The Encyclopedia of the
HARLEY-DAVIDSON

HARLEY-
DAVIDSON

The Encyclopedia of the HARLEY-DAVIDSON

Peter Henshaw & Ian Kerr

Photography by Garry Stuart

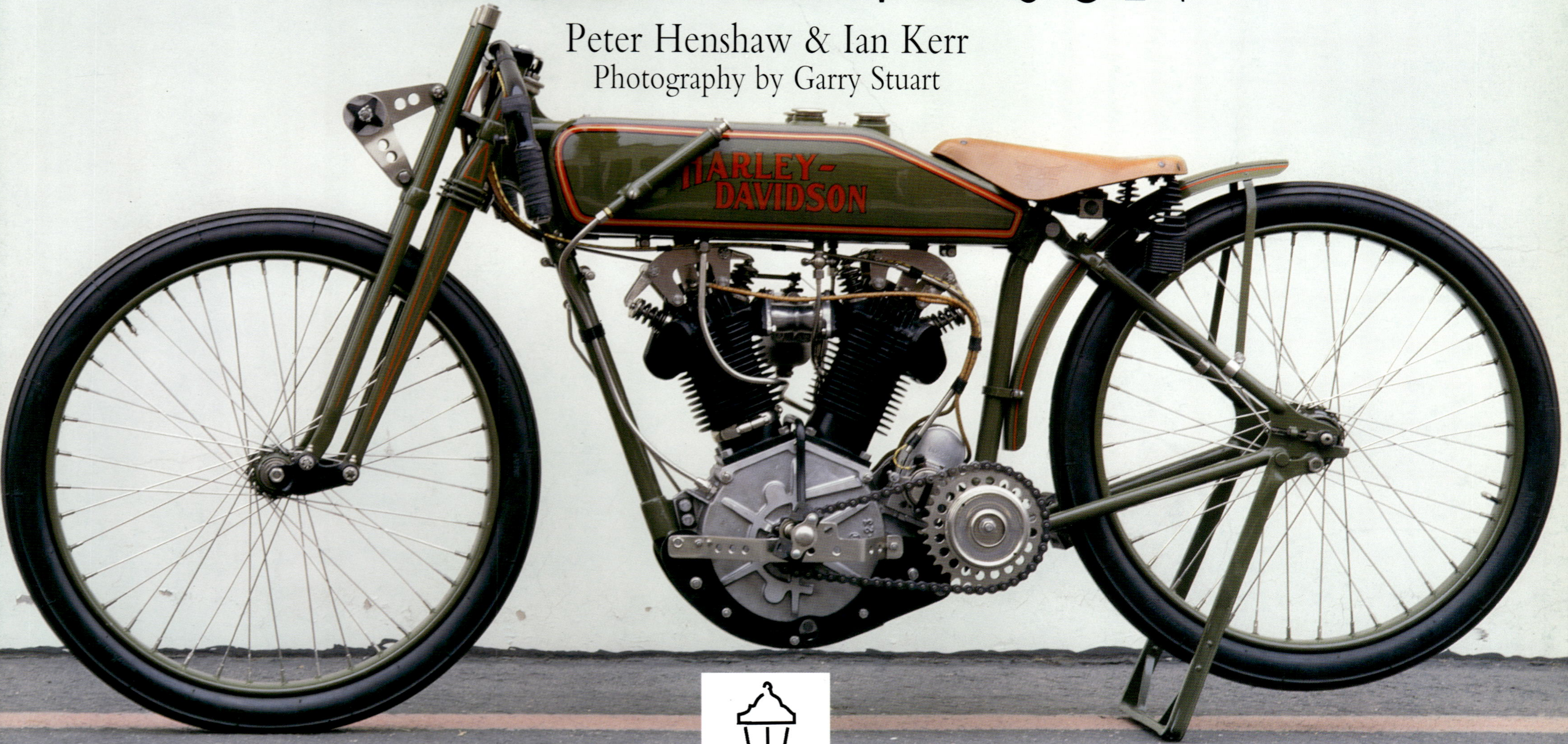

Published in 2002 by
Regency House Publishing Ltd
3 Mill Lane
Broxbourne
Hertfordshire
EN10 7AZ

ISBN 1 85361 492 0

Printed in China

PAGE 2: 1923 eight-valve racer.

TITLE PAGE: 1923 eight-valve boardtrack racer.

RIGHT: Harley-Davidson XR750.

CONTENTS

Who says you can't take a Harley off-road? An early Duo Glide takes to the water.

VR1000: Harley-Davidson's return to top-level racing wasn't a conspicuous success until the 1990s, but the work goes on.

THE HARLEY-DAVIDSON STORY (1903–2000)

1900s: Early Days

In 1903, it must have seemed to Mr and Mrs Davidson that their three sons had gone their separate ways. None had followed their father into the carpentry trade, but all were holding down jobs in engineering. William, the eldest, had already done well for himself, well established as a foreman with the Milwaukee Road railroad company; Walter, the middle one, had moved away to Kansas to work as a machinist; and young Arthur remained in Milwaukee, earning his crust as a patternmaker at the Barth Manufacturing Company. Arthur, in particular, seemed to have a bright future ahead of him: he was an outgoing, personable young man with a lot of drive, not to mention a good friend named William Harley. The Davidson brothers' parents had emigrated to the USA from Scotland in around 1871, while Bill Harley's parents had come over from the North of England. All chose to settle in Milwaukee, Wisconsin.

It was hardly surprising that Bill and Arthur were soon turning their considerable energy towards the internal combustion engine. The world they grew up in was one of optimism and belief in fast-developing technology. Steam power had transformed Europe and North America, and now electricity and the small petrol engine promised to do the same.

Where it all began: the Harley-Davidson Motor Company's first 'factory'.

After centuries of reliance on horse or steam power, mankind was on the brink of discovering a new age in transport. Maybe it was no coincidence that, in 1903, the Wright brothers tested their first powered aircraft, the Model A Ford hit the roads, and the first Harley-Davidson motorcycle was sold to a paying customer.

Some say that the young Harley and Davidson wanted to build a small petrol engine to pace cycle races; others that it was to power a row-boat, both being keen cyclists and fishermen. (Incidentally, it was by mutal agreement that their company came to be known as 'Harley-Davidson' – Davidson-Harley just didn't sound right.) Whatever the true reason, from the winter of 1900 they began to spend much of their free time designing their first engine. They weren't on their own, though, as a workmate at Barth, a German named Emile Kruger, had practical experience of the De Dion engine, and Ole Evinrude, whose name was to adorn countless outboard motors in years to come, helped them to develop a suitable carburettor. That first 10-cubic inch (165-cubic centimetre) engine wasn't a masterpiece: speed was controlled by the spark setting – but it ran

A well known portrait of four very famous men. From left to right: William A. Davidson, Walter Davidson Snr, Arthur Davidson, and William S. Harley.

and it worked, so the two young men did the obvious thing and bolted it into a bicycle frame.

Meanwhile, Arthur had been writing to his elder brother, Walter, keeping him abreast of developments. Walter was evidently interested, for in 1902 he took a job in Milwaukee and moved back home, all the better to help in the exciting new enterprise. But 'enterprise' is perhaps the wrong word. There is no evidence that the three saw their prototype motor as the basis of a business empire; all kept their full-time jobs, and it wasn't until the spring of 1903 that the first prototype putt-putted its way onto the streets of Milwaukee. For many young entrepreneurs, this would have been the cue to rush into production: after all, it ran didn't it, after a fashion? Moreover, many early motorcycle makers were all too eager to jump straight into production with machines that worked, just; they didn't stay in business long.

Harley and the Davidsons were made of sterner stuff. Only when their prototype worked consistently did they consider selling a replica. In fact, this cautious, conservative attitude was to typify the way they did business over the next four

LEFT: Even in the early days Harleys were exported. This is an F-head outfit in India.

decades. So when the tiny prototype obviously needed more power, it was discarded and a much bigger 25ci (410cc) motor was designed in its place. This one had enough power, but soon vibrated the standard bicycle frame to pieces. The only answer was a purpose-built frame, with bigger brakes, wider wheels and heftier bearings. True, it still had pedals, which you had to work like crazy to start it, it cruised at around 25mph (40km/h) and had no gears or suspension, but the motorized bicycle was becoming a true motorcycle. Better still, unlike some of its more hastily-assembled rivals, the Harley-Davidson worked, and kept working. People started to take notice, and the first orders began to trickle in.

It was time to get serious, and while William Davidson Snr was building a 10 x 8ft (3 x 2.4m) shed in the back garden (the first Harley-Davidson factory), Bill Harley was leaving Barth to study for an engineering degree. Over the winter of 1903–4 Arthur built up two bikes for paying customers, while Walter left his job to concentrate on the new venture full-time. Once word got round that the Harley-Davidson was a reliable piece of machinery, more orders flowed in. A rich uncle, James McLay, lent them enough money to build a proper factory on Chestnut Street (later renamed Juneau Avenue), where in 1906 49 bikes were produced. Harley-Davidson motorcycles emerged from that site for the next 70-odd years, and even now the famous V-twin motor is still made there. Production grew in leaps and bounds as the orders kept coming, and more staff was taken on to cope: output tripled to 152 bikes in 1907, then tripled again the following year. By now, Arthur, who was a born salesman, had left his day job and was out on the road demonstrating the bike and recruiting dealers, and his eldest brother, William, came on board as works manager. Harley-Davidson Incorporated was born.

OPPOSITE: Spot the difference? That's a 1905 single on the floor and a 1907 version on the bench.

HARLEY-
DAVIDSON
HARLEY-
DAVIDSON

LEFT: A 1905 F-head single, when bikes were still bikes. The pillion seat is non-standard and by the look of it, only for the brave!

1910s: Rapid Expansion

Meanwhile, the motorcycle had hardly been changed since 1903. It was now nicknamed the Silent Gray Fellow (it was quiet, and came in only one colour) but was much improved by the addition of sprung leading-link forks, which Bill Harley designed while at college. Incidentally, the same basic design of forks was used by the company until 1947, and was even reintroduced (in modernized form) in the late 1980s. Soon after, the 25ci (410cc) engine was enlarged to 35ci (575cc), housed in a larger, longer wheelbase frame. Walter proved the latest version by winning the Long Island Endurance Run in 1908, and consolidating it with an economy run win the week after.

But though the 5-35, as it was officially known, could top 50mph (80km/h), it still wasn't enough for the emerging motorcycle riders and the huge distances they travelled in America. Like all its rivals, Harley-Davidson needed a bigger bike.

Strange as it may now seem, Bill Harley did not invent the V-twin: when he paired two existing singles onto a beefed-up crankcase, he was merely doing the same as everyone around him. The V-twin was a relatively quick and easy way to build a more powerful engine using existing parts, though Harley's first prototype seemed hardly that. Still hampered by an atmospheric inlet valve (opened by piston suction, rather than positively via a camshaft and pushrod), it was no faster than the single – the 49-ci (803-cc) twin was quickly withdrawn from sale, and Bill went back to the drawing board.

When it reappeared as the F-type in 1911, it was clear that he'd been working

hard. It now had a mechanical inlet valve, so it could rev higher and produce more power, and the drive belt was tensioned to prevent slippage (another problem with the prototype). In some ways, the F-type, with its fixed belt drive and gravity-fed lubrication, seemed little advanced on the early pioneer motorcycles, but this was about to change. The years 1910 to 1915/16 saw a brief flurry of innovation from the American motorcycle makers which, for a while, put them ahead of the Europeans: and Harley-Davidson was right in the thick of it. A basic clutch in the rear hub allowed the rider to stop and restart without stopping and restarting the motor. That was soon overtaken by a proper multi-plate clutch and chain drive with a two-speed gear in the hub. This was a real breakthrough, which Harley followed up in

Harley-Davidson's image was somewhat different in the early days, as illustrated by this solid and reliable sidecar tug.

1915 with a three-speed gearbox and automatic engine oil pump. Also that year (they must have been working nights) was the option of electric lighting and fitted with this, the F-model became the J, using a gear-driven magneto-generator.

Harley-Davidson's success mirrored these technical leaps forward, and production continued to soar: over 3,000 bikes were sold in 1910, over 5,000 the year after and 9,000 the year after that, while over 16,000 left the factory the year war broke out in Europe and 22,000 as the decade came to a close. Only ten years after that first prototype first hit the road, Harley-Davidson had established itself as number two in the American motorcycle market; the longer-established Indian still led, with Excelsior a poor third. In fact, it was the start of an intense, sometimes bitter, rivalry between Indian and Harley-Davidson, which lasted for the next 25 years. Harley-Davidson certainly did better out of the First World War than its rival. In a fit of patriotic fervour, Indian turned over its entire production to military needs with the result that there were many disgruntled Indian dealers with no bikes to sell; Arthur Davidson lost no time in persuading them to change sides.

This Indian/Harley rivalry was never more obvious than at the racetracks which included the wooden boardtracks of the early years and the later dirt ovals. Harley and the Davidsons weren't natural race goers, but they were shrewd enough to realize that success on the racetracks could translate into sales, and engineer William Ottaway (a man central to Harley-Davidson's engineering policy for many years) developed the short wheelbase II-K racer out of the J-model. It could top 90mph (145km/h), and win races, so in 1916 Ottaway got the go-ahead for the Model 17 V-twin racer, with four valves

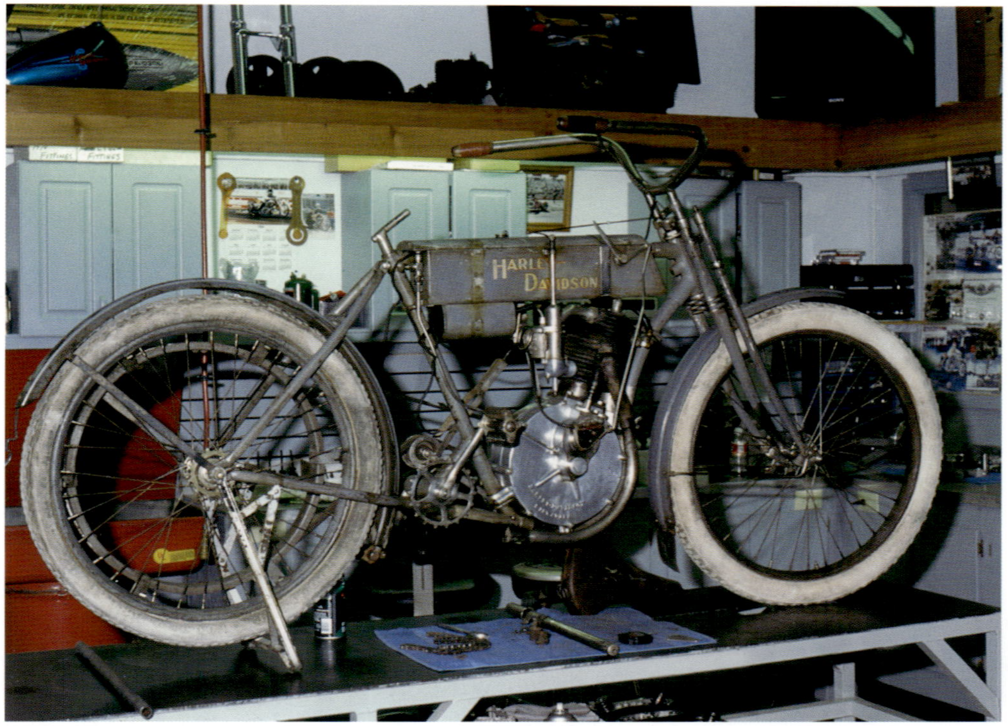

OPPOSITE: A glimmer of progress? Still with belt-drive, but sprung forks have arrived.

LEFT: A 1909 single, much as it would have left the factory.

BELOW: 1909 and the first twin, which was hardly a success.

per cylinder. With the help of cylinder-head specialist Harry Ricardo, the 8-valve produced 55hp and in the hands of the official factory team (the delightfully named 'Wrecking Crew') it was soon notching up victories all over America. However, Harley's official racing effort was short-lived as the collapse of the motorcycle market in 1920 caused the whole team to be disbanded.

Most of those racers were V-twins (though there were some four-valve singles), but that simply reflected the American motorcycle market. Harley's own V-twin had long since overtaken the little single in importance, which was drifting into the background. The availability of cheap cars (not the least of which was Henry Ford's Model T) meant that motorcycles were increasingly seen as leisure tools for sporting young men, and a few women, and they wanted power and speed. However, various motorcycle makers still believed that a quiet, reliable and cheap machine could persuade the man in the street back onto motorcycles. Harley-Davidson's contribution was the Sport Twin of 1919. The 'Sport' part of it was a misnomer, as the new bike was a docile, mild-mannered flat-twin which could barely top 50mph (80km/h). On the other hand, it was quiet and easy to ride, had an enclosed chain for practicality, a modern three-speed gearbox and the option of electric lights. Unfortunately, it failed to attract the masses back to motorcycling and was dropped after a few years. It was quite well received in Europe, owing much to the English Douglas in its design, but can be regarded as Harley-Davidson's first flop.

This was bad news, as the Sport Twin was part of a massive expansion plan based on borrowed money. Since 1903, Harley-Davidson sales had spiralled upwards year after year, to the point where it was catching Indian, the market leader. So in 1919–20, the four Founders appeared to forget their natural caution and borrowed $3 million (a huge sum at the time) to expand the factory. When it was finished, the Juneau Avenue works was the largest motorcycle plant in the world, and produced a record 28,000 machines in 1920. Now in prosperous middle age, it seemed that Arthur, Bill, Walter and William could do nothing wrong.

ABOVE LEFT: A 1913 twin, now with chain drive.

ABOVE: A 1913 single: pedals were still there to start it.

LEFT: A racer from 1914, Harley-Davidson's first official year on the track.

RIGHT: Silent Gray Fellow, quiet and lacking in colour like a retired bank clerk.

ABOVE LEFT: An unrestored 1913 V-twin.

ABOVE: A single from the same year.

LEFT: A 1914 twin with pillion seat and painted rims.

RIGHT: Pushrod-operated inlet valves were the key to the early twin's success.

ABOVE, LEFT and RIGHT: Most early Harleys were destined to haul sidecars, but some were sporting solos.

OPPOSITE LEFT: Typical 1915 riding gear: mackintosh, overtrousers and flat cap.

OPPOSITE RIGHT: Boardtrack racers were stripped to the bone. This one dates from the 1920s.

1920s: Down & Up

Although the four Founders had become accustomed to success, they had to cope with economic failure in double-quick time. Harley sales more than halved in 1921 (to 10,202), which left them with an inventory of unsold bikes and a $3 million loan to pay off. But to to have achieved the success they already had, the four had needed plenty of business acumen. They shut the newly expanded factory for a month, cut salaries by 15 per cent and abandoned the racing programme. So sudden was this decision that the racers were actually left stranded at the State Fair races at Phoenix. They had to pawn their wristwatches and borrow money from a local Harley-Davidson dealer in order to buy their train tickets home!

This abrupt belt-tightening worked: the unsold bikes found buyers and production began to recover. But it was a fluctuating recovery; the motorcycle boom time was over, thanks to cheap, affordable cars, and it took the company 20 years to match that peak of 1920. What the new generation of riders wanted was more power, and Harley-Davidson obliged with the Superpowered Twin of 1922. Although intended as a sidecar machine, the extra power from its enlarged 74-ci (1200-cc) V-twin was welcomed by keen solo riders. The engine was no more than a bigger version of the original twin, still with overhead inlet valve, side exhaust and total-loss lubrication. Like its predecessor, the bike came in magneto (FD) or electric light (JD) forms.

Harley-Davidson had always set a great deal of store by its dealers, taking time to build up a network when less patient rivals were selling bikes to anyone with cash. Now, as one of the two makes dominating the market, Harley-Davidson was able to tighten its control over dealers: factory representatives toured the country, checking up; franchise agreements were for one year only (screw up, lose your franchise); and no dealer was permitted to sell any other make alongside Harley-Davidson, even the little Cleveland, which

sold to a completely different market. Of course, if you toed the line, there were rewards to be had, as the Harley-Davidson had a good reputation and sold well; but if you didn't, you wouldn't be a Harley-Davidson dealer for very much longer.

One thing that restricted dealers' freedom was that Indian and Harley were the only two manufacturers selling bikes in large numbers. And despite the fierce rivalry between them, the two actually met regularly to fix prices. It was of course illegal, and had to be done in secret, but in 1922 Frank Weschler (of Indian) and Arthur Davidson met and agreed to sell at the same price the following year; it became an annual event in their respective diaries. But Harley-Davidson was really in better shape than Indian by the early 1920s, whether or not they were selling more bikes. It was more efficient, had a tighter grip on its dealers and a simpler model range.

Nevertheless, Indian often seemed ahead on model development: it had been first with a flat-twin (which Harley-Davidson countered with the Sport Twin) and in 1925 launched the 21ci (350cc) Prince, a single-cylinder machine in the British mould. A Prince, a 350cc BSA and a New Imperial were shipped into Milwaukee for evaluation, and soon after Harley-Davidson's own 350cc single appeared in 'A' (side-valve) and 'AA' (overhead valve – ohv) guise. The former sold well as a basic utility bike, while the AA (nicknamed 'Peashooter') did very well in racing. It also highlighted another aspect of the Harley/Indian wars: Indian was invariably first in bringing out a new model, which Harley-Davidson would soon counter. Not only the Sport Twin and

THIS PAGE: The 1922 Sport Twin was a real departure for Harley-Davidson but, apart from export sales, was not a success.

OPPOSITE: This 1922 sidecar outfit is in rare original unrestored condition.

Peashooter, but all of Harley's new bikes of the 1920s – the Two Cam, 45 and VL – were all direct responses to what Indian was already making.

The Two Cam became something of a legend. The problem for Harley riders was that Indian side-valve twins were far faster than the older ioe (inlet-over-exhaust) Harleys, even if they didn't have the same stamina. The Two Cam answered this by using a separate cam for each valve, thus allowing tuning for more light, higher compression, higher revs and more power. Available in 61ci (1000cc) and 74ci (1200cc) forms, it wasn't cheap ($50 more than the equivalent J) but it was said that a good 74 would reach 100mph (161km/h). No longer need Harley-Davidson riders be embarrassed on club runs! In fact, the Milwaukee machines, despite their earlier racing success, had gained a reputation as old men's machines – reliable and dependable, but without the performance of an Indian or Excelsior. The 45ci (750cc) Indian Scout was a particular problem for Harley dealers, and the Founders' response (which Walter Davidson had promised to shareholders in 1927) was the model D, better known as the 45. Like the Scout, it was a 45ci side-valve V-twin. Unlike the Scout, it was slow and gutless, unable to reach 60mph (97km/h) where the Scout could top 75. A carburettor kit had to be rushed into production to improve power, and to add salt to the wound, the distinctive vertically-mounted generator was prone to failure. The 45 had clearly been rushed into production without proper testing. In time, it developed into a

1922
HARLEY-
DAVIDSON
DUNLOP
H-D

thoroughly reliable machine, powering the wartime WLA and the three-wheel Servicar. But it was a long way from the care and diligence with which the first single had been developed. Was Harley-Davidson losing its way?

It must have seemed so to loyal riders and dealers when the new VL was unveiled in 1929. This replacement for the long-lived F/J models was almost all-new, the bore, stroke and capacity being all that remained the same on this side-valve V-twin for which the factory claimed a 15 to 20 per cent power improvement. Unfortunately, the real increase was just one horsepower, and to make matters worse the VL weighed a whole 120lb (54kg) more than the F/J. In a bid to overcome this lethargic performance, the engine had been given relatively small flywheels which did allow good acceleration up to 50mph (80km/h) or so, but that was about it. Top speed and hill-climbing ability were sorely lacking on this big twin, and the light flywheels also caused finger-numbing vibration. Some customers wanted the F/J and Two Cam reinstated; others wanted their money back.

As with the 45, Milwaukee had to come up with answers, and fast. After much burning of the midnight oil, the solution came in bigger, heavier flywheels and modified cams. This did the trick, but the bigger flywheels needed bigger crankcases which in turn demanded a larger frame. In other words, every VL made up to that point (there were 1,300 of them) had to be completely rebuilt. It cost Harley-Davidson $100,000 in new parts, caused much heartache to the dealers (who were expected to do the rebuilding at their own expense), and did much damage to the hard-won reputation for reliability. Better was to come, but there were more hard times to get through first.

THIS PAGE: It's 1925 and the J-model with overhead inlet, side exhaust valves has four years to run. It had electric lighting and a three-speed hand-change gearbox.

OPPOSITE: This J sidecar outfit in its original green was undeniably handsome. The substantial luggage rack was standard but the front brake was not.

HARLEY-DAVIDSON

ABOVE LEFT: By the late 1920s, Harley-Davidson could claim that its Juneau Avenue factory was the biggest in the world.

ABOVE: The 1937 National Hillclimb Championship in which Harold Seamans, competing in the 80-ci (1300-cc) Class B event, breaks the winning tape while riding a factory bike.

1930s: Slump to Recovery

Twice in its lifetime, Harley-Davidson has come close to closure. The most recent was 1985 when its major financial lender pulled out. But 50 years before, there had been talk in the boardroom of voluntarily throwing in the towel. With the benefit of hindsight, this seems odd: the company was challenging Indian for market leadership and had overcome the sales slump of 1920 and some troublesome new bikes, which were now settling down nicely. The reason was more fundamental than any of this. The slump following the Wall Street Crash of 1929 had hit motorcycle sales hard: Harley sold nearly 21,000 bikes that year, over 17,000 in 1930 but a mere 3,700-odd in 1933. It was a disaster. The four Founders were now old men; they and their families were financially secure and they'd worked hard all their lives. Why carry on? Well they did: whether it was loyalty to the workforce, some of whom had been there since the start, or simply a refusal to throw away 30 years of work. Bill Harley and the Davidsons decided to stick with it. Not only that, but they put a lot of money and time into an all-new bike that for them was a real leap into the unknown. This was the Knucklehead, a real milestone bike for the company that finally enabled it to leap ahead of Indian.

But in the meantime, men were laid off, salaries were cut by 10 per cent and the Founders reduced their own remuneration by half. The side-valve VL and 45 were gradually improved, the 45 in 1932 with a redesigned frame that did

LEFT and BELOW: 1928 was the last year for the J model, the engine shot clearly showing the pushrod inlet valve and side exhaust valves. This motor served Harley for nearly 20 years and not everybody approved of its replacement.

away with that odd vertical generator, which had led to the derogatory phrase, 'three-cylinder Harley'. Renamed the R-series, it was also given a boost when Indian temporarily dropped the Scout which had been taken over by the du Pont concern, in a move towards rationalization. The upshot was that the 45 had the middleweight market to itself. Meanwhile, the genius of tuner Tom Sifton and others was coaxing it into winning a few races, to the extent that the factory later produced its own 45 racer, the WR.

Another sign of faith in the future was the introduction of a range of bright new colours: for years, Harleys had only come in dark, drab hues that reflected their workaday image. Now, there were eye-catching reds, oranges and creams, two-tones and stylish art deco tank logos. Once again, Indian was the spur to action. The du Pont empire was based on chemicals, which gave Indian access to a whole range of paints. Harley-Davidson had no choice but to follow suit. The same went for the VLH of 1935. Indian had a big-engined 80ci (1300cc) V-twin, the Big Chief, which Harley-Davidson did not. This would never do, and the VLH had its crankpin moved outboard to allow a 4¼-in (108-mm) stroke and the biggest Harley motor yet.

But while all this was going on, the engineering department (such as it was, manpower having been depleted by the various cuts) was hard at work developing the Knucklehead. It actually took five years to reach production, partly because of the skeletal engineering department,

partly because of teething troubles with oil leaks and the new dry sump lubrication system. When it was finally unveiled in 1936, at the annual dealer conference, everyone seemed to know it would be a hit. It was new from stem to stern: the 61ci (1000cc) overhead-valve V-twin had a shorter stroke than its predecessors, so it was able to rev higher and harder than any previous Harley. It had a proper recirculating oil system in place of the outmoded total-loss system, which meant a constant supply of fresh, cool oil, better able to cope with high-speed runs. It had big valves and a high compression. It had 40bhp at 4,800rpm and could top 90mph (145km/h). There was a new four-speed constant-mesh gearbox, new clutch, new frame, new swoopy styling. The '61E' was its official title, but it was nicknamed 'Knucklehead', suggested by the shape of the rocker boxes.

Even so, the Founders, now increasingly assisted by their respective sons, tried to insist that the new bike was for limited production only, and dealers were forbidden to order them as demonstrators. Maybe they were wise, for the Knucklehead didn't have a trouble-free first year: it went on leaking oil until valve spring covers were introduced, and the springs themselves broke, but these really were teething problems. In any case, Joe Petrali set a new speed record with one of the new bikes of 136.15mph (219km/h), thanks to a tuned engine and high top gear, while police officer Fred Hamm broke the 24-hour record, riding a standard Knucklehead 1,825 miles (2937km) in 24 hours (an average of just over 76mph/122km/h, including stops). It was official, Harley-Davidson finally had a bike that was both reliable and fast.

The Knucklehead's success rubbed off on the other bikes. In 1937 the 45, 74 and

Art deco comes to Milwaukee on this 1932 VL, but Harley had Indian to thank for the new range of bright colours. The new sidecar motor proved a reliable performer after a troublesome start.

80ci (750, 1200 and 1300cc) side-valve all received the new lubrication system (which was just as well, the total-loss being a museum piece by now), and the new bike's streamlined styling. The 45 became the W-series, and the V was now the U-series.

While the Knucklehead stole the headlines, something completely different was doing good business for the company. The three-wheeled Servicar used the front end of a 45 married to a two-wheel rear end and a load-carrying box. Originally intended for garage mechanics to collect cars for repair, it proved ideal for traffic police and meter maids, becoming something of an American institution. It remained in production until 1973.

ABOVE LEFT and RIGHT: The neat 1934 '45' was more often known by the number than the official 'D'- or 'W'-series. Never fast or glamorous, it was nevertheless a tough and reliable machine that introduced thousands to motorcycling in the 1930s and 1940s.

LEFT: A late-1930s VL, ready to tour. Note the large buddy seat, panniers, buckhorn bars and luggage rack.

HARLEY-DAVIDSON

OPPOSITE: A rare VL of the early 1930s. Mirrors and buddy seat were extras.

TOP LEFT: A 1934 VL.

TOP RIGHT: The 1935 VLU was at 80ci (1300cc) the biggest Harley yet.

LEFT and ABOVE: Mid-1930s Servicars with optional large carrying box.

A study in changing attitudes: in the 1920s Harley-Davidson motorcycling was portrayed as a peaceful, almost pastoral activity. Ten years later ('The Greatest Sport of Them All'), the image is altogether more dynamic while by the 1940s power and performance had become key words, with riders' apparel taking on a distinctly militaristic flavour.

ABOVE: Four jackets from *The Enthusiast* magazine, featuring touring (1939), racing at Daytona Beach (1940), a military machine (1942) and an optimistic look towards peacetime (1944).

LEFT: A charming simple image from the 1930s.

1940s: Wartime Diversions

Just a handful of years after the four Founders contemplated closing down Harley-Davidson, the family firm was back on the crest of a wave. It had survived two major slumps and the difficult, competitive years between them, and had finally overtaken its arch-rival and transformed itself from producer of plodding side-valves into a maker of fast performance machines. The Second World War was to keep the factory busier than ever, with thousands of bikes and countless pieces of subcontract work.

But in 1940, the USA hadn't yet declared war on Germany, and plans for next year's peacetime Harleys could go ahead. The 80ci (1300cc) side-valve received aluminium cylinder heads in 1940, or better cooling, and 1941 saw a 74ci (1200cc) version of the Knucklehead, whose new centrifugally controlled oil pump finally solved the old problem of too much lubrication in some places, not enough in others. Not many of them reached civilian hands though. Also under discussion was a 45ci (750cc) overhead-valve engine, which was logical, given the success of the Knucklehead. It had been discussed for some years, and Bill Harley was a strong advocate of the idea. Three prototypes were built, one of which was taken on a 5,000-mile (8050-km) proving ride into Texas and back. There were no major problems, and the new baby Knucklehead looked promising; but it would have cost as much to build as the bigger side-valves, and the idea was dropped. It would be nearly 20 years before Harley-Davidson finally produced an overhead-valve 'baby' V-twin – the Sportster.

In any case, even by the late 1930s, the Board had had more pressing things to consider, notably the potential of big orders from the army for a standard do-everything motorcycle. Harley's solution was the WLA, basically a militarized version of the standard W-series 45. With its low compression motor, big oil bath air filter and off-road skid plate, the WLA was a truly rugged workhorse. The extra weight of all the army equipment depressed top speed to around 50mph (80km/h), but that hardly mattered unless you were being chased by BMW-mounted Germans. What did matter was that nearly 90,000 were supplied to the forces, along with a WLC version for the Canadians, which introduced thousands of servicemen to motorcycles in general and Harleys in particular.

There were plenty of other military projects at Milwaukee, not all of which saw the light of day. The XA *was* produced, basically a copy of the contemporary BMW, because that's what the army wanted. A civilian prototype of this flat-twin, shaft-drive bike was actually running in 1946, though it came to nothing. The XA motor was also mooted for a mini-Jeep, which could be parachuted

OPPOSITE: The 1942 XA was a strictly military machine. Few were made.

LEFT, BELOW LEFT and BELOW: The Knucklehead heralded a new era of style for Harley-Davidson, quite apart from being the fastest, most powerful Harley for the road yet seen. These examples all date from 1938/39 and are of 61ci (1000cc) capacity.

into battle zones, for a generator set, and as an updated modernized Servicar.

While all this was going on, a new generation of Harleys and Davidsons was taking over. William Davidson had died just before the war in Europe began; Walter Davidson died in February 1942 and Bill Harley the year after. Arthur, the sole survivor of the four Founders, was killed in a car crash in 1946. But William's son (another William) took over the helm of the company, while Harley's son controlled the engineering side: Harley-Davidson was still very much a family firm as it faced a very different post-war world.

For the first time it was to encounter a significant threat from imports. Harley-Davidson now accounted for 75 per cent of US motorcycle production, but there was a growing trickle (soon to become a flood) of British motorcycles being shipped across the Atlantic. Indian tried to meet these head-on and failed. Harley-Davidson preferred to pretend that the threat did not exist. So instead of coming up with a mid-range competitor for the Brits (a modern vertical twin, with foot gearchange and rear suspension, was discussed), the company concentrated on what it did best, big V-twins. For 1947 the Knucklehead became the Panhead (the rocker covers now looked like upturned saucepans) with aluminium cylinder heads for cooler running and hydraulic tappets for quietness. In a now-familiar scenario, the tappets proved troublesome at first, with

ABOVE LEFT: Some Knuckleheads did reach the civilian market in the 1940s.

ABOVE: A highly though possibly over-restored 1942 WL45 and 1958 Duo Glide.

fluctuating oil pressure upsetting the valve timing; but overall the Panhead was a real improvement. There was even bigger news the following year when Bill Harley's 1907 Springer forks were dropped in favour of conventional telescopic ones, and the bike was named Hydra-Glide in their honour. The first of the 'Glides' had arrived. Gradual updates followed, with foot shifting (1952), rear suspension (1958), 12-volt electrics (1964) and electric start (1965).

ABOVE RIGHT: A 1940s Knucklehead. It was necessary to go easy on the chrome as wartime shortages made themselves felt.

RIGHT: 1948 marked the first year of the Panhead, with hydraulic tappets and alloy cylinder heads.

HARLEY
DAVIDSON

So it was business as usual, except for the S-125, or Hummer, which went on sale in 1947. This little 7-ci (125-cc) two-stroke was basic in the extreme, with 3bhp, a three-speed gearbox and suspension worked by rubber bands. It was really a pre-war German DKW, offered to the US as war reparations. Indeed, BSA got it as well, as the Bantam, and there was even a Russian version. In its Harley-Davidson guise the ex-DKW was actually quite a success, at least at first. Although some traditionalist Harley dealers refused to sell what they regarded as a pipsqueak machine that didn't deserve to be called a motorcycle, it sold in steady, though not spectacular, numbers for over a decade, by which time it had grown to 10ci (165cc) and had telescopic forks.

OPPOSITE: A springer fork marks this out as an early (1948) Panhead.

LEFT: The 1946/47 Knucklehead was hardly changed from its pre-war version.

BELOW LEFT and RIGHT: By 1946 the Knucklehead had become the mainstay of the range.

1950s: A Challenge from Imports

The Hummer was all very well, but it wasn't the middleweight Harley for which dealers were clamouring. Faced with a nimble 30-ci (500-cc) Triumph or Norton, the ancient 45 looked like an overweight dinosaur. Throughout the 1950s, Harley held on to its traditional club and touring market, selling about 5–6,000 big twins a year to these riders. But it was missing out on the new sports bike market; by 1950, imports had captured 40 per cent of US sales, prompting Harley-Davidson to ask for a swingeing import tariff to be imposed. At the subsequent hearing, some unpleasant facts about how the company had restricted its dealers' freedom of action came to light and the request was refused.

Harley's answer to the imports was unveiled in 1952, and the new K-series, which finally replaced the long-serving 45, was a real enigma. Why? Well on one hand, it had an up-to-the-minute chassis, just like the imports it was designed to compete against: telescopic forks, swinging-arm rear suspension, unit-construction gearbox with foot-shift, it was all there. But its power unit was a development of the old side-valve 45, though opinion is divided among Harley historians as to how new it was; the K-series really did look like old wine in a new bottle. Side-valves were outmoded before the war, so it was mystifying that Harley-Davidson stuck with them into the 1950s. It wasn't as though the firm was short of expertise (its prototype ohv 45s had been running in 1940); nor was it short of money (it spent $3.5 million on new plant in the late 1940s); it wasn't short of time to develop a new engine; and finally, there really was an all-new prototype to meet the imports, with an all-aluminium 60-degree V-twin with ohv, twin carburettors and high cams. Whatever the reason, Harley dealers were stuck with the K-series, whose sales matched their disappointment. It could barely exceed 80mph (129km/h) where the Triumph Thunderbird could top 100. Things were improved after a couple of years when Harley boosted the capacity to 55ci (883cc), but it was still no Thunderbird beater.

Meanwhile, the whole of motorcycling in America was changing, and not just because of British imports. As we've seen, Harley's traditional club market carried steadily on, which was an organized, orderly sport, with clean-cut men and women wearing smart uniforms. But there was a new strand of biking on the way which had first come to public attention back in 1947, thanks to the supposed 'riot' at Hollister, California. The new bike gangs that were forming at the time, often composed of ex-servicemen, had a very different ethos to the orderly clubs. For them, motorcycling was at the centre of an outlaw lifestyle which rejected mainstream

OPPOSITE and ABOVE: Telescopic forks and hardtail rear end add up to the Hydra-Glide Panhead.

RIGHT: The oddball 1955 KHK had a modern chassis with a side-valve motor.

society. Their bikes reflected that. Instead of the traditional Harley 'dresser', with panniers, screen and crash-guards, the bike of a member of Satan's Sinners had every piece of superfluous equipment stripped off to cut weight and boost performance. But in time, it became a statement in itself. This was the start of the custom bike movement, and Harley-Davidson, much as it tried to distance itself from the outlaw clubs, was right at the centre of it. Ironically, it has since embraced a highly

LEFT: Harley-Davidson is keen on birthdays. This is a 50th-Anniversary Hummer from 1954.

OPPOSITE: The Hummer eventually spawned an off-road version, the forerunner of the trail bike.

stylized, sanitized version of the same thing. The difference is that new Harleys are sometimes bought by 'weekend outlaws', who wear black leather at weekends but are back in suit and tie come Monday morning.

The tourers and custom bikes are two distinct families of bikes within Harley-Davidson's current range. The third is the Sportster, which made its debut in 1957. Just like the Knucklehead of 1936, this was a milestone bike for Milwaukee. There was actually nothing radical about it, based as it was on the running gear and basic engine unit of the K-series. What it did have was a new top end with overhead valves and a 7.5:1 compression, which produced 40bhp at 5,500rpm. It wasn't enough to beat the 40ci (650cc) British twins, but it did equal them. That was taken care of the following year with the higher-compression XLH Sportster that claimed 55bhp at 6,800rpm. Once again, Harley finally had a bike that was as fast as anything else on two wheels.

The other thing it gained that year was a look we all associate with the Sportster. Its origins lay in the stripped-down XLC (no lights or battery) of which just 200 were built for competition. The road-going XLCH aped that style, with high-level staggered pipes and a tiny 2.2-gallon (10- litre) fuel tank taken from the Hummer 125. It may not have been as practical as the standard Sportster, but it looked like a lean all-engine machine, and it outsold the frumpy XL by two to one. But even with the XLCH, the Sportster wasn't an instant hit. It was actually outsold by the big twins until 1969, and throughout the early 1960s, less than one in five Harley sales was a Sportster. However, in the long run, that didn't matter, as the Sportster opened up a new market for Harley-Davidson, one which is still going strong.

423·T
MC-OHIO-51
THUNDER HUMMERS

1960s: Takeover

The Sportster had finally met the British bikes head-on, at least in straight-line speed, but there was another new threat on the horizon. Honda and Suzuki had started to export to the US, and these clean, sophisticated lightweights looked as if they would open up another market which was closed to the company. It did have the 10-ci (165-cc) Hummer, but that was now over ten years old as well as based on a pre-war design; so to meet the Japanese something else was required. It is hard to credit now, but right through the 1960s Harley-Davidson sold lots of small motorcycles, even a 3ci (50cc) moped and a scooter! These came from two sources: developments and updates of the original Hummer, and bikes built by Aermacchi in Italy.

ABOVE and OPPOSITE LEFT: The 1959 FLH Duo Glide, by then the definitive American touring motorcycle. Screen and fibreglass panniers are typical period extras.

OPPOSITE RIGHT: Balloon tyres, soft long-travel suspension – riding doesn't get much more comfortable than this.

The Italian connection was made in 1961. Two years earlier, Harley-Davidson's own development committee had recommended that the best way to market a lightweight motorcycle was to import someone else's machine and stick the Harley-Davidson badge on it. So that's what they did. Aermacchi had been making aircraft during the war, producing scooters after 1945, then a range of four-stroke singles from the mid-1950s. An 80mph (129km/h) 250, with its distinctive horizontal cylinder, fine handling and light weight was on offer by 1957 – just the thing to offer Americans who hopefully wanted to buy a miniature Harley. Instead of just importing the bikes, Harley-Davidson decided to buy Aermacchi lock, stock and barrel, and in 1961 began to import the 250 as the Harley-Davidson Sprint. It had skimpier mudguards and a smaller fuel tank for the US market, and *Cycle World* was quite impressed when it tested one. The pushrod single would happily rev to 7,500rpm, it reported, and cruised at 65–70mph (105–113km/h), though even then it was able to assert that there was nothing very high-tech about it. Sure enough, the Sprint was technically overtaken by Japanese bikes a few years later, but it nonetheless made a valuable contribution to Harley-Davidson's market share.

Not that it came without headaches. Although mechanically reliable, the Sprint was not as well made as home-grown Harleys, and the electrics in particular gave a lot of trouble. It was bad enough that

Harley-Davidson engineers and managers were trying to co-ordinate with a factory thousands of miles away that worked in a different language, but Aermacchi also bought in many of its components from other factories. From Milwaukee, getting on top of Aermacchi quality proved an almost impossible task.

In any case, head office had its own plans for the lightweight market, which slotted in under the 15ci (250cc) Sprint. First came its assault on the scooter market. The Topper, which appeared in 1960, used the Hummer's 10-ci (165-cc) two-stroke engine in restricted (5hp) or unrestricted (9hp) guise, allied with automatic transmission and full bodywork. It was not an elegant machine, lacking the perkiness of a Vespa or Lambretta, nor was the Topper cheap at $600. And despite encouraging sales in its first year (when 3,800 were sold), and a cameo role in a popular TV series, the Harley-Davidson scooter was not a success, and was dropped in 1964.

A similar fate awaited Harley-Davidson's updates on the Hummer theme; with a slightly larger 175cc engine, it became the Pacer in road form, or the Scat with off-road-style high-level exhaust and knobbly tyres. There was also the Ranger, a true off-road bike with no lights or front mudguard, higher ground clearance and lower gearing. But all three of these bikes shared that elderly two-stroke engine, and a basic frame without rear suspension. When the Japanese lightweights arrived, they underlined just how crude the Harley tiddlers were and Milwaukee dropped them in 1966.

But if Harley's home-grown lightweights were failing to set the world on fire, the Sportster was a different matter. In small towns across the continent, the favourite Saturday night pastime consisted of an impromptu drag race between Harley and Triumph/BSA/Norton. A horsepower race ensued: Triumph produced the twin carburettor Bonneville, and Norton the 45-ci (750-cc) Atlas. So Harley-Davidson upped the 'touring' XLH Sportster to 55bhp, which in standard form could run a 15.5-second quarter-mile and almost reach 100mph (161km/h). It still looked like a mini-Duo Glide, but better was to come.

By 1968, with yet more power (now up to 58bhp), new front forks and a 114-mph (183-km/h) top speed, the XLH Sportster was living up to its name; even *Cycle World* was full of praise for its handling, which three years previously it had likened to an automobile's. Also, it at last acquired the classic Sportster tank. Who cared if you needed to fill up every 80 miles?

It was at a time when the Sportster was a serious rival to the imported big bikes, particularly in the later 1960s, when its weight disadvantage disappeared beside new heavy superbikes such as the BSA Triumph triples and Honda 750 four. It wouldn't last, of course, as the performance race soon left the Sportster behind in the early 1970s, but by then the 'baby Harley' had acquired a following all of its own, just as loyal as the big twin aficionados, and for whom ultimate performance was less important than the way the Sportster looked and felt.

A baby Harley of a very different sort was the M50 moped, another import from Aermacchi which was sold in America. By the standards of the time, it was a reasonable machine, with rear suspension and an attractive price tag. And the buyers seemed to agree. It seems a sacrilege now, but in 1965 and 1966 Harley's best-selling machine was a moped. The success was short-lived, however. After a couple of years the M50 was treated to a boost to 4ci (65cc) and a big advertising campaign, yet the 65cc

ABOVE: The Aermacchi Sprint: its high-level pipe, bashplate and high-rise front mudguard indicates an off-road specification.

OPPOSITE: The Electra Glide motor, shown here in early Panhead form.

Harley never achieved the popularity of the M50.

The Sprint, on the other hand, just seemed to sell better each year, despite its relatively low-tech approach. For one thing, Aermacchi/Harley-Davidson seemed able to come up with more power and new variations on the theme year after year.

The 1966 CRS was a serious scrambles version, with 28bhp at 8,500rpm (it could rev safely to 9,500) and no lights or speedometer. In racing CR-TT form it made 35bhp at 10,000rpm and a streamlined Sprint set up a new 250cc record, at 176mph (283km/h) on the Bonneville Salt Flats.

That record, set in 1965, helped to boost Sprint sales to 3,000 a year and they rose to 5,000 in 1966 and 9,000 the year after. A 21-ci (350-cc) version appeared in 1969, though in a less frenetic state of tune, with 25bhp. Most of the Japanese opposition was faster, but the Sprint was becoming something of a niche bike, offering a different riding experience. It certainly went on selling a steady 4–4,500 a year right up to 1974. For a basic pushrod single with a four-speed gearbox and roots going back to the mid-1950s, it had a remarkable performance.

With all these small bikes coming and going, the big twins were rather left to fend for themselves in the early 1960s, but change did come mid-decade. Twelve-volt electrics in 1964 prefaced the fitting of electric start the following year, coupled with one of motorcycling's all-time classic model names – Electra Glide. Although regarded as a new model in its own right, the Electra was really just the standard FLH with push-button starting, though the name at least has proved very durable. It also added 75lb (34kg) to the weight, and to cope, the engine was updated into the Shovelhead in 1966.

As before, this was evolution, not revolution, in this case, the fitting of a Sportster-style top end to the big twin, which gave a useful power boost to 60bhp. It went up again in 1968, to 65bhp at 5,400rpm, thanks to a Tillotson carburettor and improved porting. Otherwise, it performed in much the same way as any Harley big twin before or since, happy to rumble along at 70mph (113km/h) until the tank ran dry. At the time, the Japanese had yet to come up with their own V-twin Harley lookalikes, so the Electra Glide was still offering a unique riding experience.

This was fine for the traditionalists, but Harley-Davidson was getting further and further out of the main stream. So while the motorcycle market expanded around it, the company was getting pushed into a backwater. The Aermacchi bikes had increased sales and turnover, but not profits, and the company was fast running out of money. So in 1965, for the first time since it was set up, the family firm went public. It was careful to keep control (various family members made up seven of the nine-strong Board) and the influx of cash was spent on new equipment and a big advertising campaign. But it was only putting off the inevitable. Within a couple of years, the company was back to square one and ready to accept that a takeover was the only way to survive.

There were two suitors. Bangor Punta was keen, but had a reputation as an asset-stripper. The American Machine and Foundry Company, on the other hand, had a strong engineering background; it promised that Harley's existing management would keep control, and chairman Rodney C. Gott was a Harley owner! After some courtroom wrangling, Harley-Davidson shareholders accepted the deal, and Harley-Davidson became a subsidiary of AMF in January 1969. It was a family firm no more.

TEXACO
T

OPPOSITE: All the bikes pictured here are Shovelhead Electra Glides, which sold steadily right through to the late 1960s and into the 1970s. Although by now they are in some ways outdated, for many American riders they remain the definitive touring bike. On the far left is an early 1966 version, the first year for the updated Shovelhead V-twin, while on the right are two examples showing restrained and fully loaded approaches to the 'dresser'.

RIGHT: A Shovelhead Electra Glide.

ABOVE: The 1971 FX Super Glide, with original 'boat tail' rear end. Most of these were junked and conventional rear ends substituted.

OPPOSITE LEFT: Engine detail of the FXE Super Glide. The 'E' denoted electric start, otherwise it was identical to the FX.

OPPOSITE RIGHT: A Super Glide with its later rear mudguard which allowed them to sell better.

1970s: Have We Got Trouble!

When Willie G. Davidson styled the Super Glide in 1969, he may not have known that he was opening up a whole new era of Harley history. The Super Glide was a milestone bike for the company, even though it was no bigger or faster or more advanced than any previous Harley-Davidson (let alone any other motorcycle). What it did do was to mix and match existing components to produce something new. Or at least, something that *looked* new. This worked again with the Low Rider in 1977, and enabled Harley-Davidson to get through some very difficult times when it couldn't afford, or didn't have time, to introduce genuinely new bikes.

Grandson of the original William Davidson, Willie G. had joined the family firm back in 1963 after working as a designer for Ford and Brooks Stevens. He was a professionally trained designer who happened to have grown up with motorcycles, so he was ideally placed to interpret what Harley riders wanted, and put it into practice. And the Super Glide was a perfect example of what he could do. Harley-Davidson had long tried to distance itself from the custom movement because of its associations with the outlaw element. But the reality was that thousands of Harley riders were modifying their bikes in this way, and it was a golden opportunity for Milwaukee to produce a ready-made custom bike. The Super Glide was therefore the first factory custom. The idea was simple: Harley married the frame, engine and gearbox of the touring FL with the lighter-weight forks, 19-inch (48-cm) front wheel and small headlamp of the XLCH Sportster. Despite the incongruous 'boat tail' glass fibre rear end that most owners junked, it looked long, low and handsome, and opened up a new market for the customizers who wanted an off-the-peg custom bike.

Not that the rest of the range was neglected. The XLCH Sportster was opened out to 997cc (the return of the classic 61ci engine size) though there was no attempt to supertune it. Milwaukee had accepted that it could never keep up in the horsepower race, and it was pointless to try. The buyers evidently weren't worried, and bought 20,000 Sportsters in 1973.

The small imports from Aermacchi were more of a problem. The Sprint, which had sold well for a while, had long since been overtaken by more sophisticated Japanese lightweights, and Harley-Davidson's answer was to meet these head-on with a range of two-stroke trail bikes.

Their origins lay in the Rapido, a 7-ci (125-cc) commuter bike that was unveiled in 1968. But as the TX125, it wore trendy off-road clothes, knobbly tyres and oil injection. An 11-ci (175-cc) version soon followed, and a 15-ci (250-cc) in 1975. On paper, they looked the business: oil injection two-strokes with CDI ignition, five-speed gearbox and authentic trail bike looks. And for a while they sold well with over 11,000 250s alone in 1975; but it was short-lived success. It was a measure of how fast the Japanese were progressing that the Harley-Davidson-badged trail bikes were looking a little crude and unfinished by 1976. Question marks remained over Aermacchi quality, and in 1978 Harley-Davidson pulled out of Italy altogether, selling the Varese factory to Cagiva. It has never sold a small motorcycle to the public since.

So what was AMF, the new owner of Harley-Davidson, up to all this time? It had a clear goal: the motorcycle market was booming, but Harley-Davidson wasn't making many bikes. The AMF plan was to boost production dramatically, thus generating the money needed to fund much-needed new models. They certainly succeeded in boosting production, moving bike assembly to York, Pennsylvania, and more than doubling the output of engines and gearboxes at Juneau Avenue in three years. But, as Rodny Gott later admitted, it was all pushed through too quickly with little regard to quality. As a result, AMF-era Harleys soon gained a reputation for poor quality, while a surfeit of 'top-down' management meant that the long-term Harley-Davidson employees were being ignored.

Things weren't going as planned, so AMF man Ray Tritten made a thorough study of the whole business, finding a lack of professionalism (in both engineering and marketing), complacency with regard

to the Japanese, and inefficient production.

With characteristic energy, Tritten set about cutting costs and introducing proper forecasting. He also brought in two men who were to play key roles at Harley-Davidson over the next 20 years. Jeff Bleustein was a fomer associate professor of engineering at Yale University and managed to transform the engineering department by bringing in more professionals. Vaughn Beals (already an AMF man) was put directly in charge of Milwaukee, and one of the first things he did was to sort out a long-term new model programme.

In the short term, the V-twin would be updated to see Harley-Davidson through the next few years, which appeared in the form of the Evolution motor. But for the future (and this was the radical part) there would be an all-new family of engines, 30–80ci (500–1300cc), two cylinders to six

THIS PAGE: A rare line-up of original 1971/72 Super Glides. Front drum brakes were still relied upon to stop these heavy motorcycles, but disc brakes were on the way.

OPPOSITE: A variety of Shovelhead Electra Glides, all showing the single front disc brake that Harley adopted in the early 1970s. The bike at top left has a sidecar attached which could still be ordered from the factory, though not many took advantage of the option.

(the NOVA programme). The plan was clear, a product-led recovery, funded by AMF.

But even with NOVA and Evolution going full steam ahead, Harley-Davidson still needed to survive in the meantime. Once again, Willie G. came to the rescue with two new bikes, both of them based on the Super Glide principle of mix 'n' match existing parts. One was a flop, the other a huge success. The XLCR Café Racer, with its vaguely European looks and all-black colour scheme, was just too far out from the traditional Harley look to be acceptable. It was dropped after a couple of years. The Low Rider, on the other hand, took the concept of a factory custom even further than the Super Glide. This was much closer to a real chopper, with lengthened and raked forks, low-riding 27-inch (69-cm) seat height with a king and queen seat. It was a canny blend of old and new, with tank-mounted instruments and 1903-style tank script, but with a twin disc front end and alloy wheels as well. The buying public loved it, and the Low Rider became the best- selling Harley (basic Sportster apart) over the next few years.

This was all very well, but AMF was growing impatient. For over ten years, it had been putting a great deal of money into Harley-Davidson, for very little return. Although there has been a tendency to castigate AMF as faceless big business, it really did want to make a success of Harley-Davidson. But after ten years of effort, the company was still only making small profits, and certainly not enough to fund the NOVA programme, which according to a team of consultants would need $70 million to reach production. There was a time when AMF would have signed the cheque, but not any more. Vaughn Beals made the recommendation; it was time to sell.

1980s: Like a Japanese Miracle

It was the best of times, it was the worst of times. For Harley-Davidson, the 1980s brought near-bankruptcy, a dramatic turnaround and a new-found prosperity that it still enjoys today. Not since the 1930s had it come so close to a complete close, and never had it experienced such a dramatic reversal in its fortunes. The decade had an inauspicious start: Harley's share of the US big bike market had slumped to one-third by 1980, and just 23.3 per cent by 1983. In that year, for the first time ever, it sold fewer big machines than Honda. Yet in 2000, a revitalized Harley-Davidson overtook Honda again, not only in the big bike market, but overall. For a company that until recently sold nothing smaller than the 54ci (883cc) Sportster (Honda of course, offers a complete range of bikes) it was an astonishing achievement.

But the seeds of this remarkable revival were being sown even in the darkest days of the early 1980s. Vaughn Beals' vision of a product-led recovery was showing signs of life, thanks to the revitalized engineering department led by Jeff Bleustein. First off, the FLT, launched in 1980, was almost an entirely new motorcycle, and was the first sign that Harley-Davidson was once again taking R&D (Research and Development) seriously.

The new FLT Tour Glide was Harley's response to Honda's GoldWing, a smooth and sophisticated tourer that had been steadily eating into the company's traditional market. To meet this formidable competition, the FLT had a new steel backbone frame, which rubber-mounted the venerable Shovelhead V-twin (just about the only part of the bike that *wasn't* new).

Rubber block mounts at the front and the swinging-arm pivots, plus adjustable locating links at the front of the engine and the top, were master strokes. They allowed vibration through when idling, but virtually eliminated it when the bike was on the move. One magazine road test likened it to the gentle thrumming one might feel through the deck of a steamship. It was a huge step forward, which has since been applied to nearly every Harley model and played a big part in increasing their acceptability to non-Harley riders.

The bike's geometry saw innovative thinking too: the forks were mounted off-set to the steering head, which allowed a steepish steering-head angle (for quicker steering) with long trail (for stability). There were a few changes to the Shovelhead in the form of electronic ignition and a quieter exhaust system, while the gearbox acquired a fifth speed. The twin headlamp fairing was new, as was the twin disc front end, the enclosed

All the bikes on this page and opposite demonstrate how the Evolution V-twin spread right through the range in the 1980s, and made possible Harley's product-led recovery. OPPOSITE is a Softail Custom with rear suspension hidden for that hardtail look. LEFT are two Dyna Super Glides, the lower one showing various accessories. ABOVE is a fuel-injected FLHTCUI Electra Glide.

chain and 5-gallon (22-litre) fuel tank. The FLT was really another of Harley's milestone bikes, and is still part of the range, 20 years on.

Later that same year came the Sturgis, which introduced another innovation – belt drive. It was the first production motorcycle to offer both primary and final drive by toothed belts, and the advantages were many. Belts need no lubeing and have a 20,000-mile (3220-km) life; they are quieter and smooth out the drive; they allow a dry (and thus leak-free) chain case, which gives the clutch an easier time. So successful was the system that, like the rubber-mounted motor, it has since been applied to every Harley-Davidson you can buy.

It is odd, however, that the new belt-equipped FXB-80 Sturgis was not designed as a tourer (surely the first market for a clean, long-lasting final drive) but was the

Low Rider. And in every other respect, this was a traditional Harley, with solid-mounted motor and a four-speed gearbox: sales though, convinced Harley-Davidson that belts were the way to go.

This was all very well, but the company itself was in deep trouble. AMF wanted out, and Vaughn Beals actually encouraged the move by recommending that his employer sell its troublesome subsidiary outright. And when AMF management agreed that this was a good idea, it just so happened that Beals had a management buyout proposal ready and waiting. You had to admire the man's subtlety.

The negotiations dragged on into 1981, but eventually, with the help of some massive bank loans, Harley-Davidson's management bought the company back for $80 million. To celebrate, some of them (including Willie G.) rode from York to Milwaukee, calling at jubilant dealers on the way. Once more, the advertisements ran, 'The Eagle Soars Alone'.

But they weren't out of the woods yet: market share was still slipping and quality was not what it should be. New products were on the way, but the company itself had to become more efficient. This happened by adopting Japanese business methods of which there were three in particular: just-in-time inventory cut-down on money tied up in stocks of parts; employee involvement using the skills of people on the shop floor to affect decision making; and every worker was given the means to monitor his or her own quality and output. All three, known as the 'Productivity Triad', had a tremendous effect on efficiency, productivity and quality, the problems which had bugged Harley for decades. More good news came in 1983, when Harley-Davidson's third application for tariff protection against large imported bikes was finally granted; it was for five years only, but gave the company a crucial breathing space.

But for most Harley riders, this all happened behind the scenes; what really marked the 1980s was the Evolution twin. As the name implied, it wasn't all-new (a new top-end on the old Knucklehead-based crankcase) but was as much of a leap forward as the Knucklehead had been nearly 50 years before. Aluminium-alloy cylinders were lighter and ran cooler, a narrower valve angle allowed shorter, straighter ports, the valve train was lighter, the con-rods stronger and tolerances all-round were tightened up dramatically. The result was more power than the old twin, and it was quieter, more reliable and didn't leak oil. Best of all, Harley-Davidson had finally got a new engine right first time.

It cannot be stressed enough that the Evolution V-twin saved Harley-Davidson, and made its 1980s turnaround possible.The other new model in 1983 was less noticeable, but just as significant. For once Bill Davis had built a Harley that cleverly hid its rear suspension units under the gearbox resulting in hardtail looks, but with at least a little rear suspension movement. Vaughn Beals (so the story goes) saw it, liked it and bought the patent. Since then, the Softail, as it was called, has been a hugely successful part of Harley-Davidson's range, combining the looks of a 1940s or '50s Hydra-Glide with a modicum of ride comfort. The Springer Softail of 1989 took that to its logical conclusion, combining the Softail rear end with a redesigned version of the Springer forks that Bill Harley had initiated in 1907. This was the ultimate retro machine.

But even as all these good things were happening, the company was facing collapse and rebellion. In 1985 Citicorp, one of the main financers of the buyout, announced that it wanted its money back, which would have meant bankruptcy. Only days before the plug was pulled, Richard Teerlink, Vaughn Beals and Tom Gelb managed to get new funds from Heller Financial. Meanwhile, long-suffering Harley dealers had become so disgruntled with the new regime (which insisted on lower margins, and new investment in training and marketing) that they formed their own independent association. Gradually, compromises were made and the rifts were healed.

Just as Evolution transformed the big twins, so it did the same for the Sportster. A 54-ci (883-cc) Evo Sportster replaced the old one-litre Shovelhead version in 1984, with a 67-ci (1100-cc XLH following on soon after (soon boosted to the XLH1200 we know now). Still with a four-speed gearbox and chain drive, Harley emphasized the little 883 as a loss-leader entry model. Even now, it undercuts some Japanese cruisers, though even in Evo form, it's slower and less sophisticated than any of them. If the 883 was a sensible cut-price way of boosting sales, the XR1000 did the opposite. Its engine, using modified XR750 parts, was expensively hand-assembled, but the handsome result was still a long way behind Japanese sports bikes. On the other hand, it did form the basis fo the first Buell, which led to greater things.

Only a year after its eleventh-hour flirtation with financial collapse, Harley-Davidson was visibly on the road to recovery. The Evolution had been well received, and sales were booming. It was an ideal time to go public and to raise money by selling shares on the general market. The issue was such a succes that $90 million was raised, enabling Harley to pay off many of its debts with enough left over to buy the motorhome maker Holiday Rambler. In theory, it could have used the money to resurrect the NOVA engine programme, but by then it was clear that the Evolution (originally intended as a stop-gap) would see the company through many years to come: in any case, the buying public expected a Harley-Davidson to have an air-cooled V-twin engine, not a water-cooled four.

By the end of the 1980s, the recovery was complete. The company made a $17.7 million profit in 1987 and $27 million the year after. In fact, it is instructive to compare the beginning of the 1980s to the end: productivity up by half, US market share doubled; inventory cut by 75 per cent, scrap/rework down by two-thirds and an annual profit of $59 million. Police business (once highly prized, and lost in the bad old days of poor quality) began to return and exports increased dramatically; not only that, but it must have caused some satisfaction at Milwaukee when Japan became its foremost overseas market. When Ronald Reagan visited the York plant, it was the ultimate seal of official approval; Harley-Davidson had become the American success story of the 1980s.

In the late 1980s/early '90s, Harley-Davidson nostalgia reached new heights. OPPOSITE: The Springer Softail marked the return of the springer fork. BELOW: The Heritage Softail Nostalgia, in all its black, white and chrome glory.

ABOVE: The Evo Fat Boy: who else but Harley-Davidson could get away with a name like that?

OPPOSITE: The Road King was a kind of stripped-down Electra Glide. This is a 2000 model with Twin Cam 88 engine.

1990s: Ever Upward

If the 1980s was Harley-Davidson's turnaround decade, then the 1990s saw its consolidation. The Harley bandwagon, which set off in the mid-1980s and gathered pace in the early '90s, showed no sign of stopping in 2000, or even slowing down a little. It was a measure of how far the company had come that in 1990 it could call a motorcycle 'Fat Boy' and be taken seriously. The market had changed, and there was a new breed of affluent bikers for whom motorcycling was a weekend hobby rather than a means of transport. For them, it hardly mattered that the Harley was slower, cruder and more expensive than the Japanese opposition. What did matter was that it was way ahead of all of them on high-profile imaging – a solid no-nonsense all-American motorcycle.

The way that Harleys developed through the 1990s reflected this, with more emphasis on styling than engineering. Of course, there have been engineering advances like fuel injection, balance shafts and sealed wheel bearings, but the company has chosen not to invest in anything radically new, even though, perhaps for the first time, it could easily afford to. There were rumours of a 95ci (1560cc) version of the Evo in 1989/90, an attempt to overtake the 85ci (1400cc) Japanese lookalikes. It came to nothing, as the bigger engine reportedly had unacceptable bore and piston wear. That was something else that was different about Harley-Davidson – it now knew better than to launch anything substandard.

The Fat Boy summed all this up. It was the nearest thing to a new Harley model in 1990, though really no more than a Heritage Softail with a few styling tweaks and a new paint job. Frame, tank, mudguards – almost everything was finished in silver, with the 'fat' look emphasized by solid wheels, wide and heavy valanced mudguards and wide tyres. All with a solidly-mounted engine, vibration and all, because that was what some riders wanted. An interesting contrast was the FXRT, also current in 1990. This had a rubber-mounted motor, proper modern (not stylized) fairing, air-adjustable forks and twin front disc brakes. The FXRP police version went down well, but Harley-Davidson sold a mere 304 civilian FXRTs in 1990, and over 4,000 Fat Boys. You couldn't have had a clearer indication of the reasons why people were buying Harleys. Then there was the 'Harleywood' phenomenon in which a new Harley-Davidson became the accessory of choice

LEFT: Top of the range in the early 1990s was the Ultra Classic Electra Glide.

ABOVE: The Sportster was the quiet success story of the 1980s and '90s, bringing many new riders into the fold, especially in its updated five-speed Evo 883 form.

for Hollywood names. Sylvestor Stallone, Cher, Bruce Willis and countless others were happy to be seen in the saddle of a Harley and millionaire Malcolm Forbes even formed his own club for similarly wealthy Harley riders. One less affluent owner wryly observed that in Hollywood, Harley-Davidsons had become like a certain part of the human anatomy – everyone had one!

Meanwhile, certain advances did trickle down through the range. In 1993, the 883 Sportster finally acquired belt-drive and a five-speed gearbox, both long overdue. It was still relatively cheap to buy, undercutting even some of the equivalent Japanese cruisers and played a crucial role in attracting new riders into the Harley fold. It was still doing that in 2000, seven years on from belt-drive, and 16 from the 883 Evolution motor introduction. It also retained a charming simplicity that was

lacking in some of the bigger, more expensive bikes. The Dyna Glide of 1993, on the other hand, had a new frame, but paradoxically the sole reason was to make it look like an old one! The bike was an update of the mid-range rubber-mounted FX series, but the idea here was to emulate the 1970s Low Rider, the engine of which kept its rubber mounts, but were cunningly hidden by the new frame. The names also evoked nostalgia, and the first Dyna Glide was a limited-edition Sturgis (remember 1980), followed by the raked-fork Wide Glide. Mechanically, the Dyna stuck with the 82ci (1340cc) Evolution engine, now in its tenth year and still in a very mild state of tune.

Harley-Davidson's foray into the law courts in the mid-1990s underlined how important its image had become. With the Japanese making big, reliable, good-looking V-twins, Harley became increasingly

ABOVE LEFT and RIGHT: Screamin' Eagle tuning parts could be applied to any Harley. This is a limited-edition Twin Cam Tour Glide so equipped.

LEFT: A 2000 Twin Cam Dyna Glide but, at a quick glance, it could be taken for a machine built 30 years ago.

concerned to protect those things that made it unique. The names 'Glide', 'Sportster' and 'Hog' were all registered as trademarks, and even the distinctive 'potato-potato' exhaust note (a product of the Harleys' 45-degree cylinder and a common crankpin for both con-rods) was subject to a patent application. 'The Harley-Davidson sound functions as a (trade) market and identifies Harley-Davidson alone as the source of the goods emitting that sound.' The company actually withdrew that application after a few years, but the fact that it attempted at all underlines the fact that these bikes weren't being bought for their A to B transport capability.

So strong was the nostalgia element that the touring FLs had been losing sales to the more obviously retro Heritage Softails. So Harley unveiled the FL Road King, a stripped-down tourer that harked back to the original Electra Glide, with no fairing and a detachable pillion seat. There were signs, though, that Harley-Davidson realized that all this wasn't sufficient on its own.

In 1995, fuel injection appeared on the US/Canadian-market Electra Glide Ultra, with electronic engine control according to temperature, load and so on. It didn't make the bike go any faster, but it did improve economy and cold starting as well as emissions. Even sporting pretensions were creeping back into the range: in 1998, the Sportster 1200 Sport gained more power from twin-plug heads, a higher compression and new cams to add to the twin front discs and adjustable suspension it already had.

But it was the final year of the century that saw Harley's biggest news for a long time. The new Twin Cam 88 motor was unveiled, at first for certain models only, but clearly intended to replace the Evolution in time. At 88 cubic inches (1450cc), hence the name) it was the largest production Harley engine yet, though the company had evidently decided it wasn't going to get into a cubic-inch war. Victory (92ci/1507cc), Yamaha (98ci/1602cc) and Honda (110ci/1800cc) all announced even bigger V-twins soon afterwards.

Despite its name, the Twin Cam didn't have twin overhead cams, but two cams on the shaft; in fact, it was still air-cooled and certainly contained no radical changes. But although it looked similar to the Evo, the Twin Cam was very different inside, with a bigger bore/shorter stroke, reshaped combustion chamber, new ignition and lubrication systems, oil-cooled pistons, redesigned breathing and die-cast crankcases. It was claimed to be stiffer, stronger, more powerful and more oil-tight than the old engine. Harley-Davidson also claimed that only 21 parts had been carried over from the Evolution, 'most of them screws', according to the brochure. The Twin Cam was still rubber-mounted, but for 2000 the 88B was launched, with twin contra-rotating balance shafts for what must be the smoothest Harleys yet.

So as the 21st century got underway, Harley-Davidson was facing it with a new confidence after 15 years of almost uninterrupted growth. There was record production, record sales and record profits and it seems a fair bet that it will reach the 100th anniversary of Bill Harley and Arthur Davidson's first prototype in good shape.

ABOVE: The latest incarnation of the Electra Glide, with the 88-ci (1450-cc) Twin Cam motor.

RIGHT: For those who wanted a slightly more contemporary tourer, the Tour Glide was still on offer in 2000.

The 1998 85th-Anniversary 1200 Sportster.

Heading into the 21st century, a 2000 Tour Glide.

A–Z OF HARLEY-DAVIDSON

Abresh Body Shop
Abresh, based in Milwaukee, built sidecars for Harley-Davidson for many years, starting with sidecar and van bodies made to special order. In 1936 it introduced an all-new steel-bodied sidecar, and carried on producing steel sidecars for the factory right up to 1966. The following year, Harley moved over to fibreglass bodies, laid up in the former boat-building facility at its Tomahawk plant.

Accessory Groups
To reduce costs and rationalize the optional extras business in the 1930s, Harley-Davidson reorganized its accessories into Standard or Deluxe groups, which were bought as a pack. For 1936, for example, the Standard group for the side-valve twins was a Safety Guard (engine crash bars), a side stand, a lighted speedometer with maximum-speed-attained hand, Ride Control and a steering damper. If you could afford the Deluxe group, it included dice heads on the switch panel, saddlebags and chrome plating for the fender lamp, stoplight and dice shift knob. In 1937, 90 per cent of new machines were ordered with one or the other of these groups.

Prest-o-Lite (press-to-light) reservoir for acetylene lighting. Like other manufacturers, Harley-Davidson moved to electric lighting as soon as possible.

Acetylene Lighting
Before charging systems and batteries became reliable enough to make electric lighting universal, motorcycles (Harleys included) used acetylene lights. To switch them on (though 'switch' probably isn't the right word), you turned on a tap, which dripped water onto carbide pellets which gave off carbide gas, which could be lit. The result wasn't searchlight power, but was better than nothing. In fact, a book entitled the *ABC of the Motorcycle* (published in 1910) states: 'There is only one lamp to use on a motorcycle and that is the acetylene ... Electric lights such as are used on automobiles are out of the question as they require too much extra power over that generated by the magneto.' The author was wrong: Harley-Davidson was offering electric lights on the J-model in 1915, and by 1926 all the road-going twins were so equipped.

Advertising
Harley-Davidson's advertising has become something of an art form in itself. The Milwaukee advertising firm of Klaus, Van Peterson and Dunlap obtained the Harley-Davidson contract in 1907, and incredibly, kept it for the next 50 years. Not until well after the AMF takeover was there a serious change in advertising policy, but when there was, it caused a serious rift with AMF. In 1976, AMF hired the prestigious Chicago advertising agency, Benton & Bowles, for all its recreational subsidiaries, of which Harley-Davidson was one. All

The Aermacchi era: Harley-Davidson's attempt to compete with the Japanese through the 1960s and '70s was via the Italian-made Aermacchis, which were fitted with H-D badges and sold in North America through H-D dealers. Not all dealers were pleased with this development, deriding the bikes as 'not real Harleys', which was true enough. However, they did give Harley a toehold in the small bike market, and thousands were sold each year. Whether this was worth the company's while in the long term is less clear: there were serious problems of communication with the Italian factory and quality control was a permanent headache. LEFT is a period 1970s advertisement for the 11 and 15-ci (175 and 250-cc) two-strokes, the last Aermacchis to be sold as Harley-Davidsons. Below is the neat four-stroke engine unit of the 250 Sprint, the first Aermacchi to wear an H-D badge.

Harley-Davidson national advertisements now had to adhere to strict guidelines. The trouble was that B & B had no experience of the motorcycle market, and the advertisements it produced were unsuitable. Harley-Davidson got around this by expanding other forms of publicity, such as dealer co-operative schemes, and direct mail. This was not popular with AMF, particularly when Harley-Davidson placed a full-page ad in *Easyrider* magazine for, at the time, AMF wanted to distance itself from the 'outlaw' end of the market. In the end, the parent company relented, and put Harley-Davidson back in charge of its own advertising budget.

Aermacchi

The Italian motorcycle maker, bought by Harley in 1960, and sold in 1978. Harley's reason for buying into Aermacchi was simple: by the late 1950s, it was clear the company needed to get into small motorcycles as British imports, the Triumphs, Nortons and BSAs, and to a lesser extent the Italians, were dominating the market. Although Honda had yet to secure its 'Nicest People' revolution, small motorcycles were increasingly popular, and Harley was losing market share as a result. All it could offer was the outdated 7ci (125cc) Hummer, and the quickest solution was to buy someone else's small bikes and stick Harley-Davidson badges onto them.

Aermacchi seemed like the obvious partner: it had been an aircraft manufacturer before the Second World War, but turned its attention to motorcycles in 1954. The first prototype was actually electrically-powered, but a 125cc machine, the 125N, soon followed. It was really a cross between a bike and a scooter, with a step-through fame (scooter-style) and motorcycle-type dummy fuel tank. Others followed, notably the futuristic looking 10.5-ci (172-cc) Chimera in the mid-50s, while in competition Aermacchi secured new 50cc and 75cc speed records.

But the company was small and in a weak position, despite gradually improving sales of its 11 and 15ci (175 and 250cc) four-stroke singles, with their characteristic horizontal cylinders. This was what attracted Harley: the Aermacchi singles were modern and lightweight for their time, but the Italians needed investment to carry on making them. A deal was struck: Harley-Davidson bought the whole caboodle, and Aermacchi began exporting the singles to America, complete with Harley-Davidson badges. They were successful, and despite fierce competition from the Japanese continued to sell right into the 1970s, though were looking increasingly outdated by then, despite the later addition of electric start.

But Aermacchi was not standing still, and 1967 saw the introduction of the M125

Rapido, which formed the basis of a whole new generation of two-strokes. Off-road bikes of 5 and 6ci (90 and 100cc) followed, and the much-updated TX125 trail bike in 1973. A 175 version followed, then a 250, which was fast but more expensive than the opposition. In fact, the pace of development was so fast that Aermacchi just could not keep up with the Japanese, and sales began to drop off. In 1978, faced with its own problems at home, Harley-Davidson sold the Aermacchi concern to Cagiva, a fledgling Italian bike maker which utilized the factory to good effect.

AFL-CIO
Awarded the Union Label to Harley-Davidson and four trade unions in 1988, in recognition of 'the foresight and co-operation shown by unions and management in helping to revitalize the company'. The Allied Industrial Workers Union and Machinists and Aerospace Workers shared the award.

Aircraft
There have been a couple of attempts to produce Harley-Davidson-powered light aircraft. Harvey Mummert worked for the Curtiss-Wright Aviation Company in the early 1920s, and happened to be a Harley-Davidson enthusiast. He built a wooden-framed monoplane, powered by a J-model 61ci (1000cc) V-twin. Lightly modified V-twins usually shook light aircraft to pieces due to the vibration, but Mummert had balanced his for its 1,600–1,800rpm running speed and the plane worked well, taking second place in the National Air Race. Harvey Mummert later built a small racing biplane for Sarter Tiffany, this time with the J-twin inverted and which also competed successfully.

Thereafter, most light aircraft with motorcycle engines opted for the Henderson or Indian fours, with their inherently lower vibration. An exception was the one built in the early 1930s by plane enthusiast Lester Long, who built up a flat-twin motor for aircraft from J-model cylinder assemblies and a new crankcase. The resulting 80ci (1310cc) unit was well balanced and produced 30hp, able to spin a 5ft 2-in (1.575-m) propellor at 2,750rpm. Long produced 50 of the engines himself, after Walter Davidson had shown no interest in producing it in-house.

Albrecht, Paul
A Harley-Davidson competition rider in the 1950s.

Alexander, Ronnie
A racing mechanic who worked on Ray Calborn's XR750s in the early 1970s. He was later recruited to the company's quality audit department, which had its work cut out from the late 1970s after years of quality being a poor relation to production.

Allchin, T.R.S.
A prominent English Harley-Davidson rider in the 1920s.

Allen, Irwin D.
President of the Federation of American Motorcyclists, which was voluntarily dissolved in 1919 because of declining membership and an inability to control motorcycle sport. It was replaced by the American Motorcycle Association (AMA).

Allied Industrial Workers Union
Major trade union at Milwaukee, representing production and maintenance workers.

Allis-Chalmers
Harley did much subcontracting work for this well-known maker of tractors and other agricultural machinery during the Second World War. Among the more interesting projects was a powered snow-sledge, a suitable engine being the side-valve flat-twin XA motor that Harley-Davidson was developing during the war. This itself was heavily based on the German BMW and the XA motorcycle (of which a few were built for the military) and came complete with shaft drive. Two XA power units, with three-speed and reverse gearbox, were supplied to Allis-Chalmers, but no more was heard of the project.

Alzina, Hap
An Indian dealer in Oakland, California during the 1920s. He later became Indian's west-coast distributor and was so successful that, according to author Harry Sucher, it was Alzina's sales and profits that kept Indian afloat during the difficult years from the late 1920s to the early '30s.

AMA (American Motorcycle Association)
America's governing body for motorcycle sport and manufacturing. It was founded in 1920 after the pioneering Federation of American Motorcyclists (FAM) was voluntarily wound up. Although originally intended as an independent democratic organization, the AMA later became dominated by trade interests. An example was the institution of the 21 cubic-inch racing class in the 1920s, which just happened to favour Harley-Davidson's new 21ci (350cc) single. The same thing happened in the early 1970s, when the AMA changed flat-track rules to allow overhead-valve 750s to compete and Harley's XR750 fitted the bill. What the AMA did achieve, however, was proper organization of motorcycle sport in America, which had previously been lacking.

Americade
Previously the Aspencade Motorcyclists Convention, now renamed and moved to the town of Lake George, New York. Unlike the custom-orientated Sturgis and Daytona, this one is for touring bikes, though like them, it is Harley-Davidson-dominated.

American Association of Motorcycle Road Racers
A group of racers in the 1960s which, increasingly dissatisfied with the AMA's restriction of non-approved events, formed its own organization. When big names, like Dick Mann, began to compete in these 'unofficial' events, it soon became clear that the AMA would have to move with the times and accept different forms of competition that were becoming increasingly popular with the public. It was in these exchanges, often through the pages of *Cycle World* magazine, that the AMA was increasingly accused of being dominated by Harley-Davidson and not representative of American motorcycling in general.

American Car & Foundry
The original name of the American Machine and Foundry Company (AMF), which took over Harley-Davidson in 1969. It started off in the railway rolling stock

business, and diversified into yachts, bakery and tobacco, among other things.

American Machine & Foundry Company (AMF)

AMF was the huge engineering concern that took over Harley-Davidson in 1969, selling it back to the management 12 years later. It was a big firm, based in engineering but with a desire to expand the leisure side of its business: moreover, chairman Rodney Gott was a Harley rider! In the event, AMF forestalled a hostile bid for Harley-Davidson by Bangor Punta. Since then, AMF has not had a good press, being blamed for everything that went wrong at Milwaukee in the 1970s. AMF-era Harleys were for years seen as a real low point, regarded as bad bikes produced by a profit-motivated company that had no knowledge of or interest in motorcycles. It's an unfair assessment. While it is true that AMF was motivated by the growing motorcycle market, the Harley-Davidson it acquired was under-capitalized, with outdated plant producing too few bikes. So AMF spent a lot of money re-equipping Harley's Juneau Avenue factory (built in the 1920s) and moved motorcycle assembly to its own plant at York, Pennsylvania. It also began R&D work on a radically updated V-twin and an all-new range of water-cooled machines. None of this was the work of a heartless asset-stripper.

The problem was, in its anxiety to make the best of a booming bike market and see a return on its investment, the company forced up production rates much too fast: Harley-Davidson built 27,000 bikes in 1969, and 60,000 three years later, and this was dictated from the top, with little regard to quality. Although there were assurances that moving production away from Milwaukee would not mean redundancies, some people did, however, lose their jobs. The end result was a disgruntled workforce, frustrated management and sceptical dealers, left to cope with the poor quality machines that AMF was expecting to sell. But in the final analysis, despite the mistakes AMF made and its heavy-handed approach to reorganization, there is a good case for saying that it actually saved Harley-Davidson. The company stood little chance of surviving on its own into the 1970s, having run out of money: it had nowhere else to go.

American Motorcycle Manufacturers Association

The first organization for US motorcycle makers, founded in 1914, with Arthur Davidson on the board of directors. It was instrumental in the setting up of the AMA when, at the 1919 annual meeting, it was moved that the manufacturers take over the responsibilities of FAM, which they did. This formed the nucleus of the AMA. It was later renamed the American Motorcycle and Allied Trades Association.

American Stock Exchange

For a company approaching its 100th birthday, Harley-Davidson has had very little to do with stocks and shares. The reason is simple. For the first 60 years, it was owned by the Harley and Davidson families and only went public in 1965, when in dire need of fresh capital. AMF took over the whole thing in 1969, selling the company back to the management team in 1981. Five years later, to raise more capital, Harley-Davidson finally went public again, and was floated on the American Stock Exchange; in other words, anyone with enough funds could buy a few Harley shares, and lots of people did – not

When rebadging Aermacchis as Harleys and floating itself on the Stock Market wasn't enough to keep the company going, Harley-Davidson bowed to the inevitable and sold out to the American Machine & Foundry Company. The years that followed would be seen by many as a disaster, with over-production of poor quality machines by an ownership which didn't understand the motorcycle. The reality wasn't quite as simple, and there's a case for saying that Harley-Davidson would not have survived the 1970s without the AMF takeover. For better or worse, AMF logos began to appear on Harley-Davidson fuel tanks (LEFT) and in advertisements (BELOW). However, when the company was eventually bought out by its own management in 1981, these logos were the first to go.

only traditional investors, but Harley-Davidson dealers, employees and riders. Some later admitted that they intended to sell the shares at a profit (they nearly tripled in value in four years) and buy a bike with the proceeds!

Amtorg Company
Amtorg was established by the Soviet Union and based in Washington during the Second World War, and one of its tasks was to source motorcycles and send them back to Russia. A Mr Karsov was in charge of motorcycle buying, and had bought some Indians in early 1940. He'd been doing the same job back in 1924, when Charles Cartwright, Harley-Davidson's salesman in New England, had helped him buy some Harleys on credit. It was no surprise then, when the two men met again, that the Soviet Union switched back to Harleys.

Andres, Brad
Brad Andres wasn't a factory-backed racer, but achieved great things with the side-valve KR in the 1950s. He won the National Championship in 1955 and was second to Joe Leonard the following year. The KR itself was something of a minor miracle. Despite an apparently antiquated side-valve motor (with roots going back before the Second World War), it won the American flat-track National Championship 13 times.

Andres, Leonard
Father of Brad Andres, Leonard was a Harley-Davidson dealer and tuner who took the young Ray Calborn under his wing and encouraged his early career.

Antique Motorcycle Club of America
The club caters for machines over 30 years old, and anyone interested in old bikes can join. Contact Dick Winger, Antique Motorcycle Club of America, 2008 Fitchburg Rd, Stockbridge, Massachusetts 49285.

In the context of American racing, and like its predecessor, the WR, the KR racer was developed into a very competitive bike, drawing on a couple of generations of side-valve tuning know-how. ABOVE: Notice how the bike has been made radically lighter.

Appeal Manufacturing & Jobbing Company
Held the Harley-Davidson dealer franchise in Los Angeles from 1918, but was soon replaced. The problem was that the south-west and far-west of America were developing into very significant markets, making up half of all domestic

The Antique Motorcycle Club of America's national show at Davenport. Harleys, like this Duo Glide, naturally feature prominently among club members.

sales. Distances were great, and there were many enthusiasts in the area. Unfortunately, Appeal Manufacturing lacked ambition: it refused to advertise, there was no planned sales campaign and certainly no hard, aggressive sales techniques. The result was that Indian, Excelsior and other makes sold far more bikes than Harley-Davidson.

Arthur Davidson, then sales manager, acted quickly, persuading Rich Budelier, who had made a great success of selling Harleys in San Diego, to take on the Los Angeles franchise. Another factor in boosting Harley-Davidson sales in this important market was the decision to hold stocks of machines in warehouses in Los Angeles and San Diego. Previously, Western dealers had to order specific bikes from Milwaukee, and wait.

Appel, George
Appel worked for electrical components maker Remy in 1919, and it is said that he was hired by Harley-Davidson to sort out the problems with its own electrical components. At first, the electrically equipped J-model from 1915 used the Remy generator, lighting and horn, but Milwaukee introduced its own electrical equipment on the Model 20 J in the summer of 1919. The official reason was that Remy components were no longer up to Harley standards: in reality, the in-house equipment had plentiful problems of its own, and some dealers ended up replacing it with Remy parts.

Appelle, Harvey
A senior loan officer with Citicorp Industrial Credit, the company which largely financed the management buyout of Harley-Davidson from AMF in 1981. An early meeting between Appelle, Harley-Davidson chairman Vaughn Beals and Gary Ward (the AMF man in charge of Harley-Davidson's sale) concluded that a leveraged buyout was possible.

A.R. Callow, Ltd
Harley-Davidson's early importer to South Africa, which made a slow start (39 machines in the first 18 months), though a visit from Alfred Child soon put things right. On the same trip, he sold 50 bikes to an African dictator.

Arena, Sam
A versatile rider who raced Harleys in hill climbs, TTs and flat tracks, winning races from 1934 until after 1950, when he took the classic racer's second career and became a dealer. According to author, Allan Girdler, Arena was one of the dealers who, after being shown the first XL-series Sportster, encouraged Harley to produce the stripped down XLC. That bike, with its tiny tank and (later) small

LEFT:
Harley-Davidson didn't completely ignore smaller bikes in the 1980s, though the Armstrong MT500 wasn't offered to the public. In fact, only the military qualified.

RIGHT: Art Deco styling used in the 1930s.

BELOW: The Arrow-Flite tank badge on a 1950s Duo Glide.

OPPOSITE: The atmospheric inlet valve, seen here on a 1905 single, severely limited revs and therefore power. However, a pushrod inlet valve soon replaced it.

headlamp, was the first with the classic Sportster look which survives today.

Armstrong, Erle
'Red' Armstrong was a racer in the pioneer days, riding Indians. He notably won the first race at the new board track at Tacoma, Washington, in August 1915. Average speed was 79.84mph (128.5km/h), a wheel length in front of Harley-mounted Otto Walker.

Armstrong MT500
A 30ci (500cc) off-road military motorcycle, designed by Armstrong Equipment of North Humberside, England. Harley-Davidson bought the design and manufacturing rights to it in October 1987, in an attempt to broaden its product base. The MT500 was powered by a single-cylinder Rotax engine.

Arrow-Flite Badge
Distinctive tank badge used by Harley which first appeared in 1953 and was used for the remainder of the 1950s.

Art Deco
For their first 30 years, Harley-Davidsons were hardly a byword in style. They were designed to be solid, reliable machines, and the standard colour (after the original grey) was olive green. But the 1930s heralded a new era of stylish design: more to the point, arch-rival Indian had been taken over by the du Pont chemical company, and was now offering a new range of bright colours. Rivalry between the two was so intense that Harley-Davidson had no choice but to respond, and from 1933 the side-valve Harleys came in a range of bold colours and eye-

catching graphics in the art deco style. This was a new era of speed and elegance, and Milwaukee did its best to keep up.

Artley, Roy
An early racer who was part of the famed Harley-Davidson Wrecking Crew racing team in 1920. In partnership with pioneer rider William J. Ruhle, he later took over the San Diego dealership of Hap Scherer, who himself had moved on to Los Angeles.

Artman, Bob
Appeared on a Harley in the 1933 Columbia movie, *State Trooper*.

Aspencade Motorcyclists Convention
(*See* Americade.)

Atmospheric Inlet Valve
Some early engines, including those of Harley-Davidson, relied on piston suction to open the inlet valve when, as the piston descended, the valve opened, with a light spring to close it again. It was a simple system, but limited engine speed to around 500rpm (that's 500rpm, not 5,000) which was fine for the pioneer single-cylinder machines. But as the demand for more power grew, so such lowly speeds became inadequate. This was made clear to William Harley in 1909, when his prototype V-twin produced little more power than the single from which it had derived. The answer, which appeared on the first production Harley twin in 1911, was a pushrod-operated inlet valve whose positive operation allowed higher engine speeds and thus more power.

Aurora Automatic Machine
Builder of the Thor motorcycle, the company also supplied rear hubs and brakes to Harley-Davidson from 1910–13. William Ottaway was chief engineer with Aurora, but was soon poached by Bill Harley as an able assistant engineer. (*See* Ottaway, William.)

Aykroyd, Dan
Actor Dan Aykroyd, who starred with John Belushi in the film *The Blues Brothers*, rode a Harley. After John Belushi died, Aykroyd led the funeral procession on his bike.

Bagnall, William
President of the AMA from 1965, also editor and publisher of *Motorcyclist*.

Baker, Erwin
Indian racer in the second decade of the 20th century, who also made a speciality of endurance record attempts, which he completed on Ace, Excelsior, Henderson and Indian bikes. According to author Harry Sucher, he never did the same for Harley-Davidson because Walter Davidson wouldn't agree to pay his standard fees. Also co-designed (with Everett deLong) a 90-ci (1475-cc) V4 that was tried in an Indian Powerplus frame. DeLong later built a four-cylinder prototype for Harley-Davidson, which the Board rejected.

Balcom Trading Company
Established in 1945 to import Harleys into Japan, with headquarters in Takanawa, and 35 years later, it was still doing so. The founder was Richard Child, son of Alfred, who had built up Harley-Davidson business in Japan from nothing. (*See* Child, Alfred and Richard.)

Bang, Gary
Maker of aftermarket custom parts for Harleys.

Bangor Punta
A massive company that grew out of the railway industry, it attempted a hostile takeover of Harley-Davidson in 1968. Although Harley-Davidson needed cash at the time, Bangor Punta's approaches were unwelcome, and for good reason. Roger Hull, then editor of *Road Rider* magazine, considered that the reason was down to Bangor Punta's reputation: 'You can't blame the Harley-Davidson people for getting nervous,' he said later, 'Bangor Punta had a reputation for acquiring a company, squeezing it dry and then scrapping it for the salvage.' It was Harley-Davidson's fear of being sucked dry by Bangor Punta that drove it into the arms of AMF, which took over shortly afterwards. 'I decided to play the good guy,' said Rodney Gott, boss of AMF, 'and went to Milaukee to make friends with Harley management. I showed them how to combat hostile takeovers ... Bangor Punta was the bad guy and we were the good guys, so it was pretty easy to work out a deal.' The result was that Bangor Punta offered $23 million to take over Harley-Davidson, while AMF offered a million less, and still clinched the deal.

Bank of New England
One of the four banks that helped finance the management buyout of Harley-Davidson in 1981. Citicorp Industrial Credit was the main player.

Barger, Sonny
Pictured below, the leader of the Hell's Angels in California during the 1960s. He rode a Harley. (*See* Hell's Angels.)

Barnes, Austin
A commercial photographer who, in 1938, was ordered by Walter Davidson to take pictures of a dismantled Crocker V-twin. Harley-Davidson was suspicious that the Crocker, which could outrun both Harleys and Indians at the time, was copying Harley-Davidson parts without permission. Austin's pictures showed nothing untoward.

Barr, Dave
Rode his 1972 Harley Super Glide 83,000 miles (133572km) around the world in the early 1990s. Barr had served in the US Marines and had lost both his legs below the knee while fighting in Angola. Determined to raise awareness of the plight of disabled people, he mounted his Super Glide (owned from new, and already with 150,000 miles/24140km under its wheels) and set off. Starting in Johannesburg, he rode up through Africa ('I'd never have made it through Zaire without belt drive'), where floods were the biggest problem, as there was no way of knowing how deep the water was. He was almost drowned when the bike keeled over and landed on top of him. Dave was just about able to keep his mouth above water level and hold on until a couple of villagers came along and rescued him. After Africa, Dave explored Europe and Scandanavia and crossed the Arctic circle before returning to the States. Then up to Alaska, back home to California, then down through South America, by boat to Hong Kong and by Harley through China, across the Gobi Desert and Mongolia, through Siberia, Moscow and Europe again. Dave Barr later wrote a book about his experiences on the road, entitled *Riding the Edge*, published by White Horse Press.

LEFT: Sonny Barger, outside the Oakland Hell's Angels headquarters.

OPPOSITE: Bear Butte: a familiar landmark to anyone who has made the pilgrimage to Sturgis.

Barr, Frederick
Joined Harley-Davidson's engineering department in 1908, under George Nordberg, and was a key figure there for over 20 years. Also helped Nordberg and William Davidson to expand production in 1914. He finally left in 1929 to become chief engineer to the Sankyo company in Japan. His job there was to set up production of Harleys under licence at Shinagawa, a deal which had been masterminded by Alfred Child. Harleys were initially built up from parts imported direct from Milwaukee, but by 1935 Sankyo was making all-Japanese Harley-Davidsons.

Barry's Custom Cycles
Founded by Barry Matteson in Anchorage, Alaska in 1971, where he built and sold custom and tuned Harleys. He became a fully-fledged Harley-Davidson dealer in 1976, and despite Alaskan weather (not always conducive to motorcycling) did well.

Barth Manufacturing Company
Based in Milwaukee, where Bill Harley and Arthur Davidson both worked, met, became friends and went on to build a motorcycle.

Battistini
A top custom bike builder, based in the UK, which has set the standard for custom Harley building in Europe and beyond. Rikki Battistini now runs operations from the company's new headquarters in Glasgow, Scotland.

Battle of the Twins
Contemporary race series for twin-cylinder machines, dominated by Ducati in Europe and Harley-Davidson in North America.

Bauer, Louis
Boss of Indian in the 1920s, who in 1928 teamed up with Arthur Davidson and Schwinn of Excelsior to rescue the AMA.

Beals, Vaughn
A key figure in Harley-Davidson's turnaround in the 1980s, and manager in Milwaukee for AMF from 1975. Beals knew nothing about motorcycles (at least at first) but was a highly experienced and competent manager. One of his first actions was the so-called Pinehurst Meeting (*see also* below). Until then, Harley-Davidson had no long-term strategic plan, so Beals took all the senior managers up to Pinehurst, North Carolina for a week. At the intensive brain-storming session, it was agreed that Harley-Davidson urgently needed more modern, better-quality bikes to compete with the Japanese, and a ten-year product plan was drawn up. First, the existing V-twin was to be updated (which eventually saw light of day as the Evolution V-twin in 1983). Second, a whole range of all-new water-cooled engines would be developed, the NOVA programme. Far from being an AMF placeman, it was Beals who

originally proposed that Harley-Davidson senior management buy the company back from AMF, and it was he who negotiated the deal. As chairman of the newly-independent Harley-Davidson, he oversaw the huge changes in production through the 1980s which enabled the company to survive, putting quality first, with employee involvement and just-in-time methods.

Bear Butte Café
Part of the Sturgis experience, the intriguingly named Bear Butte Café sells a lot of beer while the South Dakota festival is on.

BEARS Racing
Acronym for the British, European and American Racing Series which allowed classic Triumphs, Ducatis and Harleys to battle it out on the track, without being constantly overtaken by Japanese bikes.

Beauchamp, Rex
Early 1970s flat-track racer.

Becker, Oscar
Oscar and Sherbie Becker were among Harley-Davidson's very first employees. An early photo of three of the Founders, plus eight workers grouped around a Harley sidecar outfit, shows Sherbie sitting on the bike with Bill Harley on the sidecar chassis.

Bedell, Alan
An endurance rider in the early days, in 1917 he rode 1,000 miles (1610km) non-stop at the Ascot Park race circuit, averaging 48.3mph (77.7km) for nearly 21 hours. He also set new records for the 24 hours (48.1mph/77.4km/h), 12 hours (49.5mph/80km/h) and 500 miles (50.1mph/81km/h).

Beer
There really was such a thing as Harley-Davidson branded beer and, given Milwaukee's importance in the American brewing industry, maybe it's not so surprising the Harley-Davidson version was not brewed in the Juneau Avenue plant alongside half-finished V-twin motors, but came in cans complete with the famous Harley-Davidson logo, now collectors' items (if they haven't already been drunk).

Bell & Howell Company
Harley-Davidson did subcontract work for Bell & Howell during the Second World War, producing telescopic equipment, among other products.

Belt Drive
For the first ten years, Harley-Davidson used the leather belt drive that was almost universal on early motorcycles; then, like almost everyone else, it went over to chain drive, which was more efficient, did not slip and could handle higher power. By the late 1970s, however, a popular conversion kit for Harleys was a toothed belt drive for both primary and final transmission which smoothed out the power pulses somewhat and needed minimal maintenance. The new FXB Sturgis of 1980 was the first production motorcycle to use a toothed belt (Kawasaki later used one on its GPz305). So successful was the new set-up that 20 years later every new Harley-Davidson comes with a belt-drive transmission.

Bendix Manufacturing Co
Harley-Davidson was forced to defend itself in a legal action instigated by Bendix in 1924 when Victor Bendix claimed that the Harley-Davidson clutch design infringed his own patent. Bill Harley had designed a multi-plate clutch very similar to the Bendix, and following legal advice, the Harley-Davidson version was rapidly redesigned and $18,000 was said to have been paid in damages.

Benton & Bowles
The advertising agency used by AMF, but which was not popular with Harley-Davidson. (*See* Advertising.)

Berndt, Ralph
Independent tuner who worked for Harley-Davidson but built his own race bikes. He was instrumental in the early career of talented racer Carroll Resweber in the 1950s. (*See* Resweber, Carroll.)

Bianchi, Alfredo
Designed the Aermacchi overhead-valve single-cylinder engine that later developed into the 15 and 21ci (250 and 350cc) singles sold by Harley-Davidson under its own name.

LEFT: Harley bicycles didn't last long, but the time may now be ripe for a Harley or Buell-branded mountain bike.

BELOW LEFT: There really *was* a Harley-Davidson beer.

BELOW: The earliest belt drive, and very different from Harley's later interpretation. There was no clutch: you just jumped on and got going.

OPPOSITE
TOP LEFT: The billet look, popular in modern customizing.

TOP RIGHT: A typical Big Twin (not a Sportster).

BELOW: Battle of the Twins racing enabled Sportsters to compete again.

Bicycles
For a short time, Harley-Davidson sold bicycles bearing its own badge, intending to cash in on the booming market for cycles and the brand's own rapidly growing status. According to the advertisements, it was 'the bike red-blooded fellows choose', and 'America's finest bicycle'. The idea was that red-blooded fellows would go on to buy Harley-Davidson motorcycles when the time was right, but the idea did not survive the tough times of 1920–21, and Harley withdrew from the bicycle business.

Big Twins
Generic term referring to all the larger V-twins in each era which included the 61/74ci (1000/1200cc) twins of the 1920s and '30s, the Knucklehead, Panhead and Shovelhead up to the early '80s and the Evolution and current Twin Cam twins. Sportsters are not Big Twins.

Billets
Parts machined from solid aluminium blocks, usually by CNC machines, form a large part of the aftermarket bolt-on custom parts market. Too many of these and your bike will be labelled a 'billet barge'.

Black Hills Motor Classic
The Black Hills rally began in 1938 but is better known as Sturgis, named after the small farming town in South Dakota which serves as a centre for the week's events. Now as famous as Daytona as one of the main (Harley-dominated) events in the American motorcycle calendar.

Blake, Robert
(See *Electra Glide in Blue*.)

Bleustein, Jeff
Harley-Davidson's head of engineering, an

associate professor of engineering at Yale, who was recruited in the middle of the AMF era by Ray Tritten. He really professionalized the engineering department at Milwaukee, bringing in highly qualified people who didn't necessarily know much about motorcycles, but had good track records in other fields of engineering. This wasn't always popular (the traditional Harley way was less academic), but in the end it bore fruit. Bleustein oversaw the development of the Evolution V-twin, which was crucial to Harley's 1980s turnaround, as well as the NOVA programme of water-cooled engines, which never got beyond the prototype stage.

Blue Knights
A motorcycle club whose members all belong or have belonged to the police, fire or emergency services.

BMW-Delco
The Second World War military motorcycle, built by the Delco company but closely modelled on the contemporary BMW flat twin, with shaft drive and telescopic front forks. The US Army was most impressed with this layout, and despite Bill Harley's view that an enclosed chain was just as practical as shaft drive, insisted that Harley-Davidson build something similar. This was the XA, which like the Delco was closely based on a BMW.

Board Track Racing
An early form of racing in America, using banked tracks made up of narrow wooden planks. This had a brief popularity, with tracks being built all over the country. But although smooth compared to contemporary roads, they offered far less grip, and the splinters could produce horrific injuries when riders fell off. They were soon superseded by oval dirt tracks.

ABOVE LEFT: The Blue Knights back-patch club.

ABOVE: The XA flat-twin engine unit, based on the equivalent BMW.

LEFT: An early board-track racer

OPPOSITE:
LEFT: Bootlegged (unlicensed) accessories were a major headache for Harley-Davidson, until it called the lawyers in.

RIGHT: The Boozefighters was one of the original post-war 'outlaw' bike clubs.

Boardwalk Custom Show
An integral part of Daytona Bike Week, the Boardwalk has grown from a side event into a show in its own right. Custom bikes of all shapes, sizes and types can be seen there every year, though it's fair to say that Harleys still predominate.

Boeing Aircraft Company
Boeing was just one of the companies for which Harley-Davidson did subcontract work during the Second World War, Allis-Chalmers and Massey-Harris being others. Shell parts, aircraft parts and over 13,000 truck winches were produced alongside 88,000 motorcycles.

Bond, John R.
Worked in Harley-Davidson's engineering department from 1946–47, and later edited *Road & Track* magazine. He noted the paradox that, although senior management encouraged new and innovative prototypes, these were never incorporated into production, hence the company's conservative model policy after the war.

Bonus Bucks
A dealer incentive scheme of the 1980s, except that the Harley-Davidson dealers didn't see it that way. Their profit margins on the big twins were cut by 5 per cent, but they could earn that back by taking part in company-run schemes: demo rides, advertising, stocking levels, service training courses. Feelings ran high, and 32 dealers actually broke away to form their own dealer organization, the Dealers Alliance. Relations between the DA and Harley-Davidson were strained until Harley-Davidson modified the Bonus Bucks scheme and made its official dealer organization elective. (*See* Dealers Alliance.)

Bootlegging
Harley-Davidson is one of the strongest brands in the world, so it's hardly surprising that many companies have tried to cash in on the name with unauthorized 'branded' products. Ranging from racist T-shirts to counterfeit spare parts and pot-smoking accessories, these reflected badly on Harley-Davidson itself. So from the early 1980s the company made a big effort to regain its own trademark, winning a big court case and getting into the licensed parts and accessories business. It meant strict control of what the Harley-Davidson logo could be appended to, such as lighters, pens, toys, T-shirts, the list is endless.

Boozefighters
One of the biker groups that grew up in the American West in the late 1940s/'50s, which (together with the events at Hollister) pre-empted mass media coverage of 'motorcycle hooligans'. Harley-Davidson was careful to distance itself from the 'outlaw' culture though, ironically, its success is now based on a stylized, sanitized version of exactly that.

Boston Consulting Group
Firm of industrial analysts which reported on the state of Harley-Davidson in 1978, towards the end of the AMF era. Its conclusion was that it could be viable, but only if AMF invested $60-70 million in the NOVA range of all-new water-cooled engines. Before NOVA bore fruit, AMF would have had to have undergone a 'valley of death' where investment was bound to exceed income. AMF, as we now know, chose not to go down that road.

Brinck, Eddie
A racer in the 1910s, Eddie Brinck was one of the original Harley-Davidson Wrecking Crew, which dominated US competition from 1915–21.

British Imports
America imported British motorcycles by the thousand in the 25 years after the Second World War. These were different from Harley-Davidsons, being light, nimble 350/500cc singles that introduced a new generation to bikes. By 1950, imports had captured 40 per cent of the market, which was why Harley-Davidson tried to have an extra import tariff imposed. Not until the late 1950s, with the XL Sportster, did the company have anything really comparable with a Norton or a Triumph. Maybe it was the experience of Indian, which had tried to meet the imports head-on with small home-built bikes, but which soon succumbed to bankruptcy, that encouraged Harley-Davidson to persevere with its conservative model policy.

British Motorcycle Dealers Association
Formed to represent those who sold British bikes in America after the Second World War. They had a legal battle with Harley-Davidson in 1951, when the company sought to have the duty on imported bikes increased from 8 to 40 per cent, and imports restricted to pre-war levels – a tiny fraction of what was being sold by then. The court found against Harley-Davidson, and in the next 20 years, dealers sold ever-increasing numbers of British bikes in the US.

Brough-Superior
The Brough-Superior motorcycle contrasted interestingly with Harley-Davidson. It was hand-built in England (allegedly by technicians wearing white gloves!) and it was expensive. By the late 1920s, most V-twins in Europe were sold as exotic, sporting machinery while in America big V-twins were the standard production motorcycle.

Buddy Seat
Before the buddy seat, the only way to carry a passenger was on a separate seat on the rear mudguard (uncomfortable in those hardtail days) or in a sidecar (rapidly losing favour in the early 1930s). The buddy seat was a joint invention of William S. Harley and Frank Trispel, foreman of the experimental department. It was an elongated saddle, flaring wider at the back, which was just big enough to take a passenger and rider in close proximity. Indeed, for many, this was part of the attraction, and a set of auxiliary springs could be clipped on to cope with the extra weight. It proved very popular, and was Harley-Davidson's best-selling accessory in the 1930s.

OPPOSITE
ABOVE: Brough-Superior typified the difference between European and American V-twins. On the smaller, twistier roads of Europe, V-twins were big, exotic and expensive. In America, they were a necessity.

BELOW: This 1937 ULH shows its gigantic white buddy seat to good effect.

RIGHT: The classic WR racer, still to be seen on the track today.

Burkey, Eric
President of the IAM Lodge at the York factory. With others at Harley-Davidson, he went on a tour of the Honda factory at Marysville, Ohio to take a look at Japanese production methods. He later recalled his amazement that Harley-Davidson engineers from York were making hurried sketches (no cameras allowed) of how the Honda workers were bleeding brakes. This was the same method that had been used at Milwaukee for years, but no one at York appeared to have heard of it. This anecdote is a good illustration of one of Harley-Davidson's central problems at the time – two plants hundreds of miles from each other which didn't appear to be on speaking terms.

Buschmann, John
Worked for Harley-Davidson for just three months from late 1965–66. He was behind a stolen parts scam in which new parts were smuggled out of the factory and sold to the unsuspecting. The alarm was raised when one Milwaukee dealer was offered 10,000 Harley spark plugs at 50 per cent of the cost of production. It transpired that 29,000 plugs were missing. The FBI was called in, thousands of dollars worth of missing parts were traced, and Buschmann went to prison.

Byrnes, Ed
Ed 'Kookie' Byrnes rode a Harley-Davidson Topper scooter in the TV series *77 Sunset Strip*. In spite of this, it still wasn't a great success.

Cabbage Patch
Infamous Daytona campsite called Sopotnik's Cabbage Patch, where the main attraction during Bike Week is the Women's Coleslaw Wrestling (it is a cabbage farm after all!), where clothes are optional.

Cable, Charles
Charles Cable ran an import-export business from San Francisco in the 1920s, and part of his business was the export of Harleys to Mongolia. The factory was puzzled as to why he never ordered any spare parts, and it turned out that he was diverting machines to Japan instead. In the light of this, and because of the lacklustre performance of the current Japanese importer, Alfred Child was put in charge of a new Japanese operation, beginning an era of healthy Harley-Davidson exports to the Far East. (*See* Child, Alfred.)

California Bobber
Arguably the first real custom bike, or it at least heralded the birth of the modern custom movement. It first appeared in California in the late 1940s, and the look rapidly spread across America and then the rest of the world. The easiest way to make a big Harley or Indian lighter and faster was to dump all the superfluous bodywork, so front mudguards, screens and panniers were all removed. The rear mudguard was 'bobbed' or 'chopped' (hence 'chopper') down to the bare minimum, in keeping with the stripped-down look. As years

went by, the chopper or bobber look became ever more stylized and elaborate, popularized by Peter Fonda's Captain America chopper in the film *Easy Rider*: but the California Bobber was where it all started.

California Highway Patrol
CHiP has long been a high-profile user of Harley-Davidsons, and though the company lost the contract to Kawasaki in 1986, it has since been won back.

Cars
The advent of cheap mass-produced automobiles spelt death to many motorcycle makers. By the early 1920s there were around 400 makes of car on sale in America, and you could buy a good one for $400, the price of a sidecar outfit. Through drastic cost-cutting, Harley-Davidson managed to survive, but it was a close call, and the market had changed for good. In North America, more than in Europe, the days of the motorcycle as basic transport were over; they were increasingly becoming sporting machines for enthusiasts, and this was reflected in Harley-Davidson's concentration on big bikes.

OPPOSITE: Paintwork apart, this California Bobber demonstrates how customizing started. Note the lack of front mudguard, the 'chopped' or 'bobbed' rear guard, high bars and fat tyres. However, fancy colour schemes came later.

RIGHT: CHiP – the California Highway Patrol – was a long-term prestigious user of Harleys, until it changed to Kawasakis. After some hard work, they are back with Harley again.

Celebrity Owners

Countless celebrities are or have been Harley-Davidson owners, and there are now more than ever before. For some, it's a piece of strategic image enhancement, for others, a genuine love of the bike. Whatever the reason, it moved one rider to observe that Harleys in Hollywood are now like a certain part of the human anatomy – everyone's got one!

Music

Bryan Adams
Aerosmith
Jon Bon Jovi
Cher
Eric Clapton
David Crosby
Neil Diamond
Fabulous Thunderbirds
The Grateful Dead
Isaac Hayes
Peter Hook (Monaco)
Billy Idol
Michael Jackson
Billy Joel
Ronan Keating (Boyzone)
Tommy Lee
John Cougar Mellencamp
George Michael
Larry Mullen (U2)
Willie Nelson
Olivia Newton-John
Robert Plant
Lou Reed
Steppenwolf
Barbra Streisand
Dave Stewart (The Eurythmics)
The Stray Cats
Terence Trent D'Arby
Tonya Tucker
Eddie Van Halen
Dave Vanian (The Damned)
Stevie Ray Vaughan
Paul Young

Actors

Dan Aykroyd
Ann-Margret
Antonio Banderas
James Caan
Martin Collins
Clint Eastwood
Peter Fonda
Harrison Ford
Goldie Hawn
Don Johnson
Jay Leno
Ewan McGregor
Liam Neeson
Priscilla Presley
Micky Rourke
Kurt Russell
Arnold Schwarzenegger
Steven Segal
John Gordon Sinclair
Sylvester Stallone
Liz Taylor
Richard Thorp
Bruce Willis

Sportsmen

Eric Cantona
Chris Eubank
Paul Gascoigne
Ian Wright

Motor Racing

Johnny Herbert
Damon Hill
Eddie Irvine
Adrian Reynard
Mika Salo

Others

David Copperfield (magician)
Michael Flatley (Riverdance)
Richard Young (paparazzo photographer)

Ex-Harley Owners

Sir Malcolm Campbell
Malcolm Forbes
Clark Gable
Michael Hutchence
Charles Lindbergh
Elvis Presley
Roy Rogers

Chann, Jimmy

A successful rider for Harley-Davidson after the Second World War, winning the 1947, '48, and '49 Grand National Championships.

Child, Alfred

Alfred Child was a supersalesman who did much to spearhead Harley-Davidson's exports in the 1920s and '30s. Born in England, he worked his passage across the Atlantic at the age of 16 and after working as a butler, handyman and seller of cycle parts, joined Harley-Davidson. After acting as Arthur Davidson's assistant, he was sent to Africa, which he explored riding a J-model and sidecar. Starting in Cape Town and finishing in Cairo, he took orders for 400 bikes and galvanized many dealers into action. Then Davidson sent him to Japan, where Child was to remain until war broke out, first importing Harleys, then setting up a production facility for which drawings and machine tools were bought from Milwaukee, and the Sankyo company began to make all-Japanese two- and three-wheelers based on Harley-Davidson designs. After a disagreement, the contract was terminated and Alfred Child was again Harley-Davidson's Japanese importer; but after the Second World War, Child turned up in the USA as the importer for BSA! It was his evidence (citing his knowledge of Harley-Davidson's restrictive practices) that helped persuade the Tariff Commission to turn down Harley-Davidson's request for higher tariffs on imported bikes in 1951. Alfred Child lived to a ripe old age into the late 1970s. He had had an eventful life, to say the least.

Child, Richard

Alfred Child's son, he worked closely with his father importing Harley-Davidsons into Japan. His 400-mile (645-km) test ride of one of the first 61E Knuckleheads to arrive was not a success: it leaked oil, broke valve springs and needed frequent tappet adjustments. Despite this experience, Richard Child went on to become the official Harley-Davidson importer for Japan after the Second World War.

Chopper

A development of the original California Bobber (*see above*). Through the 1950s, '60s and '70s, the chopper grew ever more elaborate, with the movement spreading right across the world. Bars grew higher, as did seat backs, and multi-coloured metalflake paint jobs and intricate chrome work or engravings appeared. For many, the definitive chopper, with its high bars, tombstone seat and feet-forward stance, is the Panhead Harley ridden by Peter Fonda in the film *Easy Rider*.

OPPOSITE: It may seem amusing to us now, but this was regarded as the ultimate custom machine in the mid-1970s. Everything, apart from the brakeless front wheel, is to excess – forks, lights, pipes and sissy bar. It was at this point in time that the question was asked: are these motorcycles works of art or not?

Church, Gene
Dominated the Battle of the Twins series (*see above*) in the mid-1980s, winning in 1984, '85 and '86 on the Harley XR1000-powered Lucifer's Hammer. A former dirt-track rider, Church went on to race in Grand Prix.

Class C Racing
This was a racing class thought up in the early 1930s by Harley-Davidson and Indian for production machines, as opposed to expensive competition specials. The idea was to reduce the cost of racing, as both factories were finding the cost of special racers prohibitive. Instead, standard production 45ci (750cc) side-valve and 30ci (500cc) overhead-valve bikes were eligible for flat-track racing. Class C TT events (what we would now call motocross) and hill climbs also allowed 61, 74 and 80ci (1000, 1200 and 1300cc) twins to compete in Class C, which had its first season in 1933. The rules, of course, were explicitly designed to suit Harley-Davidson and Indian twins.

Colours
Harley-Davidson first offered a wider range of colours in 1933, in direct response to Indian's initiative a year or two earlier. Traditionally, Harleys came in drab grey or green, but for 1933 the twins could be had in silver and turquoise with black and gold pinstriping; black and mandarin red with gold striping; sunshine blue and white with gold striping; and police blue and white with gold striping. There was no extra charge for any of these. It was also the year of the handsome art deco bird motif on the tank. In fact, Harleys had never looked better.

Constantine, Arthur
An assistant engineer at Harley-Davidson from 1923, he was a graduate engineer fresh from Buick. He resigned the following year, after Bill Harley and Arthur Davidson rejected a 45ci (750cc) side-valve prototype he had built. Constantine went to Excelsior, where he produced the successful Super X. Harley-Davidson's own 45ci twin was brought out in direct response.

Crewe, Della
Completed a 5,400-mile (8690-km) journey across the US on a Harley-Davidson sidecar outfit in 1915, with only her dog Trouble for company. Della Crewe had had only ten days' riding experience before she left Waco, Texas on her 1914 outfit. She had considered taking the train for her cross-country tour, but rejected it as 'uninteresting'. With only dirt roads to follow, little more than rutted quagmires due to a very wet winter and spring, progress was painfully slow, particularly on the first leg of the journey to Dodge City. At one point, Della had to abandon the road and head off across the wheat fields for 4 miles before finding a useable track. The weather and roads were better as they headed north through Missouri, then crossed the Mississippi and continued on to Chicago and the Harley-Davidson factory in Milwaukee. In Indiana, Trouble was threatened with quarantine due to contact with foot-and-mouth disease, and by the time the pair reached Ohio it was November and the weather was closing in with drifting snow sending them frequently into ditches. Finally, Della and Trouble made it to New York after 5,378 miles (8655km) and no major breakdowns. 'I had a glorious trip,' said Della to reporters, 'I am in perfect health and my desire is stronger than ever to keep going.'

Crocker
A high performance V-twin built by Albert Crocker and Paul Bigsby in the late 1930s. Much to Harley-Davidson's chagrin, it was

much faster than any tuned Harley (or Indian for that matter), being capable of a genuine 100mph (161km/h). The Crocker was an expensive limited-production machine, but Walter Davidson was so incensed at its performance that he tried to get production stopped. He threatened Budd and Kelsey-Hayes (who supplied wheel rims to Crocker) and Linkert (carburettors) that if they didn't stop doing so their lucrative Harley-Davidson supply contracts would cease. They agreed, though the Crocker stayed in production for a little while longer.

Curtiss, Glenn
A pioneer aviator who rode a Harley.

The Crocker was a thorn in Harley-Davidson's side, being faster than anything out of Milwaukee, thanks to its highly tuned overhead-valve V-twin motor. Walter Davidson did his best to get production stopped.

Custom Bikes
Harley-Davidson has long been central to the custom bike movement, particularly in the USA. It grew out of the original California Bobber (*see above*), into the chopper, and later into countless variations such as street fighters and retro bikes. In America, at least, the Harley is still the bike of choice for custom builders.

Cycle Tow
Predecessor of the Servicar (*see* below), the Cycle Tow was basically a standard motorcycle with two outrigger wheels fitted to the rear of the frame. The idea behind this four-wheel motorcycle was that mechanics could ride it out to customers' cars as a two-wheeler, then hinge down the outrigger wheels, hitch the whole device to the car's rear bumper and drive back to the garage, cycle in tow. Once the car had been serviced, it would be driven back to the customer and the whole operation would be reversed. Not many Cycle Tows were sold, but the idea soon led to the immensely successful Servicar, which became a virtual American institution, remaining in production into the 1970s.

Davidson Family
Hailing from Aberdeen, Scotland, the Davidsons emigrated to America and settled in Milwaukee in 1871 or '72. William Davidson and his wife had two daughters and three sons, Arthur, Walter and William A. In the fledgling Harley-Davidson company, Arthur was the salesman, Walter the general manager (he built the first bikes) and William A. the works manager. Walter and his wife, Emma, had three sons, Gordon, Walter Jnr and Robert.

Davidson, Arthur
Co-founder of Harley-Davidson with Bill Harley, Arthur Davidson was a natural salesman who, in the words of author Herbert Wagner was 'an out-front brassy little guy. A handshaker. He loved people ... He liked a good joke and he loved telling stories.' It was his extravert personality that meshed so well with the engineering skills of his friend, Bill Harley, and played a key role in the early success of the company. While his brothers were building bikes and Bill was developing them, Arthur was out on the road recruiting reliable dealers and promoting the company through advertising. He recognized the importance of training, and founded the Harley-Davidson service school. Like Bill Harley, Arthur loved the

ABOVE: Willie G. has to be the most famous living Davidson, being the grandson of the original William A. Davidson. Not only is he the stylist behind countless bikes of the 1970s and '80s (notably the Super Glide and Low Rider), but he is a hard-working ambassador for the company, attending rallies and meeting the customers.

outdoor life. He died in December 1950, the victim of a road accident.

Davidson, Bill
Son of Willie G, and the fourth William Davidson to work for the company, he joined the 'family firm' (which of course it no longer was) in 1984 to oversee the national demo ride programme and gain experience in the advertising and promotion departments. In mid-1991 he was nominated head of the Harley Owners Group (HOG).

Davidson, Gordon
The son of Walter, he joined the sales department in 1930 after graduating from the University of Pennsylvania. Seven years later he was a company director and was appointed vice-president of manufacturing in 1942. More at home on the shopfloor than in the office, he died in 1967. His brother, Walter C., was much like their father, a natural salesman fond of company and the odd drink. Young Walter rose to become vice-president and sales manager. He died in 1974. Less is known of their cousin Allan, William A.'s son. He too worked for the family firm, but sadly died young.

Davidson, Janet
Aunt to the Davidson brothers, her connection with the Harley-Davidson story is that she painted the red pinstriping and Harley-Davidson logo onto the tank of the very first 1903 machine.

Davidson, John A.
Son of William H., he was appointed president of the company in the late 1970s, after a succession of non-family presidents failed to solve the company's problems. He resigned just before the management buyout was completed in June 1981.

Davidson, Karen
Daughter of Willie G., who trained in fashion design before joining the Harley-Davidson clothing department and becoming involved in Harley-Davidson's own MotorClothes range.

Davidson, Walter
Second of the Davidson brothers and president of the company, Walter learned his trade of machinist in the railway workshops of Milwaukee. It was he who built up the first motorcycle that his younger brother, Arthur, and William Harley designed. Often described as the most reserved of the four Founders, he turned out to be a highly competent rider, and did well in early endurance events. Walter was still president of Harley-Davidson when he died in February 1942.

Davidson, William A.
Eldest of the Davidson brothers and company works manager and vice-president, William was already foreman of a railway workshop when his younger brother and Bill Harley were building their first motorcycle. Consequently he brought a wealth of practical experience in production, metallurgy and management to the fledgling company. Nicknamed 'Big Bill' or 'Old Bill', to differentiate him from his son, William H., was a big man who loved outdoor pursuits such as hunting and fishing. Veteran employees remember him as someone who always remembered them. He died in 1937.

RIGHT: Walter Davidson, second of the Davidson brothers, pictured after winning an endurance event on an early single.

Davidson, William H.
Son of William A. Davidson, he succeeded Walter Davidson as president in 1942. He was still in the job at the time of the AMF deal, and urged shareholders to accept the takeover. He retired shortly afterwards, in 1971.

Davidson, Willie G.
Son of William H., and the man behind many key bikes in Harley's recent history, Willie G. Davidson styled the original Super Glide (1970) and Low Rider (1977), and remains head of the styling department. He trained in design, and worked for Ford and Brooks Stevens before joining the family firm. Harley's survival through the 1970s and revival in the '80s owes a great deal to Willie G., whose stylistic skills have allowed the company to produce apparently new bikes using many existing (sometimes outdated) parts. Just as important, he has remained in touch with what riders want – his Super Glide of 1970 was the first factory interpretation of what customizers were already doing. In the process, he has managed to transform himself from an earnest-looking young man with horn-rimmed spectacles and a sensible haircut into a grey-bearded biker. That in itself puts him in tune with the typical 21st-century rider of a new Harley, i.e. an affluent middle-aged professional.

Davis, Bill
Davis is a customizer who could fairly claim to be the godfather of the current line of Softail Harleys. In the early 1980s, he built a bike with the rear suspension units hidden horizontally under the gearbox, while the swinging arm and frame were reworked to give the impression of a hardtail rear end, but with more contemporary comfort. The story goes that Vaughn Beals saw this bike at a rally and was so impressed that he ordered a factory version. The official Softail appeared in 1984, proved a tremendous hit, and has been a big part of the Harley-Davidson line-up ever since.

ABOVE: The Softail was originally designed by a customizer named Bill Davis. Vaughn Beals spotted it, and the rest is history.

OPPOSITE: The reality of Daytona Beach, somewhat different from the free cruisin' image usually depicted.

Daytona Bike Week
What started as beach races at Daytona, Florida progressed to the purpose-built Daytona Speedway track and the famous Daytona 200 race. But a major festival of motorcycling has developed around the racing which, although it has seen increasing numbers of Japanese bikes in recent years, is still really Harley-Davidson-dominated. Apart from cruising the flat sandy beach in front of the town (recently restricted), there are custom shows (the Rat's Hole and Boardwalk), innumerable trade stands and Main Street, which for Bike Week is solid with Harleys.

Dealer Advisory Council
The company-sponsored association for US dealers. Until the mid-1980s, its members were appointed by the company, but they were elected by the dealers after their revolt of 1983. It still represents the vast majority of North American Harley-Davidson dealers.

Dealers Alliance
The national Harley-Davidson Dealers Alliance, formed in 1983 by 32 American Harley-Davidson dealers in protest to the new Bonus Bucks incentive. Three hundred had joined within two years, 81 per cent of whom declared they had no confidence in Harley-Davidson management; but over the years, relations did mend. The company's Dealer Advisory Council became more democratic and communication between the company and its US dealers improved. (*See* Bonus Bucks.)

Dean Witter Reynolds
Firm of investment advisors that organized Harley-Davidson's refinancing package in early 1985, a key figure being Steve Deli of the firm's Chicago office, who was able to bring together Harley-Davidson and Heller Financial, another investment bank that was interested in putting money into the company. The original financier, Citicorp, pulled out in 1984, and finding another bank was the only way the company could be saved.

Deathshead
A winged skull which appears on the Hell's Angels' backpatch and is often painted elaborately on their gas tanks. It is patented worldwide and woe betide anyone copying the design or wearing anything similar on their jackets.

De Savary, Peter
A millionaire property developer and Harley aficionado who opened up his Civil War mansion in Berkshire, England for a number of Harley rallies in the late 1980s and early 1990s. He has an impressive collection of vintage automobiles and motorcycles.

A Harley accessory? ABOVE is British entrepreneur Peter de Savary, sporting a neatly trimmed beard, while RIGHT is American customizer Rick Doss, with a rather fuller version.

OPPOSITE: The cockpit of the FLT, with all the accessories you would need for crossing the American continent. What else can you spot besides the extra mirrors and mobile phone?

Dixon, Freddy
English rider and tuner of Harley-Davidsons in the 1920s, he scored many race wins, both solo and sidecar-mounted, notably at Brooklands. He also modified an early eight-valve Harley racer to take twin carburettors.

DKW
German motorcycle maker, established in 1919, whose pre-war RT125 two-stroke was awarded to various Allied manufacturers as war reparations. BSA produced it as the Bantam, the Russians made their own version and for Harley-Davidson it became the Model S, or Hummer.

Doss, Rick
Famed American East-Coast builder and designer of custom parts under the shop name of 'Rick's Custom Harlees' (sic). He is also a consultant to the multimillion-dollar customs parts company, Custom Chrome Inc.

Dressers
Dressers are Harley-Davidsons which have been literally 'dressed' for touring with screen, panniers, spotlamps, large comfortable seat and maybe a fairing. The origins lie in 1920s and '30s America, when riders began to add practical touring accessories to cope with the long distances in that big country. It developed a look all its own, the antithesis of the stripped-down bobbers and choppers after the Second World War. But the dresser survived to become an integral part of the world of Harley-Davidson: the Electra Glide, Road Glide and Road King all draw on this tradition, being ready-made dressers (in various states of undress) available direct from the factory.

Harley-Davidson
USA

Their credo is violence...
...LUST IS THE LAW
THEY LIVE BY!
ROGER CORMAN
DEVILS ANGELS
OF ALL TIME!

Easy Rider
The film in which Peter Fonda and Dennis Hopper ride choppers across America under big blue skies, 'looking for adventure' as the Steppenwolf soundtrack goes, also featured one of the most famous Harleys of all time – Captain America. The appearance of this definitive chopper, with its high bars, raked forks and easy riding stance, made a huge impression on riders, not just in America but all over the world. The film combined late 1960s hedonism with a traditional, almost Wild West theme of two drifters in search of freedom. The only difference was, instead of horses, they rode Harley-Davidsons.

Ekins, Bud
Famous desert competition rider on Harley-Davidsons and friend of the late Steve McQueen. Ekins' collection of Harleys has often been seen in films, notably the bike ridden by James Stewart in *The Spirit of St Louis* (the Charles Lindbergh story) and the American TV series *Nichols*, in which James Garner rides a Harley-Davidson single.

Electra Glide in Blue
Another world-famous film starring a Harley. In 1971, Robert Blake played a Harley-mounted motorcycle patrolman who dreams of his perfect bike, an Electra Glide in blue. The bike's role in the film is to point up Blake's own dilemma, the choice between law enforcement and the pursuit of the road to freedom.

Electric Lighting
This arrived in 1915, along with a three-speed gearbox, to keep up with rival manufacturers and replace the soon to be outmoded acetylene lights.

Electric Starter
This arrived in 1965 on the aptly named Electra Glide. Harley-Davidson could not be accused of being late with electric start, which was only just appearing on Japanese machines at the time. It needed to be beefy to turn over the 74ci (1200cc) V-twin, and the big starter and new 12-volt electrical system together added 75lb (34kg) to the weight of the machine.

Engines
F-head Singles & Twins: 1903–1929
The name given to the first engines used by Harley-Davidson, so-called because of

OPPOSITE: Captain America: Peter Fonda's bike from *Easy Rider* inspired a whole generation of custom builders, both outside as well as within the USA.

Electric lighting on a 1925 J-model (LEFT) and electric starting on a 1965 Electra Glide (ABOVE). What's 40 years when the company has lasted for nearly 100?

the 'F-head' layout, with side-mounted exhaust valve and pushrod-operated overhead inlet valve, though very early motors had a vacuum-operated inlet valve.

Side-Valves: 1927–1948
Side-valve singles and V-twins were the basis of Harley-Davidson's engineering policy from the late 1920s to the late 1930s. A major consideration was that arch-rival Indian used side-valve engines, and they were often faster than the equivalent Harley.

Knucklehead: 1936–1947
Name given to Harley-Davidson's overhead-valve V-twin, introduced in 1936, because the rocker covers were thought to resemble a pair of knuckles. It was an engineering milestone for the company, being faster and more advanced than any of the opposition (i.e. Indian) and played a great role in transforming Harleys from reputedly being something of an old man's bike to a real sportster. It was replaced by the Panhead in 1947. (*See also* Big Twins.)

Panhead: 1947–1966
Aluminium cylinder heads for improved cooling and hydraulic tappets were added to the existing bottom end. Also cooking-pan-shaped rocker covers, hence the nickname. A clean-looking motor, but not without its teething troubles.

Shovelhead: 1966–1983
The nickname for the updated Big Twin of 1966, which adopted a Sportster-style top end to boost power. Power was increased again, to 65bhp in 1968, though capacity remained at the now familiar 74ci/1200cc. It was a long-lived engine, surviving right throught to 1983.

Evolution: 1983–1998
Substantially an updated 82ci (1340cc) big twin, which formed the bedrock of the company's turnaround in the 1980s. It was originally intended as a stop-gap before a range of all-new liquid-cooled engines were unveiled, but stayed in production as Harley's sole engine for 15 years. With aluminium-alloy cylinder barrels, a lighter

OPPOSITE
FROM TOP LEFT to RIGHT: F-Head: The first twin with overhead inlet, side exhaust valves. Knucklehead: The first ohv twin and a massive leap forward. Panhead: Hydraulic tappets introduced not only automotive technology but also teething troubles. Shovelhead: More power to cope with more weight. BELOW: The Shovelhead came with aluminium heads. The SU carburettor is non-standard.

ABOVE: The Evolution motor heralded a whole new era of reliable, oil-tight Harleys. It actually has much in common with the Shovelhead it replaced, but the initial quality was higher, as were the engineering tolerances.

ABOVE RIGHT: The Twin Cam is the current Big Twin, the largest yet at 1450cc (88ci). Like every other engine apart from the Knucklehead, it's more of an evolution than a revolution, though the 88B balance-shaft version (which appeared in 1999) is something of an innovation for Harley.

valve train and closer tolerances, it was more powerful, more oil-tight, better built and more reliable than its predecessors. Harley-Davidson owes a great deal to this engine.

Twin Cam 88: 1998–
Harley-Davidson's current Big Twin, still an air-cooled 45-degree V-twin with hydraulic tappets, but really an all-new engine compared to the Evolution, which it replaced over 1998/99. The biggest Harley yet, at 88ci (1450cc), with a bigger bore/shorter stroke, twin camshafts and a choice of carburettors or fuel injection. A balance shaft version, the 88B, was launched in 2000, though at the time of writing, the 883/1200 Sportsters are sticking with their Evolution engines.

Engraving
A customizing technique, using traditional engraving methods to decorate bare metal, usually engine crankcases and exhausts. It was pioneered by English customizer John Reed, who persuaded gun engraver Don Bloxidge to try his hand at bikes. The technique reached its apogee in the 1970s and '80s, but since then the use of heat-resistant powder coating has made possible alternative ways of decorating engine cases.

The Enthusiast
Harley's own magazine, launched in 1916 and sent out to all Harley-Davidson dealers and registered owners. Although not always a regular publication, and it could hardly be called non-partisan, it carried on for decades afterwards.

SUPER X
STURGIS
1999
BANDANNA
THE STITCHERY STUDIO
BURNSVILLE, MN

OPPOSITE: The Excelsior Super X out-performed equivalent Harleys.

ABOVE LEFT: A modern Excelsior-Henderson of 1998. Sadly, the task of competing on Harley's home ground proved more difficult than was expected.

ABOVE: The lavishly painted tank on this 1976 bicentennial bike is a wonder to behold.

Excelsior
A key early competitor of Harley-Davidson. Excelsior was formed in 1908 and later taken over by Schwinn bicycles. It produced a full range of machines but failed to survive the Depression, and was closed down by Schwinn in 1931.

Excelsior-Henderson
An American-made competitor to the Harley in the late 1990s. Harley-Davidson's turnaround in the 1980s and later success attracted some entrepreneurs into investing in a home-built competitor, notably the Polaris company (which makes the Victory) and the Hanlon brothers, who revived the Excelsior-Henderson name around a brand new 84ci (1386cc) V-twin. Production began in 1998, but at the time of writing is suspended while Excelsior-Henderson is in financial trouble. The future remains uncertain.

Factory Customs
These are factory-built Harleys (or indeed any bikes) which follow current customizing trends. For Harley-Davidson, this started in 1970 with the Super Glide, whose long low look reflected what customizers of the time were doing. It was the first time Milwaukee recognized the custom movement, despite the fact that through the 1960s there were probably more Harleys being chopped than any other bike. They even beat the Japanese on this one, who later began to produce their own 'factory customs', which were really standard bikes with higher bars and a dash of extra chrome. It was also the first major project to emerge from the styling department headed by Willie G. Davidson. The Super Glide wasn't actually a huge

success, but it did pave the way for a whole stream of factory customs from Harley-Davidson, notably the low-seated Low Rider of 1977, the Wide Glide of 1980 and the Softail of 1986. Each one followed a particular custom look, and all were successful, so much so that factory customs now play a major part in Harley-Davidson's model range. It is hard to believe now that in 1980 Harley-Davidson management was reluctant to allow the Wide Glide into production, as its wide-spaced raked forks were thought to be too strongly connected to the outlaw element of biking.

Fat Bob
A renowned Harley dealer, living in Dorset, England, and famous for his keen interest in blueberry muffins and thick shakes. In reality, a Fat Bob is the common wide style of fuel tank with two filler caps, one each side. One cap may be a fake, the style harking back to the days when gas was held in one side, oil in the other. The style is still used on Harley-Davidsons today, the left-hand (fake) cap being used to house a fuel gauge.

ABOVE: One of Arlin Fatland's custom Harleys. Note the magnificent spoked wheels.

LEFT: The Fat Bob tank harks back to the days when fuel tanks held oil as well as gas.

OPPOSITE: A familiar decorative theme on custom bikes, especially in the 1970s. This bike's so hot it's on fire!

Fatland, Arlin
Famous custom bike builder from Denver, Colorado who has been operating from his shop, Twowheelers, for almost 30 years. He still turns up at Sturgis with head-turning custom Harleys.

Flames
The flame device is the most traditional of all custom paint treatments. It harks back to the paint jobs found on hot-rod cars in California in the 1930s and 1940s.

Focus Flow Processing
Part of Harley-Davidson's transformation of its manufacturing operation in the 1980s. It had traditionally built components in batches, which had to be moved all over the plant for various operations. This meant many parts were either queueing up at production bottlenecks, or held in stock until they were needed, which wasted time and money. Reorganizing production into a continuous flow, with each employee responsible for several operations, meant different parts could be produced as needed, allowing the company to respond more efficiently and speedily to orders. Production costs and times were cut, inventories slashed, and quality improved.

Forbes, Malcolm
Millionaire Malcolm Forbes didn't start

riding until late in life, but made up for it with a Harley stationed at each of his houses, not to mention one on his private plane and another two on his yacht. He founded a club, the Capitalist Tools MC (which survives him) and had a hot-air balloon made in the shape of a Softail.

Fordyce, Skip

A popular and successful pioneer Harley-Davidson dealer in Riverside, California. He caused a scandal after the Second World War by taking on Triumph machines as well, and a horrified Harley-Davidson management threatened to withdraw his franchise. Fordyce pointed out that Harley had nothing to offer new riders in the increasingly popular 30–45-ci (500–750-cc) market, adding that he could take them to court for restraint of trade. Harley-Davidson withdrew its objection.

Forecar

Introduced in 1913 as an alternative to the conventional sidecar outfit. It replaced the motorcycle's front wheel and forks with a large load-carrying box and two steerable wheels connected by an axle. Everything aft of that was pure standard motorcycle. The Servicar which superseded it was simply the same idea back to front. (*See* Servicar.)

Forty-Five OHV (Prototype)
Although the standard long-lived Forty-Five, and even its successor the K-series, were all side-valves, there was talk of an overhead-valve version as early as 1937. Bill Harley suggested that they should use the standard bottom end but with new cylinders and heads. It would use the standard frame, in strengthened form, and produce a projected 30hp even in a mild state of tune. Bill Harley pushed hard for this model, pointing out that although it would cost about the same to make as the 74ci (1200cc) side-valve, it would also produce as much power and would be more reliable into the bargain (there were continuing piston problems with the big twins). In August, a preliminary prototype, which had been built based on a contemporary 45 ohv hill climber, with exposed valvegear but giving a good idea of how a 45 ohv might perform, gave 29hp.

However, Arthur Davidson disagreed, maintaining they should concentrate on the existing bikes and use the money saved to produce a new smaller lightweight machine. Bill countered that an all-new design would take two years to come to fruition and cost more than his idea anyway. But he was evidently on a losing streak. Walter Davidson voiced doubts that the existing 45 gearbox would be able to accept the ohv's extra power, for even in the modestly powered Forty-Five it had caused trouble. Part of the reason for this was that the primary chain was mounted on the left, and the final chain on the right, subjecting the gearbox to twisting loads. Walter concluded that it would be better to continue with the existing Forty-Five unchanged, but design an all-new 45ci or 50ci (750 or 820cc) ohv to run alongside it.

In the end, the idea was dropped because a new steel-strutted piston solved the 74's engine problems, which removed a primary reason for making the 45 ohv at all. As the board minutes recorded: 'At the time this 45 overhead-valve motor was suggested, we were apparently at our wits' end in regard to pistons in our 74 and 80 motors, and it was felt we would have to come to an overhead-valve motor to replace these side-valve motors. Since that time, a new piston has been developed which seems to be the answer to our troubles. If this proves to be correct, the necessity for overhead-valve motors to replace the side-valves is not so great. The one outsanding difference between the overhead-valve motor and the side-valve is that the overhead costs considerably more to make.'

For the 45 ohv, that sounded like the end of the matter, but the discussions went on, and William H. Davidson (son of William A., and the first new-generation Davidson to join the board) sided with Bill Harley. So a proper prototype was built, with new cylinders and heads mounted on a modified 45 bottom end, and in February 1939 it was ridden down to San Antonio in Texas on a 5,000-mile (8050-km) proving run, along with a prototype 74 ohv. There were no major problems, apart from the need for stronger crankcases on the 74 to cope with the extra power. But cost was still the stumbling block, as a 45 ohv would be relatively expensive to build, and the company preferred to concentrate on the 74 ohv and new aluminium heads for the side-valve big twins. Intended to debut in the 1939 line-up, the 45 ohv soon slipped to 1940, then 1941. Finally the war effort put paid to it; but really it was the cost of the bike that made most of the Board so lukewarm about the whole idea.

Fuller, Wyatt (See Razorback.)

Gerberick, Royal
With fellow ex-serviceman Roy Kerle, Gerberick rode a 1920 J-model sidecar outfit 19,000 miles (30,577km) through the Mid-West and Western USA, which was quite an achievement on the roads of 1922.

Gott, Rodney C.
Chief executive of AMF and the man behind the takeover of Harley-Davidson in 1969. AMF's business was mostly based in industrial products, and Gott wished to expand into the leisure industry to balance the company's investment. He was an enthusiast of Harley-Davidson motorcycles and sincerely wanted the company to succeed, but was convinced that rapidly expanding production and non-family management were the answer. As a motorcyclist, however, he did understand the market, something that his successor Tom York (who came from an accounting background) did not.

Goulding
Supplied sidecars, such as the Litecar, to Harley-Davidson for the Forty-Five, which was offered from 1929. Although Harley-Davidson bought sidecars from several sources, and eventually brought their production in-house, it considered at one time buying all of them from Goulding.

ABOVE: Goulding supplied wheels as well as sidecars.

OPPOSITE: Ignoring the fact that this is an Indian bike, it does, however, have a Goulding sidecar, as also supplied to Milwaukee for factory fitment.

Indian

Hadfield, Al
Set a new record for the 'Three Flags' run (Canada to Mexico non-stop), beating Cannonball Baker's run on an Indian. Hadfield, without factory support, covered the 1,689 miles (2718km) on a standard 61ci (1000cc) Harley in 51 hours 22 minutes, a record that was broken by Wells Bennett on an Excelsior the following year. However, Hadfield managed to retrieve it that November, in a time of 49 hours exactly.

Hamm, Fred
Held a number of endurance records for Harley-Davidson before the Second World War. Hamm was a motorcycle police officer from Pasadena who broke records in his spare time. He smashed the Canada–Mexico Three Flags run in 1936, his time of 28 hours 7 minutes on a 74ci (1200cc) Harley-Davidson knocking 10 hours off the existing record. Also notable was his non-stop 24-hour run the following year, using a 61E Knucklehead. A 5-mile circle was marked out with flags and, for night-time running, flares were used on the smooth, dry bed of Lake Muroc. Despite 100-degree daytime temperatures and below-freezing nights, Fred Hamm lapped the circle continuously, covering 1,825 miles (2937km) in the 24 hours. He stopped only for fuel and fruit juice, and the Knucklehead's only problem was needing a new chain, fitted at the 1,400th mile.

Hamsters
Formed over 20 years ago by Arlen Ness, Arlin Fatland and others, the Hamsters are an élite group of Harley custom bike builders. They meet at Sturgis each year and can be spotted by the vivid yellow T-shirts they all wear.

Hardtail
The term for any motorcycle without rear suspension, though it is usually applied to custom Harleys, hence the current Harley-Davidson 'Softail' range, which conceals the rear suspension to give the appearance of a hardtail.

Harley, William A.
Son of the original Bill Harley who succeeded his father as chief engineer on his death in 1943. However, he started with a lowly job in the shipping department, at his father's behest, and learned every aspect of the production process by practical experience. William A. was also an enthusiastic rider, and did some test riding during the development of the Knucklehead. As head of engineering, he supervised Harley-Davidson's post-war model programme, the 7ci (125cc) Hummer, the Panhead and Sportster; but

according to author Harry Sucher, he was a cautious and conservative man when it came to engineering. He retained his job (not to mention the vice-presidency) after AMF took over in 1969 but sadly died in August 1971. His son John was parts and accessories manager.

Harley, William S.
Bill Harley was sometimes described as the 'fourth Davidson', so close was his working relationship with the three brothers. He was a boyhood friend of Arthur Davidson in particular, and the two shared a passion for the new internal combustion engine. William was trained as a draughtsman, so he was instrumental in designing, not only the first Harley-Davidson, but also the bikes in the first 30 years. Naturally enough, he became chief engineer. Bill Harley was a quiet and introverted man, but like the eldest Davidson, he was also a lover of the outdoors and later developed a talent for sketching wildlife as much as for shooting it. He died in September 1943, aged 66.

Harley Owners Group (HOG)
A company-sponsored group of Harley-Davidson riders, it is the first such group in the motorcycle industry and probably the most successful. It was formed in 1983 as a means of communication between Harley-Davidson and its customers via ride-outs, newsletters and details of events. So successful has it been that there are now branches of HOG all over the world. The first year's membership is free with a new bike, though HOG's development as a social group is underlined by the numbers who stay in it after that first year. Seventeen years on, HOG is the world's largest manufacturer-sponsored owners' group, with over half a million members worldwide.

OPPOSITE LEFT
Emblem of the Hamsters motorcycle club.

RIGHT
An aftermarket hardtail frame, just as unforgiving as the original.

LEFT: They get everywhere! A HOG member of the Tenerife Chapter.

Heller Financial
The investment bank that refinanced Harley-Davidson in 1985 after the original financer of the buyout (Citicorp) decided to pull out. According to author Chris Reid, Heller was the only firm that would consider supporting Harley at the time, and without that support, Harley-Davidson would have faced bankruptcy. Heller was impressed by the strength of the brand name, the recovering motorcycle market and the steps already taken to turn the company around. One of Heller's top managers was also a Harley rider, which probably helped.

Hell's Angels
Often associated with Harley-Davidson, especially in their early days, this was a fact that Harley-Davidson not always viewed with relish The first Hell's Angels club was formed in 1948 by a breakaway group from the Pissed Off Bastards in Berdoo, San Bernardino, part of the general growth of 'outlaw' groups in the late 1940s and early '50s, particularly in the wake of events at Hollister, California. Possibly one of the original (certainly the most notorious) motorcycle associations of all time, the Angels now consist of 1,600 members in 100 chapters all over the world. The Hell's Angels thrived as part of the post-war counter-culture, but as a

company, Harley-Davidson did its best to dissociate itself from all the back-patch clubs, seeing it as detrimental to the image of motorcycling which, quite apart from anything else, was bad for business. Instead, Harley publicity material showed clean-cut, clear-eyed, smartly turned out folk riding its bikes: their uniforms would even have passed muster on a parade ground. Not until the late 1970s did Harley riders begin to wear denim and beards in official advertisements. Now, of course, a vaguely Hell's Angel look is *de rigueur* at every HOG meeting and HOG is organized in chapters, just like the Angels.

Sonny Barger was probably the most famous of them all, founding the Oakland, California chapter in 1957. Known as the Chief to his fellow members, he was to spend 21 spells in prison (a total of 13 years) for drunk driving, assaults with deadly weapons, kidnapping, drug dealing and racketeering. He was also, like many Angels, devoted to Harleys, and at 61 years of age still rides one today, a Road King on which he covered nearly 12,000 miles (19312km) in the first three months. He explained his mileage rate was down because of the demands of a book tour to promote his autobiography! (*See also* Hollister.)

High-Flo Muffler
A certain style of tail-pipe fitted in 1934, with the pipe exiting above the lower mudguard stay, and not parallel to the ground, as was traditional.

Hill Climbing
A popular sport in America from the very early years of motorcycling, and particularly vibrant in the 1920s and '30s. Competitors had to ride up an extremely steep slope, and the higher they got, the more likely they were to win, and it was better still if they actually broached the summit. Harley-Davidson was in there almost from the beginning, as big torquey V-twins were ideally suited to this demanding task. Moreover, it had its masters of the art with riders such as Joe Petrali. So popular was it that the factory built many special hill climb bikes in the late 1920s and early '30s. Hill climbing remains a popular sport, though the most successful machines are now lighter-weight off-roaders rather than the big V-twins.

Holiday Rambler
Maker of motorhomes, and bought by Harley-Davidson in December 1986. Harley-Davidson had just gone public, and now had the access to finance (not to

LEFT: Hell's Angels have an extensive network. Here are two English members at the Boothill Saloon.

ABOVE: The High-Flo muffler was a popular fitment in the 1930s.

OPPOSITE: A very American sport: this XR750 has been radically adapted for hill climbing.

FOR 300 MILES
Harley-Davidson Wins Dodge City 300-Mile International Grand Prize Race
Breaks all World's Records
Ride a Winner
HARLEY-DAVIDSON
HOG

mention managerial freedom) to expand by acquisition. Holiday Rambler promised to even out Harley-Davidson's business by not holding all its eggs in the one motorcycle basket, and by broadening its customer base. That it was able to buy out a company for $155 million, only a year after near-bankruptcy, underlines how rapid Harley-Davidson's turnaround was.

Hollister
The now infamous events at Hollister, a small town in California, on 4 July 1947, have become part of biking folklore. The 'riot' by '4,000 members of a motorcycle club' (*Life* magazine) was undoubtedly blown out of all proportion by the media. More to the point, for the purpose of our story, many (if not most) of the riders arrived on chopped Harleys and Indians. The reporting of this event actually helped kickstart the customizing movement, in which a stripped down, chopped Harley became a symbol of rebellion. It was the 'one percenters' image, which Harley-Davidson did its best to distance itself from over the succeeding 25 years, until it realized that the image (if not the reality) was actually good for business.

OPPOSITE: This clean two-tone Hydra-Glide has a tasteful and restrained amount of extras.

ABOVE: This Hydra, of about the same age, is in its more original form, but has a buddy seat.

Honda
Of all the Japanese manufacturers, it was probably Honda which affected Harley-Davidson most, and for two reasons. One, its advertising campaigns of the 1960s ('You Meet the Nicest People on a Honda') opened up a whole new market for motorcycles, which Harley-Davidson and many of its dealers didn't really understand, and led to the company's retreat from the mainstream into a niche market. Two, Honda's GoldWing struck directly at Harley-Davidson's traditional touring market. It was smooth, quiet and refined, and the fact that Honda later began building it at a plant in Marysville, Ohio shows just how important it was to the US market.

'Hospitals'
An ironic in-house reference to the areas set aside at the York Pennsylvania plant in the 1970s for bikes that needed fixing. At one point, half had missing parts, while others had simply been badly put together. Chief executive Vaughn Beals later recalled that the first 100 XLCR Café Racers off the line in 1976 needed $100,000-worth of work before they could be shipped out.

Hub Gears
Hub gears, mounted in the rear-wheel hub, were used before the gearbox and chain drive as we know them appeared. The earliest Harley-Davidsons had a single-speed direct-belt drive; a hub-mounted clutch was introduced in 1912 (so you didn't have to restart the motor every time the machine was stopped) and a two-speed hub was unveiled in 1914, though Bill Harley had patented a two-speeder four years earlier. But this was only a stop-gap, as a three-speed countershaft gearbox, with chain primary and final drive, came in the following year. The basic engine/transmission layout for Harley-Davidsons would not change for the next 65 years.

Hydra-Glide
Official name given to the first production Harley-Davidson with telescopic front forks, for model year 1949. Although it was no more than manufacturers all around the world were doing at the time, it was a radical step for Harley-Davidson, which had used the same basic design of leading link springer fork since 1908. The new hydraulically-damped forks, with their long travel, soft springing and a soft, wide 5-inch (12.7-cm) section front tyre, added a new dimension of comfort, though diehard traditionalists complained that they lacked the steering precision of the originals. There were also some instances of riders falling off in tight turns as the forks could compress sufficiently to allow the end of the crash bar to hit the ground. A factory kit with firmer damping was subsequently produced which cured the fault.

Hydraulic Valve-Gear
Used by all Harley-Davidsons since the Panhead of 1949. The valves are opened by hydraulic pressure rather than solid pushrods, the advantages being quieter running and lower maintenance.

'I Make the Eagle Fly'
A late-1970s attempt to involve the workforce using exhortations to make them value quality. There was a slight improvement, but at great cost (overtime went up and production dropped), and real quality improvements didn't come until employees were given genuine involvement a few years later.

Import Tariffs
On three occasions after the Second World War, Harley-Davidson asked for higher import tariffs to be slapped on imported motorcycles. The first was in 1950, in response to the increasing numbers of British bikes being sold in America. At the hearing before the US Trade Commission, Harley-Davidson's charge of unfair

HARLEY-DAVIDSON
STOP

competition was scuppered by Alfred Child (*see above*), who revealed Harley's then current restrictive practices towards its dealers. As a result, the application was turned down. The same thing happened in 1978 when Harley-Davidson applied for protection against Japanese imports. This time, dealers testified that the Aermacchi-built machines were outdated and uncompetitive against the Japanese. Only in 1983 did the company succeed in getting new tariffs, imposed on imports of over 700cc for five years: Harley had a stronger case this time, and the Ronald Reagan administration was more sympathetic to higher tariffs. In a brilliant piece of PR, Harley-Davidson was able to request that the 1983 tariff be lifted a year earlier than its five-year term.

Indian Dispatch-Tow
Indian's three-wheeled response to the Servicar, which at first sight was seen to have advantages of its own. Even Harley dealers pointed out that the Dispatch-Tow was a handier size than the Servicar, and had an instantly accessible fold-down towbar while the Servicar's had to be stored in the box and bolted on. Harley fitted a fold-down towbar in 1936.

Industrial Engines
Harley-Davidson made stationary industrial engines during the early 1930s. During 1931–32, the plant was operating at a mere 10 per cent of its capacity, and the company desperately needed something other than motorcycles to keep it occupied. Industrial engines promised entry into a whole new market, while making use of existing machine tools. A complete range of singles, horizontally-opposed twins and V-twins was brought out, but sales were poor and the line was dropped after less than two years.

Jack Pine Enduro
A gruelling cross-country competition of around 500 miles (805km), featuring rough tracks to be negotiated and fast-flowing streams to be forded. It seems to have been traditional for younger members of the Davidson and Harley families to take part, and William H. Davidson, later to become president of Harley-Davidson, won the event in 1930.

Japanese Production Methods
(*See* Productivity Triad.)

Jelly Bean System
Nothing to do with confectionary, rather a production method introduced in the 1980s which meant that every different model and colour of Harley would come down the production line, one after another. This direct response to demand from dealers was more efficient than the old method of making the same model in large batches, then storing them until they were sold.

John Brown Motolamp
Name given to the single 7-in (18-cm) headlamp which replaced the twin smaller headlamps in 1913.

Jones Brothers
The first dealer to sell Harley-Davidsons in New Zealand, after a visit by Arthur Davidson in 1920.

Juneau Avenue
Address of the Harley-Davidson headquarters in Milwaukee, Wisconsin.

Kennedy, Pat
Kennedy is best known for his extreme Swedish-style long-forked choppers which he creates at his base in Tombstone, Arizona.

OPPOSITE: The famous Juneau Avenue works, where Harley-Davidson has been based since 1905, has certainly seen some changes.

ABOVE: The Dispatch-Tow was Indian's answer to the Servicar.

RIGHT: Pat Kennedy specializes in long-forked choppers.

Kilbourn Finance Company
A subsidiary of Harley-Davidson, set up in the late 1930s as a finance company for Harley dealers and customers.

Koto Trading Company
Unofficially imported Harley-Davidsons to Japan in the early 1920s, via trader Charles Cable. The bikes were supposedly bound for Mongolia, but Cable thought it more profitable to divert them to Japan, undermining the official importer. (*See also* Child, Alfred.)

Laconia Bike Week
First started in 1923, the event traditionally takes place during the beginning of June. It was based on racing (like Daytona), which had actually been taking place since 1917.

LBO
Leveraged buyout: a technical term for the management buyout of Harley from AMF in 1981. 'Management buyout' is not strictly correct, as most of the finance actually came from banks, not the management itself.

Leading Link Forks
An important early innovation from Harley-Davidson, designed by Bill Harley in 1907. Until then, motorcycle pioneers had mainly used unsprung cycle-type forks, which as speeds rose gave a poor ride and handling; they could also fracture. Harley came up with the solution in his final year at university: two sets of fork legs were connected at the bottom by a pair of forged steel links, which formed a pivot joint. Springing came courtesy of four long coil springs mounted inside the moving set of fork legs – the upper springs for load, the lowers for rebound. The wheel's axle was mounted just in front of the forks' lower attachment (hence leading link). For the low-powered machines of the day, they were rigid, had adequate travel, and a constant wheelbase. And they remained Harley-Davidson's standard front suspension for the next 40 years. They were even reintroduced (albeit in redesigned form) in 1990!

Leather & Lace
A US club and motorcycle association for female Harley riders.

Leno, Jay
Probably the most famous face on American TV, Leno is also well known for his bike collection, which contains a number of Harleys. He regularly acts as 'Road Captain' for the 'Love Ride' each November in California, which raises millions of dollars for charity.

Leonard, Joe
An independent rider in the 1950s, he was not backed by the factory, but nevertheless won the US National Championship on a KRTT. In fact, while the KR was Harley's official racer, the championship was won more often by private riders (notably Carroll Resweber and Brad Andres as well as Leonard) than by factory-backed ones. Between them, Leonard, Resweber and Bart Markel won the title ten times.

Leopold, B.F.
Leopold was instrumental in forcing Harley-Davidson to build its own sidecars. He bought up the Rogers sidecar company in late 1925, Rogers having been Harley's

main supplier, and refused to reserve his entire production for the company, to the point of stopping production altogether. Faced with the sudden withdrawal of its main supplier, Harley-Davidson had no choice but to start making sidecars itself.

Liberty Editions
Special-edition Harleys built in 1985 in honour of the Statue of Liberty and to celebrate its centennary. For each of the 1,750 Liberty Edition bikes sold (there were three versions), Harley-Davidson donated $100 to the Statue of Liberty/Ellis Island Foundation Inc. The company also organized a 'Ride for Liberty', in which Vaughn Beals started from the West Coast and Willie G. took a southern route. They met in Washington and led thousands of riders to Liberty State Park, NJ, where Beals presented the foundation with a cheque for $250,000.

Lightweight Motorcycles
(*See* Aermacchi.)

Lubrication
In the early days, engine lubrication was taken care of either by a drip feed, by which oil was dripped through a needle valve at a rate that could be adjusted by the rider, or by a hand pump, which one pumped a couple of times every few miles to keep the motor happy. The problem with the latter was that most riders, paranoid that the engine wasn't receiving enough oil, would overdo things and tended to over-oil it instead. The result was oil blow-by past the piston rings, carbon build-up, overheating and premature wear. The only cure was an automatic oil pump, which Harley-Davidson introduced for 1915. Driven off the magneto drive shaft, the pump fed through a sight glass so that nervous riders could reassure themselves that all was well. The pump delivered 4psi, laughably low by modern standards, though at the time Harley-Davidson claimed it would feed against a resistance of up to 70psi. By the standards of the day, it was a precision piece of equipment, and if any pump was faulty, dealers were asked to return it direct to Milwaukee for repair.

Lucifer's Hammer
Famous Harley-powered race bike, which won the Battle of the Twins series three years running, 1984–86.

Magazines
In the early 1970s, the growing custom bike scene spawned a number of lifestyle magazines on the subject, the most famous of which was and still is *Easyrider*.

Magri, Armando
Described by author David Wright as 'the patriarch of Harley-Davidson dealers', he came from Chico, California where he grew up buying, selling and racing bikes when he had the money. He secured a job with Frank Murray's Harley-Davidson dealership in Sacramento in July 1937 and eventually took it over in 1949. His ability to mix with all levels of riders (from the Harley and Davidson families to Hell's Angels) helped business and the dealership grew. Despite the chance to take on Japanese and British franchises, Magri stayed loyal to Harley-Davidson, and retired in the early 1980s, aged 72, leaving a large, healthy business behind.

Marshall & Ilsley Bank
Funded much of Harley-Davidson's early growth via loans, notably a major factory expansion (to 80,000sq ft/7200sq m) in 1911 and again in 1918, when the four Founders reputedly borrowed $3 million from the bank to pay for a huge factory expansion, this time to nearly 600,000sq ft (55,000sq m). Building started in late 1918 and was completed by April 1920. Harley-Davidson had probably the largest motorcycle factory in the world, though to build it the Founders had to overcome their natural distrust of bank loans and the associated interest rates. However, the Marshall & Ilsley Bank was impressed: at the new factory's dedication ceremony, the bank's president declared that Walter Davidson's credit rating was such that he could obtain a million-dollar loan at any time, just by signing his name.

Materials As Needed
System by which only materials immediately needed for production were held in the factory. Harley-Davidson introduced this in the early 1980s in order to cut inventories and costs.

McCann, Jeff
A well respected Californian custom painter, he has worked for many of the top custom bike builders including Arlen Ness, and Ron Simms.

Memorabilia
High prices can be commanded for the most obscure remnants of Harley-Davidson history and there is a large collectibles market for them all over the world.

OPPOSITE
LEFT: Leading link forks on a 1925 J-model – standard Harley equipment until 1948.

CENTRE: A Leather & Lace member.

RIGHT: Jay Leno is one of many celebrity Harley riders.

BELOW: Harley-Davidson memorabilia includes just about everything imaginable.

Military Tricycle (Prototype)
Before the advent of the Jeep, the US military was looking around for something to replace the conventional sidecar outft, and in the summer of 1939 was testing two rival tricycles, both based on motorcycle technology. The Delco company supplied a three-wheeler based on a BMW front end and automotive rear axle, while an Indian trike had already been undergoing testing for about a year. The army was evidently impressed by the shaft-driven Delco, as it afterwards asked Harley-Davidson to submit a bid to build 50 shaft-drive trikes. Bill Harley would have preferred chain drive, and proposed a trike with twin $^{5}/_{8}$-in (16mm) chain drives, but this was ignored. But he didn't give up, and later proposed an enclosed rear chain for a 45-based trike, which would overcome all the army's objections to high-maintenance chains.

Nevertheless, the army wanted shaft drive, so that's what it got, and a small batch of 61ci (1000cc) ohv V-twin trikes, with shaft drive and automotive axles, was built in early 1940. But when the first one arrived at Camp Holabird for testing, several problems arose. It vibrated so badly that the fuel line fractured four times; the engine leaked oil, lacked power in top gear and stalled in thick mud; the brakes had to be adjusted often and the whole device was tail-heavy, with a tendency to pop unwanted wheelies, though the latter was acknowledged to be an endemic problem with the tricycle layout. The trike was brought back to the factory and fitted with a slightly bigger 69ci (1130cc) engine, built by modifying a prototype 74 ohv; it was also found that the frame had cracked, so the rear end was strengthened. In the end, the project came to nothing as the army found that the Willys Jeep was far more suited to its needs than any three-wheeler. However, Harley did build an XA-powered trike which was destined to become the new Servicar in 1945.

Miller, Theodore
Assistant sales manager to Arthur Davidson in the early 1920s, and in charge of domestic sales.

Morrow, Carl
Top engine tuner and speed guru. Carl's Speedshop in Daytona is a state-of-the-art high-tech workshop, turning out high-performance parts for road and track Harleys.

Motorcycles As Needed
Dealers protesting against the Bonus Bucks programme in 1983 used this as a parody of Harley's 'Materials As Needed' scheme. They would only order bikes they could sell very quickly, refusing to hold stocks of machines imposed by the factory.

Muhammad Ali (Cassius Marcellus Clay)
Despite once warning Joe Frazier that 'motorcycles is for crazy people', Muhammad Ali later bought a Harley.

ABOVE: Jeff McCann, a leading custom painter.

RIGHT: The 'Mustang' scooter.

Muscular Dystrophy Association
Harley-Davidson's favourite charity, which benefited in particular from the company's 85th birthday to which anyone was invited, as long as they contributed $10 to the MDA. In all, the weekend event raised $500,000 for the charity.

Mustang
Small American moped, the tank of which was used on choppers in the 1970s, coining the phrase 'Mustang-style gas tank'.

Ness, Arlen
The godfather of US customizing, Arlen Ness has been designing and building custom Harleys for around 30 years. He has concentrated on the Low Rider style and also builds outrageous bikes to showcase both his skills and his burgeoning custom parts business, 'Two Bad', for example, being powered by two supercharged 61ci (1000cc) Sportster V-twins. Arlen Ness parts are now an integral part of the custom movement. But though Arlen Ness (and his son Cory, who is part of the business) is thoroughly grounded in Harleys, there are signs of diversification, and the latest addition to the catalogue is a set of Ness-branded race leathers.

New York Stock Exchange
Harley-Davidson went public on the New York Stock Exchange in June 1986, raising $90 million in stock and subordinated debt. Selling company shares on the general market allowed Harley-Davidson to buy Holiday Rambler motorhomes and secured its independence from long-term debt.

Arlen Ness has progressed through just about every custom style, but always comes up with his own ideas. Who else could have thought up this two-wheeled Ford Thunderbird?

Nordberg, George
Production assistant in the early days under William Davidson, Nordberg joined Harley-Davidson in 1909.

NOVA
A new engine programme, launched in 1976, which aimed at producing a complete range of all-new water-cooled engines to power a new generation of modern Harleys. The plan was for a modular system in which a range of engines (including a 30ci/500cc twin, 61ci/1000cc four and 92ci/1500cc six) would all share vital components to cut costs and valves, pistons and gears would be common to all. Parent company AMF seemed willing to underwrite the cost (it was to put $10 million into NOVA), and a contract was awarded to Porsche to develop the range as Harley-Davidson's own engineering department had enough to do, keeping the existing V-twin up with emissions legislation, and developing the upcoming Evolution V-twin. Over the next few years, development continued, and when the Evolution was unveiled in 1983, a few selected dealers were shown a running prototype of the V4 NOVA. If everything had gone to plan, we might have seen water-cooled V4 and V6 Harley-Davidsons on the market by 1985–86.

Unfortunately, AMF's attitude to Harley-Davidson's long-term future changed as the 1970s drew to a close, particularly after new chief executive Tom York came on board. A team of consultants recommended that NOVA get the final go-ahead, at a cost of $60–$80 million. York refused, and instead began to look for ways of getting rid of Harley-Davidson altogether. NOVA was effectively dead from that point on. Ironically, the air-cooled pushrod Evolution proved such a success that the need for NOVA receded

HARLEY-DAVIDSON
XR750
1
HOG
HOG
GOODYEAR
EAGLE

into the background anyway. Only now, nearly 25 years later, is Harley-Davidson thought to be once again working on a new range of water-cooled engines.

O'Brien, Dick
Harley's racing director from 1957, he succeeded Hank Syvertson, being only the third man to hold the post. He had worked for a Harley-Davidson dealer before the war and volunteered for pilot training when war broke out. O'Brien returned to his job as service manager afterwards, before joining Harley-Davidson's racing department which, under his leadership, enjoyed a run of success through the 1960s until 1969, when the AMA rules were changed to allow larger 45ci (750cc) ohv machines to compete with the side-valves. O'Brien's response was the XR750, a short-stroke version of the Sportster V-twin. It was a disaster, but the completely redesigned XR of 1972 had more success, going on to dominate American flat-track racing for the next quarter-century. Dick O'Brien retired in 1983.

OPPOSITE: The XR750, in classic flat-track guise. Once it had been developed and the initial problems overcome, the XR750 came to dominate American flat-track racing. The VR1000 (which you can just see in the left background) faces a tougher task on the circuits.

RIGHT: A 1920s eight-valve racer. After several years of dimissing racing as a waste of good money, the four Founders plunged into competition wholeheartedly with this purpose-built machine.

Ottaway, William
Next to the four Founders, Ottaway was a figure central to Harley-Davidson's racing and engineering policy. He joined in 1913 as assistant engineer to William Harley. He was an instinctive engineer, and in four years designing Aurora Thor machines had made a name for himself in racing circles; Bill Harley realized that Harley-Davidson needed some of Ottaway's natural flair. One of his first jobs was to improve the efficiency of the standard Harley-Davidson 61ci twin, and he also spent much time reducing vibration levels. Ottaway is probably best known for his work on Harley racing machines, notably the II-K (using a tuned version of the 61-ci/1000-cc side-valve twin) and the Model 17 8-valve racer. The latter (after refinement by cylinder-head specialist Harry Ricardo) produced 55hp, plenty to win races in 1915. These technical advances, plus tight team organization from Ottaway, allowed Harley-Davidson's all-conquering Wrecking Crew to sweep the racing boards in the late teens. It was also William Ottaway who warned Bill Harley that his clutch design might infringe Bendix patents and was proved right when Bendix took Harley-Davidson to court. He increasingly concentrated on road bikes after Harley-Davidson dropped its big racing programme in 1921 and Joe Petrali took on the development of race machines. When William Davidson died in 1937, Ottaway took over as interim production manager and agreed to allow trade unions into the plant for the first time. In 1943 he took over as chief engineer when Bill Harley died, though this function was soon passed on to Bill's son, William.

Pacific Motors
This is a story of shady goings-on in 1920s Los Angeles, where Indian dealer C. Will Risdon had the upper hand, particularly in the lucrative and high profile police sales. He decided to retire in 1925, and looked around for someone to buy his well-established, well-equipped dealership. An offer came in from A.F. Van Order but Risdon was suspicious, knowing that Van Order worked for Harley dealer Rich Budelier; but Van Order's story was that he had been sacked, and wanted to set up in competition with his old boss. Risdon was persuaded, and his dealership passed to Van Order's holding company, named Pacific Motors. He immediately increased the fire insurance coverage to its maximum, and the following month the entire premises burnt down! Nothing was ever proved in court, but on the night of the fire Van Order was 'out of town' and Budelier was found in bed, fully clothed and with his shoes on.

Parker, Scott
An official Harley-Davidson racer in the 1980s/90s and the youngest rider ever to earn an AMA expert licence. He was team-mate to Jay Springsteen in the official racing team for 1986 and won the National Championship himself in 1990 and '91.

Parkhurst, Leslie
'Red' Parkhurst was a central member of the Harley-Davidson Wrecking Crew, and a highly successful and popular racer on both side-valve and eight-valve Harleys. He started racing Harley-Davidsons privately, serving his apprenticeship on the notorious pine-planked Motordromes (nicknamed 'Murderdromes') but caught the attention of the factory team when racing at Milwaukee. He was recruited to the team and, being recently married, abandoned the Motordromes for safer dirt tracks. The following year, after many wins, he had a serious accident at Rockford, Illinois, but carried on racing until the USA entered the First World War. Despite declaring he would never return to racing, Bill Ottaway managed to persuade him back with the lure of a ride on the eight-valve racer for the 1919 season; he won the 200 Mile Internal Road Race Championship at Marion, Indiana and in early 1920 set up new speed records (although not officially accepted) of 112.61mph (181.2km) over the mile on an eight-valve and 82.09mph (132.1km) on a sidecar outfit. But by this time Red Parkhurst was one of the oldest racers on the track and he decided to retire for good, citing his responsibility to his wife and child.

Perkins, Dudley
An early Harley-Davidson dealer in San Francisco, who helped Tom Sifton obtain his sub-dealership in 1929.

Pershing, General John
Led the first US military operation to use motorized units, including Harley-Davidson solos and sidecars in the 1916 attempt to track down the army of General Francisco Villa in Mexico.

Petrali, Joe
Almost single-handledly took on Harley-Davidson's minimal competition strategy

Scott Parker, US National Champion in 1990 and '91 on an XR750.

from the late 1920s. He was a terrific rider on both tracks and hill climbs (Hill Climb Champion 1932–38), a practical mechanic, and was central to the development of the 61E Knucklehead. Born in 1904 in San Francisco, the son of Italian immigrants, Petrali made his name racing in the 21ci (350cc) class, and was taken on by Harley-Davidson to 'racerize' its own 21ci Peashooter. While helping develop the Knucklehead, he warned the company that the bike's oil leaks and valve spring breakage problems needed sorting out before production. He also set up speed records with the Knucklehead. With Hank Syvertson, he built a streamlined bike with twin carburettors, higher compression, different cams and magneto, plus a high top gear that allowed a theoretical top speed of 160mph (257km). On the first run, the bike went into a serious tank slapper as Petrali approached 100mph. He came in and ordered that all the aerodynamic bodywork be removed, then went out again and captured the mile speed record at 136.15mph (219.1km). This was disappointing for Petrali, as Walter Davidson had promised him a $1,000 bonus if he got to 150mph! Joe was evidently given many of the odd jobs at Harley-Davidson, and in 1938 supervised the dismantling of a Crocker V-twin, which Milwaukee suspected of encroaching on Harley-Davidson patents. He resigned from the company that year after Walter Davidson refused him a job in the experimental department and went on to work for Howard Hughes, later taking to the speaker circuit where he was able to air his reminiscences of Hughes and Harley-Davidson. Joe Petrali died in 1974 as a result of heatstroke suffered at Bonneville Speed Week.

Pinehurst Meeting

This was a seminal week in Harley history. Vaughn Beals, the man brought in by AMF to turn Harley-Davidson around, decided in April 1976 to take all Harley-Davidson senior managers on a week-long brainstorming session to Pinehurst, North Carolina where the long-term project strategy could be thoroughly explored. What emerged was a two-point plan. First was to improve the existing V-twin gradually and this was eventually unveiled as the Evolution engine, which carried Harley-Davidson right through the 1980s and '90s. Second was the NOVA family of all-new engines. (*See* NOVA.)

Poetry

Martin Jack Rosenblum may well be the only Harley-Davidson poet in the world. He has a doctorate from the School of Objectivist Poets, and in his 40s achieved a lifelong ambition to own a Harley. While recovering from a near-fatal illness, he made a complete career change from teaching poetry at university level to helping students lacking in college skills. He now works as Harley-Davidson's official archivist at Juneau Avenue. Here is a selection of his work, all taken from his book *The Holy Ranger*, and all are produced with the kind permission of the man himself.

The Name Is Harley

standing there with the look
of invention in front of that
original shack new ideas took
hold which built a reputation
all the way into the American ideas

that govern individuality & style:
William S. Harley has lent his name

to the modern horsemen whose
apocalyptic ride strengthens
open trail visions that make

this photographic action take place
this sunrise in April as I take the
keys to my Harley machine then walk

the irregular stone path to
the garage contemplating in
poetic anticipation exactly
what will
be coming
within an
afternoon –
upon this
Iron Name

Dr Martin Rosenblum is probably the world's only Harley-Davidson poet. He's certainly the only one sanctioned by the factory!

Evolution Amulet

the 30th Anniversary XLH sat in the dining
room
during winter months & once after a late
movie
when all slept & the first real snow landed
I got on it & turned the lights toward a
front window shining into the blanket
being tossed upon the birch tree
as winter's comfort & signalling
with yellow flashes reflecting
on an empty television screen:
the first spring rain bringing
this Sportster down the ramp
& across the lawn & sidewalk
to the garage where it
fires up once the carb
swallows ample passion
but the rain keeps any
travel in the alley to
just beads upon fresh
wax or upon the chrome my reflection
while I get down nearer the exhausts
to hear what winter kept in silence

Harley-Davidson 85th Anniversary Homecoming Contemplation

out of the Kenosha holding strip
onto 94 with all areas alongside

covered by raised fists & victory
then on p43 of the HD magazine I

find my Harley racing orange front
fender in that photo coming around

the bend in back & my beard sprouting
out from the black helmet – well being

there in June & here now in my office
between appointments just from buying

this to read & opening my napkin
to keep sandwich crumbs off that

front fender it coming at me
as I was upon it with cycles

behind it columns roaring
& such lovely ladies with

leather fringe sparkling
upon the interstate path

into Milwaukee Iron
coming home I rode:

a pack of Strokers
at my side & front

to rear passion
burning V-twins

in the mystical
gas tank logos'

signatures that
chant the poems

I cannot write

Mid-Summer on a Harley-Davidson

Wisconsin cornfields strangled
without water

& gravel shoulders coughing
dust in the wake

of that eighteen-wheeler ahead
so the downshift & then faster

onto this county trunk
as blackbirds' chatter

stick in the windless
hum of tires going to

the broken yellow
lines as a center

for the holy
morning ride

History's American Springer

all the imports hid behind plastic
fairings and their riders crouched
on top of them wearing bright
tennis shoes
in the later
eighties when exposed satisfaction
aesthetically sprung onto those
Softail front ends in chromed
memories of the
future here in an
American reply with
our riders deep inside
of these machines such
that every element
upon Milwaukee
street poetry was concretely observed
through that
Springer expanding & contracting
anytime the rider's heart would &
blood pulsed through the sparkling
coils straight from homemade poetic

Potato Patent
Harley-Davidson has been famously litigious over the use of its trademark, most notably of all over the attempt to patent the 'potato-potato' exhaust note. This is a distinctive sound, a product of the 45-degree V-twin which because it's not produced by any other motorcycle is instantly recognizable as emanating from a Harley. The company reasoned that this too could be copied by a rival manufacturer, and so began exploring whether it was legally possible to patent a noise, even one as distinctive as this. The legal process went on for several years, and in 2000 an Australian musician by the name of Bill Cook was actually to claim that he'd taken out an American copyright on potato-potato when making an album called *Steel Stallions* in 1993. According to Cook, legal counsel for Harley-Davidson had been trying to reclaim ownership of the patent through the US Patent and Trademark Office, despite opposition from rival manufacturers. The musician cheekily maintained that he was prepared to hand the copyright over if Harley-Davidson gave away a copy of his album with every bike it sold, for which he would charge $15 a time. It's true that Harley-Davidson gave up its attempt to patent the sound in 2000, but was it really because of Cook? 'It's a nice story,' said one Harley-Davidson official, 'pity it's not true.'

Presley, Elvis
The King was a confirmed Harley rider, a fact which made it to the cover of *The Enthusiast* (Harley's own magazine) in May 1956. According to the text, he had started out on a 165 and was by then onto his third Harley, a brand new red-and-white KH. It was, of course, great publicity for the company, so it must have been disappointed by the1964 film *Roustabout*, in which Elvis rode a Honda!

Price Fixing
Price fixing was a feature of the relationship between Harley-Davidson and Indian between the World Wars. It seems astonishing now that these two firms, which were such bitter rivals and which indulged in cut-throat competitive techniques, should also meet once a year to agree on common prices. The first meeting was in 1922 between Arthur Davidson and Indian's general manager, Frank Weschler. In the words of author Harry Sucher, these long-term rivals shared 'crêpes Suzette and lobster thermidor in an atmosphere of somewhat strained conviviality'. It was agreed that henceforth, similar competing models from Harley-Davidson and Indian

Despite riding a Honda in the film *Roustabout*, Elvis owned a Harley. It was an early XL Sportster, with screen, and appeared (as seen here) on the cover of *The Enthusiast* magazine.

THE Enthusiast
A MAGAZINE FOR MOTORCYCLISTS
153
Elvis Presley
Audubon Drive
Memphis
Harley-Davidson
Solo
56
2
Memphis Harley Davidson Co.
Memphis
Tenn.

would carry the same prices. Typical was the 1931 dinner meeting at New York's Waldorf Astoria Hotel: the Harley-Davidson Big Twin and Indian Chief would both cost $320, the Harley-Davidson 45 and Indian Scout, $295, and so on. At one point, Indian president, Paul du Pont, turned up unannounced, and warned of the consequences if the contents of these secret meetings ever leaked out. Price fixing was of course illegal.

Productivity Triad

Three things crucial to Harley-Davidson's survival and turnaround in the 1980s were employee involvement, just-in-time production and allowing workers to control quality for themselves. In 1981, Harley-Davidson's central problem wasn't a lack of demand for its bikes, but the way they were made; in other words, inefficiently, expensively and of poor quality. The answer was to use Japanese production methods, of which the Productivity Triad was Harley-Davidson's interpretation.

1) Employee Involvement: There was a wealth of knowledge and experience on the shopfloor in Milwaukee and York, but it was being ignored. This came to a head in the AMF days, when production was forced up despite what those on the shopfloor said. Quality circles had already been tried in a half-hearted fashion, but this time they worked, because they (and other employee involvement techniques) had the commitment of management, and employees realized that they really were being heard and could have a genuine influence.

2) Just-In-Time (JIT) Production: Traditionally, like much US industry at the time, Harley-Davidson had relied on large batch production of both certain components and particular models of bike. This meant holding large stocks of raw materials and parts, sometimes for months until they were needed, which tied up a great deal of capital and slowed down the production process. Instead, JIT delivered only those parts to the assembly line which were needed to build each bike (*see* Jelly Bean System). Meanwhile, departments were reorganized to enable them to rapidly switch between making small batches of different parts, rather than large batches. In the ten years after JIT was introduced, Harley-Davidson's inventory was reduced by 75 per cent, scrap and rework was down by 68 per cent, and productivity up by 50 per cent.

3) Statistical Operator Control: A much simpler idea than it sounds. Employees were given the training to monitor the quality of their own work. Once they were proficient in this, problems could be rectified at source as soon as they arose because they were detected immediately by the person doing the job, rather than by a quality inspector further down the line after 500 faulty parts had been made or (worse) by no one at all.

These three methods were all inter-related, and it is doubtful that any one of them could have worked without the other two. Together, they transformed the efficiency and effectiveness of Harley-Davidson's production process, and turned the company around.

Quality

(*See* Productivity Triad.)

Rat's Hole

Big Daddy Rat holds these world-famous custom bike shows at Daytona Bike Week in March and the Sturgis Rally in August. He also puts on an annual event at the Essen Motor Show in Germany.

Razorback

A make of custom Harley during 1993–94.

OPPOSITE: A selection of customs at the Rat's Hole Show, at Daytona in 1989. These sleek 'bodywork bikes' are a phenomenon of the 1990s.

RIGHT: Rat's Hole, 1990, and Big Daddy was there in person.

Built by Wyatt Fuller (who has been riding for 30 years but only customizing for three), the Razorback Motor Works specialized in clean-looking single-colour bikes, with the engine exposed but all the nooks and crannies smoothed out and filled in. So successful was the look that Wyatt was bought out by Harley-Davidson in 1994 and the man himself was contracted to design accessories.

Reagan, Ronald
As well as agreeing to a new import tariff in 1983 to protect Harley-Davidson, President Ronald Regan toured the York assembly plant in May 1987. All politicians like to be associated with success, and Reagan was no exception, particularly as the tour followed Harley-Davidson's high profile request that the tariff be dropped a year ahead of schedule. It was a symbolic moment: at one point, Reagan was invited to start a newly-built Sportster: 'This thing won't take off on me, will it?' He never was a motorcyclist.

Red River Run
On the annual memorial weekend, the Vietnam Veterans Motorcycle Club gather in Red River, Mexico to ride, party and remember their fallen comrades at the Vietnam memorial chapel in the mountains.

Remy
Remy electrical equipment was introduced on the 61ci (1000cc) J-model in 1916, in

the form of battery and lights. In 1919, the factory dropped these in favour of its own electrics, though it was over a year before they were acceptably reliable.

Resweber, Carroll
An official factory rider in the late 1950s who won the National Championship four times, in 1958, '59, '60 and '61. Incredibly, he did all this while working a full-time day job for Mercury Marine. Not surprisingly, his home town of Cedarburg, Wisconsin, held a 'Carroll Resweber day'. He was on course for a fifth title when he ran into Jack Gholson and Dick Klamforth at the Lincoln, Illinois track in 1962 where the dust was so thick that he reputedly had his eyes shut at the time. He was seriously injured, lying unconscious for nine days, and was left with one arm and leg shorter than the others. But Resweber survived and went on to work in the Harley-Davidson race department. One of the better Resweber anecdotes came out of a race in late 1950s Laconia. It was early days for the racing K-model, and the gearbox was turning out to be too weak. Resweber didn't want to race that day, having a huge hangover, but being a factory rider he had to do so. So he determined to break the K's weak gearbox and retire early. However, and despite full-throttle clutchless heavy-footed gearchanges, it ran the full distance.

Ricardo, Harry
An English cylinder-head specialist, whose pioneering work on four-valve combustion chambers inspired generations of engineers. At Joe Petrali's suggestion, he was brought in to develop Harley-Davidson's eight-valve racer for 1915. Walter Davidson refused at first, faced with the cost of Ricardo's steamship ticket and his fee, but relented when Petrali

SALOON
THE WOODS
Satellite
Big Screen T.V.
PACKAG
LIQUOR
OPEN

reminded him that $25,000 had already been spent on the the eight-valve. Ricardo duly worked his magic and the bike was transformed into a race winner. Less successful was the side-valve Model A 21ci (350cc) single that Harry Ricardo had had a hand in and which had a tendency to seize if ridden hard for any length of time.

Rikuo
Model name for Harley-Davidsons built under licence in Japan in the 1930s. (*See* Child, Alfred.)

Robert Baird & Co
The Wall Street firm of brokers which launched Harley-Davidson shares on the stock market in 1965. Until then, much of the company was still directly owned by the Harley and Davidson families, 50 members of the two clans owning 53 per cent of the voting shares and in effect controlling the company's destiny. But by the mid-1960s, Harley-Davidson was losing market share and desperately needed money. One way to raise it was to sell shares on the open market. Robert Baird & Co was later retained to advise Harley-Davidson on the AMF takeover.

OPPOSITE: Scene at the Red River Run.

RIGHT: A Shriner Servicar.

Robinson, Dot
National president of Motor Maids of America in the 1940s and wife of Harley-Davidson dealer Earl Robinson.

Robinson, Earl
Detroit Harley-Davidson dealer from the 1930s, who set transcontinental records solo and with his wife Dot. He rode the 3,000 miles (4828km) from New York to Los Angeles on a tuned 45 in 3 days 6½ hours in 1935, then made the same run with Dot, setting a sidecar record.

Rogers Company
Supplied Harley-Davidson with sidecars up to 1925, when founder B.F. Rogers died and the company was bought by Eros Manufacturing. Rogers always resisted Harley-Davidson's demands for it to make sidecars for it exclusively, and carried on selling to other manufacturers such as Ace and Excelsior. However, Harley-Davidson did persuade Rogers to build a special wide-bodied sidecar which would carry two adults. Its 56½in (143.5cm) track was the same as the standard US car track, though it was designed more with exports in mind rather than the home market.

Rourke, Mickey
Yet another high-profile celebrity Harley rider of the 1980s. Screen actor Rourke spoke of his fascination with Harley-Davidson: 'It's a personal thing that can't be described. It's part of you.' Not that Hollywood Harleys were a new idea: in the 1930s and '40s, stars like Clark Gable, Robert Young and Tyrone Power were all happy to be photographed astride big V-twins. On the other hand, the Harley as a celebrity fashion accessory may have gone a little too far, for it's been said that in Hollywood, Harley-Davidsons are just like a certain part of the human anatomy – everyone had one.

Scherer, Julian
'Hap' Scherer was Harley's publicity manager in the early days, having joined the company in 1915. He was a particularly strong exponent of the Sport Twin, realizing that the whole American motorcycle industry was in danger of only catering to the limited enthusiast market, with big V-twins. Scherer, along with other industry salesmen, wanted to expand the market with easier to ride machines like the Sport Twin. In 1920, he actually rode one from Denver to Chicago in 48 hours, but sales fell off after he left the company. He was sacked by Walter Davidson after allegedly co-operating with Harley-Davidson's main rivals. Scherer later went to work for Rich Budelier's new Harley-Davidson dealership in Los Angeles.

Schwinn, Ignaz
The bicycle magnate who bought up Excelsior, for some time the number three American manufacturer, after Harley-Davidson and Indian, and kept it going until 1931.

Servicar
The three-wheeled Servicar used the front end of a 45 married to a two-wheel rear end and a load-carrying box. Originally intended for mechanics to collect vehicles for repair, it proved ideal for traffic police and meter maids, becoming something of an American institution.

Shriners
Believe it or not, there is a strong association between that peculiarly American phenomenon, the Shriners, and Harley-Davidson. Drill teams have become common, and perform intricate, highly

disciplined manoeuvres at outdoor shows. Many of them are based in the US's south-eastern corner, and the Al Menah team from Nashville is typical. It has 16 members mounted on identical police-style Harleys (without the sirens or red lights) and was formed in 1958. Even by the early 1980s, it had still won more top three trophies than any other team. The competitions aren't just about bike control; there's a strong militaristic aspect to the Shriners in their demand for cleanliness, uniformity and a good turn-out. Inspections are said to be just as tough as those for the Marines!

Side Van

Introduced in 1915 as an adjunct to the successful sidecar line. It was basically a wooden box on a sidecar chassis, with a hinged top for loading, and proved very popular with commercial users. Among the factory options was your own

ABOVE: Before the Servicar, side vans (commercial sidecars) were popular with businesses that couldn't afford a Ford van.

LEFT: Silent Gray Fellow undergoes silent contemplation: very few of these bikes are now unrestored.

OPPOSITE: Big man, big bike: Ron Simms of Bay Area Custom Cycles.

company's signwritten name on the box, at 10 cents per letter, and a folding top and side curtains to protect the rider from inclement weather.

Sifton, Tom
An almost legendary tuner of Harleys, Tom Sifton started out as a sub-dealer in San Francisco in 1929 and became a full dealer in 1933: but he increasingly spent his time tuning Harley-Davidsons for competition. With detailed work on the porting and valves, he was able to make the unpromising 45 into a race winner, but just how he did it remained a secret from fellow tuners and even from the factory's racing department. In fact, Sifton went on making side-valve Harleys go faster for decades, working his magic on the WR racer in the late 1940s and the KR in the 1950s and 1960s.

Silent Gray Fellow
The nickname given to the early Harley-Davidson single, owing to its standard colour and quiet running. It was a typical machine of the time, with belt drive, inlet-over-exhaust head and atmospheric inlet valve. Top speed was around 45–50mph (72–80km/h) and the engine's peak was about 500rpm: only when more positive valve operation was used could engine speeds rise. But of the thousands of similar machines which appeared in both America and Europe in the first decade of the 20th century, the Silent Gray Fellow was one of the few survivors, for no better reason than that it was well made and reliable.

Simms, Ron
Has operated as Bay Area Custom Cycles in San Francisco for 30 years. His tough 'gangster'-style customs with distinctive skeleton/skull paint jobs are instantly recognizable.

Spiegelhoff, Johnny
A racer in the 1930s and '40s, and though better known as a Indian devotee, Spiegelhoff was a committed Harley rider in the 1930s who worked as an inspector on the Milwaukee assembly line. But he switched allegiance when the Harley-Davidson racing department refused to make the changes he wanted to his bike. So he left his job, bought an Indian, and scored several race wins, notably at Daytona in 1947. He even took on an Indian dealership in Milwaukee, right on Harley's home turf.

Springsteen, Jay
A legendary racer on dirt or pavement, his career stretches back over 20 years and continues to this day.

Sturgis Bike Week
(See Black Hills Classic.)

Styling
The styling of Harley-Davidsons – the way they look – is now central to their appeal, and just as important as the trademark V-twin. But a styling department wasn't established until 1963, when president William H. Davidson phoned his son Willie G. (then working for an industrial design company) and invited him back to the family firm to run it. Even so, it was eight years before Willie G's first seminal bike appeared, the Super Glide, using an Electra Glide engine and frame with a lighter-weight Sportster front end. It was a nod in the direction of contemporary customizers, and the first true factory custom. The Super Glide was followed by the less successful Café Racer (whose European look was too remote from the traditional Harley) and the very successful FXS Low Rider, with raked, lengthened forks, low seat and fat rear tyre. The Low Rider sold by the thousands, but the clever part was that, like most of Willie G's designs, it used mainly existing components to save money and time. Since then, other styling milestones have been the Wide Glide (with wide-spaced forks – another custom favourite) and the Softail (with hidden rear suspension to ape the authentic hardtail look). The Springer Softail used an updated version of Bill Harley's original Springer front forks and the Fat Boy (another good seller) has solid wheels and a single colour. As the 1990s ended, Harley-Davidson seemed to have reached the point where the way its bikes looked was far more important to buyers than any technical considerations.

SuperRide
A nationwide demonstration ride programme in 1985 which took place over three weekends when 40,000 potential customers took 90,000 test rides, though not enough bought bikes to justify the $3 million that SuperRide cost to set up. But since then, Harley-Davidson has become one of the foremost companies for offering demo rides at shows, events and dealerships.

Syvertson, Hank
Long-time member of the Harley-Davidson racing department, who assisted Joe Petrali in the development of the Knucklehead and in building Joe's record-breaking version of the same bike.

LEFT: They don't come much more experienced than this. Here Jay Springsteen is lost in contemplation before a race.

OPPOSITE: The XLCR Café Racer was one of Willie G's rare flops. Call it a European-type Sportster or what you will, but few were sold.

Teerlink, Richard
Harley-Davidson's chief financial officer at the time of the 1984 crisis with Citicorp. Citicorp, Harley's original financial backer from the 1981 buyout, wanted its money back; so Teerlink, together with Vaughn Beals, Tom Gelb and others, managed to negotiate a refinancing deal with Heller Financial. It was finally signed on 31 December 1985, just weeks before the company would have needed to file for protection against bankruptcy, for which a legal team had already been hired. Richard Teerlink later succeeded Vaughn Beals as chief executive officer.

Terminator
Who else but Arnold Schwarzenegger could plays an android, returned to our time from the future in order to protect the future saviour of the human race? And, of course, he rides an all-black Harley-Davidson. As he says to the prone biker he's pointing a large gun at: 'Give me your clothes, and your motorcycle.'

Thompson, Charley
One-time president of Harley-Davidson, who once said that the biggest job for dealers was to find things for customers to do with their bikes once they'd bought them. He was reflecting the way the market was changing into a leisure industry, with accessories, clothing and dealer-organized events, like ride-outs, being at least as important as the motorcycles themselves.

Tipton, Ken
Builder of the Harley-Davidson specials from the 1950s, who went on to produce his own proprietary tuning parts.

Tritten, Ray
Brought in by AMF in 1973 to tackle Harley-Davidson's continuing problems. The moving of bike assembly to a factory in York, Pennsylvania had created its own problems, while quality and market share were heading downhill. He made some small improvements to production and forecasting, but probably his biggest contribution was to bring in Jeff Bleustein (*see* above) to head up a new, more professional engineering department and to appoint Vaughn Beals as AMF's man in Milwaukee. Beals was a particularly good choice, as this finally put a senior AMF person in day-to-day contact with Harley-Davidson management. Previously, AMF had been attempting to control both Milwaukee and York plants from its own remote headquarters.

Triumph
Triumph, the famous manufacturer based in the English Midlands, was a thorn in Harley-Davidson's side in the 1940s, '50s and '60s. It was probably the most successful of the makes imported into the US after the Second World War, notably with the twin-cylinder Thunderbird and Bonneville. Although smaller and lighter than any Harley, the Triumphs were faster and more nimble, and by 1950 imports had captured 40 per cent of the American motorcycle market. Harley-Davidson's response to Triumph and the other imports

was slow to come, but arrived in the form of the side-valve K, overhead-valve Sportster, and later with the smaller Aermacchi bikes. Harley-Davidson's official view was that none of these imports were direct competitors to its own big V-twins, a short-sighted view that nearly led to its demise. Ironically, Triumph and its British contemporaries used much the same argument when confronted with Japanese competition.

T-Shirts
T-shirt slogans featuring Harleys have become something of an art form. Here is a selection:
'God Rides a Harley'
'Harley-Davidson: If I had to explain, you wouldn't understand'
'Harley's Best, **** the Rest'
'I'd rather see my sister in a whorehouse than my brother on a Honda'
'On the Seventh Day, God Created the Harley-Davidson (And on the eighth, he traded it in for something lighter and faster)'

Tucker, Mike
One of the first people to drag race a Harley. Tucker used a more or less standard FLH with a tuned engine and all the superfluous equipment removed.

Vietnam Vets (*See also* Red River Run)
A US national club whose members all served in the Vietnam conflict from 1964–73. They all sport patches commemorating their fallen brothers and others declared missing in action.

Von Gumpert, Eric
A long-standing export manager from 1913, Eric von Gumpert negotiated the first exports of Harley-Davidsons to Britain, and was involved in many of the early export markets, including Europe and Japan. As a veteran employee, he prepared Harley-Davidson's submission to the Tariff Commission in 1951, applying for higher import duties on bikes brought into the USA.

Walker, Otto
Captained the Harley-Davidson racing team in 1915, winning the 300-mile (483-km) race through Los Angeles' closed streets at 68.31mph (110km/h). Team mate Red Parkhurst came second.

Warbird
A special body kit designed by Mike Corbin of Hollister, California, which can be fitted to big twins or sportsters giving them the appearance of fighter aircraft and a racier image. They come unpainted, the builders, in this case Arlin Fatland, getting individual paint jobs designed.

Warr, Fred
A famous Harley-Davidson dealer in London, who later became the official importer for Britain. He is still in business as F.H. Warr & Co.

Watson, Duncan
Official Harley-Davidson importer for Britain from April 1914, though business only really got underway after the First World War, in 1919. He was also Lord Mayor of London and was knighted by George V. Watson made a good job of establishing Harley-Davidson in Britain, but withdrew in 1925 when a new 33 per cent import tax made the American bikes prohibitively expensive. It would be many years before Harley-Davidson had another UK importer.

Web Sites
There are probably thousands of web sites associated with Harley-Davidson, but here are four that are official:
www.harley-davidson.co.uk
www.harley-davidson.com
www.hog.com
www.buell.com

The Wild One
Stanley Kubrick's 1954 film, starring Marlon Brando, was supposedly a dramatization of the events of 1947 in Hollister, California (*see* above), which reinforced the bad boy biker image in the minds of the general public. The film was actually banned for a time, until 1970 in England, but the image stuck. It really didn't matter to most people that Brando rode a Triumph – in the eyes of the bikers, Lee Marvin's rival gang all rode chopped Harleys, so it followed that they must be the *real* outlaw bikes.

The Wrecking Crew
The almost legendary Harley-Davidson racing team of lthe late 1910s and early '20s, which at that time swept the board in America. Under the direction of Bill Ottaway, a hugely talented group of riders was brought together, with strict team

tactics and pit-stop techniques adding to the success. Prominent riders were Jim Davids, Ralph Hepburn, Walter Higley, Fred Ludlow, Otto Walker and Ray Wishaar.

Zimmerman, Fred
A dealer in Rochester in the 1920s, Fred Zimmerman enjoyed the disinction of selling new Indians under the 'Rochester Harley-Davidson' name. Fred obtained a Harley-Davidson franchise for the area in 1922, despite being a Jew. (The Founders were basically anti-Semitic, and had a policy of excluding Jews from any part of the business.) Zimmerman's new dealership did well, particularly with the advent of the fast Two Cam Harleys. It was said that he used to fettle Two Cams used by the booze runners in those Prohibition days, as well as those used by the state police to chase them! The story goes that a business rival informed Harley-Davidson of Fred Zimmerman's ethnic background and factory support for his dealership was gradually withdrawn. In the end, Zimmerman cancelled his franchise and took one with Indian instead, but kept the name 'Rochester Harley-Davidson'. His right to do so was upheld in court.

OPPOSITE:
LEFT: Triumph twins (this is a flat-track racer) seriously threatened Harley-Davidson in the 1950s and '60s.

RIGHT: However, some very interesting people ride Harleys nowadays.

RIGHT:
Corbin Warbird, as painted by Arlin Fatland.

YEAR ON YEAR

All years refer to Model Years.
Listed are main models for sale only.
Prices are retail for US market.
Main specifications and changes only are listed,
not minor year-to-year changes

1903

Model: 24.7ci (405cc) F-head single
Price: $200
Production: 3
Colour: Black with Gold pinstriping (by Aunt Janet Davidson)
Specification: 24.7ci F-head single, 3 x $3\frac{1}{2}$in (76 x 89mm) bore x stroke, auto inlet valve, belt transmission, loop frame, 28in-wheels (711mm), wheelbase 51in (130cm), weight 185lb (84kg)

1904

Model: Model 0, 24.74ci (405cc) F-head single
Price: $200
Production: 8
Colour: Black with Gold pinstriping
Specification: As 1903

1905

Model: Model 1, 24.74ci (405cc) F-head single
Price: $200
Production: 16
Colour: Black with Gold pinstriping
Specification: As 1903

1906

Model: Model 2, 26.8ci (440cc) F-head single

A 1905 single. Few changes occurred that year, but production doubled.

Price: $210
Production: 50
Colours: Black with Gold pinstriping, Renault Grey with Red pinstriping
Specification: 26.8ci (440cc) F-head single, $3\frac{1}{8}$in x $3\frac{1}{2}$in (79 x 89mm) bore x stroke, optional hand crank for engine, otherwise as 1903

1907

Model: Model 3, 26.8ci (440cc) F-head single
Price $210
Production: 150
Colours: Renault Grey with Red pinstriping, Black
Specification: Sager front fork used, otherwise as 1906

1908

Model: Model 4, 26.8ci (440cc) F-head single
Price: $210
Production: 450
Colours: As 1907
Specification: As 1907

1909

The Range
Model: Model 5 - 30ci (**500cc**) F-head single, 28in (711mm) wheels, battery ignition
5A - 30ci F-head single, 28in wheels, magneto
5B - 30ci F-head single, 26in (660mm) wheels, battery ignition
5C - 30ci F-head single, 26in wheels, magneto
5D - 50ci (819cc) F-head V-twin, magnet
Prices
5 - $210
5A - $250
5B - $210
5C - $250
5D - $325

LEFT: There was little difference between the 1905 (front) and 1907 (rear) singles.

TOP: The 1907 Model 3 with Sager front fork.

ABOVE: A rare 1909 5D V-twin, only produced in that year.

Production
5 - 864
5A - 54
5B - 168
5C - 36
5D - 27
TOTAL - 1,149
Colours: As 1907
Specification:
5-5C - 30ci F-head single, $3^{5/16}$in x $3^{1/2}$in (78 x 89mm) bore x stroke, auto inlet valve, Bosch magneto on A and C.
5D - 50ci (820cc) F-head V-twin, 3 x $3^{1/2}$in (76 x 89mm) bore x stroke, auto inlet valve, drip-feed lubrication, 7hp, Bosch magneto, belt drive, loop frame, 28in (711mm) wheels.

1910

The Range
6 - 30ci (**500cc**) F-head single, 28in (711mm) wheels, battery ignition
6A - 30ci F-head single, 28in wheels, magneto
6B - 30ci F-head single, 26in (660mm) wheels, battery ignition
6C - 30ci F-head single, 26in wheels, magneto
6E - Racer, 30ci F-head single
Prices
6- $210
6A - $250
6B - $210
6C - $250
6E - $275
Production
6 - 2,302
6A - 334
6B - 443
6C - 88
6E - 1
TOTAL - 3,168
Colours: As 1907
Specification
Belt idler fitted, and wider $1^{3/4}$in (44mm) drive belt, otherwise as 1909.

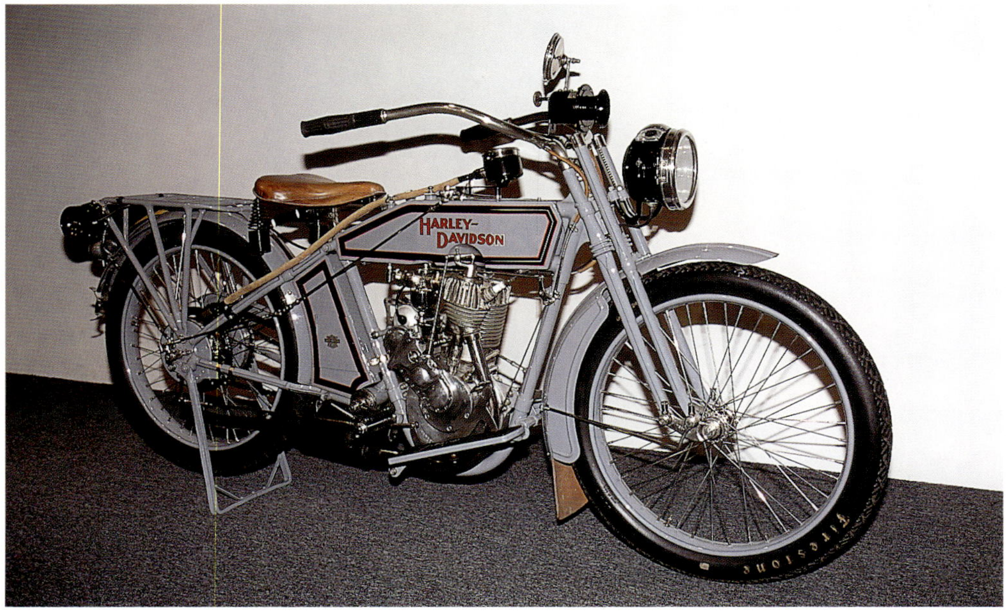

TOP: The 1910/11 single: an option of 26- or 28-inch wheels allowed different seat heights.

ABOVE: The 1912 X8A combined a 30ci (500cc) single with magneto and rear-wheel clutch..

1911

The Range
7 - 30ci (**500cc**) F-head single, 28in (711mm) wheels, battery ignition
7A - 30ci F-head single, 28in wheels, magneto
7B - 30ci F-head single, 26in (660mm) wheels, battery ignition
7C - 30ci F-head single, 26in wheels, magneto
7D - 50ci (820cc) F-head V-twin, 28in (711mm) wheels, magneto
Prices
7 - $225
7A - $250
7B - $225
7C - $250
7D - $300
Production
Total 5,625 (details not available)
Colours
Renault Grey with pinstriping in red, gold, blue, dark grey or black
Specifications
7D - 50ci F-head V-twin, 3 x $3^{1/2}$in (76 x 89mm) bore x stroke, mechanical inlet valve, drive belt tensioner

1912

The Range
8 - 30ci (**500cc**) F-head single, battery ignition
X8 - 30ci F-head single, battery ignition, rear wheel clutch
8A - 30ci F-head single, magneto
X8A - 30ci F-head single, magneto, rear wheel clutch
8D - 50ci (**820c**) F-head V-twin, magneto
X8D - 50ci F-head V-twin, magneto, rear wheel clutch
X8E - 61ci (1000cc) F-head V-twin, magneto, rear wheel clutch, chain drive
Prices
8 - $200
X8 - $210

8A - $225
X8A - $235
8D - $275
X8D - $285
X8E - $285
Production
Total 3,852 (details not available)
Colours
As 1911
Specifications
8-X8D - Mostly as 1911, but with free-wheel clutch in rear hub of 'X'-models, operated by handle on fuel tank. Also new frame and Troxel saddle suspension system on all bikes.
X8E - 61ci flathead V-twin, 3⁵⁄₁₆in x 3½in (84 x 89mm) bore x stroke, mechanical inlet valve, lubrication by hand-operated oil pump (replaces drip-feed as main system), free-wheel clutch, chain final drive.

1913

The Range
9A - 35ci (**575cc**) F-head single, belt drive
9B - 35ci F-head single, chain drive
9E - 61ci (**1000cc**) F-head V-twin, chain drive
9F - 61ci F-head V-twin, chain drive
9G - 61ci F-head V-twin Forecar Delivery Van
Prices
9A - $290
9B - $290
9E - $350
9F - n/a
9G - n/a
Production
9A - 1,510
9B - 4,601
9E - 6,732
9F - 49
9G - 63
TOTAL - 12,955
Colours
As 1911

TOP and ABOVE: There was now chain drive for the single, in 1913 (top) and 1914 (above) form. Harley-Davidson built over 6,000 singles in 1914, but little more than half that the following year.

Specifications
9A/B - 35ci F-head single, 3⁵⁄₁₆in x 4in (84 x 102mm) bore x stroke, mechanical inlet valve, 5hp (hence also known as 5-35)
9G - 61ci F-head V-twin with front-mounted delivery box

1914

The Range
10A - 35ci (**575cc**) F-head single, belt drive
10B - 35ci F-head single, chain drive
10C - 35ci F-head single, two-speed
10E - 61ci (**1000cc**) F-head V-twin, chain drive
10F - 61ci F-head V-twin, two-speed
10G - Delivery Van, two-speed
Prices
10A - $200
10B - $210
10C - $245
10E - $250
10F - $285
10G - n/a
Production
10A - 316
10B - 2,034
10C - 877
10E - 5,055
10F - 7,956
10G - 171
TOTAL - 16,409
Colours
As 1911
Specification
10A/B/C - Kick starter, two-speed rear hub (C), H-D's own brakes and hubs (B/C)
10E/F - Kick starter, two-speed rear hub (F), H-D's own brakes and hubs

1915

The Range
11B - 35ci (**575cc**) F-head single
11C - 35ci F-head single, two-speed
11E - 61ci (**1000cc**) F-head V-twin, single speed

11F - 61ci F-head V-twin, three-speed
11G - Delivery Van
11H - 61ci F-head V-twin, single-speed, electric lighting
11J - 61ci F-head V-twin, three-speed, electric lighting
11K - Racer, 61ci F-head V-twin

Prices
11B - $200
11C - $230
11E - $240
11F - $275
11H - $275
11J - $310
11K - $250

Production
11B - 670
11C - 545
11E - 1,275
11F - 9,855
11G - 98
11H - 140
11J - 3,719
11K - 188
TOTAL - 16,490

Colours
As 1911

Specifications
Singles - Automatic oil pump
Twins - Three-speed gearbox (11F/J), automatic oil pump, full electrical system (generator, lights, horn, ignition (11H/J)

1916

The Range
16B - 35ci (**575cc**) F-head single
16C Solo - 35ci F-head single, three-speed
16E - 61ci (**1000cc**) F-head V-twin, single-speed
16F - 61ci F-head V-twin, three-speed
16J - 61ci F-head V-twin, three-speed, electrics
16S Stripped Stock Racer - 35ci F-head single
16R - Roadster Racer - 61ci F-head V-twin
16T Track Racer - 61ci F-head V-twin

Prices
B - $200
C - $230
E - $240
F - $265
J - $295
Production
B - 292
C - 862
E - 252
F - 9,496
J - 5,898
R/S/T - 123
Delivery van - 7
TOTAL - 16,930
Colours
Renault Grey with red, black, dark blue or gold pinstriping
Specification
Kickstart (C), common frame, footboards and brake rod on all bikes

1917

The Range
17B - 35ci (**575cc**) F-head single
17C - 35ci F-head single, three-speed
17E - 61ci (**1000cc**) F-head V-twin, single-speed
17F - 61ci F-head V-twin, three-speed
17J - 61ci F-head V-twin, three-speed, electrics
17S Stripped Stock Racer - 35ci F-head single
17R Roadster Racer - 61ci F-headV-twin
17T Stripped Stock Racer - 61ci F-head V-twin
Prices
B - $215
C - $240
E - $255
F - $275
J - $310
S -
R - $280
T - $280

OPPOSITE: The 11F 61-ci (1000-cc) twin of 1915, with sidecar (a common fitment at the time). 'F' denotes a 3-speed gearbox.

ABOVE: An original unrestored 11F. Note the pushrods for the overhead inlet valves. The exhaust valves are side-mounted.

Production
B - 124
C - 605
E - 68
F - 8,527
J - 9,180
RST - 18
TOTAL - 18,522
Colours
Olive Green with Coach Green and gold pinstriping
Specifications
Singles - Dixie magneto
Twins - Four-lobe cam, revised valve timing, Dixie magneto

1918

The Range
18B - 35ci (**575cc**) F-head single, magneto
18C Solo - 35ci F-head single, magneto, three-speed
18E - 61ci (**1000cc**) F-head V-twin, single-speed, magneto
18F Solo - 61ci F-head V-twin, three-speed, magneto
18J Solo - 61ci F-head V-twin, three-speed, electrics
18R Roadster Racer - 61ci F-head V-twin, three-speed
18FUS - 61ci F-head V-twin, Presto-lite equipped, supplied to Govt. only
Prices
B - $235
C - $260
E - $275
F - $290
J - $320
Production
B - 19
C - 251
E - 5
F - 11,764
J - 6,571
R - 3

FUS - 8,095
TOTAL - 26,708
Colours
Olive Green, Coach Green, Pea Green and black pinstriping
Specifications
Singles - Berling magneto
Twins - Remy generator, many minor changes

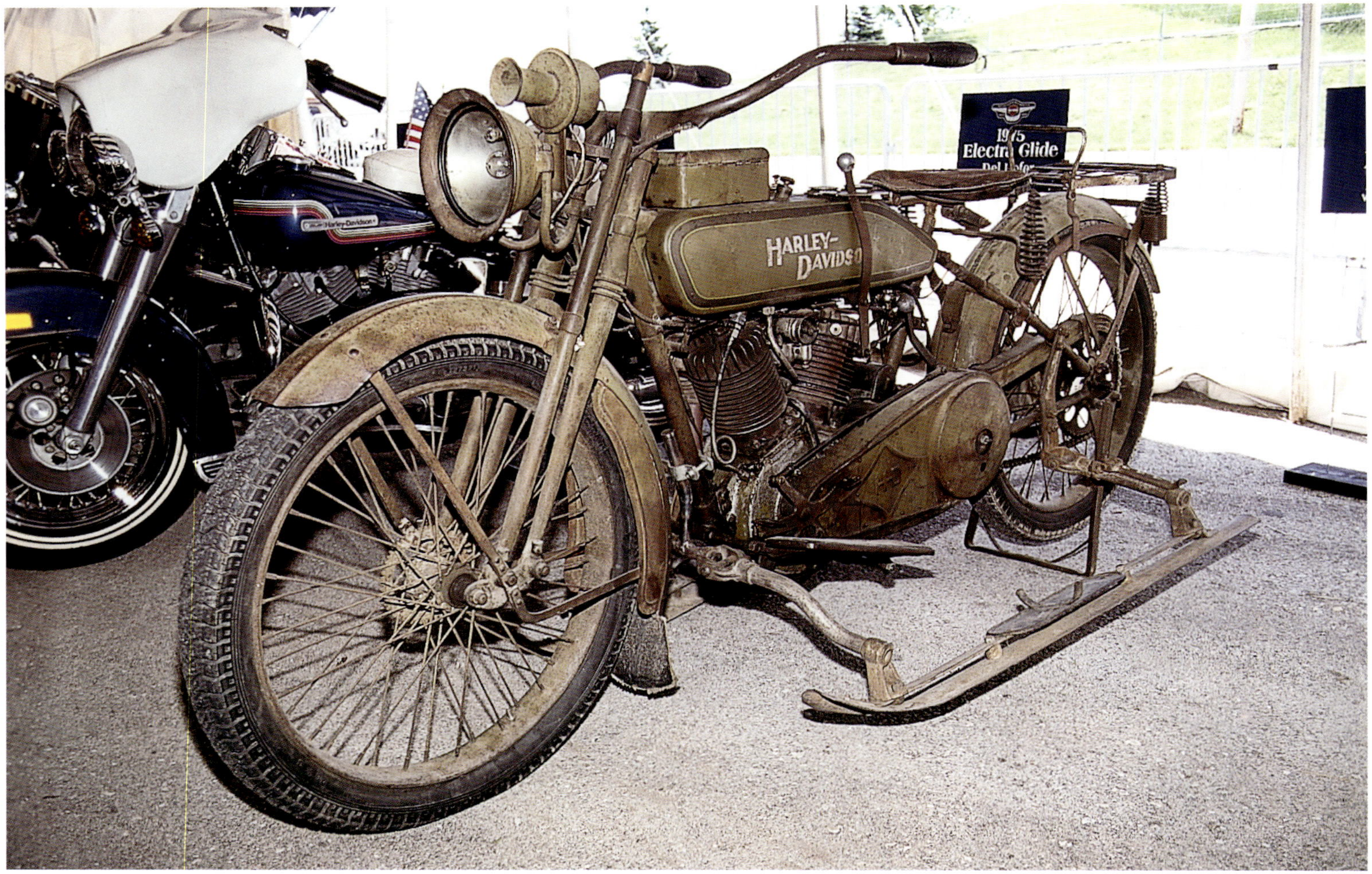

1919

The Range
19WF Sport Twin - 35ci (**575cc**) ho twin, three-speed, magneto
19F - 61ci (**1000cc**) F-head V-twin, three-speed, magneto
19FS - 61ci F-head V-twin 'sidecar motor', three-speed
19J - 61ci F-head V-twin, three-speed, electrics
19JS - 61ci F-head V-twin 'sidecar motor', three-speed, electrics
Prices
WF - $335
F - $350
FS - $350
J - $370
JS $370
Bicycles - $30–$45
Production
W - 753
F - 5,064
J - 9,941
FS - 7,521
TOTAL - 23,279
Colours
As 1918
Specifications
W Sport Twin - 35ci horizontally-opposed twin, 2³/₄in x 3in (70 x 76mm) bore x stroke, 3.75:1 compression, roller-bearing crank 6hp, unit-construction three-speed gearbox, enclosed drive chain
FJ Twins - 2¹/₂in (6.4cm) wider handlebars

1920

The Range
20WF Sport Twin - 35ci (**575cc**) ho twin, three-speed, magneto
20WJ Sport Twin - 35ci ho twin, three-speed, electrics
20F - 61ci (**1000cc**) F-head V-twin, three-speed, magneto
20FS - 61ci F-head V-twin, three-speed, magneto, sidecar gearing
20J - 61ci F-head V-twin, three-speed, electrics
20JS - 61ci F-head V-twin, three-speed, electrics, sidecar gearing
Prices
WF - $335
WJ - n/a
F - $370
FS - $370
J - $395
JS - $395
Production
WF - 4,459
WJ - 810
F - 7,579
J - 14,192
TOTAL - 27,040
Colours
As 1918
Specifications
WJ - Full electrics, with H-D's own generators and coils, plus lights and horn

There was electric lighting for this 1919 twin, with unusual ski attachment for bad weather. There were no fair-weather bikers then!

1921

The Range

21CD Commercial - 37ci (606cc) F-head singel, three-speed, magneto
21WF - 35ci (**575cc**) ho twin, three-speed, magneto
21WJ - 35ci ho twin, three-speed, electrics
21F - 61ci (**1000cc**) F-head V-twin, three-speed, magneto
21FS - 61ci F-head V-twin, three-speed, magneto, sidecar gearing
21J - 61ci F-head V-twin, three-speed, electrics
21JS - 61ci F-head V-twin, three-speed, electrics, sidecar gearing
21FD - 74ci (**1200cc**) F-head V-twin, three-speed, magneto
21FDS - 74ci F-head V-twin, three-speed, magneto, sidecar gearing
21JD - 74ci F-head V-twin, three-speed, electrics
21JDS - 74ci F-head V-twin, three-speed, electrics, sidecar gearing

Prices

CD - $430
WF - $415
WJ - $445
F - $50
FS - $450
J - $485
JS - $485
FD - $485
FDS - $485
JD - $520
JDS - $520

Production

WF - 1,100
WJ - 823
F - 2,413
FD - 277
J - 4,526
JD - 2,321

TOTAL - 11,460

Colours

As 1918

Specifications

CD - New single for commercial use, using one cylinder of the 74ci V-twin, chassis as on V-twin
FD/JD - 74ci F-head V-twin, bored and stroked (3 7/16in x 4in/87 x 102mm) version of 61ci, 1 1/4in (32mm) Schebler carburettor, 18hp.

1922

The Range

22CD Commercial - 37ci (606cc) F-head single, three-speed
22WF Sport Twin - 35ci (**575cc**) ho twin, three-speed, magneto
22WJ Sport Twin - 35ci ho twin, three-speed, electric s
22F - 61ci (**1000cc**) F-head V-twin, three-speed, magneto
22FS - 61ci F-head V-twin, three-speed, magneto, sidecar gearing
22J - 61ci F-head V-twin, three-speed electrics
22JS - 61ci F-head V-twin, three-speed, electrics, sidecar gearing
22FD - 74ci (**1200cc**) F-head V-twin, three-speed, magneto
22FDS - 74ci F-head V-twin, three-speed, magneto, sidecar gearing
22JD - F-head V-twin, three-speed, electrics
22JDS - F-head V-twin, three-speed, electrics, sidecar gearing

Prices

CD - $315
WF - $310
WJ -$315
F - $335
FD - $360
FS - $335
FDS - $360
J - $365
JS - $365
JD - $390
JDS - $390
Bicycles - $35-$45

Production

CD - 39
WF - 38
WJ - 455
F - 1,824
J - 3,183
FD - 909
JD - 3,988

TOTAL - 10,436

Colours

Brewster Green with gold pinstriping

Specifications

All big Twins - Roller-bearings for male con-rod, longer-life valve springs, double-plunger oil pump

In 1922, the FD could be yours for $360 (without the sidecar). Or there was a Harley-Davidson-badged bicycle for $35.

1923

The Range

23WF Sport Twin - 35ci (**575cc**) ho twin, three-speed, magneto
23WJ Sport Twin - 35ci ho twin, three-speed, electrics
23F - 61ci (**1000cc**) F-head V-twin, three-speed, magneto
23FS - 61ci F-head V-twin, three-speed, magneto, sidecar gearing
23J - 61ci F-head V-twin, three-speed, electrics
23JS - 61ci F-head V-twin, three-speed, electrics, sidecar gearing
23FD - 74ci (**1200cc**) F-head V-twin, three-speed, magneto
23FDS - 74ci F-head V-twin, three-speed, magneto, sidecar gearing
23JD - 74ci F-head V-twin, three-speed, electrics
23JDS - 74ci F-head V-twin, three-speed, electrics, sidecar gearing

Prices

WF - $275
WJ - $295
F - $285
FD - $310
FS - $285
FDS - $310
J - $305
JS - $305
JD - $330
JDS - $330

Production

WF - 614
WJ - 481
F - 2,822
FD - 869
J - 4,802
JD - 7,458
TOTAL - 17,046

Colours

As 1922

Specifications

FD/JD - Improved cylinder-head cooling, larger generator (JD/JDS)

1924

The Range

24FE - 61ci (**1000cc**) F-head V-twin, alloy pistons, three-speed, magneto
24FES - 61ci F-head V-twin, alloy pistons, three-speed, magneto, sidecar gearing
24JE - 61ci F-head V-twin, alloy pistons, three-speed, electrics
24JES - 61ci F-head V-twin, alloy pistons, three-speed, electrics, sidecar gearing
24FD - 74ci (**1200cc**) F-head V-twin, cast-iron pistons, three-speed, magneto
24FDS - 74ci F-head V-twin, cast-iron pistons, three-speed, magneto, sidecar gearing
24JD - 74ci F-head V-twin, cast-iron pistons, three-speed, electrics
24JDS - 74ci F-head V-twin, cast-iron pistons, three-speed, electrics, sidecar gearing
24FDCA - 74ci F-head V-twin,alloy pistons, three-speed, magneto
24FDSCA - 74ci F-head V-twin, alloy pistons, three-speed, magneto, sidecar gearing
24JDCA - 74ci F-head V-twin, alloy pistons, three-speed, electrics
24JDSCA - 74ci F-head V-twin, alloy pistons, three-speed, electrics, sidecar gearing

Prices

FE - $300
FES - $300
JE - $320
JES - $320
FD - $315
FDS- $315
JD - $335
JDS - $335
FDCA - $325
FDSCA - $325
JDCA - $345
JDSCA -$345

Production

FE - 2,708
FD - 502

FDCB - 90
FDCA - 351
JE - 4,994
JD - 2,955
JDCB - 3,034
JDCA - 3,014
TOTAL - 17,648
Colours
Olive Green, with maroon, black and gold pinstriping
Specifications
FE/JE - Aluminium-alloy pistons with four rings, Alemite grease fittings
FD/JD twins. Revised flywheels with different balance factor, Alemite grease fittings

1925

The Range
25FE - 61ci (**1000cc**) F-head V-twin, iron-alloy pistons, three-speed, magneto
25FES - 61ci F-head V-twin, iron-alloy pistons, three-speed, magneto, sidecar gearing
25JE - 61ci F-head V-twin, iron-alloy pistons, three-speed, electrics
25JES - 61ci F-head V-twin, iron-alloy pistons, three-speed, electrics, sidecar gearing
25FDCB - 74ci (**1200cc**) F-head V-twin, iron-alloy pistons, three-speed, magneto
25FDCBS - 74ci F-head V-twin, iron-alloy pistons, three-speed, magneto, sidecar gearing
25JDCB - 74ci F-head V-twin, iron-alloy pistons, three-speed, electrics
25JDCBS - F-head V-twin, iron-alloy pistons, three-speed, electrics, sidecar gearing
Prices
FE - $295
FES - $295
JE - $315
JES - $315
FDCB - $315

FDCBS- $315
JDCB - $335
JDCBS - $335
Production
F - 1,318
FD - 433
J - 4,114
JD - 9,506
TOTAL - 15,371
Colours
As 1924
Specifications
All Twins - Wider frame (lower saddle 3in/76mm), iron-alloy pistons with narrower and deeper rings, 20in (508mm) wheels, many detail changes

OPPOSITE
LEFT: The V-twin racer at the height of the board-track craze.

RIGHT: One of Harley's first purpose-built racers, the eight-valve twin (1923).

LEFT: A 1925 J-model, with the electric equipment that most buyers now preferred. This bike also has the optional speedometer.

1926

The Range

26A - 21ci (**350cc**) flathead single, three-speed, magento

26B - 21ci flathead single, three-speed, electrics

26AA - 21ci OHV single, three-speed, magento

26BA - 21ci OHV single, three-speed, electrics

26S Racer - 21ci OHV single, magneto

26J - 61ci (**1000cc**) F-head V-twin, three-speed, electrics

26JS - 61ci F-head V-twin, three-speed, electrics, sidecar gearing

26JD - 74ci (**1200cc**) F-head V-twin, three-speed, electrics

26JDS - 74ci F-head V-twin, three-speed, electrics, sidecar gearing

Prices

A - $210
B- $235
AA - $250
BA - $275
S - $300
J - $315
JS - $315
JD - $335
JDS - $335

Production

A - 1,128
AA - 207
B- 5,979
BA - 676
F - 760
FD - 232
J - 3,749
JD - 9,544
TOTAL - 22,275

Colours

As 1924

Specifications

A/B singles - 21ci 2$^7/_8$in x 3$^1/_4$in (73 x 83mm) bore x stroke, flat-head (A/B) with 8hp or OHV (AA/BA) with 12hp, Ricardo cylinder-heads, 26in (660mm) wheels, three-speed gearbox

J/JD - 26 x 3$^1/_4$in (660 x 84mm) balloon tyres, higher output generator, bigger battery

TOP and ABOVE: Two different versions of the 1927 21-ci (350-cc) single: the utilitarian side-valve B (top) and the ohv Peashooter (above).

1927

The Range

27A - 21ci (**350cc**) flathead single, three-speed, magento

27B - 21ci flathead single, three-speed, electrics

27AA - 21ci OHV single, three-speed, magento

27BA - 21ci OHV single, three-speed, electrics

27S Racer - 21ci OHV single, magneto

27J - 61ci (**1000cc**) F-head V-twin, three-speed, electrics

27JS - 61ci F-head V-twin, three-speed, electrics, sidecar gearing

27JD - 74ci (**1200cc**) F-head V-twin, three-speed, electrics

26JDS - 74ci F-head V-twin, three-speed, electrics, sidecar gearing

Prices

A - $210
B- $235
AA - $250
BA - $275
S - $300
J - $310
JS - $310
JD - $320
JDS - $320

Production

A - 444
AA - 98
B- 3,711
BA - 524
F - 246
FD - 209
J - 3,561
JD - 9,691
TOTAL - 18,484

Colours
As 1924
Specifications
A/B Singles - Reinforced frame, new Ricardo cylinder-heads

1928

The Range
28B - 21ci (**350cc**) flathead single, three-speed, electrics
28BA - 21ci OHV single, three-speed, electrics
28J - 61ci (**1000cc**) F-head V-twin, three-speed, electrics
28JS - 61ci F-head V-twin, three-speed, electrics, sidecar gearing
28JX Sport - 61ci F-head V-twin, three-speed, electrics
28JXL Special Sport - 61ci F-head V-twin, Dow-metal pistons, three-speed, electrics
28JH Two Cam - 61ci F-head V-twin, Dow-metal pistons, three-speed, electrics
28JD - 74ci (**1200cc**) F-head V-twin, three-speed, electrics
28JDS - 74ci F-head V-twin, three-speed, electrics, sidecar gearing
28JDX Sport - 74ci F-he74 V-twin, three-speed, electrics
28JDXL Special Sport - 61ci F-head V-twin, Dow-metal pistons, three-speed, electrics
28JDH Two Cam - 74ci F-head V-twin, Dow-metal pistons, three-speed, electrics
Prices
B - $235
BA - $255
J - $310
JS - $310
JX - $310
JXL - $325
JD - $320
JDS - $320
JDX - $320
JDXL - $335
JH - $360

JDH - $390
Production
A - 519
AA - 69
B - 3,483
BA - 639
F - 141
FD - 131
J - 4,184
JD - 11,007
TOTAL - 20,173
Colours
Singles - As 1924
Twins - Main colour as 1924, but options are Black, Fawn Grey, Azure Blue, Maroon, Coach Green, White, Cream or Police Blue
Specifications
B/BA Singles - Dow-metal pistons, lighter flywheels and con-rods, revised gearing
J/JD Twins - 25in (635mm) wheels, front-wheel brake
JX/JXL/JDX/JDXL - Sports version of J/JD, with Two Cam slimmer chassis and tuned J/JD twin (bigger valves, Dow-metal pistons)
JH/JDH Two Cam Twins - Sports version of existing J/JD twins, same capacity but with direct-action valve gear, Dow-metal domed pistons, narrower fuel tank, 18in (457mm) wheels, throttle-controlled oil pump. lower, slimmer profile

ABOVE: The almost legendary JDH Two Cam of 1928.

ABOVE RIGHT: 1929 was the first year for the Forty-Five.

RIGHT: All bikes had twin headlamps for 1929.

1929

The Range

29B - 21ci (**350cc**) flathead single, three-speed, electrics
29BA - 21ci OHV single, three-speed, electrics
29C - 30.5ci (**500cc**) flathead single, three-speed, electrics
29D - 45ci (**750cc**) flathead V-twin, three-speed, electrics
29DL - 45ci flathead V-twin, high compression, three-speed, electrics
29J - 61ci (**1000cc**) F-head V-twin, three-speed, electrics
29JS - 61ci F-head V-twin, three-speed, electrics, sidecar gearing
29JH Two Cam - 61ci F-head V-twin, Dow-metal pistons, three-speed, electrics
29JD - 74ci (**1200cc**) F-head V-twin, three-speed, electrics
29JDS - 74ci F-head V-twin, three-speed, electrics, sidecar gearing
29JDH Two Cam - 74ci F-head V-twin, Dow-metal pistons, three-speed, electrics

Prices

B- $235
BA - $255
C - $255
D - $290
DL - $290
J - $310
JS - $310
JD - $320
JDS - $320
JDH - $370

Production

A - 197
AA - 26
B- 1,592
BA - 404
C - 1,570
D - 4,513
DL - 2,343
F - 191
FD - 73
J - 2,886
JD - 10,182
TOTAL - 23,977

Colours

As 1928

Specifications

C Single - 30.5ci flathead single, 3 3/32in x 4in (79 x 102mm) bore x stroke, 10hp, used Model B's chassis
D Twins - 45ci (750cc) flat-head V-twin, 2 3/4in x 3 13/16in (70 x 97mm) bore x stroke, Dow-metal pistons, Ricardo cylinder-heads, battery/coil ignition, three-speed gearbox

1930

The Range

30B - 21ci (350cc) flathead single, three-speed
30C - 30.5ci (500cc) flathead single, three-speed
30D - 45ci (**750cc**) flathead V-twin, low-compression, three-speed

LEFT: By 1930, Harley-Davidson was well into its side-valve gear. This one has a four-tube muffler and twin headlights.

OPPOSITE
ABOVE: A 1930 VL, a high-compression 74ci (1200cc) side-valve with sidecar.

BELOW: The 30.5-ci (500-cc) single of 1931 was not a great success.

30DS - 45ci flathead V-twin, low-compression, three-speed, sidecar model
30DL Sport - 45ci flathead V-twin, Dow-metal pistons, three-speed
30DLD Special Sport - 45ci flathead V-twin, high compression, three-speed
30V - 74ci (**1200cc**) flathead V-twin, medium-compression, three-speed
30VL Sport - 74ci flathead V-twin, high-compression, Dow-metal pistons
30VS - 74ci flathead V-twin, Dow-metal pistons, three-speed, sidecar model
30VC Commercial - 74ci flathead V-twin, nickel-iron pistons, three-speed

Prices
B- $235
BA -$255
C -$260
D -$310
DS -$310
DL - $310
DLD -$310
V (all models) - $340

Production
B - 577
BA - 767
C - 1,627
D - 2,000
DS - 213
DL -3,191
DLD - 206
V - 1,960
VS - 3,612
VL -3,246
VC - 1,174

TOTAL - 18,573

Colours
Singles - Olive Green with vermilion, maroon and gold pinstriping
Twins - As singles, with options of Maroon, Cream, Azure Blue, Fawn Grey, Coach Green or Police Blue. Any combination of two colours could be ordered as well.

Specifications
C - Now uses Model D's new frame (lower saddle, more ground clearance), I-beam front forks, automatic primary chain oiling
D - New frame (lower saddle, more ground clearance), I-beam front forks, auto primary chain oiling
V - New flat-head (side-valve) Big Twin to replace the J Model. 74ci flathead V-twin, 3 7/16in x 4in (87 x 102mm) bore x stroke, Ricardo cylinder heads, magnesium-alloy pistons, total loss lubrication, three-speed gearbox, dual headlights (1930 only)

1931

The Range
31B - 21ci (350cc) flathead single - now for export only
31C - 30.5ci (500cc) flathead single, three-speed
31D - 45ci (**750cc**) flathead V-twin, low-compression, three-speed
31DS - 45ci flathead V-twin, low-compression, three-speed, sidecar gearing
31DL Sport - 45ci flathead V-twin, high-compression, three-speed
31DLD Special Sport - 45ci flathead V-twin, high-compression, three-speed
31V - 74ci (**1200cc**) flathead V-twin, medium-compression, three-speed
31VL Sport - 74ci flathead V-twin, high-compression, Dow-metal pistons
31VS - 74ci flathead V-twin, medium-compression, three-speed, sidecar gearing
31VC Commercial - 74ci flathead V-twin, low-compression, three-speed

Prices
C - $260
D - $310
DS - $310
DL - $310
DLD - $325
V (all models) - $340

Production
C - 1,059
D - 715

DS - 276
DL - 1,306
DLD - 241
V - 825
VS - 1,994
VL - 3,477
TOTAL - 9,893

Colours

Olive Green with vermilion, maroon and gold pinstriping, with options of White, Maroon/Cream, Black/Vermilion, Olive Green/Vermilion, or Police Blue/Fawn Grey.

Specifications

D Twins - Reinforced frame (as per Tom Sifton's recommendation), Schebler De Luxe carburettor, new generator drive gears, new clutch assembly

1932

The Range

32B - 21ci (350cc) flathead single, three-speed
32C - 30.5ci (500cc) flathead single, three-speed
32R - 45ci (**750cc**) flathead V-twin, low-compression, three-speed
32RS - 45ci flathead V-twin, low-compression, three-speed, sidecar model
32RL Sport - 45ci flathead V-twin, Dow-metal pistons, three-speed
32RLD Special Sport - 45ci flathead V-twin, high-compression, three-speed
32V - 74ci (**1200cc**) flathead V-twin, medium-compression, three-speed
32VL Sport - 74ci flathead V-twin, high-compression, Dow-metal pistons
32VS - 74ci flathead V-twin, medium-compression, three-speed, sidecar gearing
32VC Commercial - 74ci flathead V-twin, low-compression, three-speed
32G Servicar - 45ci flathead V-win, three-speed
32GD Servicar - 45ci flathead V-twin, three-speed, large compartment
32GE Servicar - 45ci flathead V-twin, three-speed, air tank

Prices

B - $195
C - $235
R - $295
RS - $295
RL - $295
RLD - $310
V (all models) - $320
G Servicar - $450

Production

B - 535
C - 540
R - 410
RS - 111
DL - 628
DLD - 98
V - 478
VS - 1,233
VL - 2,684
VC - 239
G/GA Servicar - 219
TOTAL - 7,175

Colours

Main colour as 1931, with options of Delft Blue/Turquoise, Vermilion, Vermilion/Black, White, Police Blue or Delft Blue

Specifications

R Twins - Replaces D twins. 45ci flathead V-twin with stronger valve springs, more air space between cylinders, larger flywheels, new crankcases, aluminium pistons, heavy-duty oil pump, Linkert carburettor

ABOVE and TOP: Both these 1932/33 VLs show the art deco bird motif, used on all models in 1933.

V Twins - More air space between cylinders, improved oil pump, new generator (more consistent current, better cooling), heavy-duty front forks
G Servicars - Basically a three-wheeler version of the Model R, for commercial use, single drum brake inside rear axle casing

1933

The Range
33B - 21ci (350cc) flathead single, three-speed
33CB - 30.5ci (**500cc**) flathead single, three-speed
33C - 30.5ci flathead single, three-speed
33R - 45ci (**750cc**) flathead V-twin, low-compression, three-speed
33RS - 45ci flathead V-twin, low-compression, three-speed, sidecar model
33RL Sport - 45ci flathead V-twin, high compression, three-speed
33RLD Special Sport - 45ci flathead V-twin, high-compression, three-speed
33V - 74ci (**1200cc**) flathead V-twin, medium-compression, three-speed
33VL Sport - 74ci flathead V-twin, high-compression, three-speed
33VS - 74ci flathead V-twin, medium-compression, three-speed, sidecar gearing
33VLD - 74ci flathead V-twin, high compression, magnesium-alloy pistons, three-speed
33VC Commercial - 74ci flathead V-twin, low-compression, three-speed
33G Servicar - 45ci flathead V-win, three-speed+reverse
33GD Servicar - 45ci flathead V-twin, three-speed, large compartment
33GE Servicar - 45ci flathead V-twin, three-speed, air tank
Prices
B - $187.50
C - $225
R - $280
RS - $280
RL - $280
RLD - $290
V (all except VLD) - $310
VLD - $320
Production
B - 123
C - 183
R - 162
RS - 3
RL - 264
RLD - 68
V - 233
VS - 164
VL - 886
VLD - 780
VC - 109
UFS - 499
G/GA Servicar - 182
TOTAL - 3,656
Colours
Singles & Servicars - Silver/Turquoise with Black/Gold strips and bird motif
CB Single - Olive Green/Black with bird motif
Twins (except Servicar) - Sunshine Blue/White with gold striping, Silver/Turquoise, Olive/Brilliant Green, Black/Mandarin Red or Police Blue/White, all with bird motif.
Specifications
CB - New model, using 30.5ci single in Model B (21ci) frame).
Otherwise, no major changes

1934

The Range
34B - 21ci (350cc) flathead single, three-speed
34CB - 30.5ci (**500cc**) flathead single, three-speed
34C - 30.5ci flathead single, three-speed
34R - 45ci (**750**cc) flathead V-twin, low-compression, three-speed
34RS - 45ci flathead V-twin, low-compression, three-speed, sidecar model
34RL Sport - 45ci flathead V-twin, three-speed
34RLD Special Sport - 45ci flathead V-twin, high-compression, three-speed
34VD - 74ci (**1200cc**) flathead V-twin, low-compression, three-speed
34VDS - 74ci flathead V-twin, low-compression, three-speed, sidecar gearing
34VFDS - 74ci flathead V-twin, low-compression, nickel-iron pistons, three-speed, sidecar gearing
34VLD Special Sport - 74ci flathead V-twin, high compression, three-speed+reverse
34G Servicar - 45ci flathead V-win, three-speed+reverse
34GD Servicar - 45ci flathead V-twin, three-speed+reverse, large compartment
34GE Servicar - 45ci flathead V-twin, three-speed, air tank

1934, and the tank loses that beautiful art deco bird in favour of a winged diamond. Harley-Davidson was weathering the Depression, with very low sales, and closure was on the agenda. To save money, plain black details replaced chrome or nickel plate.

Prices
B - $187.50
C - $225
CB - $197.50
R - $280
RL - $280
RLD - $290
V (all models) - $310
G Servicar - $430
GD Servicar - $430
GE Servicar - $485
Production
B - 424
C - 220
CB - 310
R - 450
RS - 302
RL - 743
RLD - 240
VD - 664
VDS - 1,029
VL - 886
VLD - 4,527
VFD/VFDS - 1,335
Servicars - 229
TOTAL - 11,359
Colours
Singles - Silver/Red Teak or Olive Green/Black with Art Deco logo
CB Single - As 1933, with Art Deco logo
Twins - Silver/Red Teak, Red Teak/Black, Black/Orlando Orange, Seafoam Blue/Silver, Black/Olive Green, Vermilion/Copper, all with Art Deco logo
Specifications
R Twins - Aluminium-alloy pistons, new oil pump, new clutch
V Twins - Aluminium-alloy pistons, High-flo muffler
Servicars - Straight-bore cylinders with T-slot pistons

$310 would buy you a brand new VL in 1934, but few could afford it.

In 1935, chromium began to make a tentative appearance on handlebars and headlamps, but it would be several years before sales fully recovered to pre-Depression levels.

1935

The Range

35R - 45ci (**750cc**) flathead V-twin, low-compression, three-speed
35RS - 45ci flathead V-twin, low-compression, three-speed, sidecar model
35RL - 45ci flathead V-twin, high-compression, three-speed
35RLD Special Sport - 45ci flathead V-twin, three-speed
35RLDR Competition Special - 45ci flathead V-twin, three-speed
35VD - 74ci (**1200cc**) flathead V-twin, low-compression, three-speed
35VDS - 74ci flathead V-twin, low-compression, three-speed, sidecar gearing
34VFDS - 74ci flathead V-twin, low-compression, nickel-iron pistons, three-speed, sidecar gearing
35VLD Special Sport - 74ci flathead V-twin, high compression, three-speed
35VLDJ Competition Special - 74ci flathead V-twin, three-speed
35VLDD Sport Solo - 80ci (**1300cc**) flathead V-twin, high compression, three-speed, solo bars
35VDDS - 80ci flathead V-twin, three-speed, sidecar model
35G Servicar - 45ci flathead V-win, three-speed+reverse
35GD Servicar - 45ci flathead V-twin, three-speed+reverse, large compartment
35GE Servicar - 45ci flathead V-twin, three-speed+reverse, air tank

Prices

R - $295
RL - $295
RS - $295
RLD - $305
RLDR - $322
VD - $320
VLD - $320
VDS - $320
VLDJ - $333.50
VLDD - $347
VDDS - $347
G Servicar - $440
GD Servicar - $440
GE Servicar - $495

Production

R - 543
RS - 392
RL - 819
RLD - 177
RLDR - 29
VD - 585
VDS - 1,189
VL - 886
VLD - 3,963
VLDJ - 102
VFDS - 327
VLDD -179
Servicars - 567
TOTAL - 9,758

Colours

Egyptian Ivory/Regent Brown, Teal Red/Black, Olive Green/Black, Verdant Green/Black, Venetian Blue/Silver, Silver/Black (police only), Cameo Cream/Dawn Grey, Potomac Blue

Specifications

All models - Carburized braked drums and harder linings
R Twins - Constant-mesh gearbox, quickly-detachable rear wheel
V Twins - New 80ci flathead V-twin, 3.422 x 3.125in (87 x 79mm) bore x stroke, same chassis as 74ci. Four-speed transmission option for sidecar use

1936

The Range

36R - 45ci (**750cc**) flathead V-twin, low-compression, three-speed
36RS - 45ci flathead V-twin, low-compression, three-speed, sidecar gearing
36RL - 45ci flathead V-twin, high-compression, three-speed
36RLD Special Sport - 45ci flathead V-twin, high compression, three-speed
36RLDR Competition Special - 45ci flathead V-twin, three-speed
36E - 61ci (**1000cc**) OHV V-twin, medium-compression, four-speed
36EL Special Sport - 61ci OHV V-twin, high-compression, four-speed
36ES - 61ci OHV V-twin, medium-compression, four-speed, sidecar gearing
36VD - 74ci (**1200cc**) flathead V-twin, low-compression, three-speed
36VDS - 74ci flathead V-twin, low-compression, three-speed, sidecar gearing
36VLH Sport - 80ci (**1300cc**) flathead V-twin, high compression, three-speed, solo bars
36VHS - 80ci flathead V-twin, three-speed, sidecar gearing
36G Servicar - 45ci flathead V-win, three-speed+reverse
36GD Servicar - 45ci flathead V-twin, three-speed+reverse, large compartment
36GE Servicar - 45ci flathead V-twin, three-speed+reverse, air tank

Prices

R - $295
RL - $295
RS - $295

RLD - $305
RLDR -$320
E -$380
EL - $380
ES - $380
VD - $320
VLD - $320
VDS - $320
VLH - $340
VHS - $340
G Servicar - $440
GD Servicar - $440
GE Servicar - $495

Production
R - 539
RS - 437
RL - 355
RLD - 540
RLDR - 79
E - 152
EL - 1,526
ES - 26
VD - 176
VDS - 623
VL - 886
VLD - 1,577
VLH - 2,046
VHS - 305
VFHS - 35
VFDS - 600
Servicars - 648
TOTAL - 10,550

Colours
Sherwood Green/Silver, Dusk Grey/Royal Buff, Teak Red/Black, Venetian Blue/Croydon Cream, Maroon/Nile Green

Specifications
R 45ci Twins - Deeper finning on heads and cylinders, redesigned combustion chambers, 1.25in (32mm) carburettor (RL/RLD), many minor changes
E 61ci Twins - All-new OHV 'Knucklehead' V-twin, 61ci, 3 5/16in x 3 1/2in 84 x 89mm) bore x stroke, dry sump lubrication, four-speed gearbox, heavy duty clutch, saddle-type twin fuel tanks, 18in wheels
V 74ci/80ci Twins - Deeper finning on heads and cylinders, new combustion chamber, four-speed gearbox optional, many minor changes
Servicars - Deeper finning on cylinders, redesigned combustion chambers, many minor changes

1937

The Range
37W - 45ci (**750cc**) flathead V-twin, low-compression, three-speed
37WS - 45ci flathead V-twin, low-compression, three-speed, sidecar gearing
37WL - 45ci flathead V-twin, high-compression, three-speed
37WLD Special Sport - 45ci flathead V-twin, high compression, three-speed
37WLDR Competition Special - 45ci flathead V-twin, three-speed
37E - 61ci (**1000cc**) OHV V-twin, medium-compression, four-speed
37EL Special Sport - 61ci OHV V-twin, high-compression, four-speed
37ES - 61ci OHV V-twin, medium-compression, four-speed, sidecar gearing
37U - 74ci (**1200cc**) flathead V-twin, low-compression, four-speed
37UL Special Sport - 74ci flathead V-twin, high-compression, four-speed
37US - 74ci flathead V-twin, low-compression, four-speed, sidecar gearing
37UH - 80ci (**1300cc**) flathead V-twin, medium-compression, four-speed
37ULH Sport - 80ci flathead V-twin, high compression, four-speed
37UHS - 80ci flathead V-twin, medium-compression, four-speed, sidecar gearing
37G Servicar - 45ci flathead V-win, three-speed+reverse
37GD Servicar - 45ci flathead V-twin, three-speed+reverse, large compartment
37GE Servicar - 45ci flathead V-twin, three-speed+reverse, air tank

1936 was the first year of the 80-ci (1300-cc) side-valve twin – the biggest Harley yet.

Prices
W - $355
WL - $355
WS - $355
WLD - $355
WLDR - $380
E - $435
EL -$435
ES - $435
U - $395
UL - $395
US - $395
UH - $415
ULH -$415
UHS - $415
G Servicar - $515
GD Servicar - $515
GE Servicar - $570
Production
W - 539
WS - 232
WL - 560
WLD - 581
WLDR - 145
E - 126
EL - 1,829
ES - 70
U - 612
US - 1,080
UL -2,861
UH - 185
ULH - 1,513
UHS - 400
Servicars - 816
TOTAL - 11,549
Colours
Teal Red/Black, Bronze Brown/Delphine Blue, Police Silver/Black, Delphine Blue/Teal Red, Olive Green/Black
Specifications
W 45ci Twins - redesigned R-series, with styling from the E-series Knucklehead and a redesigned 45ci flathead V-twin with dry sump lubrication, roller bearings, saddle fuel tank, tanktop instrument panel
E 61ci Twins - No major changes
U 74ci/80ci Twins - redesigned V-series, with styling from the E-series Knucklehead and redesigned engines with dry sump lubrication, roller bearings, E-model four-speed gearbox, frame and running gear, saddle fuel tank, tanktop instrument panel, automatic primary chain oiling
Servicars - Used styling from E-series Knucklehead and redesigned 45ci V-twin from W-series (see above)

1938

The Range
38WL - 45ci (**750cc**) flathead V-twin, high-compression, four-speed
38WLD Sport - 45ci flathead V-twin, high compression, four-speed
38WLDR Competition Special - 45ci flathead V-twin, four-speed
38E - 61ci (**1000cc**) OHV V-twin, medium-compression, four-speed
38EL Special Sport - 61ci OHV V-twin, high-compression, four-speed
38ES - 61ci OHV V-twin, medium-compression, four-speed, sidecar gearing
38U - 74ci (**1200cc**) flathead V-twin, medium-compression, four-speed
38UL Special Sport - 74ci flathead V-twin, high-compression, four-speed
38US - 74ci flathead V-twin, medium-compression, four-speed, sidecar gearing
38UH - 80ci (**1300cc**) flathead V-twin, medium-compression, four-speed
38ULH Special Sport - 80ci flathead V-twin, high compression, four-speed
38UHS - 80ci flathead V-twin, medium-compression, four-speed, sidecar gearing
38G Servicar - 45ci flathead V-win, three-speed+reverse
38GD Servicar - 45ci flathead V-twin, three-speed+reverse, large compartment
Prices
WL - $355
WLD - $355
WLDR - $380
E - $435
EL -$435
ES - $435
U - $395
UL - $395
US - $395
UH - $415
ULH - $415
UHS - $415
G Servicar - $515
GD Servicar - $515
Production
W - 302
WS - 247
WL - 309
WLD - 402
WLDR - 139
EL - 2,289
ES - 189
U - 504
US - 1,193
UL - 1,099
UH - 108
ULH - 579
UHS - 132
Servicars - 525
TOTAL - 8,017
Colours
Hollywood Green with gold striping, Teak Red/Black, Venetian Blue/White, Police Silver/Black (police), Silver Tan/Sunshine Blue, Olive Green/Black
Specifications
W 45ci Twins - Many minor changes
E 61ci Twins - Redesigned gearbox, many minor changes
U/UH 74ci/80ci Twins - Many minor changes
Servicars - Many minor changes

1938 and the Servicar gets a helping hand. The three-wheeler was to be Harley-Davidson's longest-running model.

1939

The Range

39WL - 45ci (**750cc**) flathead V-twin, high-compression, four-speed
39WLD Sport - 45ci flathead V-twin, high compression, four-speed
39WLDR Competition Special - 45ci flathead V-twin, four-speed
39EL Special Sport - 61ci (**1000cc**) OHV V-twin, high-compression, four-speed
39ES - 61ci OHV V-twin, medium-compression, four-speed, sidecar gearing
39U - 74ci (**1200cc**) flathead V-twin, medium-compression, four-speed
39UL Sport - 74ci flathead V-twin, high-compression, four-speed
39US - 74ci flathead V-twin, medium-compression, four-speed, sidecar gearing
39UH - 80ci (**1300cc**) flathead V-twin, medium-compression, four-speed
39ULH Special Sport - 80ci flathead V-twin, high compression, four-speed
39UHS - 80ci flathead V-twin, medium-compression, four-speed, sidecar gearing
39G Servicar - 45ci flathead V-win, three-speed+reverse
39GD Servicar - 45ci flathead V-twin, three-speed+reverse, large compartment

Prices

WL - $355
WLD - $355
WLDR - $380
EL - $435
ES - $435
U - $395
UL - $395
US - $395
UH - $415
ULH - $415
UHS - $415
G Servicar - $515
GD Servicar - $515

Production

W - 260
WS - 170

WL - 212
WLD - 326
WLDR - 173
EL - 2,695
ES - 214
U - 421
US - 1,327
UL - 902
UH - 92
ULH - 384
UHS - 109
Servicars - 525
TOTAL - 7,810

Colours

Black with Ivory panel, Airway Blue/White, Teak Red/Black, Police Silver/Black

Specifications

W 45ci Twins - Revised pistons/rings, longer-life valve springs, 'cat's-eye' streamlined instrument panel
E 61ci Twins - Revised pistons/rings, improved valve springs, optional three-speed gearbox, cat's-eye instrument panel
U 74ci/80ci Twins - Revised piston rings, longer inlet manifold (to cool carburettor), thinner cylinder walls (80ci), new four-speed gearbox, cat's-eye instrument panel
Servicars - As W 45ci Twins

1940

The Range

40WL Sport - 45ci (**750cc**) flathead V-twin, high-compression, four-speed
40WLD Special Sport - 45ci flathead V-twin, aluminium cylinder heads, high compression
40WLDR Competition - 45ci flathead V-twin, aluminium heads
40EL Special Sport - 61ci (**1000cc**) OHV V-twin, high-compression, four-speed
40ES - 61ci OHV V-twin, medium-compression, four-speed, sidecar gearing
40U - 74ci (**1200cc**) flathead V-twin, medium-compression, four-speed
40US - 74ci flathead V-twin, medium-compression, four-speed, sidecar gearing
40UL Special Sport - 74ci flathead V-twin, four-speed
40UH - 80ci (**1300cc**) flathead V-twin, medium-compression, four-speed
40ULH Special Sport - 80ci flathead V-twin, high compression, four-speed
40UHS - 80ci flathead V-twin, medium-compression, four-speed, sidecar gearing
40G Servicar - 45ci flathead V-win, three-speed+reverse
40GD Servicar - 45ci flathead V-twin, three-speed+reverse, large compartment

Prices

WL - $350
WLD - $350
WLDR -$395
EL - $430
ES - $430
U - $385
UL - $385
US - $385
UH - $410
ULH - $410
UHS - $410
G Servicar - $515
GD Servicar - $515

Production

W - 439
WS - 202
WL - 569
WLD - 567
WLDR - 87
EL - 3,893
ES - 176
U - 260
US - 1,516
UL - 822
UH - 187
ULH - 672
UHS - 163
Servicars - 908
TOTAL - 10,461

OPPOSITE: The Knucklehead (this is a 1939 example) heralded a new era of high performance for Harley-Davidson.

RIGHT: This 1940 WL45 is in civilian trim, but most of those built over the next few years would be to a military specification. It was at this time the smallest bike Harley made, and the workhorse of the range.

FAR RIGHT: A 1940 Knucklehead.

Colours
Black with Flight Red strip, Clipper Blue/White, Squadron Grey/Bittersweet, Flight Red/Black, Police Silver/Black (police only)

Specifications
W 45ci Twins - Aluminium heads, deep-finned cylinders, larger carburettor (WLD, WLDR), 5 x16in (127 x 406mm) tyres optional
E 61ci Twins - New piston rings, rear crankcase baffle removed, larger main-bearing, larger intake ports, manifold and 1½in (38mm) carburettor, many minor changes
U 74ci/80ci Twins - Larger main-bearing, constant-mesh four-speed gearbox, aluminium cylinder heads (80ci), many minor changes
Servicars - As W 45ci Twins

1941

The Range
41WL - 45ci (**750cc**) flathead V-twin, four-speed
41WLD - Sport - 45ci flathead V-twin
41WLDR - Special Sport - 45ci flathead V-twin, aluminium heads
41WLA - 45ci flathead V-twin, US Army
41WLC - 45ci flathead V-twin, Canadian Army
41EL Special Sport - 61ci (**1000cc**) OHV V-twin, high-compression, four-speed
41ES - 61ci OHV V-twin, medium-compression, four-speed, sidecar gearing
41U - 74ci (**1200cc**) flathead V-twin, medium-compression, four-speed
41US - 74ci flathead V-twin, medium-compression, four-speed, sidecar gearing
41UL Special Sport - 74ci flathead V-twin, high-compression, four-speed
41FL Special Sport - 74ci OHV V-twin, high-compression, four-speed
41FS - 74ci OHV V-twin, medium-compression, four-speed, sidecar gearing
41UH - 80ci (**1300cc**) flathead V-twin, medium-compression, four-speed
41ULH Special Sport - 80ci flathead V-twin, high compression, four-speed
41UHS - 80ci flathead V-twin, medium-compression, four-speed, sidecar gearing
41G Servicar - 45ci flathead V-win, three-speed+reverse
41GD Servicar - 45ci flathead V-twin, three-speed+reverse, large compartment

Prices
WL - $350
WLD - $365
WLDR - $385
E - $425
EL - $425
ES - $430
FL - $465
FS - $465
U - $385
UL - $385
US - $385
UH - $410
ULH - $410
UHS - $410
G Servicar - $515
GD Servicar - $515

Production
W - 4,095
WL - 4,277
WLA - 2,282
WLC - 149

WLD - 455
WLDR -171
EL - 2,280
ES - 261
FL - 2,452
FS - 156
U - 884
US - 1,888
UL - 715
UH - 126
ULH - 420
UHS - 1,112
Servicars - 1,159
TOTAL - 22,882
Colours
Brilliant Black, Skyway Blue, Flight Red, Cruiser Green, Police Silver (police only)
Specifications
W 45ci Twins - Many minor changes
E/F 61/74ci Twins - New F-series 74ci OHV V-twin is enlarged version of 61ci, with $3^{7}/_{16}$in x $3^{31}/_{32}$in (87 x 101mm) bore x stroke, heavier flywheels. All E/F twins had redesigned crankcases, clutch, revised geometry (29-degree rake), 16in (406mm) wheels
U/UH 74/80ci Twins - Many minor changes
Servicars - As W 45ci Twins

1942
The Range
42WL - 45ci (**750cc**) flathead V-twin, four-speed
42WLD Special Sport - 45ci flathead V-twin, aluminium heads, four-speed
42WLA - 45ci flathead V-twin, US Army
42WLC - 45ci flathead V-twin, Canadian Army
42E - 61ci (**1000cc**) OHV V-twin, high-compression, four-speed
42EL Special Sport - 61ci OHV V-twin, medium-compression, four-speed
42U - 74ci (**1200cc**) flathead V-twin,

ABOVE: The 80-ci (1300-cc) side-valve twin was soon to be replaced by the 74-ci (1200-cc) Knucklehead. Even as war clouds gathered, the evolution of the range continued.

OPPOSITE: War Child: Harley-Davidson sold thousands of WLA 45s to the military, though few of them survived.

medium-compression, four-speed
42UL Special Sport - 74ci flathead V-twin, high-compression, four-speed
42F - 74ci OHV V-twin, medium-compression, four-speed
42FL Special Sport - 74ci OHV V-twin, high-compression, four-speed
42XA - 45ci flathead ho twin, US Army
42G Servicar - 45ci flathead V-win, three-speed+reverse

Prices

WL - $350
WLD - $365
E - $425
EL - $425
F -$465
FL - $465
U - $385
UL - $385
UH - $410
ULH - $410
UHS - $410
G Servicar - $525
GD Servicar - $510

Production

WLA - 13,051
WLC - 9,820
WL - 142
WLD - 133
XA - 1,011
EL - 620
ELA - 8
ELC - 45
ES - 164
FL - 799
FS - 107
U - 421
US - 978
UL - 405
Servicars - 399
TOTAL - 28,103

Colours

As 1941

Specifications

No major changes

1943

The Range

43WLA - 45ci (**750cc**) flathead V-twin, US Army
43WLC - 45ci flathead V-twin, Canadian Army
43E - 61ci (**1000cc**) OHV V-twin, medium-compression, four-speed
43EL Special Sport - 61ci OHV V-twin, high-compression, four-speed
43U - 74ci (**1200cc**) flathead V-twin, medium-compression, four-speed
43UL - 74ci flathead V-twin, high-compression, four-speed
43F - 74ci OHV V-twin, medium-compression, four-speed
43FL Special Sport - 74ci OHV V-twin, high-compression, four-speed
43XA - 45ci flathead ho twin, US Army
43G Servicar - 45ci flathead V-win, three-speed+reverse

Prices

E - $425
EL - $425
F - $465
FL - $465
U - $385
UL - $385

Production

WLA - 24,717
WLC - 2,647
EL - 53
ES - 105
FL - 33
FS - 112
U - 493
US - 1,315
UL - 11
Servicars - 135
TOTAL - 29621

Colours

Grey or Silver

Specifications

No major changes

1944

The Range

42WLA - 45ci (**750cc**) flathead V-twin, US Army
43WLC - 45ci flathead V-twin, Canadian Army
44E - 61ci (**1000cc**) OHV V-twin, medium-compression, four-speed
44EL Special Sport - 61ci OHV V-twin, high-compression, four-speed
44U - 74ci (**1200cc**) flathead V-twin, medium-compression, four-speed
44UL - 74ci flathead V-twin, high-compression, four-speed
44F - 74ci OHV V-twin, medium-compression, four-speed
44FL Special Sport - 74ci OHV V-twin, high-compression, four-speed
44G Servicar - 45ci flathead V-win, three-speed+reverse

Prices

E - $425
EL - $425
F - $465
FL - $465
U - $385
UL - $385

Production

WLA - 11,351
WLC - 5,356
WL - 57
EL - 116
ES - 105
FL - 172
FS - 67
U - 580
US - 206
UL - 1366
Servicars - 57
TOTAL - 19,433

Colours

As 1943

Specifications

W 45ci Twins - No major changes
E/F 61/74ci Twins - Synthetic rubber tyres
U 74ci Twins - Synthetic rubber tyres
Servicars - Synthetic rubber tyres

1945

The Range

45WLA - 45ci (**750cc**) flathead V-twin, four-speed, US Army
45WLR - 45ci flathead V-twin, four-speed, Russian Army
45WL - 45ci flathead V-twin, four-speed
45E - 61ci (**1000cc**) OHV V-twin, medium-compression, four-speed
45EL Special Sport - 61ci OHV V-twin, high-compression, four-speed
45ES - 61ci OHV V-twin, medium-compression, four-speed, sidecar gearing
45U Commercial - 74ci (**1200cc**) flathead V-twin, medium-compression, four-speed
45UL - 74ci flathead V-twin, high-compression, four-speed
45US - 74ci flathead V-twin, medium compression, four-speed, sidecar gearing
45F - 74ci OHV V-twin, medium-compression, four-speed

ABOVE: The FL was the enlarged 74-ci (1200-cc) version of the Knucklehead, though the 61-ci (1000-cc) EL continued alongside it.

OPPOSITE: 1946, and peacetime and chrome returns with a vengeance to this FL, on silencer, wheel rims, front hub and headlamp rim.

45FL Special Sport - 74ci OHV V-twin, high-compression, four-speed
45FS - 74ci OHV V-twin, four-speed, sidecar gearing
45G Servicar - 45ci flathead V-win, three-speed+reverse

Prices
WL - $395.97
E - $463.67
EL - $463.67
ES - $463.67
F - $465
FL - $465
FS - $465
U - $427.25
UL - $427.25
US - $427.25
Servicar - $580.33
XA - $500

Production
WLA - 8,317
WL - 1,357
EL - 398
ES - 282
FL - 619
FS - 131
U - 513
US - 217
UL - 555
Servicars - 86
TOTAL - 12,475

Colours
Grey

Specifications
W 45ci Twins - Aluminium cylinder heads optional
E/F 61/74ci Twins - No changes
U 74ci Twins - Aluminium cylinder heads optional

1946

The Range
46WL - 45ci (750cc) flathead V-twin, three-speed

46E - 61ci (**1000cc**) OHV V-twin, medium-compression, four-speed
46EL Special Sport - 61ci OHV V-twin, high-compression, four-speed
46ES - 61ci OHV V-twin, medium-compression, four-speed, sidecar gearing
46U - 74ci (**1200cc**) flathead V-twin, medium-compression, four-speed
46UL - 74ci flathead V-twin, high-compression, four-speed
46US - 74ci flathead V-twin, medium compression, four-speed, sidecar gearing
46F - 74ci OHV V-twin, medium-compression, four-speed
46FL Special Sport - 74ci OHV V-twin, high-compression, four-speed
46FS - 74ci OHV V-twin, medium-speed, four-speed, sidecar gearing
46G Servicar - 45ci flathead V-win, three-speed+reverse

Prices
WL - $395.97
E - $463.67
EL - $463.67
ES - $463.67
F - $465
FL - $465
FS - $465
U - $427.25
UL - $427.25
US - $427.25
Servicar - $593.93

Production
WL - 4,410
EL - 2,098
ES - 244
FL - 3,986
FS - 418
U - 670
US - 1,052
UL - 1,800
Servicars - 1,444
TOTAL - 16122

Colours
Grey or Flight Red

Specifications
No major changes

1947

The Range
47WL - 45ci (750cc) flathead V-twin, three-speed
47E - 61ci (**1000cc**) OHV V-twin, medium-compression, four-speed
47EL Special Sport - 61ci OHV V-twin, high-compression, four-speed
47ES - 61ci OHV V-twin, medium-compression, four-speed, sidecar gearing
47U - 74ci (**1200cc**) flathead V-twin, medium-compression, four-speed
47UL - 74ci flathead V-twin, high-compression, four-speed
47US - 74ci flathead V-twin, medium compression, four-speed, sidecar gearing
47F - 74ci OHV V-twin, medium-compression, four-speed
47FL Special Sport - 74ci OHV V-twin, high-compression, four-speed
47FS - 74ci OHV V-twin, medium-speed, four-speed, sidecar gearing
47G Servicar - 45ci flathead V-win, three-speed+reverse

Prices
WL - $490
E - 590
EL - $590
ES - $590
F - $605
FL - $605
FS - $465
U - $545
UL - $545
US - $545
Servicar - $710

Production
WL - 3,338
EL - 4,117
ES - 237
FL - 6,893
FS - 401
U - 422
US - 1,267

UL - 1,243
Servicars - 2,177
TOTAL - 20,095
Colours
Brilliant Black, Flight Red, Skyway Blue, Police Silver (police only), Olive Green (export)
Specifications
No major changes

1948

The Range
48S - 7ci (125cc) two-stroke single, three-speed
48WL - 45ci (750cc) flathead V-twin, three-speed
48E Sport - 61ci (**1000cc**) OHV V-twin, medium-compression, four-speed
48EL Special Sport - 61ci OHV V-twin, high-compression, four-speed
48ES - 61ci OHV V-twin, medium-compression, four-speed, sidecar gearing
48U Sport - 74ci (**1200cc**) flathead V-twin, medium-compression, four-speed
48UL Sport Special - 74ci flathead V-twin, high-compression, four-speed
48US - 74ci flathead V-twin, medium compression, four-speed, sidecar gearing
48F Sport - 74ci OHV V-twin, medium-compression, four-speed
48FL Special Sport - 74ci OHV V-twin, high-compression, four-speed
48FS - 74ci OHV V-twin, medium-speed, four-speed, sidecar gearing
48G Servicar - 45ci flathead V-win, three-speed+reverse
Prices
WL - $535
E - $635
EL - $635
ES - $635
F - $650
FL - $650
FS - $650
U - $590
UL - $590

US - $590
Servicar - $755
Production
S - 10,117
WL - 2,124
EL - 4,321
ES - 198
FL - 8,071
FS - 334
U - 401
US - 1,006
UL - 970
Servicars - 1,778
TOTAL - 29,320
Colours
Brilliant Black, Flight Red, Azure Blue, Police Silver (police only)

Specifications
S 7ci Single - 125cc two-stroke, 2 1/16in x 2 29/32in (52 x 60mm) bore x stroke, 6.6:1, 1.7hp, three-speed gearbox, rigid frame, girder forks, six-volt electrics, 3 1/4 x 19in (82.5 x 483mm) tyres
W 45ci V-twin - No major changes
E/F 61/74ci Twins - Revised 'Panhead' motor with new top end (hydraulic tappets and aluminium heads), new wishbone frame
U 74ci Twins - New wishbone frame
Servicars - No major changes

1949

The Range
49S - 7ci (125cc) two-stroke single, three-speed
49WL - 45ci (750cc) flathead V-twin, three-speed
49E Hydra-Glide - 61ci (**1000cc**) OHV V-twin, medium-compression, four-speed
49EL Hydra-Glide Sport - 61ci OHV V-twin, high-compression, four-speed
49ES - 61ci OHV V-twin, medium-compression, four-speed, sidecar gearing
49EP - 61ci OHV V-twin, medium-compression, four-speed, spring fork, sidecar gearing
49ELP Sport - 61ci OHV V-twin, medium-compression, four-speed, spring fork
49F Hydra-Glide - 74ci (**1200cc**) OHV V-twin, medium-compression, four-speed
49FL Hydra-Glide - 74ci OHV V-twin, high-compression, four-speed

OPPOSITE: The 1948 Servicar was little different to the 1938 version, with the same 45-ci (750-cc) side-valve twin and 3-speed hand-change gearbox.

RIGHT: The first Panhead of 1948.

BELOW: The Hydra-Glide of 1949/50.

49FLP Sport - 74ci OHV V-twin, high-compression, four-speed
49FS - 74ci OHV V-twin, medium-compression, four-speed, sidecar gearing
49FP - 74ci OHV V-twin, medium-compression, four-speed, spring fork sidecar gearing
49G Servicar - 45ci flathead V-win, three-speed+reverse

Prices
S - $325
WL - $590
E - $735
EL - $735
ES - $735
F - $750
FL - $750
FS - $750
Servicar - $860

Production
S - 7,291
WL - 2,289
WLA - 436
EL - 3,419
ELP - 99
ES - 177
FL - 8,014
FLP - 486
FS - 490
Servicars - 1,039
TOTAL - 23,740

Colours
Single - Black
Twins - Brilliant Black, Peacock Bue, Burgundy, Police Blue, Metallic Congo Green

Specifications
S 7ci Single - No major changes
W 45ci V-twin - No major changes
E/F 61/74ci Twins - Hydra-Glide hydraulically-damped forks (includes larger front brake and headlamp, redesigned bars), spring forks still optional
Servicars - No major changes

1950

The Range

50S - 7ci (125cc) two-stroke single, three-speed
50WL - 45ci (750cc) flathead V-twin, three-speed
50E Hydra-Glide - 61ci (**1000cc**) OHV V-twin, medium-compression, four-speed
50EL Hydra-Glide Sport - 61ci OHV V-twin, high-compression, four-speed
50ES - 61ci OHV V-twin, medium-compression, four-speed, adjustable-rake forks, sidecar gearing
50F Hydra-Glide - 74ci (**1200cc**) OHV V-twin, medium-compression, four-speed
50FL Hydra-Glide - 74ci OHV V-twin, high-compression, four-speed
50FS Hydra-Glide - 74ci OHV V-twin, medium-compression, four-speed, adjustable-rake forks, sidecar gearing
50G Servicar - 45ci flathead V-win, three-speed+reverse

Prices

S - $325
WL - $590
E - $735
EL - $735
ES - $735
F - $750
FL - $750
FS - $750
Servicar - $860

Production

S - 4,708
WL - 1,108
WLA - 15
EL - 2,046
ES - 177
FL - 7,407
FS - 544
Servicars - 1,003
TOTAL - 17,008

Colours

Single - Brilliant Black, Flight Red, Sportsman Yellow, Riviera Blue
Twins - Brilliant Black, Riviera Blue, Ruby Red, Police Silver (police only), White, Flight Red, Metallic Green, Azure Blue

Specifications

S 7ci Single - No major changes
W 45ci V-twin - No major changes
E/F 61/74ci Twins - Redesigned cylinder heads (10% power increase)
Servicars - No major changes

1951

The Range

51S - 7ci (125cc) two-stroke single, three-speed
51WL - 45ci (750cc) flathead V-twin, three-speed
51EL Hydra-Glide Sport - 61ci (**1000cc**) OHV V-twin, high-compression, four-speed
51ELS - 61ci OHV V-twin, medium-compression, four-speed, sidecar gearing
51FL Hydra-Glide Sport - 74ci (**1200cc**) OHV V-twin, high-compression, four-speed
51FLS Hydra-Glide - 74ci OHV V-twin, medium-compression, four-speed, sidecar model
51G Servicar - 45ci flathead V-win, three-speed+reverse

Prices

S - $365
WL - $730
EL - $885
ELS - $885
F - $750
FL - $900
FLS - $900
Servicar - $1,095

Production

S - 5,101

The 1952 WR racer, the only one in existence. Thanks to the tuning genius of Tom Sifton and others, it was a competitive machine for several years.

WL - 1,044
WLA - 1
EL - 1,532
ES - 76
FL - 6,560
FS - 135
Servicars - 778
TOTAL - 15,227
Colours
Single - Persian Red, Sportsman Yellow, Rio Blue, Metallic Blue, Metallic Green
Twins - Brilliant Black, Rio Blue, Persian Red, Police Silver (police only), White, Metallic Blue, Metallic Green
Specifications
S 7ci Single - Roller-bearing small-end bearing, new speedometer, other minor changes
W 45ci V-twin - Fixed jet Linkert carburettor
E/F 61/74ci Twins - Several minor changes
Servicars - Hydraulic rear brake, 16in (406mm) rear wheels

1952

The Range
52S - 7ci (125cc) two-stroke single, three-speed
52K Sport - 45ci (750cc) flathead V-twin, four-speed
52EL Hydra-Glide Sport - 61ci (**1000cc**) OHV V-twin, four-speed
52ELS - 61ci OHV V-twin, four-speed, sidecar model
52ELF - 61ci OHV V-twin, four-speed, foot-change
52FL Hydra-Glide Sport - 74ci (**1200cc**) OHV V-twin, four-speed
52FLS Hydra-Glide - 74ci OHV V-twin, four-speed, sidecar model
52FLF Hydra-Glide - 74ci OHV V-twin, four-speed, foot-change
52G Servicar - 45ci flathead V-win, three-speed+reverse
Prices
WL - $730
K - $865
EL - $955
ELS - $955
ELF - $955
F - $750
FL - $970
FLS - $970
FLF - $970
Servicar - $1175
Production
S - 4,576
K - 1,970
EL - 918
ES - 42
FL - 5,554
FLS - 186
Servicars - 1,047
TOTAL - 14,293
Colours
Single - Tropical Green, Rio Blue, Persian Red, Marine Blue Metallic
K - Black, Rio Blue, Persian Red, Bronco Bronze
E/F Twins - Brilliant Black, Rio Blue, Tropical Green, Persian Red, Police Silver (police only), Marine Blue Metallic, Bronco Bronze Metallic, White
Specifications
S 7ci Single - Many minor changes
K 45ci V-twin - New 45ci flathead V-twin, replacing W-series, 2.75in x 3.813in (70 x 97mm) bore x stroke, unit-construction, aluminium-alloy cylinder heads, alloy pistons, 30hp, four-speed gearbox, foot-change, hand-clutch, swinging-arm rear suspension, hydraulic front forks, six-volt electrics, 19in (483mm) wheels
E/F 61/74ci Twins - Foot-change optional, many minor changes
Servicars - No major changes

1953

The Range
53ST - 10ci (165cc) two-stroke single, three-speed
53K Sport - 45ci (**750cc**) flathead V-twin, four-speed
53KK - 45ci flathead V-twin, four-speed
53FL Hydra-Glide Sport - 74ci (**1200cc**) OHV V-twin, four-speed
53FLF Hydra-Glide - 74ci OHV V-twin,, four-speed, foot-change
53FLE Hydra-Glide - 74ci OHV V-twin, four-speed, sidecar model
53FLEF Hydra-Glide - 74ci OHV V-twin, four-speed, foot-change, sidecar model
53G Servicar - 45ci flathead V-win, three-speed+reverse
Prices
ST - $405
K - $875
FL - $1,000
FLF - $1,000
FLE - $1,000
FLEF - $1,000
Servicar - $1,190
Production
ST - 4,225
K - 1,723
FL - 1,986
FLF - 3,351
Servicars - 1,146
TOTAL - 12,431
Colours
Single - Pepper Red, Glacier Blue, Forest Green, Glamour Green
Twins - Pepper Red, Forest Green, Silver (police only), Glacier Blue, Brilliant Black, White, Cavalier Brown, Glamour Green, Cadillac Grey/Azure Blue (Motor Maids special)
Specifications
ST 10ci Single - Replaces S-series, 165cc two-stroke, .375in x 2.281in (9.5 x 58mm) bore x stroke, Alemite grease fittings, revised gearing, 5hp restrictor kit available for those with restricted licence
Many minor changes
K 45ci V-twin - KK has KR racing spec cams, ports and ball-bearing crankshaft
F 74ci Twins - Redesigned hydraulic tappets for more consistent valve timing, FLE 74ci motor replaces 61ci, with milder cam and smaller carburettor
Servicars - No major changes

1954

The Range
54ST - 10ci (**165cc**) two-stroke single, three-speed
54STU - 10ci two-stroke single, three-speed
54KH Sport - 55ci (900cc) flathead V-twin, four-speed
54FL Hydra-Glide Sport - 74ci (**1200cc**) OHV V-twin, four-speed
54FLF Hydra-Glide - 74ci OHV V-twin, four-speed, foot-change
54FLE Hydra-Glide - 74ci OHV V-twin, four-speed, sidecar model
54FLEF Hydra-Glide - 74ci OHV V-twin, four-speed, foot-change, sidecar model
54G Servicar - 45ci flathead V-win, three-speed+reverse
Prices
ST - $405
STU - $405
KH - $925
FL - $1,015
FLF - $1,015
FLE - $1,015
FLEF - $1,015
Servicar - $1,240
Production
ST - 2,835
KH - 1,579
FL - 4,757
Servicars - 1,397
TOTAL - 10,568
Colours
Pepper Red, Daytona Ivory, Glacier Blue, Anniversary Yellow, Forest Green, Black, White (police only), Silver (police only), Cadillac Grey/Azure Blue (Motor Maids special)

The little 7-ci (125-cc) Hummer was based on a German DKW design. This 1954 bike was built in Harley-Davidson's 50th Anniversary year.

Specifications
ST 10ci Single - STU is 5hp restricted version
KH 55ci V-twin - Replaces 45ci K, 55ci flathead V-twin, 2.745in x 4.562in (70 x 116mm) bore x stroke, 6.8:1 compression ratio, 38hp (12% power increase), aluminium pistons, gearbox and clutch redesigned to suit, revised geometry
F 74ci Twins - No major changes
Servicars - No major changes

1955

The Range
55B Hummer - 7ci (125cc) two-stroke single, three-speed
55ST- 10ci (**165cc**) two-stroke single, three-speed
55STU - 10ci two-stroke single, three-speed
55KH Sport - 55ci (**900cc**) flathead V-twin, four-speed
55KHK Super Sport - 55ci
55FL Hydra-Glide Sport - 74ci (**1200cc**) OHV V-twin, four-speed
55FLF Hydra-Glide Sport - 74ci OHV V-twin, four-speed, foot-change
55FLE Hydra-Glide - 74ci OHV V-twin, four-speed, sidecar model
55FLEF Hydra-Glide - 74ci OHV V-twin, four-speed, foot-change, sidecar model
55FLH Hydra-Glide Super Sport - 74ci OHV V-twin, four-speed
55FLHF Hydra-Glide Super Sport - 74ci OHV V-twin, four-speed, foot-change
55G Servicar - 45ci (750cc) flathead V-win, three-speed+reverse

Prices
ST - $405
STU - $405
KH - $925
KHK - $995
FL - $1,015
FLF - $1,015
FLE - $1,015
FLEF - $1,015
FLH - $1,083
FLHF - $1,083
Servicar - $1,240

Production
ST - 2,263
B - 1,040
KH - 1,065
KR Racers -135
FL - 5,142
Servicars - 1,041
TOTAL - 10,686

Colours
Black, Pepper Red, Aztec Brown, Anniversary Yellow, Atomic Blue, Silver (police only), White (police only), Hollywood Green

Specifications
B 7ci Single - New model, 125cc two-stroke single, $2\frac{1}{16}$ x $2\frac{9}{32}$in (52 x 60mm) bore x stroke, 6.6:1 compression, 1.7hp, three-speed gearbox, multi-plate clutch
ST/STU 10ci Single - No major changes

KH 55ci V-twin - KHK Super Sport with sports cams and polished cylinder heads
F 74ci Twins - FLH Super Sport, new model, 8:1 compression ratio, polished and flowed intake ports, many minor changes
Servicars - No major changes

1956

The Range

56B Hummer - 7ci (125cc) two-stroke single, three-speed
56ST- 10ci (**165cc**) two-stroke single, three-speed
56STU - 10ci two-stroke single, restricted, three-speed
56KH Sport - 55ci (**900cc**) flathead V-twin, four-speed
56KHK Super Sport - 55ci
56FL Hydra-Glide Sport - 74ci (**1200cc**) OHV V-twin, four-speed
56FLF Hydra-Glide Sport - 74ci OHV V-twin, four-speed, foot-change
56FLE Hydra-Glide - 74ci OHV V-twin, four-speed, sidecar model
56FLEF Hydra-Glide - 74ci OHV V-twin, four-speed, foot-change, sidecar model
56FLH Hydra-Glide Super Sport - 74ci OHV V-twin, high-compression, four-speed
56FLHF Hydra-Glide Super Sport - 74ci OHV V-twin, high-compression, four-speed, foot-change
56G Servicar - 45ci (750cc) flathead V-win, three-speed+reverse

Prices

B - $320
ST - $405
STU - $405
KH - $935
KHK - $1,003
FL - $1,055
FLF - $1,055
FLE - $1,055
FLEF - $1,055
FLH - $1,123
FLHF - $1,123
Servicar - $1,240

Production

ST - 2,219
B - 1,384
KH - 539
KHK - 714
KR Racers - 60
FL/FLH - 5,806
Servicars - 1,203
TOTAL - 11,925

Colours

B Single - Pepper Red, Atomic Blue
ST Singles - Pepper Red, Atomic Blue, Champion Yellow, Tangerine
Twins - Pepper Red, Black, Atomic Blue, Champion Yellow, Silver (police only), White (police only), Flamboyant Metallic Green

Specifications

B 7ci Single - No major changes
ST/STU 10ci Single - No major changes
KH 55ci V-twin - Lower saddle, redesigned oil pump, redesigned gearbox
F 74ci Twins - Higher lift cams for FLH, many minor changes
Servicars - Some minor changes

1955, and the Big Twin has yet to acquire rear suspension or a genuine hardtail. But Hydra-Glide forks on big soft tyres and a sprung saddle were enough for some.

1957

The Range

57B Hummer - 7ci (125cc) two-stroke single, three-speed
57ST - 10ci (**165cc**) two-stroke single, three-speed
57STU - 10ci two-stroke single, three-speed
57XL Sportster - 54ci (883cc) OHV V-twin, four-speed
57FL Hydra-Glide Sport - 74ci (**1200cc**) OHV V-twin, four-speed
57FLF Hydra-Glide Sport - 74ci OHV V-twin, four-speed, foot-change
57FLH Hydra-Glide Super Sport - 74ci OHV V-twin, high-compression, four-speed
57FLHF Hydra-Glide Super Sport - 74ci OHV V-twin, high-compression, four-speed, foot-change
57G Servicar - 45ci (750cc) flathead V-win, three-speed+reverse

Prices

B - $356
ST - $445
STU - $445
XL - $1,103
FL - $1,167
FLF - $1,167
FLH - $1,243
FLHF - $1,243
Servicar - $1,367

Production

ST - 2,401
B - 1,350
KH - 90
KR Racers - 29
XL - 1,983
FL/FLH - 5,616
Servicars - 1,192
TOTAL - 12,661

Colours

Singles - Pepper Red, High Fire, Skyline Blue, Birch White

ABOVE LEFT: The 1956 KH was substantially a new bike, with modern chassis and four-speed foot-change gearbox, but a side-valve motor.
ABOVE : Duo Glide brought rear suspension at last.

Twins - Pepper Red, Black, Skyline Blue, Birch White, Metallic Midnight Blue
Servicars - Pepper Red, Black, Skyline Blue, Silver (police only), Metallic Midnight Blue

Specifications

Hummer 7ci Single - No major changes
ST/STU 10ci Single - No major changes
XL 54ci Twins - New model, 54ci V-twin, 3in x 3.812in (76 x 97mm) bore x stroke. 7.5:1 compression ratio, unit construction, dry sump, cast-iron cylinders and heads, four-speed gearbox, foot-change, 18in (457mm) wheels, telescopic front forks, rear swinging arm
F 74ci Twins - Some minor changes
Servicars - Some minor changes

1958

The Range

58B Hummer - 7ci (125cc) two-stroke single, three-speed
58ST - 10ci (**165cc**) two-stroke single, three-speed
58STU - 10ci two-stroke single, restricted, three-speed
58XL Sportster - 54ci (**883cc**) OHV V-twin, medium-compression, four-speed
58XLH Sportster - 54ci OHV V-twin, high-compression, four-speed
58XLC Sportster Racing - 54ci OHV V-twin, high-compression, four-speed
58XLCH Sportster Super Sport - 54ci OHV V-twin, high-compression, four-speed
58FL Duo Glide Sport - 74ci (**1200cc**) OHV V-twin, four-speed
58FLF Duo Glide Sport - 74ci OHV V-twin, four-speed, foot-change
58FLH Duo Glide Super Sport - 74ci

OHV V-twin, high-compression, four-speed
58FLHF Duo Glide Super Sport - 74ci OHV V-twin, high-compression, four-speed, foot-change
58G Servicar - 45ci (750cc) flathead V-win, three-speed+reverse

Prices
B - $375
ST - $465
STU - $465
XL - $1,155
FL - $1,255
FLF - $1,255
FLH - $1,320
FLHF - $1,320
Servicar - $1,450

Production
ST - 2,445
B - 1,677
KR Racers - 9
XL - 579
XLH - 711
XLCH - 239
FL/FLH - 6,038
Servicars - 1,192
TOTAL - 12,890

Colours
Singles - Skyline Blue, Tropical Coral, Calypso Red, Black
XL Sportster - Skyline Blue, Sabre Grey Metallic, Calypso Red, Black
Twins - Calypso Red, Black, Skyline Blue, Sabre Grey Metallic, Police Silver, Birch White (police only)
Servicars - Calypso Red, Black, Skyline Blue, Birch White, Police Silver

Specifications
B 7ci Single - No major changes
ST/STU 10ci Single - No major changes
XL 54ci Twins - XLC/XLCH new models, off-road versions of XL/SLH, with 2.25gal (1 litre) 'peanut' tank, magneto ignition, no silencers, no lights. Otherwise, no major changes
F 74ci Twins - No major changes
Servicars - Hydra-Glide front forks standard

1959

The Range
59B Hummer - 7ci (125cc) two-stroke single, three-speed
59ST - 10ci (**165cc**) two-stroke single, three-speed
59STU - 10ci two-stroke single, restricted, three-speed
59XL Sportster - 54ci (**883cc**) OHV V-twin, medium-compression, four-speed
5XLH Sportster - 54ci OHV V-twin, high-compression, four-speed
59XLC Sportster Racing - 54ci OHV V-twin, high-compression, four-speed
59XLCH Sportster Super Sport - 54ci OHV V-twin, high-compression, four-speed
59FL Duo Glide Sport - 74ci (**1200cc**) OHV V-twin, four-speed
59FLF Duo Glide Sport - 74ci OHV V-twin, four-speed, foot-change
59FLH Duo Glide Super Sport - 74ci OHV V-twin, high-compression, four-speed
59FLHF Duo Glide Super Sport - 74ci OHV V-twin, high-compression, four-speed, foot-change
59G Servicar - 45ci (750cc) flathead V-win, three-speed+reverse

Prices
B - $385
ST - $475
STU - $475
XL - $1,175
XLH - $1,200
XLCH - $1,285
FL - $1,280
FLF - $1,280
FLH - $1,345
FLHF - $1,345
Servicar - $1,500

Production
ST - 2,311
B - 1,285
KR Racers - 33
XL - 42
XLH - 947
XLCH - 1,059
FL/FLH - 5,767
Servicars - 812
TOTAL - 12,256

Colours
Singles - Skyline Blue, Calypso Red, Hi-Fi Turquoise, Hi-Fi Red
XL Sportster - Skyline Blue, Calypso Red, Black, Hi-Fi Turquoise, Hi-Fi Red
Twins - Calypso Red, Black, Skyline Blue, Hi-Fi Turquoise, Hi-Fi Red, Police Silver, Birch White (police only)
Servicars - Calypso Red, Black, Skyline Blue, Birch White, Police Silver

Specifications
B 125cc Single - No major changes
ST/STU 165cc Single - No major changes
XL Sportsters - XLH has high-lift intake cams, XLCH now road legal, with lights, battery/generator electrics, 2-into-1 exhaust with silencer
FL Duo Glides - Duo Glide new model, swinging-arm rear suspension with twin hydraulic shock absorbers, redesigned frame, hydraulic rear brake, many minor changes
Servicars - No major changes

The Sportster: This is a 1959 XLCH with the small tank (from the Hummer) and high-level pipes and was the start of a new and highly successful family of Harley-Davidsons.

1960

The Range
60A Topper - 10ci (**165cc**) two-stroke single, automatic transmission
60AU Topper - 10ci two-stroke single, 5hp restricted, automatic
60BT Super 10 - 10ci two-stroke single, three-speed
60BTU Super 10 - 10ci two-stroke single, 5hp restricted, three-speed
60XLH Sportster - 54ci (**883cc**) OHV V-twin, four-speed
60XLCH Sportster - 54ci OHV V-twin, four-speed
60FL Duo Glide Sport - 74ci (**1200cc**) OHV V-twin, four-speed
60FLF Duo Glide Sport - 74ci OHV V-twin, four-speed, foot-change
60FLH Duo-Glide Super Sport - 74ci OHV V-twin, high-compression, four-speed
60FLHF Duo Glide Super Sport - 74ci OHV V-twin, high-compression, four-speed, foot-change
60G Servicar - 45ci (750cc) flathead V-win, three-speed+reverse

Prices
Topper AH - $430
TopperAU - $430
Super 10 BT - $455
Super 10 BTU - $465
Sportster XLH - $1,225
Sportster - $1,310
Duo Glide FL - $1,310
Duo Glide FLF - $1,310
Duo Glide FLH - $1,375
Duo Glide FLHF - $1,375
Servicar - $1,530

Production
Topper - 3,801
Super 10 - 2,488
Sportster - 2,765
Duo Glide - 5,967
Servicars - 707
TOTAL - 15,728

Colours
Topper - Pepper Red, Strato Blue, Granada Green
Super 10 - Pepper Red, Hi-Fi Red, Hi-Fi Green, Hi-Fi Blue
Sportster - Skyline Blue, Black, Hi-Fi Red, Hi-Fi Green, Hi-Fi Blue
Twins - Black, Skyline Blue, Hi-Fi Red, Hi-Fi Green, Hi-Fi Blue, Police Silver, Birch White (police only)
Servicars - Black, Skyline Blue, Birch White, Police Silver

Specifications
A/AU Topper scooter - New model, 165cc two-stroke single, 2.375in x 2.281in (60 x 58mm) bore x stroke, 6.6:1 compression, reed intake valve, aluminium cylinder head, rubber mountings, automatic belt drive and chain final drive, leading-link front suspension, swinging-arm rear suspension, recoil starter, AU is 5hp restricted version
Super 10 - New model update of B single, now 10ci two-stroke single, as Topper, 6hp (BT) and 5hp (BTU)
XL Sportster - No major changes
FL Duo Glides - New headlamp nacelle, some minor changes
Servicars - New headlamp nacelle

ABOVE: Not many people know that Harley-Davidson made a scooter. The 10-ci (165-cc) Topper, this one from 1960, was a belated attempt to cash in on the growing scooter market. However it had none of the style of a Vespa or Lambretta, but the Topper is a very rare beast today.

OPPOSITE: A 1960 Duo Glide, set up for touring with screen and panniers or saddlebags.

1961

The Range

61AH Topper - 10ci (**165cc**) two-stroke single, automatic transmission

61AU Topper - 10ci two-stroke single, 5hp restricted, automatic

61BT Super 10 - 10ci two-stroke single, three-speed

61BTU Super 10 - 10ci two-stroke single, 5hp restricted, three-speed

61C Sprint - 15ci (246cc) OHV single, four-speed

61XLH Sportster - 54ci (**883cc**) OHV V-twin, four-speed

61XLCH Sportster - 54ci OHV V-twin, four-speed

61FL Duo Glide Sport - 74ci (**1200cc**) OHV V-twin, four-speed

61FLF Duo Glide Sport - 74ci OHV V-twin, four-speed, foot-change

61FLH Duo Glide Super Sport - 74ci OHV V-twin, high-compression, four-speed

61FLHF Duo Glide Super Sport - 74ci OHV V-twin, high-compression, four-speed, foot-change

61G Servicar - 45ci (750cc) flathead V-win, three-speed+reverse

Prices

Topper AU - $445
Super 10 BT - $465
Super 10 BTU - $465
Sportster XLH - $1,250
Sportster XLCH - $1,335
Duo Glide FL - $1,335
Duo Glide FLF - $1,335
Duo Glide FLH - $1,400
Duo Glide FLHF - $1,400
Servicar - $1,530

Production

Topper - 1,341
Super 10 - 1,527
Sportster - 2,014
Duo Glide - 4,927
Servicars - 628
TOTAL - 10,437

Colours

Topper - Strato Blue
Super 10 - Skyline Blue, Hi-Fi Red, Hi-Fi Green, Hi-Fi Blue, Birch White
Sportster - Pepper Red, Black, Hi-Fi Red, Hi-Fi Green, Hi-Fi Blue
Duo Glide - Black, Pepper Red, Hi-Fi Red, Hi-Fi Green, Hi-Fi Blue, Police Silver, Birch White (police only)
Servicars - Black, Skyline Blue, Birch White, Pepper Red, Police Silver

Specifications

AH/AU Topper - AH has high-compression 9hp motor, sidecar and utility box optional
Super 10 - 16in (406mm) wheels
Sprint - New model, Aermacchi Sprint made in Italy, rebadged as Harley-Davidson. 15ci OHV four-stroke single, 2.6 x 2.83in (66 x 72mm), 18bhp/7,500rpm, four-speed, motor hung from single strut steel frame
XL Sportster - No major changes
FL Duo Glides - 'Single-fire' ignition (separate points and coil for each cylinder)
Servicars - No major changes

ABOVE LEFT: The Super 10 was a 10-ci (165-cc) development of the Hummer, and still sold reasonably well despite the age of the basic design.

ABOVE: A 1961 Duo Glide festooned with period extras, just the sort of thing the average long distance rider would fit: glassfibre panniers (starting to replace traditional leather), screen, pillion grab handle, whitewall tyres and screen. In other words, a dresser.

1962

The Range

62AH Topper - 10ci (**165cc**) two-stroke single, automatic transmission
62AU Topper - 10ci two-stroke single, 5hp restricted, automatic
62BT Pacer - 11ci (**175cc**) two-stroke single, three-speed
62BTF Ranger - 10ci two-stroke single, three-speed
62BTH Scat - 11ci two-stroke single, three-speed
62BTU Pacer - 10ci two-stroke single, 5hp restricted, three-speed
62C Sprint - 15ci (**246cc**) OHV single, four-speed
62H Sprint H - 15ci OHV single, high compression, four-speed
62XLH Sportster - 54ci (**883cc**) OHV V-twin, four-speed
62XLCH Sportster - 54ci OHV V-twin, four-speed
62FL Duo Glide Sport - 74ci (**1200cc**) OHV V-twin, four-speed
62FLF Duo Glide Sport - 74ci OHV V-twin, four-speed, foot-change
62FLH Duo Glide Super Sport - 74ci OHV V-twin, high-compression, four-speed
62FLHF Duo Glide Super Sport - 74ci OHV V-twin, high-compression, four-speed, foot-change
62G Servicar - 45ci (750cc) flathead V-win, three-speed+reverse

Prices

Topper AH - $445
Topper AU - $445
Pacer BT - $465
Ranger BTF - $440
Scat BTH - $475
Pacer BTU - $465
Sportster XLH - $1,250
Sportster XLCH - $1,335
Duo Glide FL - $1,335
Duo Glide FLF - $1,335
Duo Glide FLH - $1,400
Duo Glide FLHF - $1,400
Servicar - $1,555

Production

Pacer/Ranger/Scat - 1,983
Sportster - 1,998
Duo-Glide - 5,184
Servicars - 703
TOTAL - 9,868

Colours

Topper - Birch White, Tango Red, Granada Green
Pacer/Ranger/Scat - Tango Red, Skyline Blue, Hi-Fi Red, Hi-Fi Purple
Sportster - Black, Tango Red, Hi-Fi Red, Hi-Fi Purple, Hi-Fi Blue
Duo Glide - Black, Tango Red, Hi-Fi Red, Hi-Fi Purple, Hi-Fi Blue
Servicars - Black, Birch White, Tango Red, Skyline Blue, Police Silver (police only)

Specifications

AH/AU Topper - No major changes
BT Pacer - New update of Super 10, 11ci two-stroke single, longer stroke, higher compression (7.65:1), BTH Scat is on/off-road version, BTF Ranger off-road, BTU Pacer 5hp restricted - BTU and BTF continue with 10ci single
Sprint H - New model based on Sprint. 9.2:1 compression, 19.5bhp/6,700rpm, off-road styling with short mudguards, high-level exhaust, semi-knobbly tyres
XL Sportster - No major changes
FL Duo Glides - No major changes
Servicars - No major changes

1963

The Range

63AH Topper - 10ci (**165cc**) two-stroke single, automatic transmission
63AU Topper - 10ci two-stroke single, 5hp restricted, automatic
63BT Pacer - 11ci (**175cc**) two-stroke single, three-speed
63BTH Scat - 11ci two-stroke single, three-speed
63BTU Pacer - 10ci two-stroke single, 5hp restricted, three-speed
63C Sprint - 15ci (**246cc**) OHV single, four-speed
63H Sprint H - 15ci OHV single, high-compression, four-speed
63XLH Sportster - 54ci (**883cc**) OHV V-twin, four-speed
63XLCH Sportster - 54ci OHV V-twin, four-speed
63FL Duo Glide Sport - 74ci (**1200cc**) OHV V-twin, four-speed
63FLF Duo Glide Sport - 74ci OHV V-twin, four-speed, foot-change
63FLH Duo Glide Super Sport - 74ci OHV V-twin, high-compression, four-speed
63FLHF Duo Glide Super Sport - 74ci OHV V-twin, high-compression, four-speed, foot-change
63G Servicar - 45ci (750cc) flathead V-twin, three-speed+reverse

Prices

Topper AH - $460
Topper AU - $460
Pacer BT - $485
Scat BTH - $495
Pacer BTU - $485
Sportster XLH - $1,270
Sportster XLCH - $1,355
Duo-Glide FL - $1,360
Duo-Glide FLF - $1,360
Duo-Glide FLH - $1,425
Duo-Glide FLHF - $1,425
Servicar - $1,590

Production

Topper - 978
Pacer/Ranger/Scat - 1,740
Sportster - 1,433
Duo Glide - 4,246
TOTAL - 8,397

Colours

Topper - Black, Tango Red, White
Pacer/Scat - Tango Red, Black, Hi-Fi Red, Hi-Fi Turquoise
Sportster - Black, Tango Red, Horizon Metallic Blue, Hi-Fi Turquoise, Hi-Fi Red, Hi-Fi Purple
Duo Glide - Black, Tango Red, Horizon Metallic Blue, Hi-Fi Red, Hi-Fi Purple, Hi-Fi Turquoise
Servicars - Black, Birch White, Tango Red, Horizon Metallic Blue, Police Silver (police only)

Specifications

AH/AU Topper - No major changes
BT/BTH/BTU - Swinging-arm rear suspension, BTU has restricterd 11ci single
Sprint H - Power increase to 25bhp/8,700rpm, 1.06in (27mm) Dell'Orto carburettor, megaphone exhaust
XL Sportster - No major changes
FL Duo-Glides - No major changes
Servicars - No major changes

1964

The Range

64AH Topper - 10ci (**165cc**) two-stroke single, automatic transmission
64AU Topper - 10ci two-stroke single, 5hp restricted, automatic
64BT Pacer - 11ci (**175cc**) two-stroke single, three-speed
64BTH Scat - 11ci two-stroke single, three-speed
64BTU Pacer - 10ci two-stroke single, 5hp restricted, three-speed
64C Sprint - 15ci (**246cc**) OHV single, four-speed
64H Sprint H - 15ci OHV single, high-compression, four-speed
64XLH Sportster - 54ci (**883cc**) OHV V-twin, four-speed
64XLCH Sportster - 54ci OHV V-twin, four-speed
64FL Duo Glide Sport - 74ci (**1200cc**) OHV V-twin, four-speed
64FLF Duo Glide Sport - 74ci OHV V-

twin, four-speed, foot-change
64FLH Duo Glide Super Sport - 74ci OHV V-twin, high-compression, four-speed
64FLHF Duo Glide Super Sport - 74ci OHV V-twin, high-compression, four-speed, foot-change
64GE Servicar - 45ci (750cc) flathead V-win, three-speed+reverse

Prices
Topper AH - $470
Topper AU - $470
Pacer BT - $495
Scat BTH - $505
Pacer BTU - $495
Sportster XLH - $1,295
Sportster XLCH - $1,360
Duo-Glide FL - $1,385
Duo-Glide FLF - $1,385
Duo-Glide FLH - $1,450
Duo-Glide FLHF - $1,450
Servicar - $1,628

Production
Topper - 825
Pacer/Scat - 2,760
Sportster - 1,433
Duo Glide - 5,500
Servicar - 725
TOTAL - 11,243

Colours
Topper - Black, Fiesta Red
Pacer/Scat - Black, Fiesta Red
Sportster - Black, Fiesta Red, Hi-Fi Red, Hi-Fi Blue
Duo Glide - Black, Fiesta Red, Hi-Fi Red, Hi-Fi Blue
Servicars - Black, White, Fiesta Red, Police Silver (police only)

Specifications
AH/AU Topper - No major changes
BT/BTH/BTU - No major changes
XL Sportster - Full-width aluminium drum front brake
FL Duo Glides - No major changes
Servicars - Electric starter, 12-volt electrics

1965

The Range
65M 50 - 3ci (49cc) two-stroke single, three-speed
65AH Topper - 10ci (165cc) two-stroke single, automatic transmission
65BT Pacer - 11ci (**175cc**) two-stroke single, three-speed
65BTH Scat - 11ci two-stroke single, three-speed
65C Sprint - 15ci (**246cc**) OHV single, four-speed
65H Sprint H - 15ci OHV single, high-compression, four-speed
65XLH Sportster - 54ci (**883cc**) OHV V-twin, four-speed
65XLCH Sportster - 54ci OHV V-twin, four-speed
65FLB Electra Glide Sport - 74ci (**1200cc**) OHV V-twin, four-speed
65FLFB Electra Glide Sport - 74ci OHV V-twin, four-speed, foot-change
65FLHB Electra Glide Super Sport - 74ci OHV V-twin, high-compression, four-speed
65FLHFB Electra Glide Super Sport - 74ci OHV V-twin, high-compression, four-speed, foot-change
65GE Servicar - 45ci (750cc) flathead V-win, three-speed+reverse

Production
Topper - 500
Pacer/Scat - 1,250
Sportster - 3,770
Electra Glide - 6,930
Servicar - 625
TOTAL - 13,075

Colours
M-50 - Holiday Red
Topper - Pacific Blue, Holiday Red
Pacer/Scat - Black, Holiday Red, Pacific Blue
Sportster - Black, Holiday Red, Hi-Fi Red, Hi-Fi Blue
Electra Glide - Black, Holiday Red, Birch White (police only), Police Silver (police only), Hi-Fi Red, Hi-Fi Blue
Servicars - Black, Birch White, Holiday Red, Police Silver (police only)

Specifications
M-50 - New model, Aermacchi-built two-stroke moped, three-speed, motorcycle styling
AH Topper - No major changes
BT/BTH - No major changes
XL Sportster - 12-volt electrics, auto-advance ignition
FL Electra Glides - Electric start, 12-volt electrics, 5-gal (22.7-litre) tank, auto-advance ignition
Servicars - No major changes

1966

The Range
66M 50 - 3ci (**49cc**) two-stroke single, three-speed
66MS 50 Sport - 3ci two-stroke single, three-speed
66BTH Bobcat - 11ci (175cc) two-stroke single, three-speed
66C Sprint - 15ci (**246cc**) OHV single, four-speed
66H Sprint H - 15ci OHV single, high-compression, four-speed
66XLH Sportster - 54ci (**883cc**) OHV V-twin, four-speed
66XLCH Sportster - 54ci OHV V-twin, four-speed
66FLB Electra Glide Sport - 74ci (**1200cc**) OHV V-twin, four-speed
66FLFB Electra Glide Sport - 74ci OHV V-twin, four-speed, foot-change
66FLHB Electra Glide Super Sport - 74ci OHV V-twin, high-compression, four-speed

66FLHFB Electra Glide Super Sport - 74ci OHV V-twin, high-compression, four-speed, foot-change

Prices
Bobcat - $515
Sportster XLH - $1,411
Sportster XLCH - $1,415
Electra-Glide FLB - $1,545
Electra-Glide FLFB - $1,545
Electra-Glide FLHB - $1,610
Electra-Glide FLHFB - $1,610

Production
Bobcat - 1,150
Sportster -4,800
Electra Glide - 7,800
Servicar - 625
TOTAL - 14,375

Colours
Bobcat - Holiday Red, Indigo Metallic, Sparkling Burgundy
Electra Glide - Black, Indigo Metallic, Sparkling Burgundy, Birch White (police only), Hi-Fi Blue
Servicars - Black, Birch White, Holiday Red

Specifications
MS50 Sport - Restyled version of M50, racing-style fuel tank, flat seat
BTH Bobcat - Restyle of BT Pacer, mechanically unchanged
XL Sportster - 15% power increase, Tillotson carburettor
FL Electra Glides - New 'Shovelhead' V-twin, new cylinders, aluminium cylinder heads, Sportster-style rocker boxes, Linkert, then Tillotson carburettor
Servicar - No major changes

1967

The Range
67M 50 - 3ci (49cc) two-stroke single, three-speed
67M 65 - 5ci (**65cc**) two-stroke single, three-speed
67MS 65 - 5ci two-stroke single, three-speed
67H Sprint H - 15ci (**246cc**) OHV single, four-speed
67SS Sprint SS - 15ci OHV single, four-speed
67XLH Sportster - 54ci (**883cc**) OHV V-twin, four-speed
67XLCH Sportster - 54ci OHV V-twin, four-speed
67FLB Electra Glide Sport - 74ci (**1200cc**) OHV V-twin, four-speed
67FLFB Electra Glide Sport - 74ci OHV V-twin, four-speed, foot-change
67FLHB Electra Glide Super Sport - 74ci OHV V-twin, high-compression, four-speed
67FLHFB Electra Glide Super Sport - 74ci OHV V-twin, high-compression, four-speed, foot-change
67GE Servicar - 45ci (750cc) flathead V-twin, three-speed+reverse

Prices
Sportster XLH - $1,650
Sportster XLCH - $1,600
Electra Glide FLB - $1,735

ABOVE: Electra Glide: Harley's best known name was born in 1965 with the unveiling of an electric-start Duo Glide. Apart from that and a 12-volt system to cope with the rigours of turning over 74-cubic inches of compression on a winter's morning, it was no different. However, the extra electrics added a massive 75lb (34kg) to the weight of the machine. The Panhead motor (seen here) was only offered in the Electra Glide for 1965.

OPPOSITE: 1964, and the Duo Glide is still the standard American touring motorcycle, still in kickstart form only.

Electra Glide FLFB - $1,735
Electra Glide FLHB - $1,800
Electra Glide FLHFB - $1,800
Servicar - $1,930
Production
Sportster - 4,500
Electra Glide - 7,750
Servicar - 600
TOTAL - 12,850
Colours
Electra Glide - Black, Crystal Blue, Sparkling Burgundy, Hi-Fi Blue
Servicar - Holiday Red, Birch White
Specifications
Sprint SS - Replaces Sprint (only minor changes), Sprint H continues
XL Sportster - Electric start, revised shock absorbers
FL Electra Glides - No major changes
Servicar - Fibreglass box

The Sportster XLCH of 1968, the cheapest twin in the range, and actually listed at a slightly lower price than the basic XLH. It had higher compression but less equipment, though the XLCH, far from being the 'competition' model it was originally intended to be, would soon be no more than the standard XL with different styling.

1968

The Range
68MS 50 Sport - 3ci (49cc) two-stroke single, three-speed
68M 65 - 4ci (**65cc**) two-stroke single, three-speed
68MS 65 Sport - 4ci two-stroke single, three-speed
68M125 Rapido - 7ci (124cc) two-stroke single, four-speed
68H Sprint H - 15ci (**246cc**) OHV single, four-speed
68SS Sprint SS - 15ci OHV single, four-speed
68XLH Sportster - 54ci (**883cc**) OHV V-twin, four-speed
68XLCH Sportster - 54ci OHV V-twin, four-speed
68FLB Electra Glide Sport - 74ci (**1200cc**) OHV V-twin, four-speed
68FLFB Electra Glide Sport - 74ci OHV V-twin, four-speed, foot-change
68FLHB Electra Glide Super Sport - 74ci OHV V-twin, high-compression, four-speed
68FLHFB Electra Glide Super Sport - 74ci OHV V-twin, high-compression, four-speed, foot-change
68GE Servicar - 45ci (750cc) flathead V-twin, three-speed+reverse
Prices
Sportster XLH - $1,650
Sportster XLCH - $1,600
Electra Glide FLB - $1,735
Electra Glide FLFB - $1,735
Electra Glide FLHB - $1,800
Electra Glide FLHFB - $1,800
Servicar - $1,930
Production
Sportster - 6,875
Electra Glide - 6,950
Servicar - 600
TOTAL - 14,425
Colours
Electra Glide - Black, Jet Fire Orange, Sparkling Burgundy, Hi-Fi Sparkling Blue
Servicar - Birch White
Specifications
XL Sportster - Longer travel front forks
FL Electra Glides - Redesigned wet clutch
Servicar - No major changes

1969

The Range
69M 65 - 3ci (**65cc**) two-stroke single, three-speed
65MS 65 Sport - 3ci two-stroke single, three-speed
69ML 125 Rapido - 7ci (124cc) two-stroke single, three-speed
69SS Sprint SS - 15ci (**246cc**) OHV single, four-speed
69ERS Sprint Scrambler - 15ci single, four-speed
69XLH Sportster - 54ci (**883cc**) OHV V-twin, four-speed
69XLCH Sportster - 54ci OHV V-twin, four-speed
69FLB Super Sport - 74ci (**1200cc**) OHV V-twin, four-speed
69FLFB Super Sport - 74ci OHV V-twin, four-speed, foot-change
69FLHB Super Sport - 74ci OHV V-twin, high-compression, four-speed
69FLHFB Super Sport - 74ci OHV V-twin, high-compression, four-speed, foot-change
69GE Servicar - 45ci (750cc) flathead V-twin, three-speed+reverse
Prices
Sportster XLH - $1,765
Sportster XLCH - $1,698
Electra Glide FLB - $1,885
Electra Glide FLFB - $1,885
Electra Glide FLHB - $1,900
Electra Glide FLHFB - $1,900
Servicar - $2,065
Production
Sportster - 7,800
Electra Glide - 7,300
Servicar - 475
TOTAL - 15,575

Colours
Sportster/Electra Glide - Black, Jet Fire Orange, Birch White, Sparkling Burgundy, Sparkling Blue, Sparkling Gold
Servicar - Birch White
Specifications
Sprint Scrambler - Replaces Sprint H, Sprint SS continues
XL Sportster - Stacked silencers
FL Electra Glides - No major changes
Servicar - No major changes

1970

The Range
70MS 65 Sport Leggero - 4ci (65cc) two-stroke single, three-speed
70MSR Baja - 6ci (98cc) two-stroke single, four-speed
70MLS Rapido - 7ci (124cc) two-stroke single, four-speed
70SS Sprint - 21ci (**346cc**) OHV single, four-speed
70ERS Sprint Scrambler - 21ci OHV single, four-speed
70XLH Sportster - 54ci (**883cc**) OHV V-twin, four-speed
70XLCH Sportster - 54ci OHV V-twin, four-speed
70FLP Super Sport - 74ci (**1200cc**) OHV V-twin, four-speed
70FLPF Super Sport - 74ci OHV V-twin,, four-speed, foot-change
70FLH Super Sport - 74ci OHV V-twin, high-compression, four-speed
70FLHF Super Sport - 74ci OHV V-twin, high-compression, four-speed, foot-change
70GE Servicar - 45ci (750cc) flathead V-twin, three-speed+reverse
Production
Sportster - 8,560
Electra-Glide - 7,615
Servicar - 494
TOTAL - 16,669
Colours
All models - Birch White, Sparkling Blue

LEFT: The 1966/67 Electra Glide marks the debut of the Shovelhead motor, with the familiar Panhead bottom end and capacity, with new cylinder heads. For many buyers, this was just a taste of their own personal dresser.

BELOW LEFT: At the other end of the range, the 7-ci (125-cc) Rapido was an import from Aermacchi in 1969. It was more up to date than the Hummer/Bobcat it replaced, but was still not as sophisticated as the small Japanese bikes.

Specifications
Baja 100 - New model from Aermacchi, 6ci two-stroke single, petroil lubrication, pure off-road machine
MLS Rapido - Now has off-road styling
Sprint SS/Scrambler - New model, enlarged 21ci OHV single replaces 15ci
XL Sportster - Optional 'boat tail' seat/mudguard unit, XLCH has XL's points/coil ignition
FL Electra Glides - Alternator replaces generator, redesigned crankcases
Servicar - No major changes

1971

The Range
71MS 65 Sport Leggero - 4ci (65cc) two-stroke single, three-speed
71MSR Baja - 6ci (98cc) two-stroke single, four-speed

LEFT: Diversification: in the 1970s, Harley-Davidson's search for new markets led it to experiment with snowmobiles; but they failed to stand the test of time.

RIGHT: Meanwhile, the traditional touring market provided far more reliable bread-and-butter income. FLT seen here didn't arrive until 1980, by which time the touring sector had already shrunk to just one piece of a much bigger motorcycle market.

71MLS Rapido - 7ci (124cc) two-stroke single, four-speed
71SS Sprint SS - 21ci (**346cc**) OHV single, four-speed
71SX Sprint SX - 21ci OHV single, four-speed
71XLH Sportster - 54ci (**883cc**) OHV V-twin, four-speed
71XLCH Sportster - 54ci OHV V-twin, four-speed
71FX Super Glide - 74ci (**1200cc**) OHV V-twin, four-speed
71FLP Super Sport - 74ci OHV V-twin, four-speed
71FLPF Super Sport - 74ci OHV V-twin, four-speed, foot-change
71FLH Super Sport - 74ci OHV V-twin, high-compression, four-speed
71FLHF Super Sport - 74ci OHV V-twin, high-compression, four-speed, foot-change
71GE Servicar - 45ci (750cc) flathead V-twin, three-speed+reverse

Production
Sportster - 10,775
Super Glide - 4,700
FL/FLH - 6,675
Servicar - 500
TOTAL - 22,650

Colours
All models - Birch White, Sparkling Blue

Specifications
Sprint SX - Replaces Sprint Scrambler
XL Sportster - Wet clutch with single spring
FX Super Glide - New model, combining FLH 74ci OHV V-twin (but without electric start), frame and running gear with Sportster XLH front end, boat tail rear end, 16in (406mm) front wheel, 19in (483mm) rear, 3.5-gal (16-litre) dual fuel tank
FL/FLH - Bendix/Zenith carburettor, 10in (254mm) front disc brake
Servicar - No major changes

1972

The Range
72MS 65 Sport Leggero - 4ci (**65cc**) two-stroke single, three-speed
72MC 65 Shortster - 4ci two-stroke single, three-speed
72MSR Baja - 6ci (98cc) two-stroke single, four-speed
72MLS Rapido - 7ci (124cc) two-stroke single, four-speed
72SX Sprint SX - 21ci (346cc) OHV single, four-speed
72XLH Sportster - 61ci (**1000cc**) OHV V-twin, four-speed
72XLCH Sportster - 61ci OHV V-twin, four-speed
72FX Super Glide - 74ci (**1200cc**) OHV V-twin, four-speed
72FLP Super Sport - 74ci OHV V-twin, four-speed
72FLPF Super Sport - 74ci OHV V-twin, four-speed, foot-change
72FLH Super Sport - 74ci OHV V-twin, high-compression, four-speed
72FLHF Super Sport - 74ci OHV V-twin, high-compression, four-speed, foot-change
72GE Servicar - 45ci (750cc) flathead V-twin, three-speed+reverse

Production
Sportster - 18,150
Super Glide - 6,500
FL/FLH - 9,700
Servicar - 400
TOTAL - 34,750

Colours
All models - Birch White, Sparkling Blue

Specifications
Shortster 65 - New model, based on M65 with mini-bike styling.
Baja 100 - Lighting option (model Baja 100L)
XL Sportster - Bigger 61ci V-twin, bored-out version of 54ci
FX Super Glide - Boat tail dropped, oil pump updated
FL/FLH - No major changes
Servicar - No major changes

1973

The Range
73Z 90 - 5.5ci (**90cc**) two-stroke single
73X 90 - 5.5ci two-stroke single
73SR 100 - 6ci (98cc) two-stroke single, four-speed
73TX125 - 7ci (124cc) two-stroke single, four-speed
73SS Sprint SS - 21ci (**346cc**) OHV single, four-speed
73SX Sprint SX - 21ci OHV single, four-speed

Three years of Super Glides, the bike that did so much to prop up Harley through the difficult years. It was a clever combination of existing parts to create what looked like a brand-new bike. The 1971 (far left, top) and 1972 (far left, bottom) versions sport the original 'boat tail' rear end, which certainly looked different but horrified the traditionalists. Most of the glassfibre rears were junked in favour of conventional mudguards, as a result of which they are now collectors' items. What most buyers wanted was something more like the 1973 version (left) with a more recognizably Harley rear end. The solo seat is non-standard on this one, as are the twin front disc brakes, though the Super Glide did get a single front disc at the factory that year.

73XLH Sportster - 61ci (**1000cc**) OHV V-twin, four-speed
73XLCH Sportster - 61ci OHV V-twin, four-speed
73FX Super Glide - 74ci (**1200cc**) OHV V-twin, four-speed
73FL - 74ci OHV V-twin, four-speed
73FL - 74ci OHV V-twin, high-compression, four-speed
73GE Servicar - 45ci (750cc) flathead V-twin, three-speed+reverse

Production
Sportster - 20,700
Super Glide - 7,625
FL/FLH - 8,775
Servicar - 425
TOTAL - 37,525

Colours
All models - Birch White

Specifications
SR100 - Now has automatic lubrication
TX125 - Replaces Rapido, similar specification with trail-bike styling
Sprint SS/SX - Electric start
XL Sportster - Disc brakes front and rear, Kayaba front forks
FX Super Glide - Disc brakes front and rear, firmer suspension
FL/FLH - Rear disc brake
Servicar - Rear disc brake

1974

The Range
74Z 90 - 5.5ci (**90cc**) two-stroke single
74X 90 - 5.5ci two-stroke single
74SR100 - 6ci (98cc) two-stroke single, four-speed
74SX 125 - 7ci (124cc) two-stroke single, four-speed
74SX 175 - 11ci (174cc) two-stroke single, five-speed
74SS Sprint SS - 21ci (**346cc**) OHV single, four-speed
74SX Sprint SX - 21 ci OHV single, four-speed
74XLH Sportster - 61ci (**1000cc**) OHV V-twin, four-speed
74XLCH Sportster - 61ci OHV V-twin, four-speed
74FX Super Glide - 74ci (**1200cc**) OHV V-twin, four-speed
74FXE Super Glide - 74ci OHV V-twin, four-speed
74FL Police - 74ci OHV V-twin, four-speed
74FLH - 74ci OHV V-twin, four-speed
74FLHF - 74ci OHV V-twin, four-speed

Production
Sportster - 23,830
Super Glide - 9,233
FLH - 5,166
FLHF - 1,310
FL Police - 791
XR750 - 100
TOTAL - 40,430

Colours
All models - Birch White, Vivid Black, Sunburst Burgundy, Sunburst Blue

Specifications
SX125 - Replaces TX125
SX175 - New model from Aermacchi, similar specification to Yamaha DT-1 trail bike
XL Sportster - Return spring on throttle
FX Super Glide - FXE Super Glide has electric start, with system from FLH. Higher first gear, two-cable throttle, other minor changes
FL/FLH - Keihin carburettors, alarm system

1975

The Range

75Z 90 - 5.5ci (**90cc**) two-stroke single
75X 90 - 5.5ci two-stroke single
75SX 125 - 7ci (**124cc**) two-stroke single, four-speed
74RC 125 - 7ci two-stroke single, four-speed
75SX 175 - 11ci (174cc) two-stroke single, five-speed
75SS 250SS - 15ci (**242cc**) two-stroke single, five-speed
75SX 250SX - 15ci two-stroke single, five-speed
75XLH Sportster - 61ci (**1000cc**) OHV V-twin, four-speed
75XLCH Sportster - 61ci OHV V-twin, four-speed
75FX Super Glide - 74ci (**1200cc**) OHV V-twin, four-speed
75FXE Super Glide - 74ci OHV V-twin, four-speed
75FL Police - 74ci OHV V-twin, four-speed
75FLH - 74ci OHV V-twin, four-speed
75FLHF - 74ci OHV V-twin, four-speed

Production

Sportster - 19,410
Super Glide - 12,410
FLHF - 1,535
FL Police - 900
TOTAL - 34,255

Colours

All models - Birch White, Vivid Black, Sunburst Burgundy, Sunburst Blue, Sunburst/Vivid Orange

Specifications

250SS/SX - New model from Aermacchi, 15ci two-stroke single, five-speed, automatic lubrication, indicators, SS road bike, SX trail bike
XL Sportster - Left-foot gearchange/right-foot brake pedal
FL/FLH - Two-cable throttle

The short-lived Z90 was one of the smaller Aermacchi imports, this one with unshrouded forks and high-level silencer. The high-level pipe was there to imply off-road ability for, at the time, trail bikes were the latest big thing, and were a useful means of giving small machines extra sales appeal in the big bike-centred US market. The Z90 was only listed for one year, however, and within three years the whole Aermacchi liaison had been brought to an end. It would be over 20 years before Harley-Davidson sold another sub-54ci (883cc) machine.

1976

The Range

76SS 125 - 7ci (**124cc**) two-stroke single, four-speed
76SXT 125 - 7ci two-stroke single, four-speed
76SS 175 - 11ci (174cc) two-stroke single, five-speed
76SS 250 - 15ci (**242cc**) two-stroke single, five-speed
76SX 250 - 15ci two-stroke single, five-speed
76XLH Sportster - 61ci (**1000cc**) OHV V-twin, four-speed
76XLCH Sportster - 61ci OHV V-twin, four-speed
76FX Super Glide - 74ci (**1200cc**) OHV V-twin, four-speed
76FXE Super Glide - 74ci OHV V-twin, four-speed
76FLH-1200 - 74ci OHV V-twin, four-speed

Production

Sportster - 18,082
Super Glide - 17,695
FLH-1200 - 11,891
TOTAL - 47,668

Colours

All models - Birch White, Vivid Black, Sunburst Burgundy, Sunburst/Vivid Orange, Vivid Blue, Vivid Red, Vivid Brown, Champagne Silver

Specifications

Liberty Editions of all models, to celebrate USA's bicentennary. Black metalflake main colour, with red, white and blue detailing

1977

The Range

77SS 125 - 7ci (**124cc**) two-stroke single, four-speed
77SXT 125 - 7ci two-stroke single, four-speed
77SS 175 - 11ci (174cc) two-stroke single, five-speed
77SS 250 - 15ci (**242cc**) two-stroke single, five-speed
77SX 250 - 15ci two-stroke single, five-speed
77XLH Sportster - 61ci (**1000cc**) OHV V-twin, four-speed
77XLCH Sportster - 61ci OHV V-twin, four-speed
77XLT Sportster - 61ci OHV V-twin, four-speed
77XLCR Café Racer - 61ci OHV V-twin, four-speed
77FX Super Glide - 74ci (**1200cc**) OHV V-twin, four-speed

77FXE Super Glide - 74ci OHV V-twin, four-speed
77FXS Low Rider - 74ci OHV V-twin, four-speed
77FLH-1200 Electra Glide - 74ci OHV V-twin, four-speed
77FLHS Electra Glide - 74ci OHV V-twin, four-speed

Production
Sportster - 17,915
Café Racer - 1,923
Super Glide - 11,449
FLH/FLHS - 9,226
Low Rider - 3,742
TOTAL - 44,255

Colours
All models - Birch White, Vivid Black, Sunburst Burgundy, Sunburst/Vivid Orange, Vivid Blue, Vivid Red, Vivid Brown, Charcoal Silver, Midnight Blue Metallic, Arctic White

Specifications
XLT Sportster - New model, touring version of XLH, with thicker seat, 3.5-gal (16-litre) tank, touring handlebars, Electra Glide-type saddle bags, windscreen, higher gearing
XLCR Café Racer - New model, combining Sportster power train with redesigned European-style chassis and looks. Redesigned frame, cast aluminium wheels, wrinkle-black finish on engine, low bars, twin front discs, bikini fairing, rearset footpegs
FX/FXE Super Glides - Showa front forks, cast aluminium rear wheel optional
FXS Low Rider - New model, restyled version of FXE Super Glide. Lower frame, raked forks, deeply dished king & queen seat, 3.5-gal (16-litre) dual tanks (instruments on tank console), shorter rear shocks, cast aluminium wheels (19in /483mm) front, 16in/406mm rear), flat drag-style handlebars with risers, highway (set forward) pegs
FLH-1200 - Revised gearbox
FLHS Electra Glide Sports - limited edition of FLH, brown with tan trim

1978

The Range
78SX 250 - 15ci (242cc) two-stroke single, five-speed
78XLH Sportster - 61ci (**1000cc**) OHV V-twin, four-speed
78XLCH Sportster - 61ci OHV V-twin, four-speed
78XLT Sportster - 61ci OHV V-twin, four-speed
78XLCR Café Racer - 61ci OHV V-twin, four-speed

ABOVE LEFT: The RR250 racer was really an Aermacchi, but badged as a Harley-Davidson brought some high-profile racing success to Milwaukee. This 15-ci (250-cc) water-cooled twin won the 250 World Championship three times, from 1974–76, with Walter Villa riding. But it was a last blaze of glory for Aermacchi, which was sold to Cagiva soon afterwards.

ABOVE: The SX250 of 1976, was the final incarnation of the two-stroke Aermacchis with Harley badges. It was fast, and on paper looked the part to meet the Japanese, with a five-speed gearbox and automatic lubrication. But it cost too much, and build quality wasn't up to Honda standards. From 1979, Harley-Davidson accepted that it could (or would have to) survive without making small bikes.

78FX Super Glide - 74ci (**1200cc**) OHV V-twin, four-speed
78FXE Super Glide - 74ci OHV V-twin, four-speed
78FXS Low Rider - 74ci OHV V-twin, four-speed
78FLH-1200 Electra Glide - 74ci OHV V-twin, four-speed
78FLH-80 Electra Glide - 82ci (1340cc) OHV V-twin, four-speed

Sample Prices
XLCH Sportster - $3,370
FLH-80 Electra Glide - $4,905

Production
Sportster - 16,360
Café Racer - 1,201
Super Glide - 10,088
FLH-1200/80 -9,406
Low Rider - 9,787
TOTAL - 46,842

Colours
All models - Birch White, Vivid Black, Vivid Red, Chestnut Brown, Charcoal Silver, Arctic White, Black Cherry

Specifications
XL/XLCH/XLT Sportsters - Twin front discs, electronic ignition, siamezed exhaust system, cast wheels optional, 75th Anniversary version of XLH (Black & Gold with leather seat)
XLCR Café Racer - Dual seat optional
FX/FXE Super Glides - Electronic ignition, cast-iron valve guides, revised intake manifold
FXS Low Rider - Electronic ignition, cast-iron valve guides, revised intake manifold
FLH-1200 - Electronic ignition, cast-iron valve guides, revised cam and intake manifold, cast wheels optional, 75th Anniversary version (Black & Gold with leather seat)
FLH-80 - New 82ci V-twin, bored and stroked version of 74ci twin

The 1978 XLH Sportster, a British-registered bike. Harley-Davidson had reason to be grateful to the Sportster: throughout the 1970s, it was the best-selling bike in the range and in the best year (1974) nearly 24,000 were built. It also introduced many riders to Harleys, who went on to buy bigger machines. The exception was the XLCR Café Racer, which was just too radical for most tastes. Gradual updates (disc brakes, bigger capacity etc) kept many interested in the Sportster through the 1970s.

OPPOSITE:
ABOVE LEFT: Tour Glide was an attempt to radically update the standard touring Harley.

ABOVE RIGHT: Wide Glide married a standard low-riding rear end with widely-spaced front forks. This is a 1980 Shovelhead version.

1979

The Range
79XLH Sportster - 61ci (**1000cc**) OHV V-twin, four-speed
79XLCH Sportster - 61ci OHV V-twin, four-speed
79XLS Sportster - 61ci OHV V-twin, four-speed
79FXE-1200 Super Glide - 74ci (**1200cc**) OHV V-twin, four-speed
79FXS-1200 Low Rider - 74ci OHV V-twin, four-speed
79FXS-80 Low Rider - 82ci (**1340cc**) OHV V-twin, four-speed
79FXEF-1200 Fat Bob - 74ci OHV V-twin, four-speed
79FXEF-80 Fat Bob - 82ci OHV V-twin, four-speed
79FLH-1200 Electra Glide - 74ci OHV V-twin, four-speed
79FLH-80 Electra Glide - 82ci OHV V-twin, four-speed
79FLH-1200 Police Electra Glide - 74ci OHV V-twin, four-speed

79FLH-80 Police Electra Glide - 82ci OHV V-twin, four-speed
79FLHC Electra Glide Classic - 82ci OHV V-twin, four-speed

Production
Sportster - 11,789
Café Racer - 9
Super Glide - 9,942
FLH-1200/80 - 11,442
Low Rider - 13,260
TOTAL - 46,442

Colours
All models - Birch White, Vivid Black, Brilliant Red, Chestnut Brown, Concord Blue, Classic Tan, Classic Creme, FXS Silver, Charcoal Grey

Specifications
XL/XLCH Sportsters - Minor changes
XLS Sportster - New model, based on Sportster with Low Rider styling - raked forks, 16-in (406-mm) rear wheel, drag bars, highway pegs etc
XLCR Café Racer - Discontinued
FXE Super Glide - Twin front discs
FXS Low Rider - Minor changes
FXEF1200/80 Fat Bobs - New models based on Super Glide, with twin tanks, bobbed rear mudguard and buckhorn (higher) bars
FLH Electra Glide Classic - New model, with cast wheels and Tour Pak fairing, saddlebags, crashbars and luggage rack, Tan & Cream colour scheme

1980

The Range
80XLH Sportster - 61ci (**1000cc**) OHV V-twin, four-speed
80XLS Sportster - 61ci OHV V-twin, four-speed
80FXE-1200 Super Glide - 74ci (**1200cc**) OHV V-twin, four-speed
80FXS-1200 Low Rider - 74ci OHV V-twin, four-speed
80FXS-80 Low Rider - 82ci (**1340cc**) OHV V-twin, four-speed
80FXEF-80 Fat Bob - 82ci OHV V-twin, four-speed
80FXEF-80 Fat Bob - 82ci OHV V-twin, four-speed
80FXWG Wide Glide - 82ci OHV V-twin, four-speed
80FXB Sturgis - 82ci OHV V-twin, four-speed
80FLH-1200 Electra Glide - 74ci OHV V-twin, four-speed
80FLH-80 Electra Glide - 82ci OHV V-twin, four-speed
80FLHS Electra Glide - 82ci OHV V-twin, four-speed
80FLHC Electra Glide Classic - 82ci OHV V-twin, four-speed
80FLT Tour Glide - 82ci OHV V-twin, five-speed

Sample Prices
XLH Sportster - $5,867
FXWG Wide Glide - $5,683
FLT Tour Glide - $6,013
FXB Sturgis - $5,687

Production
Sportster - 14,747
Super Glide - 7,942
FLH Electra Glide - 7,512
FLT Tour Glide - 4,480
Low Rider - 5,925
FXWG Wide Glide - 6,085
TOTAL - 46,691

Colours
All models - Birch White, Vivid Black, Rich Red, Saddle Brown, Bright Blue, FXS Silver, Charcoal Grey, Candy Red

Specifications
XL Sportsters - Optional Hugger with lower seat (shorter shocks and thinner seat)
FXE Super Glide - No changes
FXS Low Riders - Motorola electronic ignition, 2-into-1 exhaust
FXB Sturgis - New model, based on FXS-80 Low Rider, with belt primary and final drive, 2-in (51-mm) extended forks, oil cooler, highway pegs, cast wheels
FXEF1200/80 Fat Bobs - Electronic ignition, cast or spoke wheels
FLT Tour Glide - New model, using 82ci V-twin, three-point rubber-mounted to isolate vibration, five-speed gearbox, rectangular steel backbone frame, frame-mounted fairing, enclosed rear chain, twin 10-in (254-mm) front discs, single 12-in (305-mm) rear disc, 16-in (406-mm) wheels, separate (not tank-mounted) instruments
FLH Electra Glide Classic - Electronic ignition
FXWG Wide Glide - New model, using 82ci V-twin and four-speed gearbox, extended wide-spaced front forks, 21-in (533-mm) front wheel, wider front axle, twin fuel tanks (5-gal/22.5-litre) rigidly-mounted engines, low stepped seat (26 1/4-in/67-cm), buckhorn handlebars, staggered shorty dual exhaust system, black colour scheme with orange flames on tank

1981

The Range
81XLH Sportster - 61ci (**1000cc**) OHV V-twin, four-speed
81XLS Sportster - 61ci OHV V-twin, four-speed
81FXE Super Glide - 82ci (**1340cc**) OHV V-twin, four-speed
81FXS Low Rider - 82ci OHV V-twin, four-speed
81FXEF Fat Bob - 82ci OHV V-twin, four-speed

81FXWG Wide Glide - 82ci OHV V-twin, four-speed
81FXB Sturgis - 82ci OHV V-twin, four-speed
81FLH Electra Glide - 82ci OHV V-twin, four-speed
81FLHS Electra Glide - 82ci OHV V-twin, four-speed
81FLH Electra Glide Heritage - 82ci OHV V-twin, four-speed
81FLHC Electra Glide Classic - 82ci OHV V-twin, four-speed
81FLT Tour Glide - 82ci OHV V-twin, five-speed
81FLTC Tour Glide - 82ci OHV V-twin, five-speed

Production
Sportster - 10,102
Super Glide - 3,085
Fat Bob - 3,691
FLH Electra Glide - 6,001
FLT Tour Glide - 2,793
Low Rider - 7,223
FXWG Wide Glide - 5,166
FXB Sturgis - 3,543
TOTAL - 41,604

Colours
All models - Birch White, Vivid Black, Classic Tan, Classic Creme, Candy Root Beer, Metallic Blue, Light Red, Dark Carmine, Orange, Olive, Light Charcoal Metallic, Dark Silver

Specifications
XLH Sportster - Shorter front forks, options include cast wheels, 16-in (406-mm) rear wheel, 3.3-gal (15-litre) tank
XLS Sportster - Buckhorn handlebars, shorty dual exhaust system, 3.3-gal tank, options include cast wheels, 16-in rear wheel, 2.2-gal (10-litre) tank
All 82ci V-twins - V-Fire II electronic ignition, lower 7.4:1 compression ratio, longer valve guides, extra oil drain lines from rocker boxes, revised valve guide seals, Girling rear brake caliper

FAR LEFT:
ABOVE: Even in 1981, some people still opted for a sidecar. This is a Heritage edition of the standard FLH.

BELOW: At a glance, this bike could have been made at any time during the last 50 years. However, a closer look at the Shovelhead motor narrows it down to between 1966 and 1982.

LEFT: The 1982 FXB Sturgis brought belt drive to a new Harley for the first time, though unlike present belt-drive Harley-Davidsons, this one had a primary belt as well. It was a big step forward by the engineering department, needing no lubrication or adjustment, helping to smooth out the transmission and keeping the back of the bike cleaner.

FLH Electra Glide Heritage - New model, FLH with retro styling, fringed leather saddlebags, windscreen, suspended seat
FLT Tour Glides - Revised exhaust, FLTC had special colour scheme

1982

The Range
82XLH Sportster - 61ci (**1000cc**) OHV V-twin, four-speed
82XLS Roadster - 61ci OHV V-twin, four-speed
82FXE Super Glide - 82ci (**1340cc**) OHV V-twin, four-speed
82FXS Low Rider - 82ci OHV V-twin, four-speed
82FXWG Wide Glide - 82ci OHV V-twin, four-speed
82FXB Sturgis - 82ci OHV V-twin, four-speed
82FXR Super Glide II - 82ci OHV V-twin, five-speed
82FXRS Super Glide II - 82ci OHV V-twin, five-speed
82FLH Electra Glide - 82ci OHV V-twin, four-speed
82FLT Tour Glide - 82ci OHV V-twin, five-speed
82FLTC Tour Glide - 82ci OHV V-twin, five-speed

Production
Sportster - 5,947
Super Glide - 1,617
Super Glide II - 6,255
FLH Electra Glide - 6,185
FLT Tour Glide - 2,029
Low Rider - 1,816
FXWG Wide Glide - 2,348
FXB Sturgis - 1,833
TOTAL - 28,030

Colours
All models - Birch White, Vivid Black, Classic Tan, Classic Creme, Metallic Blue, Light Red, Dark Carmine, Orange, Brown, Emerald Green, Red, Pewter, Maroon

Specifications
XLH Sportster /XLS Roadster - Lighter frame, lower 8:1 compression ratio
FXB Sturgis - Higher bars, gold cast wheels
FLT Tour Glide - New handlebars, seat, primary chain oiler, higher output alternator, locks for luggage
FXR/FXRS Super Glide II - New model, using 81ci V-twin, rubber-mounted, bar-mounted instruments, new frame, twin 10-in (254-mm) front discs, 19-in (483-mm) front wheel, 16-in (406-mm) rear wheel, wire wheels. FXRS has sissy bar, stepped seat, highway pegs, two-tone paint and cast wheels

ABOVE LEFT and ABOVE
In 1982, the FX family was in full swing. Little did Willie G. realize (or maybe he did) that the original FX Super Glide would spawn a whole new range of Harleys, slotting between the Sportsters on one hand and the FLH/FLT tourers on the other. They made a virtue out of custom styling, of which the Super Glide was a pioneer for factory machines. The 1978 Low Rider was the first (labelled 'the thinking man's chopper' by *Cycle World* magazine), and was an instant hit. Fat Bob, Sturgis and Wide Glide soon followed.

OPPOSITE: The sportiest Harley of its day. The XR1000 was a seriously souped-up XLX, with aluminium XR-type high-compression cylinder heads, twin carburettors, twin disc brakes and cast wheels. It was fast but expensive, and at the time arguably the sort of indulgence Harley could well have done without.

1983

The Range

83XLX Sportster - 61ci (**1000cc**) OHV V-twin, four-speed
83XLH Sportster -61ci OHV V-twin, four-speed
83XLS Roadster - 61ci OHV V-twin, four-speed
83XR1000 Sportster - 61ci OHV V-twin, four-speed
83FXE Super Glide - 82ci (**1340cc**) OHV V-twin, four-speed
83FXWG Wide Glide - 82ci OHV V-twin, four-speed
83FXSB Low Rider - 82ci OHV V-twin, four-speed
83FXDG Disc Glide - 82ci OHV V-twin, four-speed
83FXR Super Glide II - 82ci OHV V-twin, five-speed
83FXRS Super Glide II - 82ci OHV V-twin, five-speed
83FXRT Sport Glide - 82ci OHV V-twin, five speed
83FLH Electra Glide - 82ci OHV V-twin, four-speed
83FLHS Electra Glide Sport - 82ci OHV V-twin, four-speed
83FLHT Electra Glide - 82ci OHV V-twin, five-speed
83 FLHTC Electra Glide Classic - 82ci OHV V-twin, five-speed
83FLT Tour Glide - 82ci OHV V-twin, five-speed
83FLTC Tour Glide Classic - 82ci OHV V-twin, five-speed

Production

Sportster - 9,756
Super Glide - 1,215
Super Glide II - 4,810
FLH Electra Glides - 6,485
FLT Tour Glide - 1,078
Low Rider - 3,277
Wide Glide - 2,873

TOTAL - 29,494

Colours

All models - Birch White, Vivid Black, Classic Tan, Classic Creme, Pewter, Maroon, Indigo Blue, Claret Red, Slate Grey Metallic

Specifications

XLH Sportster - High bars, vacuum-advance ignition, 3.3-gal (15-litre) tank, new seat, less restrictive exhaust system
XLX Sportster - New model, a stripped-down version to sell for $3,995. Lower bars, solo seat, peanut fuel tank, cast wheels, no tachometer, in black only
XR1000 Sportster - New model based on XLX, with all-new top end – aluminium XR-style cylinder heads, ported and polished by tuner Jerry Branch, 9:1 compression ratio, V-Fire III electronic ignition, twin 1.4-in (36-mm) Dell'Orto carburettors, megaphone silencers, cast wheels
FLH Electra Glide - Belt final drive
FLHT/C Electra Glide - New models, combining FLT chassis with FLH-style frame-mounted fairing, 16-in (406-mm) rear wheel, larger luggage
FLT Tour Glides - Lower seat, stiffer suspension
FXDG Disc Glide - New model, based on Wide Glide with aluminium disc rear wheel, belt final drive and black shorty dual exhaust
FXSB Low Rider - New model, based on FXS Low Rider with belt final drive and 21-in (533-mm) front wheel
FXRT Sport Glide - New model based on FXRS chassis, classified as sports-tourer – aerodynamic frame-mounted fairing, anti-dive front forks, enclosed rear chain, adjustable air suspension, frame-mounted panniers, low touring handlebars, adjustable pillion pegs, 4.2-gal (19-litre) tank, various touring options

1984

The Range

84XLX Sportster - 61ci (**1000cc**) OHV V-twin, four-speed
84XLH Sportster - 61ci OHV V-twin, four-speed
84XLS Roadster - 61ci OHV V-twin, four-speed
84XR1000 Sportster - 61ci OHV V-twin, four-speed
84FXE Super Glide - 82ci (**1340cc**) OHV V-twin, four-speed
84FXST Softail - 82ci Evolution V-twin, four-speed
84FXWG Wide Glide - 82ci OHV V-twin, four-speed
84FXSB Low Rider - 82ci twin, four-speed
84FXRDG Disc Glide - 82ci Evolution V-twin, four-speed
84FXRS Low Glide - 82ci Evolution V-twin, five-speed
84FXRT Sport Glide - 82ci Evolution V-twin, five speed
84FXRP Police - 82ci Evolution V-twin, five-speed
84FLH Electra Glide - 82ci OHV V-twin, four-speed
84FLHS Electra Glide Sport - 82ci OHV V-twin, four-speed
84FLHX Electra Glide - 82ci OHV V-twin, four-speed
84FLHTC Electra Glide Classic - 82ci Evolution V-twin, five-speed
84FLTC Tour Glide Classic - 82ci Evolution V-twin, five-speed

Sample Prices

FXRS Low Glide - $7,560
FXDRG Disc Glide - $8,199
FLHTC Electra Glide Classic - $8,799

Production

Sportster - 10,620
Super Glide - 2,107
FXRS/T/DG - 5,703
Softail - 5,413

FLH Electra Glide - 7,569
FLT Tour Glide - 1,334
Low Rider - 2,877
Wide Glide - 2,227
TOTAL - 37,850
Colours
All models - Birch White, Vivid Black, Classic Tan, Classic Creme, Pewter, Maroon, Indigo Blue, Claret Red, Candy Purple, Rose Metallic, Slate Grey Metallic
Specifications
XLH/XLX Sportsters - Single front disc, diaphragm clutch, alternator
XR1000 Sportster - Improved brakes, orange/black racing colours optional
FXRS/T/P - Evo V-twin, diaphragm clutch – FXRS Low Glide has shorter rear shocks and front forks
FXRDG - One-year limited edition based on FXRS with disc rear wheel, different paint and graphics
FXST Softail - New model, based on Wide Glide with hidden rear shocks for hardtail look - belt drive, rigidly-mounted Evo V-twin, four-speed gearbox, triangulated swinging arm, 21-in (533-mm) front wheel, 16-in (406-mm) rear wheel, buckhorn handlebars, wire wheels
FXWG Wide Glide - Belt drive, chain primary, later 1984 models may have used Evo V-twin
FLH/S/X Electra Glide - Diaphragm clutch, staggered dual exhausts, highway pegs
FLHTC Electra Glide - Evo V-twin, diaphragm clutch, air shocks, air front forks with anti-dive, enclosed rear chain
FLTC Tour Glide Classic - Evo V-twin, belt drive, five-speed gearbox, diaphragm clutch

1985

The Range
85XLX Sportster - 61ci (**1000cc**) OHV V-twin, four-speed
85XLH Sportster - 61ci OHV V-twin, four-speed
85XLS Roadster - 61ci OHV V-twin, four-speed
85FXEF Fat Bob - 82ci (**1340cc**) Evo V-twin, five-speed
85FXST Softail - 82ci Evo V-twin, five-speed
85FXWG Wide Glide - 82ci Evo V-twin, four-speed
85FXSB Low Rider - 82ci Evo V-twin, four-speed
85FXRS Low Glide - 82ci Evo V-twin, five-speed
85FXRT Sport Glide - 82ci Evo V-twin, five-speed
85FXRC Low Glide Custom - 82ci Evo V-twin, five-speed
85FLHTC Electra Glide Classic - 82ci Evo V-twin, five-speed
85FLTC Tour Glide Classic - 82ci Evo V-twin, five-speed
Sample Prices
XLX Sportster - $4,695
FXST Softail - $8,399
Production
Sportster - 6,514
FXRS/T/DG - 8,095
Softail - 4,529
FLH Electra Glide - 4,793
FLT Tour Glide - 1,847
Low Rider - 2,359
Wide Glide - 4,171
TOTAL - 32,308
Colours
All models - Birch White, Vivid Black, Classic Tan, Classic Creme, Candy Blue, Slate Grey Metallic
Specifications
XLH/XLX Sportsters - No changes
FXRS/T - Belt final drive
FXRC Low Glide Custom - One-year limited edition based on FXRS, chrome engine covers, wire wheels, Candy Orange
FXEF Fat Bob – Evo V-twin, wet-multiplate clutch, belt final drive
FXSB Low Rider - Evo V-twin, wet multiplate clutch, belt final drive
FXWG Wide Glide - Evo V-twin, wet multiplate clutch, belt final drive

1986

The Range
86XLH 883 Sportster 54ci (883cc) Evo V-twin, four-speed
86XLH 1100 Sportster - 67ci (1100cc) Evo V-twin, four-speed
86FLST Heritage Softail - 82ci (**1340cc**) Evo V-twin, five-speed
86FXR Super Glide - 82ci Evo V-twin, five-speed
86FXRS Low Glide - 82ci Evo V-twin, five-speed
86FXRT Sport Glide - 82ci Evo V-twin, five-speed
86FXRD Sport Glide Grand Touring - 82ci Evo V-twin, five-speed
86FXST Softail - 82ci Evo V-twin, five-speed
86FXSTC Softail Custom - 82ci Evo V-twin, five-speed
86FXWG Wide Glide - 82ci Evo V-twin, four-speed
86FLHTC Electra Glide Classic - 82ci Evo V-twin, five-speed
86FLTC Tour Glide Classic - 82ci Evo V-twin, five-speed
Sample Prices
XLH 883 Sportster - $3,995
XLH 1100 Sportster - $5,199
FLHTC Electra Glide Classic - $10,224
FLTC Tour Glide Classic & sidecar - $13,423
Production
Sportsters - 14,379
FXR/S/T/D - 8,680
Softails - 8,694
FLH Electra Glide - 4,456
FLT Tour Glide - 1,442
Wide Glide - 1,199
TOTAL - 38,850

OPPOSITE: 1985 Electra Glide Classic with full touring kit.

ABOVE: The 1987 Softail Custom used the now ubiquitous Evolution 82-ci (1340-cc) V-twin, as did the whole range by then. The Evo was a hugely important engine for Harley-Davidson, originally intended as a stopgap before an all-new range of water-cooled motors was ready. It was only a little more powerful than the Shovelhead, but was more reliable and better built, with closer tolerances. This engine carried the company through the next 15 years, until replaced by the Twin Cam in 1998. The bike illustrated is also significant in being a Softail. It was a master stroke of marketing to launch this hardtail-look rear end just as nostalgia was becoming big business. This popular model promised traditional looks with at least some modern comfort, though suspension travel was limited compared to a conventional rear suspension.

Colours
All models - Birch White, Vivid Black, Candy Red, Slate Grey Metallic, Candy Blue, Candy Pearl, Signal Red
Specifications
XLH 883 Sportster - Major update of XLH Sportster, with smaller 54ci version of Evo V-twin (3 x 3.81in/76 x 97mm) bore x stroke), aluminium heads and cylinders, 1.3-in (33-mm) Keihin carburettor, four-speed gearbox, new frame, cast wheels (19in/483mm) front, 16in/406mm rear), low bars, solo seat, peanut tank – upgrade package with different paint and trim
XLH 1100 Sportster - As 883 Sportster with 67ci (1100cc) Evo (3.35 x 3.8in/85.1 x 96.5mm bore x stroke)
FXR Super Glide - New model, stripped-down version of FXRS
FXRS Low Glide - Revised indicator switch, optional Sport package with longer-travel suspension and twin front discs
FXRD Sport Glide Grand Touring - Based on FXRT with air rear suspension, rubber-mounted bars, high-spec stereo, full luggage, 2-into-1 exhaust
FXST Softail - Five-speed gearbox
FLST Heritage Softail - New model, based on Softail with fork shrouds, 16-in (406-mm) wire wheels and circular air cleaner; aims to imitate style of 1949–57 Hydra Glide

1987

The Range
87XLH 883 Sportster - 54ci (883cc) Evo V-twin, four-speed
87XLH 1100 Sportster - 67ci (1100cc) Evo V-twin, four-speed
87FLST Heritage Softail - 82ci (**1340cc**) Evo V-twin, five-speed
87FLSTC Heritage Softail Special - 82ci Evo V-twin, five-speed
87FXR Super Glide - 82ci Evo V-twin, five-speed
87FXRS Low Rider - 82ci Evo V-twin, five-speed
87FXRS-SP Low Rider Sport - 82ci Evo V-twin, five-speed
87FXLR Low Rider Custom - 82ci Evo V-twin, five-speed
87FXRT Sport Glide - 82ci Evo V-twin, five-speed
87FXST Softail - 82ci Evo V-twin, five-speed
87FXSTC Softail Custom - 82ci Evo V-twin, five-speed
87FLHS Electra Glide Sport - 82ci Evo V-twin, five-speed
87FLHTC Electra Glide Classic - 82ci Evo V-twin, five-speed
87FLTC Tour Glide Classic - 82ci Evo V-twin, five-speed

Sample Prices
FXRS Low Rider - $8,449
FXSTC Softail Custom - $9,499
FLHTC Electra Glide Classic - $10,395

Production
Sportsters - 13,794
FXR/S/T - 8,000
Softails - 7,686
FLH Electra Glide - 6,452
FLT Tour Glide - 225
TOTAL - 36,157
Specifications
FLHS Electra Glide Sport - Same spec as FLHTC, detachable screen, no fairing, top box or sissy bar
FXR Super Glide - Belt final drive
FXRT Sport Glide - Improved stereo
FLST Heritage Softail - Special package includes windscreen, passing lamps, leather saddlebags, two-piece seat

1988

The Range
88XLH 883 Sportster - 54ci (**883cc**) Evo V-twin, four-speed
88XLH 883 Sportster Hugger - 54ci Evo V-twin, four-speed
88XLH 883 Sportster Deluxe - 54ci Evo V-twin, four-speed
88XLH 1200 Sportster - 73ci (1203cc) Evo V-twin, four-speed
88FLST Heritage Softail - 82ci (**1340cc**) Evo V-twin, five-speed
88FLSTC Heritage Softail Special - 82ci Evo V-twin, five-speed
88FXR Super Glide - 82ci Evo V-twin, five-speed
88FXRS Low Rider - 82ci Evo V-twin, five-speed
88FXRS-SP Low Rider Sport - 82ci Evo V-twin, five-speed
88FXLR Low Rider Custom - 82ci Evo V-twin, five-speed
88FXRT Sport Glide - 82ci Evo V-twin, five-speed
88FXST Softail - 82ci Evo V-twin, five-speed
88FXSTC Softail Custom - 82ci Evo V-twin, five-speed
88FXSTS Springer Softail - 82ci Evo V-twin, five-speed
88FLHS Electra Glide Sport - 82ci Evo V-twin, five-speed
88FLHTC Electra Glide Classic - 82ci Evo V-twin, five-speed
88FLTC Tour Glide Classic - 82ci Evo V-twin, five-speed
Sample Prices
XLH1200 Sportster - $5,875
FXST Softail - $9,375
FLHTC Electra Glide Classic - $10,545
Production
Sportsters - 16,533
FXR/S/T - 7,119
Softails - 15,408
FLH Electra Glides - 7,314
FLT Tour Glides - 848
TOTAL - 47,222
Colours
Single-tones: Vivid Black, Bright Cobalt Candy Blue, Bright Candy Plum, Candy Brandywine, Candy Crimson,
Two-tones: Brandywine & Crimson, Bright Cobalt Blue & Brilliant Silver, Candy Bronze & Creme, Vivid Black & Creme, Metallic Blue & Cadet Blue Metallic, Creme & Champagne Gold

ABOVE and RIGHT: More examples of the late 1980s Softail line. The standard Softail (1987, above) and Softail Custom (1988, below) were actually very similar, but the range was soon expanded to include the Springer Softail as well. Both these bikes are pictured in typical rally scenes.

Specifications
XLH883 Sportsters - 1.6-in (41-mm) Keihin carburettor, longer rear shocks, longer swinging arm, 1.5-in (38-mm) front forks, solo seat, cast wheels. Hugger package includes lower seat and buckhorn bars. Deluxe package includes two-tone paint, buckhorn bars, wire wheels, dual seat, extra chrome, tachometer
XLH1200 Sportster - Replaces XLH1100 with 73ci Evo V-twin (3.498 x 3.812in/88.8 x 96.8mm bore x stroke), 1.5-in (38-mm) front forks, otherwise as 883 Deluxe
FXR/S/LR/RT/ST - 39-mm front forks
FXSTS Springer Softail - New model, based on FXST with exposed springer front forks, much updated version of original 1903 spring fork

1989

The Range
89XLH 883 Sportster - 54ci (**883cc**) Evo V-twin, four-speed
89XLH 883 Sportster Hugger - 54ci Evo V-twin, four-speed
89XLH 883 Sportster Deluxe - 54ci Evo V-twin, four-speed
89XLH 1200 Sportster - 73ci (1203cc) Evo V-twin, four-speed
89FLST Heritage Softail - 82ci (**1340cc**) Evo V-twin, five-speed
89FLSTC Heritage Softail Special - 82ci Evo V-twin, five-speed
89FXR Super Glide - 82ci Evo V-twin, five-speed
89FXRS-CONV Low Rider Convertible - 82ci Evo V-twin, five-speed
89FXRS-SP Low Rider Sport - 82ci Evo V-twin, five-speed
89FXLR Low Rider Custom - 82ci Evo V-twin, five-speed
89FXRT Sport Glide - 82ci Evo V-twin, five-speed
89FXST Softail - 82ci Evo V-twin, five-speed
89FXSTC Softail Custom - 82ci Evo V-twin, five-speed
89FXSTS Springer Softail - 82ci Evo V-twin, five-speed
89FLHS Electra Glide Sport - 82ci Evo V-twin, five-speed
89FLHTC Electra Glide Classic - 82ci Evo V-twin, five-speed
89FLHTCU Electra Glide Classic Ultra - 82ci Evo V-twin, five-speed
89FLTC Tour Glide Classic - 82ci Evo V-twin, five-speed
89FLTCU Tour Glide Classic Ultra - 82ci Evo V-twin, five-speed

Sample Prices
FXRS Convertible - $9,475
FXSTS Springer Softail - $10,759

Production
Sportsters - 16,967
FXR/S/T/P/CONV - 7,015
Softails - 19,756
FLH Electra Glides - 10,148
FLT Tour Glides - 1,185
TOTAL - 55,071

Colours
Single tones: Vivid Black, Metallic Blue, Candy Brandywine, Bright Candy Plum,
Two-tones: Brandywine & Crimson, Metallic Blue & Brilliant Silver, Vivid Black & Creme, Candy Bronze & Creme, Metallic Blue & Cadet Blue Metallic, Creme & Champagne Gold, Vivid Black & Brilliant Silver

Specifications
XLH883/1200 Sportsters - Aluminium intake manifold, offset gudgeon pins, new air cleaner
FXR/S/T/P - New starter, one-piece pinion shaft and right flywheel
FXRS-CONV - New model, based on FXRS with quickly-detachable screen and leather saddlebags, highway pegs and sissy bar
FLST Heritage Softail - 32-amp alternator, one-piece pinion shaft and right flywheel
FLHS/T Electra Glides - 32-amp alternator, one-piece pinion shaft and right flywheel
FLHTCU Electra Glide Classic Ultra - As FLHTC with cruise control, CB radio, intercom, front and rear sound systems, fairing lowers
FLTC Tour Glide - 32-amp alternator, one-piece pinion shaft and right flywheel
FLTCU Tour Glide Classic Ultra - As FLTC with cruise control, CB radio, intercom, front and rear sound systems, fairing lowers

1990

The Range
90XLH 883 Sportster - 54ci (**883cc**) Evo V-twin, four-speed
90XLH 883 Sportster Hugger - 54ci Evo V-twin, four-speed
90XLH 883 Sportster Deluxe - 54ci Evo V-twin, four-speed
90XLH 1200 Sportster - 73ci (1203cc) Evo V-twin, four-speed
90FLST Heritage Softail - 82ci (**1340cc**) Evo V-twin, five-speed
90FLSTC Heritage Softail Classic - 82ci Evo V-twin, five-speed
90FLSTF Fat Boy - 82ci Evo V-twin, five-speed
90FXR Super Glide - 82ci Evo V-twin, five-speed
90FXRS Low Rider - 82ci Evo V-twin, five-speed
90FXRS-CONV Low Rider Convertible - 82ci Evo V-twin, five-speed
90FXRS-SP Low Rider Sport - 82ci Evo V-twin, five-speed
90FXLR Low Rider Custom - 82ci Evo V-twin, five-speed
90FXRT Sport Glide 82ci Evo V-twin, five-speed
90FXST Softail - 82ci Evo V-twin, five-speed
90FXSTC Softail Custom - 82ci Evo V-twin, five-speed
90FXSTS Springer Softail - 82ci Evo V-twin, five-speed
90FLHS Electra Glide Sport - 82ci Evo V-twin, five-speed
90FLHTC Electra Glide Classic - 82ci Evo V-twin, five-speed
90FLHTCU Electra Glide Classic Ultra - 82ci Evo V-twin, five-speed
90FLTC Tour Glide Classic - 82ci Evo V-twin, five-speed
90FLTCU Tour Glide Classic Ultra - 82ci Evo V-twin, five-speed

Sample Prices
XLH883 Sportster - $4,250
FLSTF Fat Boy - $10,995
FLHTCU Electra Glide Classic Ultra - $13,695

Production
Sportsters - 15,163
FXR/S/T/P/CONV - 8,485
Softails - 24,138
FLH Electra Glides - 10,313
FLT Tour Glides - 1,108
TOTAL - 59,207

Colours
Single tones: Vivid Black, Bright Candy Plum, Bright Candy Ruby, Bright Candy Hi-Fi Blue, Fine Silver Metallic, Dark Candy Ruby.
Two-tones: Bright Candy Hi-Fi Blue & Silver, Bright & Dark Candy Ruby, Vivid Black & Black Pearl, Vivid Black & Creme, Creme & Champagne Gold, Bright & Dark Candy Hi-Fi Blue

Specifications
XLH883/1200 Sportsters - 1.57-in (40-mm) Keihin CV carburettor
FXR/S/T/P/CONV - 40-mm Keihin CV carburettor, one-piece right-hand flywheel, redesigned diaphragm clutch
FLSTS Fat Boy - New model, based on FLST Heritage Softail - 16-in (406-mm) disc wheels, wide FLH bars, dual shotgun exhaust system, redesigned clutch, Silver

with Dark Yellow highlights
FXST/FLST - 40-mm Keihin CV carburettor, one-piece right-hand flywheel, redesigned clutch
FLTC/U - 40-mm Keihin CV carburettor, one-piece right hand flywheel, redesigned clutch
FLHTC/U - 40-mm Keihin CV carburettor, one-piece right-hand flywheel, redesigned clutch

1991

The Range
91XLH 883 Sportster - 54ci (**883cc**) Evo V-twin, five-speed
91XLH 883 Sportster Hugger - 54ci Evo V-twin, five-speed
91XLH 883 Sportster Deluxe - 54ci Evo V-twin, five-speed
91XLH 1200 Sportster - 73ci (1203cc) Evo V-twin, five-speed
91FLSTC Heritage Softail Classic - 82ci (**1340cc**) Evo V-twin, five-speed
91FLSTF Fat Boy - 82ci Evo V-twin, five-speed
91FXR Super Glide - 82ci Evo V-twin, five-speed
91FXRS Low Rider - 82ci Evo V-twin, five-speed
91FXRS-CONV Low Rider Convertible - 82ci Evo V-twin, five-speed
91FXRS-SP Low Rider Sport - 82ci Evo V-twin, five-speed
91FXLR Low Rider Custom - 82ci Evo V-twin, five-speed
91FXRT Sport Glide - 82ci Evo V-twin, five-speed
91FXSTC Softail Custom - 82ci Evo V-twin, five-speed
91FXSTS Springer Softail - 82ci Evo V-twin, five-speed
91FXDB Dyna Glide Sturgis - 82ci Evo V-twin, five-speed
91FLHS Electra Glide Sport - 82ci Evo V-twin, five-speed
91FLHTC Electra Glide Classic - 82ci Evo V-twin, five-speed
91FLHTCU Electra Glide Classic Ultra - 82ci Evo V-twin, five-speed
91FLTC Tour Glide Classic - 82ci Evo V-twin, five-speed
91FLTCU Tour Glide Classic Ultra - 82ci Evo V-twin, five-speed

Sample Prices
XLH1200 Sportster - $6,245
FXRS Low Rider - $9,760
FXDB Sturgis - $11,520

Production
Sportsters - 17,725
FXR/S/T/P/CONV/DB - 10,077
Softails - 26,322
FLH Electra Glides - 10,098
FLT Tour Glides - 787
TOTAL - 65,009

Colours
Single tones: Vivid Black, Bright Sapphire Metallic, Wineberry Pearl, Vivid Yellow, Bright Candy Ruby, Candy Sapphire Sun-glo
Two-tones: Bright and Dark Sapphire Metallic, Bright and Dark Candy Ruby, White and Vivid Yellow, White and Turquoise, White and Yellow

Specifications
XLH883 Sportsters - Five-speed gearbox,

ABOVE: A standard Softail from 1990. In the late 1980s, the Softails took over from Sportsters as the best-selling Harleys, which shows just how far the nostalgia market had developed. Over 26,000 were made in 1991.

OPPOSITE: A 1991 Dyna Glide Daytona. The Dyna Glide didn't look very new, but it was. The new steel backbone frame allowed the big V-twin to be rubber-mounted, substantially reducing vibration levels. Otherwise it was the usual canny mix of existing components, with the 82-ci (1340-cc) Evo twin and FXRS front end.

belt drive (Deluxe only), rubber-mounted pegs, self-cancelling indicators, several minor engine changes
XLH1200 Sportster - Five-speed gearbox, belt drive, self-cancelling indicators, several minor engine changes
FXR/S/T/P/CONV - Kevlar base gasket, graphite head gasket, self-cancelling indicators, four-sided fuel inlet valve
FLST/FXST/FLSTF - Kevlar base gaskets, graphite head gaskets, four-sided fuel inlet valve, self-cancelling indicators, other minor changes, black and chrome engine (FLSTC)
FXDB Dyna Glide Sturgis - New model, Evo 82ci V-twin, new frame with rubber mounts using large-diameter rectangular backbone welded to cast steering head, rubber engine mounting system with new engine tilt and location, FXRS forks and front end, single front disc brake, 19-in (483-mm) front wheel, 16-in (406-mm) rear, all engine changes as FX/FL series
FLTC/U – Kevlar base gaskets, graphite head gaskets, four-sided fuel inlet valve
FLHTC/U - Kevlar base gaskets, graphite head gaskets, four-sided fuel inlet valve

1992

The Range
92XLH 883 Sportster - 54ci (**883cc**) Evo V-twin, five-speed
92XLH 883 Sportster Hugger - 54ci Evo V-twin, five-speed
92XLH 883 Sportster Deluxe - 54ci Evo V-twin, five-speed
92XLH 1200 Sportster - 73ci (1203cc) Evo V-twin, five-speed
92FLSTC Heritage Softail Classic - 82ci (**1340cc**) Evo V-twin, five-speed
92FLSTF Fat Boy - 82ci Evo V-twin, five-speed
92FXR Super Glide - 82ci Evo V-twin, five-speed
92FXRS-CONV Low Rider Convertible - 82ci Evo V-twin, five-speed
92FXRS-SP Low Rider Sport - 82ci Evo V-twin, five-speed
92FXLR Low Rider Custom - 82ci Evo V-twin, five-speed
92FXRT Sport Glide - 82ci Evo V-twin, five-speed
92FXSTC Softail Custom - 82ci Evo V-twin, five-speed
92FXSTS Springer Softail - 82ci Evo V-twin, five-speed
92FXDB Dyna Glide Daytona - 82ci Evo V-twin, five-speed
92FXDC Dyna Glide Custom - 82ci Evo Vtwin, five-speed
92FLHS Electra Glide Sport - 82ci Evo V-twin, five-speed
92FLHTC Electra Glide Classic - 82ci Evo V-twin, five-speed
92FLHTCU Electra Glide Classic Ultra - 82ci Evo V-twin, five-speed
92FLTC Tour Glide Classic - 82ci Evo V-twin, five-speed
92FLTCU Tour Glide Classic Ultra - 82ci Evo V-twin, five-speed
Sample Prices
XLH883 Sportster Hugger - $5,075
FLHTC Electra Glide Classic - $15,350
FXDB Daytona - $12,120
Total Production - 76,000
Colours
Vivid Black, True Pearl etc
Specifications
XLH883/1200 Sportsters - New grease fittings, halogen headlight, new horn, other minor changes
FXDB Dyna Glide Daytona - New limited-edition model based on Dyna Glide Sturgis; twin front discs, buckhorn bars, gold wheels, two-tone Indigo Blue Metallic & Gold Pearlglo colour scheme
FXDC Dyna Glide Custom - New model based on Dyna Glide Sturgis; twin front discs, two-tone Silver and Black, silver frame, unpainted engine/gearbox
FXR/S/T/P/CONV - New grease fittings, new horn, caged needle bearing for camshaft
FLST/FXST/FLSTF - New grease fittings, new horn, caged needle bearing for camshaft
FLHS/TC/U - New grease fittings, new horn, caged needle bearing for camshaft, lower seat (FLHS), glossy luggage interior
FLTC/U - New grease fittings, new horn, caged needle bearing for camshaft, glossy luggage interior

1993

The Range
93XLH 883 Sportster - 54ci (**883cc**) Evo V-twin, five-speed
93XLH 883 Sportster Hugger - 54ci Evo V-twin, five-speed
93XLH 883 Sportster Deluxe - 54ci Evo V-twin, five-speed
93XLH 1200 Sportster - 73ci (1203cc) Evo V-twin, five-speed
93FLSTC Heritage Softail Classic- 82ci (**1340cc**) Evo V-twin, five-speed
93FLSTN Heritage Softail Nostalgia - 82ci Evo V-twin, five-speed
93FLSTF Fat Boy - 82ci Evo V-twin, five-speed
93FXR Super Glide - 82ci Evo V-twin, five-speed
93FXRS-CONV Low Rider Convertible - 82ci Evo V-twin, five-speed
93FXRS-SP Low Rider Sport - 82ci Evo V-twin, five-speed
93FXLR Low Rider Custom - 82ci Evo V-twin, five-speed
93FXSTC Softail Custom - 82ci Evo V-

twin, five-speed
93FXSTS Springer Softail - 82ci Evo V-twin, five-speed
93FXDL Dyna Low Rider - 82ci Evo V-twin, five-speed
93FXDWG Dyna Wide Glide - 82ci Evo V-twin, five-speed
93FLHS Electra Glide Sport - 82ci Evo V-twin, five-speed
93FLHTC Electra Glide Classic - 82ci Evo V-twin, five-speed
93FLHTCU Electra Glide Classic Ultra - 82ci Evo V-twin, five-speed
93FLTC Tour Glide Classic - 82ci Evo V-twin, five-speed
93FLTCU Tour Glide Classic Ultra - 82ci Evo V-twin, five-speed

Sample Prices
XLH883 Sportster Deluxe - $5,820
FLSTN Heritage Nostalgia - $13,000
FXDL Dyna Low Rider - $11,800
FLHTCU Electra Glide Classic Ultra (90th Anniversary) - $16,100

Total Production - 82,000

Colours
Single tones: Vivid Black, Mandarin Orange, Scarlet Red, Victory Red Sun-Glo, Bright Aqua Sun-Glo, 90th Anniversary, Bright Wineberry Sun-Glo
Two-tones: Victory Red Sun-Glo & Aqua Sun-Glo, Black & Scarlet Red

Specifications
XLH883 Sportsters - Belt drive on all 883s
FFXR/S/T/P/CONV - No major changes
FXDWG Dyna Wide Glide - New model, using Dyna Glide chassis with wide-set front forks, high bars and 21-in (533-mm) front wheel of earlier Wide Glide, bobbed rear mudguard, single front disc brake, one-piece Fat Bob fuel tank
FXDL Dyna Low Rider - New model using Dyna Glide chassis - removable pillion seat, twin front discs, black and chrome engine/gearbox, staggered shorty dual exhaust system, black cast wheels
FLSTN Heritage Softail Nostalgia - Limited edition based on FLST with two-tone Birch White & Black, small leather saddle bags
FLST/FXST/FLSTF - No major changes
FLHS/TC/U - Remote oil reservoir mounted under engine, battery under seat, 15% bigger saddlebags
FLTC/U - Remote oil reservoir mounted under engine, battery under seat, 15% bigger saddlebags

1994

The Range
94XLH 883 Sportster - 54ci (**883cc**) Evo V-twin, five-speed
94XLH 883 Sportster Hugger - 54ci Evo V-twin, five-speed
94XLH 883 Sportster Deluxe - 54ci Evo V-twin, five-speed
94XLH 1200 Sportster - 73ci (1203cc) Evo V-twin, five-speed

ABOVE: Some police departments stayed loyal to Harley-Davidson through the 1980s. This is a 1992 FXRP (denoting a police-spec Dyna Glide) in service in Daytona.

LEFT: Harley wasn't backward in celebrating its own long history, and this is a 90th-Anniversary FLH in 1993. Handlebar fairing and panniers were standard, but the top box was probably a factory extra. Note the generous passenger backrest.

OPPOSITE
LEFT: The FLHR Road King (this is the original 1994 version) was no more nor less than the latest Electra Glide, stripped of all the touring equipment apart from spotlights and panniers.

RIGHT: Some still use the big Harleys for camping trips, though the rider has left little room for a passenger. Twin front discs were standard by this time, but were universally criticized for not being effective in stopping a bike the weight of the Electra Glide.

94FLSTC Heritage Softail Classic- 82ci (**1340cc**) Evo V-twin, five-speed
94FLSTN Heritage Softail Special - 82ci Evo V-twin, five-speed
94FLSTF Fat Boy - 82ci Evo V-twin, five-speed
94FXR Super Glide - 82ci Evo V-twin, five-speed
94FXDS-CONV Low Rider Convertible - 82ci Evo V-twin, five-speed
94FXLR Low Rider Custom - 82ci Evo V-twin, five-speed
94FXSTC Softail Custom - 82ci Evo V-twin, five-speed
94FXSTS Springer Softail - 82ci Evo V-twin, five-speed
94FXDL Dyna Low Rider - 82ci Evo V-twin, five-speed
94FXDWG Dyna Wide Glide - 82ci Evo V-twin, five-speed
94FLHR Electra Glide Road King - 82ci Evo V-twin, five-speed
94FLHTC Electra Glide Classic - 82ci Evo V-twin, five-speed
94FLHTCU Electra Glide Classic Ultra - 82ci Evo V-twin, five-speed
94FLTC Tour Glide Classic - 82ci Evo V-twin, five-speed
94FLTCU Tour Glide Classic Ultra - 82ci Evo V-twin, five-speed

Total Production - 96,000

Colours

Vivid Black, Aqua Pearl, Victory Red Sun-Glo, Two-tone Aqua Pearl & Silver, two-tone Victory Red Sun-Glo & Platinum

Specifications

XLH883/1200 Sportsters - No major changes
FXR/LR/DWG/DL/STC/STS - Larger engine sprocket
FLSTN Heritage Softail Special - '93 FLSTN in two-tone Birch and Silver

FLSTF/STC - Larger engine sprocket
FLHR Electra Glide Road King - New model based on FLHTC, without fairing. Quickly-detachable windscreen and pillion seat, air suspension, large headlight in nacelle, twin front disc brakes, cast wheels
FLHTC/U - Larger engine sprocket, sealed cruise switch (U only)
FLTC/U - Larger engine sprocket, sealed cruise switch (U only)

1995

The Range

95XLH 883 Sportster - 54ci (**883cc**) Evo V-twin, five-speed
95XLH 883 Sportster Hugger - 54ci Evo V-twin, five-speed
95XLH 883 Sportster Deluxe - 54ci Evo V-twin, five-speed
95XLH 1200 Sportster - 73ci (1203cc) Evo V-twin, five-speed
95FLSTC Heritage Softail Classic - 82ci (**1340cc**) Evo V-twin, five-speed
95FLSTN Heritage Softail Special - 82ci Evo V-twin, five-speed
95FLSTF Fat Boy - 82ci Evo V-twin, five-speed
95FXD Dyna Super Glide - 82ci Evo V-twin, five-speed
95FXDS-CONV Low Rider Convertible - 82ci Evo V-twin, five-speed
95FXSTSB Bad Boy - 82ci Evo V-twin, five-speed
95FXSTC Softail Custom - 82ci Evo V-

twin, five-speed
95FXSTS Springer Softail - 82ci Evo V-twin, five-speed
95FXDL Dyna Low Rider - 82ci Evo V-twin, five-speed
95FXDWG Dyna Wide Glide - 82ci Evo V-twin, five-speed
95FLHT Electra Glide Standard - 82ci Evo V-twin, five-speed
95FLHR Electra Glide Road King - 82ci Evo V-twin, five-speed
95FLHTC Electra Glide Classic - 82ci Evo V-twin, five-speed
95FLHTCU Electra Glide Classic Ultra - 82ci Evo V-twin, five-speed
95FLHTCUI Electra Glide Classic Ultra - 82ci Evo V-twin fuel injection, five-speed
95FLTCU Tour Glide Classic Ultra - 82ci Evo V-twin, five-speed

Sample Prices

XL883 Sportster Deluxe - $6,120
FXD Dyna Super Glide - $9,995
FXSTB Bad Boy - $13,850
FLHTCU Electra Glide Classic Ultra - $16,050

Total Production - 105,000

Colours

Single tones: Vivid Black, Aqua Pearl, Victory Red Sun-Glo
Two-tones: Aqua Pearl & Silver, Nugget Yellow & Birch White, Charcoal Satinbrite & Vivid Black, Vivid Black & Silver, Birch White & Gold, Burgundy Pearl & Scarlet Red, Vivid Black &Burgundy Pearl

Specifications

XLH883/1200 Sportsters - Electronic speedometer
FXD Dyna Super Glide - New base-model Super Glide, using Dyna Glide chassis. Staggered shorty dual exhaust, cast wheels, electronic speedometer
FXDS-CONV/DWG/DL - Electronic speedometer
FXSTSB - New model based on FXSTS. Black steel springer forks, staggered shorty dual exhaust, spoked front wheel, slotted disc rear wheel, floating brake calipers
FLHT Electra Glide Standard - new base-model of FLHTC. Cast wheels, tachometer, fuel gauge, voltmeter, oil-pressure gauge, air suspension, fork-mounted fairing, lockable saddlebags, luggage rack, accessory plug
FLHTCUI - Sequential port fuel injection

1996

The Range

96XLH 883 Sportster - 54ci (**883cc**) Evo V-twin, five-speed
96XLH 883 Sportster Hugger - 54ci Evo V-twin, five-speed
96XLH 1200 Sportster - 73ci (**1203cc**) Evo V-twin, five-speed
96XL 1200C Sportster Custom - 73ci Evo V-twin, five-speed
96XL 1200S Sportster Sport - 73ci Evo V-twin, five-speed
96FLSTC Heritage Softail Classic- 82ci (**1340cc**) Evo V-twin, five-speed
96FLSTN Heritage Softail Special - 82ci Evo V-twin, five-speed
96FLSTF Fat Boy - 82ci Evo V-twin, five-speed
96FXD Dyna Super Glide - 82ci Evo V-twin, five-speed
96FXDS-CONV Low Rider Convertible - 82ci Evo V-twin, five-speed
96FXSTSB Bad Boy - 82ci Evo V-twin, five-speed
96FXSTC Softail Custom - 82ci Evo V-twin, five-speed
96FXSTS Springer Softail - 82ci Evo V-twin, five-speed
96FXDL Dyna Low Rider - 82ci Evo V-twin, five-speed
96FXDWG Dyna Wide Glide - 82ci Evo V-twin, five-speed
96FLHT Electra Glide Standard - 82ci Evo V-twin, five-speed
96FLHR Electra Glide Road King - 82ci Evo V-twin, five-speed
96FLHRI Electra Glide Road King - 82ci Evo V-twin fuel injection, five-speed
96FLHTC Electra Glide Classic - 82ci Evo V-twin, five-speed
96FLHTCI Electra Glide Classic - 82ci Evo V-twin fuel injection, five-speed
96FLHTCU Electra Glide Classic Ultra - 82ci Evo V-twin, five-speed
96FLHTCUI Electra Glide Classic Ultra - 82ci Evo V-twin fuel injection, five-speed
96FLTCUI Tour Glide Classic Ultra - 82ci Evo V-twin fuel injection, five-speed

Sample Prices

XL883 Sportster - $5,095
XL1200S Sportster Sport - $7,910
FXDL Dyna Low Rider - $13,030
FXDWG Wide Glide - $14,030
FLTCUI Tour Glide Classic Ultra - $17,410

Total Production - 119,000

Colours

Single tones: Vivid Black, Patriot Red Pearl, States Blue Pearl, Platinum Silver, Mystique Green Metallic, Violet Pearl
Two-tones: Victory Sun-glo & Platinum, Violet & Red Pearl, Platinum Silver & Black, Mystique Green &Black,
FXSTSB - Vivid Black with yellow, purple or turquoise graphics

Specifications

XLH883/1200 Sportsters - 3.3-gal (15-litre) fuel tank
XL1200S Sportster Sport - New model, based on XLH1200. Adjustable front forks, adjustable rear shocks, 3.3-gal tank
XL1200C Sportster Custom - New model, based on XLH1200. Staggered shorty dual exhaust, spoked front wheel, disc rear wheel
FXSTC/STS/B - Electronic speedometer
FLSTN/STC - Electronic speedometer

OPPOSITE: The 1996 FLHT Electra Glide Standard was a new 'entry-level' version of the big tourer, less expensive than the Classic and Ultra versions, but still with the handlebar fairing and luggage. However, it was a starting point for any serious tourist.

ABOVE: By 1997, fuel injection was an option on the Electra Glide, though this one is the carburettored version. The object of injection was not increased power, but cleaner emissions and more consistent running, especially from cold. The Electra Glide had long since ducked out of the power game, assuming it had ever been in it.

1997

The Range

97XLH 883 Sportster - 54ci (**883cc**) Evo V-twin, five-speed
97XLH 883 Sportster Hugger - 54ci Evo V-twin, five-speed
97XLH 1200 Sportster - 73ci (**1203cc**) Evo V-twin, five-speed
97XL 1200C Sportster Custom - 73ci Evo V-twin, five-speed
97XL 1200S Sportster Sport - 73ci Evo V-twin, five-speed
97FLSTC Heritage Softail Classic - 82ci (**1340cc**) Evo V-twin, five-speed
97FLSTS Heritage Springer - 82ci Evo V-twin, five-speed
97FLSTF Fat Boy - 82ci Evo V-twin, five-speed
97FXD Dyna Super Glide - 82ci Evo V-twin, five-speed
97FXDS-CONV Low Rider Convertible - 82ci Evo V-twin, five-speed
97FXSTSB Bad Boy - 82ci Evo V-twin, five-speed
97FXSTC Softail Custom - 82ci Evo V-twin, five-speed
97FXSTS Springer Softail - 82ci Evo V-twin, five-speed
97FXDL Dyna Low Rider - 82ci Evo V-twin, five-speed
97FXDWG Dyna Wide Glide - 82ci Evo V-twin, five-speed
97FLHT Electra Glide Standard - 82ci Evo V-twin, five-speed
97FLHR Electra Glide Road King - 82ci Evo V-twin, five-speed
97FLHRI Electra Glide Road King - 82ci Evo V-twin fuel injection, five-speed
97FLHTC Electra Glide Classic - 82ci Evo V-twin, five-speed
97FLHTCI Electra Glide Classic - 82ci Evo V-twin fuel injection, five-speed
97FLHTCU Electra Glide Classic Ultra - 82ci Evo V-twin, five-speed
97FLHTCUI Electra Glide Classic Ultra - 82ci Evo V-twin fuel injection, five-speed
97FLTCUI Tour Glide Classic Ultra - 82ci Evo V-twin fuel injection, five-speed
Total Production - 132,000

1998

The Range

98XLH 883 Sportster - 54ci (**883cc**) Evo V-twin, five-speed
98XLH 883 Sportster Hugger - 54ci Evo V-twin, five-speed
98XL53C Sportster Custom 53 - 54ci Evo V-twin, five-speed
98XLH 1200 Sportster - 73ci (**1203cc**) Evo V-twin, five-speed
98XL 1200C Sportster Custom - 73ci Evo V-twin, five-speed
98XL 1200S Sportster Sport - 73ci Evo V-twin, five-speed
98FLSTC Heritage Softail Classic - 82ci (**1340cc**) Evo V-twin, five-speed
98FLSTF Fat Boy - 82ci Evo V-twin, five-speed
98FLSTS Heritage Springer - 82ci Evo V-twin, five-speed
98FXD Dyna Super Glide - 82ci Evo V-twin, five-speed
98FXDS-CONV Low Rider Convertible - 82ci Evo V-twin, five-speed
98FXSTC Softail Custom - 82ci Evo V-twin, five-speed
98FXSTS Springer Softail - 82ci Evo V-twin, five-speed
98FXSTB Night Train - 82ci Evo V-twin, five-speed
98FXDL Dyna Low Rider - 82ci Evo V-twin, five-speed
98FXDWG Dyna Wide Glide - 82ci Evo V-twin, five-speed
98FLHT Electra Glide Standard - 82ci Evo V-twin, five-speed
98FLHR Road King - 82ci Evo V-twin, five-speed
98FLHRCI Road King Classic - 82ci Evo V-twin fuel injection, five-speed
98FLHP Road King Police - 82ci Evo V-twin, five-speed
98FLHPI Road King Police EFI - 82ci Evo V-twin, fuel injection, five-speed
98FLTR Road Glide - 82ci Evo V-twin, five-speed
98FLTRI Road Glide EFI - 82ci Evo V-twin fuel injection, five-speed
98FLHTC Electra Glide Classic - 82ci Evo V-twin, five-speed
98FLHTCI Electra Glide Classic - 82ci Evo V-twin fuel injection, five-speed
98FLHTCUI Electra Glide Classic Ultra - 82ci Evo V-twin fuel injection, five-speed
98FLHTPI Electra Glide Police EFI - 82ci Evo V-twin fuel injection, five-sped
Total Production - 150,000

Colours

Single tones: Vivid Black, Lazer Red Pearl, Sinister Blue Pearl, Violet Pearl
Two-tones: Lazer Red & Black Midnight, Red & Champagne Pearl, Sinister Blue

Pearl & Platinum Silver, Mystique Green & Black
Specifications
XLH883/1200 Sportsters - New rear light, new oil pump (increased capacity), electronic speedometer (XLH883 only), integrated timer sensor/ignition module (except 1200S), 1200S gets 15% torque increase from 4-pole coil, 2-plug heads, 10:1 compression, less restrictive exhaust, new camshaft, new ignition
XL53C Sportster Custom - New model, based on basic 883 with chrome details, bullet headlight, disc 16-inch (406-mm) rear wheels and spoked 21-inch (533-mm) front wheel, twin seat
FXD/DL/CONV/WG - Nine-plate clutch, new tail light and neutral switch
FXSTC/STS/STF/FLSTS - Nine-plate clutch, neutral switch
FXSTB Night Train - New model, based on FXSTC, Europe-only for the first year. All-black Evo V-twin, black theme for oil tank, belt guard, mirrors etc, drag bars, forward control, minimal seat, wide front forks, 21-inch front wheel, 16-inch disc back rear wheel
FLHCRI Road King Classic - New model, based on FLHR, with leather saddlebags and slant-cut silencers
FLTRI Road Glide - New model to replace Tour Glide, comes in stripped-down form with no Tour-Pak, saddlebag guard rails or fender bumpers, new fairing, low windscreen, new seat, cast wheels, new sound system, gauges for fuel, ambient temp, volt, oil pressure
All FLs - Nine-plate clutch, upgraded cruise control, neutral switch, new sound system
95th Anniversary models - XL1200C, FXDWG, FLSTF, FLSTS, FLHTCI, TLHTCUI, FLHRCI, FLTRI - special graphics, individually numbered plates, Two-tone Midnight Red/Champagne Pearl

1999

The Range
99XLH 883 Sportster - 54ci (**883cc**) Evo V-twin, five-speed
99XLH 883 Sportster Hugger - 54ci Evo V-twin, five-speed
99XL53C Sportster Custom 53 - 54ci Evo V-twin, five-speed
99XLH 1200 Sportster - 73ci (**1203cc**) Evo V-twin, five-speed
99XL 1200C Sportster Custom - 73ciEvo V-twin, five-speed
99XL 1200S Sportster Sport - 73ci Evo V-twin, five-speed
99FLSTC Heritage Softail Classic - 82ci (**1340cc**) Evo V-twin, five-speed
99FLSTF Fat Boy - 82ci Evo V-twin, five-speed
99FLSTS Heritage Springer - 82ci Evo V-twin, five-speed
99FXD Dyna Super Glide - 88ci (**1450cc**) Twin Cam V-twin, five-speed
99FXDX Dyna Super Glide Sport - 88ci Twin Cam V-twin, five-speed
99FXST Softail Standard - 88ci Twin Cam V-twin, five-speed
99FXSTB Night Train - 82ci Evo V-twin, five-speed
99FXSTS Springer Softail - 82ci Evo V-twin, five-speed
99FXDL Dyna Low Rider - 88ci Twin Cam V-twin, five-speed
99FXDWG Dyna Wide Glide - 88ci Twin Cam V-twin, five-speed
99FLHT Electra Glide Standard - 88ci Twin Cam V-twin, five-speed
99FLHR Road King - 88ci Twin Cam V-twin, five-speed

99FLHRCI Road King Classic - 88ci Twin Cam V-twin fuel injection, five-speed
99FLTRI Road Glide EFI - 88ci Twin Cam V-twin fuel injection, five-speed
Total Production - 170,000
Colours
Single tones: Vivid Black, Lazer Red Pearl, Aztec Orange Pearl, Cobalt Blue Pearl, Diamond Ice Pearl, Nugget Yellow Pearl, Scarlet Red Pearl, White Pearl
Two-tones: Sinister Blue & Diamond Ice, Aztec Orange & Diamond Ice, Mystic Green & Black, Lazer Red & Black
Specifications
XLH883/1200 Sportsters - No major changes
XL53C - Longer front forks, drag bars and forward controls
XL1200C - Drag bars, forward controls
FXD/DL/WG - Twin-Cam 88 V-twin, 3.74 x 4in (95 x 102mm), 88ci, carburettor, rubber-mounted, claimed more power/torque, better reliability, stiffer engine/gearbox joint, redesigned lubrication system, quieter
FXDX - New model, based on FXD. Dyna chassis with Twin Cam 88 V-twin in all-black form, black XR750-style handlebars, twin front discs, spoked wheels, raked forks
FXSTS/STB/FLSTF/STS/FTC - New speedometer to allow on-board engine diagnostics
FLHR/CI/CUI/FLTRI - Twin Cam 88 V-twin for all the tourers, integrated rear brake reservoir and master cylinder, speedometer to allow on-board engine diagnostics (FLHR/CI only)

2000

The Range
00XL883 Sportster - 54ci (**883cc**) Evo V-twin, five-speed
00XL883 Sportster Hugger - 54ci Evo V-twin, five-speed
00XL53C Sportster Custom 53 - 54ci Evo V-twin, five-speed
00XL1200C Sportster Custom - 73ci (**1203cc**) Evo V-twin, five-speed
00XL1200S Sportster Sport - 73ci Evo V-twin, five-speed
00FLSTC Heritage Softail Classic - 88ci (**1450cc**) Twin Cam 88B V-twin, five-speed
00FLSTF Fat Boy - 88ci Twin Cam 88B V-twin, five-speed
00FXD Dyna Super Glide - 88ci Twin Cam V-twin, five-speed
00FXDX Dyna Super Glide Sport - 88ci Twin Cam V-twin, five-speed
00FXDL Dyna Low Rider - 88ci Twin Cam V-twin, five-speed
00FXDWG Dyna Wide Glide - 88ci Twin Cam V-twin, five-speed
00FXST Softail Standard - 88ci Twin Cam 88B V-twin, five-speed
00FXSTB Night Train - 88ci Twin Cam 88B V-twin, five-speed
00FXSTD Softail Deuce - 88ci Twin Cam V-twin, five-speed
00FLHR Road King - 88ci Twin Cam V-twin, five-speed
00FLTRI Road Glide - 88ci Twin Cam V-twin fuel injection, five-speed
00FLHRCI Road King Classic - 88ci Twin Cam V-twin fuel injection, five-speed
00FLHTCUI Electra Glide Classic Ultra - 88ci Twin Cam V-twin fuel injection, five-speed
Total Production - 203,000 (projected from Sept. 2000)
Colours
Single tones: Vivid Black, Chrome Yellow, Aztec Orange Pearl, Cobalt Blue Pearl, Luxury Rich Red, Bronze Pearl, Diamond Ice Pearl, Sinister Blue, Concord Purple, Suede Green Pearl, White Pearl,
Two-tones: Luxury Rich Red/Vivid Black, Sinister Blue/Diamond Ice, Aztec Orange/ Diamond Ice, Vivid Black/Chrome Yellow, Suede Green/ Vivid Black
Specifications
XLH883/1200 Sportsters - Four-pot front disc brake, sealed wheel bearings, sealed battery, stiffer crankshaft, redesigned tappet blocks, new transmission internals (smoother gearchange). XL53C and 1200C have 3.7-gal (16.8-litre) Fat Bob fuel tank
FXD/DL/WG - Four-pot front disc brake, sealed wheel bearings, sealed battery, new drive belt, new transmission internals, redesigned belt guard, swinging arm, fuse system
FXDX - As above, plus cartridge-type front forks with adjustable pre-load, rebound and compression damping, gas-charged rear shocks with dual rate springs and adjustment for pre-load and rebound damping, wider tyres, redesigned seat, cast alloy wheel option, new sidestand
All Softails - New Twin Cam 88B V-twin, with twin counter-rotating balancer shafts to reduce vibration
FXST/B/D/FLSTF/STC - Four-pot front disc brake, sealed wheel bearings, sealed

OPPOSITE:
LEFT: The 1999 Electra Glide Ultra Classic, complete with its new Twin Cam V-twin. Despite the name, the latest Harley twin does not have twin overhead cams, though it is a clear advance on the Evo. A balance shaft version is soon to follow.

RIGHT: King of the Retros: the Springer Softail (this is a 1988 version) really was the ultimate nostalgia machine, combining the Softail rear end with an updated version of Bill Harley's original 1905 springer front fork. It didn't work as well as a modern telescopic fork, but does that really matter?

BELOW: The five-speed 2001 FXSTS Springer Softail with 88ci (1450-cc) Twin Cam 88B.

battery
FXSTD - New model, based on FXST. Stretched 4-gal (18-litre) tank, chrome front end, disc rear wheel, new rear mudguard
FLSTF - Repositioned gear lever, redesigned exhaust (easier removal) and rear mudguard
FLHR/CI/TCUI/FLTRI - Four-pot front disc brake, sealed wheel bearings, sealed battery, new drive belt, new transmission internals, 9-spoke cast alloy wheels, new frame (easier access to engine top end)

2001

The Range
01XL883 Sportster - 54ci (**883cc**) Evo V-twin, five-speed
01XL883 Sportster Hugger - 54ci Evo V-twin, five-speed
01XL53C Sportster Custom 53 - 54ciEvo V-twin, five-speed
01XL1200C Sportster Custom - 73ci (**1203cc**) Evo V-twin, five-speed
01XL1200S Sportster Sport - 73ci Evo V-twin, five-speed
01FLSTC Heritage Softail Classic - 88ci (**1440cc**) Twin Cam 88B V-twin, five-speed
01FLSTCI Heritage Softail Classic EFI - 88ci Twin Cam 88B V-twin fuel injection, five-speed
01FLSTS Heritage Springer - 88ci Twin Cam 88B V-twin, five-speed
01FLSTF Fat Boy - 88ci Twin Cam 88B V-twin, five-speed
01FLSTFI Fat Boy EFI - 88ci Twin Cam 88B V-twin fuel injection, five-speed
01FXD Dyna Super Glide - 88ci Twin Cam V-twin, five-speed
01FXDX Dyna Super Glide Sport - 88ci Twin Cam V-twin, five-speed
01FXDXT Dyna Super Glide T-Sport - 1,449cc Twin Cam V-twin, five-speed
01FXDL Dyna Low Rider - 88ci Twin Cam V-twin, five-speed
01FXDWG Dyna Wide Glide - 88ci Twin Cam V-twin, five-speed
01FXST Softail Standard - 88ci Twin Cam 88B V-twin, five-speed
01FXSTB Night Train - 88ci Twin Cam V-twin 88B, five-speed
01FXSTD Softail Deuce - 88ci Twin Cam 88B V-twin, five-speed
01FXSTDI Softail Deuce EFI - 88ci Twin Cam 88B V-twin fuel injection, five-speed
01FXSTS Springer Softail - 88ci Twin Cam 88B, five-speed
01FLHR Road King - 88ci Twin Cam V-twin, five-speed
01FLTRI Road Glide - 88ci Twin Cam V-twin fuel injection, five-speed
01FLHRCI Road King Classic - 88ci Twin Cam V-twin fuel injection, five-speed
01FLHT Electra Glide Standard - 88ci Twin Cam V-twin, five-speed
01FLHTCUI Electra Glide Classic Ultra - 88ci Twin Cam V-twin fuel injection, five-speed

Colours
Single tones: Vivid Black, Luxury Blue Pearl, Jade Sunglo Pearl, White Pearl, Luxury Rich Red Pearl, Chrome Yellow Pearl, Real Teal Pearl, Concord Purple Pearl, Bronze Pearl, Diamond Ice Pearl, Suede Green Pearl
Two-tones: Luxury Blue & Diamond Ice, Luxury Rich Red & Black, Vivid Black & Chrome Yellow, Concord Purple & Diamond Ice, Real Teal & Birch White, Vivid Black & Chrome Yellow, Suede Green & Black

Specifications
All Sportsters - New oil pump, Spinylok cylinder liners to reduce oil consumption
XL883 Sportsters - 7% power increase from revised induction/exhaust systems
All Dyna Glides - New fuse panel, easier to use side stand, improved gearbox internal shifters, factory fitted security/immobilizer system, Dunlop H-D tyres
FXDL Low Rider - Lower seat (from shorter rear shocks), low-rise bars on pull-back risers
FXDWG Wide Glide - New seat and backrest
FXDXT T-Sport - New model, using Twin Cam 88B balance-shaft V-twin, twin front discs, adjustable front and rear suspension, fork-mounted fairing, adjustable screen, ballistic nylon saddlebags
All Softails - New engine diagnostics, factory fitted security/immobiliser systems, EFI option on Fat Boy, Deuce and Heritage Softail Classic
FLH/T - Improved gearbox internal shifters, factory fitted security/immobilizer system, new engine diagnostics, Dunlop H-D tyres, metal air dam below steering head removed to improve hot weather comfort

ABOVE: The latest 2001 Springer Softail, with an updated version of Bill Harley's original 1905 springer front fork. This doesn't work as well as a modern telescopic fork, but the buyers don't seem to mind.

OPPOSITE:
Who else but Harley could get away with a bike called Fat Boy?

SALOON
PROUD TO BE YOUR BUD.
BULL O' THE WOODS
Satellite
Big Screen T.V.

HARLEY-DAVIDSON LIFESTYLE

What then is the work of Life? What the business of great men, that pass the stage of the world in seeming triumph as these men, we call heroes, have done? Is it to grow great in the mouth of fame and take up many pages in history? Or is their business rather to add virtue and piety to their glory, which alone will pass them into eternity and make them truly immortal? Thus wrote Daniel Defoe, 'On the Instability of Human Glory'.

It is unlikely when William S. Harley and Arthur Davidson of Milwaukee, Wisconsin produced their first motorcycle in 1903, that they had any notion of how famous they would eventually become, particularly as there is a debate as to their initial intentions. Did they produce the bike entirely for themselves, or was it to be the first of many, and the start of a serious business venture?

Whatever the truth, even when they began to receive requests for bikes, they would have had no idea that they were founding a dynasty and creating a legend that would become a 21st-century icon.

That first Harley-Davidson, like those that were to follow, was built to be strong and reliable on roads that were far from ideal for rudimentary forms of motorized transport. However, that initial policy of strength and reliability was to serve them in good stead and enable Harley and the Davidsons to stay in business as other firms appeared and then went to the wall.

LEFT: Riding into the wind.

OPPOSITE: Sturgis: riding out to Wyoming and the Devil's Tower.

In 1909 when the first V-twin was added, the beginning of the legend was consolidated. 'The Motor Company', as it was known in-house, has been faithful to the same basic format ever since, through thick and thin, good times and bad, because it has stuck to its initial principles. As a result the motorcyclists, Harley's buying public, have remained faithful in their turn.

In this way, a bond has been created and a fierce brand loyalty has developed that has spilled out onto the social scene and society in general, creating a unique

lifestyle to evolve around the name Harley-Davidson, which is probably one of the best known the world over. People have it tattooed across their bodies; they hang posters on their walls; and apply the emblem wherever there is space. They even buy books on the subject!

And this doesn't only apply to motorcyclists. There are those with no interest whatsoever in bikes with whom the name Harley-Davidson will strike an immediate chord and, if asked to name a bike, is the first that would spring to mind. Watch any disinterested bystander when a Harley booms into view: they turn, they look and stand, mouth open, as the bike heads out of focus into the far horizon. This does not happen with a Honda step-through.

Harley-Davidson has come to stand for the American way of life as it is lived on two wheels. It may only be a bike, but to most it offers a sense of freedom that many feel is missing from their busy lives. Maybe it is this factor more than anything else that has led to the advent of the weekend bikers, or 'rebels with a cause'.

On first sight, fake henna tattoos and stick-on ponytails may make them look different, but a second glance will confirm that they are also wearing smart tailored leathers: they do not want their true status and affluence to go unnoticed. The Harley has become a style icon for the 'rubbies', RUBs or rich urban bikers, to give them their full title. They may drive an executive car from Monday to Friday, but once the weekend arrives, a top-of-the-range Harley will suddenly appear and the transformation is complete, from city slicker to weekend rebel at the lift of a garage door.

However, at the other end of the scale is what many would call the true biker.

LEFT: The Harley Owners Group (HOG) is their officially sanctioned club, with chapters existing all over the world.

OPPOSITE: Do you feel brave enough to ask this man what he thinks of the latest Honda FireBlade?

These are associated with the patch clubs and would have you believe they are one of the 'one percenters'. For these riders, their Harley is not a leisure tool to be kept under covers during the week; it is an important part of their life, something which defines them and connects them to their associates. This is the motorcycle lifestyle which has been fully embraced.

Both of these groups are at the extreme limits of the Harley lifestyle, one so diverse that it encompasses every facet of society and its products. There are goods, clothing and other items in such abundance that one wonders what their connection is with the world's most famous motorcycle, apart from earning the royalties that tumble into the corporate coffers.

While it is an easy matter to trace the origins of the Harley-Davidson and to chart its progress by means of particular models, race successes, or company decisions, what is far more difficult is to understand how the Harley lifestyle phenomon came into being.

Opinions differ as to what the initial impetus was, but there is one factor which has helped consolidate the Harley image: namely the media. Perhaps the best illustration of this is the infamous Hollister incident of 1947, which many claim to be

HARLEY-DAVIDSON
BIKE WEEK

the beginning of the back-patch gangs and anti-motorcycle propaganda, and which led to one of the film industry's most famous productions, *The Wild One*, starring Marlon Brando.

The truth is that the events at Hollister had their roots in the Second World War, or at least its ending. During the war, many US servicemen learned to ride on a Harley WLA. Freshly demobbed in the mid-1940s, many suddenly found themselves abruptly returned to 'normal' life, and deprived of the comradeship and element of danger that had become a way of life. Those unrestricted by jobs or families naturally drifted, and becoming a bike gang member was something that echoed many of their wartime experiences: the kudos associated with being something of an outsider, and being surrounded with people with whom you could completely identify.

Meanwhile, the American Motorcycle Association (AMA) had decided to resurrect its Gypsy Tour, a race series current throughout the US that took in flat tracks and hill climbs that had been popular prior to the war. With cheap gasoline and nothing much to tie them to any one place, these groups of riders headed for the races to watch and compete. There was little in the way of hotels, the motel concept having not yet caught on, so accommodation and public facilities in general were sparse apart from the big cities. Riders would often camp out or sleep rough in small towns that simply couldn't cope with the influx.

So it was with Hollister, California, a town of just 4,000 inhabitants at the time, whose only claim to fame was that it was America's biggest producer of garlic. On the Fourth of July weekend in 1947 they were playing host to an AMA-sanctioned

ABOVE: A rally enthusiast: every badge tells a story.

RIGHT: Pets are allowed at Sturgis, so now you can see the point of those broad Fat Bob tanks.

OPPOSITE: Flawless tans and big hair feature prominently at Sturgis.

Cruising Daytona: it doesn't have to be along the beach.

had been heavy-handed hardly mattered. Something big was going on, and when the editor of the *San Francisco Chronicle* got wind of it a reporter and photographer were sent out by chartered plane. (Some say he was tipped off by a police radio dispatcher, while according to others, they arrived merely by chance!)

However, by the time they arrived it was all over, but having put in so much effort to get there, they persuaded one of the revellers to pose, beer bottle in hand, astride his Harley outside Johnnie's bar, which had been the centre of the action. It was to become one of the most famous shots ever when it appeared all over the US, in newspapers, magazines and eventually in books. Following the report and picture appearing in the *Chronicle* it was also picked up by *Life* magazine who used it as a front cover shot. All this served to exaggerate the events at Hollister which began by being described as a riot, and ended with a quote from the police chief saying 'It's just one hell of a mess', which somewhat strengthened the story.

Within a week, everyone riding a motorcycle was typecast a fat overweight, bleary-eyed 'Oakland Boozefighter', out to

LEFT: Many people still ride to Daytona, stopping for fuel and refreshment along the road.

OPPOSITE
LEFT: Cruising on a hand-change, foot-clutch Harley in Sundance, Wyoming.

RIGHT: A Harley was the first prize at this casino in Deadwood.

capture every good-looking women they encountered, mounted, of course, on Harley Davidsons: so the biker image was born and was associated with Harley ever after.

This was the true incident that Stanley Kubrick's film *The Wild One* further immortalized in 1953, a young Marlon Brando playing an introspective Johnny Strabler, the leader of an outlaw gang.

In fact, Harleys did not feature prominently in the film (Brando was riding a Triumph), which compared motorcyclists to cowboys. The unfortunate effect was to make the words biker and outlaw synonymous, with Harleys as instruments of terror.

Moreover, this image continued after Hollister, with numerous other reported gatherings of bikers all being tarred with the same brush. The fact that the word 'outlaw' had been applied to them caused the AMA secretary at the time, Lin A. Kuchler to utter a much quoted comment: 'The disreputable cyclists were possibly one percent of the total number of motorcyclists, only one percent are hoodlums and troublemakers.'

This statement, despite it being made with the best of intentions to support the average rider, caused resentment among many. As a result, the 1% symbol came to be worn as a badge of honour and gangs like the Hell's Angels, Road Rats and Satan's Slaves united under this banner.

The AMA continued with its damage-limitation policy, but by now it was too late, and the anti-motorcycle lobby was in full cry, while some of these 'one percenters' continued to fuel the fire with anti-social behaviour and reckless riding. Meanwhile, Harley-Davidson attempted to distance itself from the trouble by issuing statements, but privately the company knew that this somewhat negative interest

LEFT: Ready to go!

ABOVE: There are no age limits when it comes to riding.

OPPOSITE: Hunlett, Wyoming, where the queue for the Ponderosa outstrips the queue of bikes.

LIQUOR
PONDEROSA
CAFE BAR
Coors

in motorcycles was doing it no harm at all, and was helping it to sell bikes at a time when business was slack.

Japanese manufacturer Honda eventually spent a great deal of money on an advertising campaign claiming that 'You meet the nicest people on a Honda'. This did something towards changing public opinion and softened the motorcycling image to a certain degree. It also caused Harley to reassess its position. As the 1960s progressed, a new breed of rider was appearing who liked the style, but required a clean image rather than a criminal record.

Now to be seen outside the smartest cafés and shopping in town for designer clothes, motorcycle pilots began to wonder why it was necessary to take long swinging kicks to get a magneto to breathe life into a cold iron engine. Instead they required a quick press of a button to avoid breaking into a sweat or creasing their designer jeans. They wanted to be able to 'head out on the highway looking for adventure', as Steppenwolf would have said, but stay in smart hotels and motels along the way. This was the start of the weekend biker, and from this time forward Harley had, not only to market a bike, but an image as well. What follows is just a small look at what that involved and how the bikes came to appear along with all the other associated items and events that make up the 'Harley Lifestyle'.

Cruising is an integral part of big events like Sturgis and Daytona. Of course, people go there to meet old friends, make new ones, see the displays and watch the racing. But for many, the main reason for going is the cruise, not only to pose, but also to watch others do the same thing. It's the ultimate in audience participation, for anyone can climb on their bike and be an instant part of the spectacle.

LEFT: Just to prove you were there!
ABOVE: This is what they came to see: Devil's Tower, Wyoming.

LEFT: An Angel rides his Harley chop through an English autumn landscape.

BELOW: Stay cool and look mean.

OPPOSITE
LEFT: Sonny Barger with his own FXRS.

RIGHT: Don't be fooled, by Monday morning he'll be back in suit and tie.

You can almost smell the surf! The gathering of the faithful on Daytona beach.

MANGIA
PIZZA
248-2510
SICILIAN PIZZA
ITALIAN SUBS
Outlaws
mc
Milwaukee
Florida
Sunset

OPPOSITE: One percenters from Florida, at Daytona.

LEFT and ABOVE: Conflicting images at Daytona: sexist tattos can exist alongside women riding their own bikes and being anything but submissive.

200 miles (322km) took place in 1937, with 120 riders racing on the part-road, part-beach track of 3.2miles (5km). While the field was liberally laced with Harleys, it was actually one of their rivals, an Indian, that took the honours in that inaugural meeting. It was such a success, that it became an annual event, shelved during the war, then restarted in 1947. The only problem was that races could only be run according to the vagaries of the tide, so in 1960 the last beach race was run (won by Brad Andres on a Harley) and the event moved 10 miles (16km) inland to a purpose-built stadium covering 450 acres (180 hectares) on the outskirts of the city. The first race here was also won on a

Harley ridden by Roger Reiman, whose name was to become synonymous with the marque.

Times were changing, however, and Harley was only ever to take the chequered flag another five times before the Japanese brands began to dominate, not only the classic 200-mile race, but also all the other races at Daytona.

But the coming of the Japanese was far from being the end of Harley-Davidson influence at Daytona – quite the reverse. While the racing had migrated inland, 'Bike Week', a massive seven-day meeting of like-minded bikers, remained on the narrow strip of land adjacent to the beach. It became an independent motorcycle scene, grew up, became one big party for bikers in general, and centred itself on Main Street, right in the centre of the spit. It hardly mattered that the beach racing had stopped: riders from all over the US would make it their home for a week, even while much of the rest of the US was covered in snow and ice. The vast majority never even set foot in either the main banked tri-oval circuit or, for that matter, the Municipal Stadium further down the road to watch the short-track racing.

Instead this was a chance for many to ride the 12-mile (19-km) stretch of the beach, hang out on Main Street, sink a few beers and just chill out. In 1973 the embryo 'Rat's Hole' custom show started and a year later 15,000 turned up to see the bikes on display, at a time when the police chief was expecting a third of that number.

This was also the first year that any trouble was recorded after the publication of the new motorcycle regulations for the city of Daytona. When these somewhat draconian laws were enforced, a minor backlash occurred resulting in a number of arrests, which the police were quick to play down.

Meanwhile, manufacturers like Harley-Davidson were realizing that Daytona was becoming much bigger than a mere race week and looked to setting up shop. Along Main Street, customizers and accessory suppliers took up short-term leases and shops and the crowds rolled in big time. In 1976, Harley took over a hotel for its exhibition and what with Harley drag races just down the road, the city was teeming with evidence of the marque: Daytona had become a second home for Harley-Davidson. And it had become an international event, with Harley riders arriving from all over the world. But matters were getting somewhat out of hand: the police turned a blind eye to the odd spot of impromptu beach racing, but in 1977 gang warfare broke out and a few shootings took place.

The following year all was quiet, but from then onward everything began to slide downhill. There were shoot-outs in bars and the one percenter element was rife. Gradually the police adopted a zero tolerance policy as visitors were encouraged to stay out of town by an element intent on breaking the week's annual monopoly of the American motorcycle scene. Many riders already felt they were being ripped off by hotels and traders in Main Street and needed little encouragement to stay away or move to the outlying towns. Custom bikes had to be hidden as they contravened the local laws, forcing the custom shows underground, and the whole Bike Week phenomenon was heading for collapse.

This occurred when a biker failed to stop for a police road block and was shot and killed. In court it emerged that the police were being trained and commanded to 'shoot to kill' bikers if there were any signs of misbehaviour, and the revelation resulted in the police chief and district attorney having to resign.

The new police chief adopted a different approach: he met the locals, worked out a plan of action and Bike Week returned to its former life as a party venue, but one which was now under control. However, much of its spontaneity was lost: it became too organized, even orchestrated, and was certainly over-commercialized.

But all this has not deterred over half a million flocking to Daytona every year for some winter sun.While there are now a few other makes appearing in the massive line-up and parades of bikes (even some Japanese sports bikes), Daytona is still very much Harley-dominated during Bike Week.

Standard, customized, chopped, it matters little, the parking lots and Main

OPPOSITE: Out for a quiet Sunday ride: a procession of riders at Daytona.

ABOVE: The classic Daytona image, and a classic bike to go with it.

Street reverberate to the thunder of tons and tons of Harley metal glistening in the sun as it rumbles by.

Policing is now a low-key affair that manages to keep a lid on things without upsetting the visitors. Arrests and citations are few and the main problem now revolves around traffic flow and the increasing number of accidents. This is inevitable considering the vast number of bikes that now criss-cross the city and run up and down the state visiting the various other attractions.

While racing may have been the start of the Daytona happening, it is now just a sideshow along with the other events that go to make up the week. For Harley-Davidson, meanwhile, Daytona is still a big event in the calendar, and it has taken over the massive Ocean centre for putting on displays, where test rides are run, and the end parade finishes at Harley Heaven. This is on the last Sunday of the week and signals the end of the partying before the town is inundated by the 'spring breakers' (vacationing students), who the police now regard as more troublesome than the motorcyclists. How times change!

RIGHT: A night scene: Harley-Davidson doesn't exactly dominate Daytona over Bike Week, but it comes very close to it.

OPPOSITE LEFT and RIGHT: Sturgis, where Main Street serves the same purpose as Daytona beach. It's the place to see and where one can be seen. But ride a couple of blocks away, and you can forget the massive bike event a short distance away.

Sturgis Bike Week

The Sturgis Bike Week, or to give it its correct title, the Black Hills Motor Classic, is probably the biggest bike event in American, after Daytona. It did not, however, grow out of racing, so it is probably the biggest bike gathering of its type anywhere. In fact, the town has the honorary title of Bike City USA – for the month of August at least!

It's certainly the best known the world over and is seen by many as a predominantly Harley gathering; but when you look back at its origins, Harley-Davidson was not part of the initial scene at all. A local Indian motorcycle dealer by the name of Clarence J.C. 'Pappy' Hoel started it in 1938, along with friends from his motorcycle club, the Jack Pine Gypsies.

Motorcycle stunts predominated in those early meetings, Hoel being something of a performer himself. These ran alongside AMA-sanctioned races, which were part of the reformed Gypsy Tour mentioned earlier. From this evolved several small events all around the locality, which all intertwined and became part of a ten-day happening that now sees 400,000-plus people arriving in the romantic Black Hills of South Dakota.

Although given the generic name of Sturgis, while the event was on, the actual town of Sturgis was the domain of the AMA breed of clean-cut racing types. Down at Deadwood, a small town 10 miles (16km) away, a different type of visitor was setting camp, and the town became the preserve of outlaw gangs: eventually, however, the two factions met and merged as they sought out things to do and places to visit. What with Harley-Davidson basing itself in Rapid City just down the road, the whole scene started to expand, with riders doing the rounds of

COLE'S
CENEX
STOP

Smiley's Tap is another of the perennial favourites at Daytona.

The alarmingly named Road Kill Café at Sturgis, next door to the while-U-wait tattoo and body-piercing parlour.

towns in the area until the event became known as the Sturgis Rally.

In the beginning, however, it was very much a local event, quite a small biker meeting in the summer and almost an invitation-only gathering. It remained this way until the end of the 1970s, when a major article in the bike press gave it a glowing report: TV cottoned on, featured Sturgis and the rest is history with thousands flocking to be part of it, which in turn encouraged Harley-Davidson to establish an official presence.

South Dakota has some of the best biking roads in the US, so this all dovetailed nicely. Harley put its efforts into displays at Rapid City where there was a big civic centre at its disposal. This gave riders a 30-mile (48-km) blast along the Interstate 90 from Sturgis and they began to make the event their own by running events and promotions.

But like Daytona there was a period of big trouble, which took some of the shine off the event. Again, like Daytona, things are much calmer these days, but the legacy remains of a heavy law enforcement presence that is firm but fair. Now, the greatest problem is overcrowding, pure and simple. In 1947, 400 people took part in the Black Hills – now it is more like 400,000, and people often have to stay out of town once the local accommodation and campsites have been filled. Like Daytona there is Main Street where bikes are parked five-deep down its half-mile length. Parties take place at virtually every campsite and there are unofficial competitions to go with the official hill climb that is still run as a throwback to those early days. Drag racing, as one might imagine, is rife, both organized and unofficial!

For many, however, it is the strange mix of outlaw bikers and weekend rebels that attracts them to Sturgis. While the lid

OPPOSITE: Cruising down Main Street is one-way, and done very slowly.

ABOVE: People, bikes and sideshows sum up the Sturgis scene.

ABOVE RIGHT: Just married: this well dressed couple at Sturgis have just tied the knot – on a bike!

remains firmly on, some of the partying, which continues day and night, is bordering on the obscene. Nothing seems to be too crazy or out of the way and people do not seem to mind cameras or videos recording their antics!

One thing is for sure: there is plenty for everyone, with 124 events recently recorded over the ten days. These are watched by visitors from 50 US states and 17 foreign countries, and take place day and night, emphasizing the fact that there is no time whatsoever to sleep.

Whatever the locals may privately think, Sturgis is big business and brings in a level of currency that would be hard to replace by means of normal tourism, no matter how picturesque the area is.

Bike City USA is safe for the foreseeable future and will predominantly remain the province of the fans of Harley-Davidson.

Laconia

Once upon a time the name Laconia stood alongside Daytona and Sturgis. Now efforts are being made to return the New England town to popularity with the Harley fraternity with the heavy promotion of Laconia Bike Week. First started in 1923, the event traditionally takes place during the beginning of June. It was based on racing (like Daytona), which had been actually been taking place since 1917. These were primitive races and events and it was not until the 1930s that AMA-sanctioned races began.

In 1990 the New Hampshire International Speedway Stadium was built near the town of Loudon, which confuses some, as the Laconia event is also sometimes known as Loudon. Similarly, the hill climb run at Mount Belknap also comes under the generic Laconia title. In the heart of the town (pop. 16,000) is Wiers Beach, a dilapidated run-down amusement mile on the shore of the lake. Here during Bike Week it becomes home to the bikers and unfortunately has been the scene of some quite serious rioting over the years. This stems back to the time of Hollister when it was hushed up, but it has led to the cancellation of the event at least once, in the 1960s. However, after some serious aggravation took place during the 1980s, the police took their revenge at the start of the 1990s without actually stopping the event.

Heavily restrictive traffic orders were implemented and deliberately caused

As night falls, it's business as usual on Main Street. This is one balmy evening in 1998.

If it gets too crowded in town, Sturgis is surrounded by undulating roads filled with other bikes.

massive traffic jams around the area. It became an offence to switch off engines, punishable by on-the-spot fines, with the result that engines overheated and died. Having proved their point, the authorities concentrated their efforts into organizing events to keep visitors happy and deter them from contemplating making trouble.

It has worked and now 200,000-plus make the trip to watch concerts, take part in competitions, attend bike shows and customs and watch the races in a peaceful, convivial atmosphere. Maybe Laconia can still sit alongside Sturgis and Daytona as one of the big three events of the year.

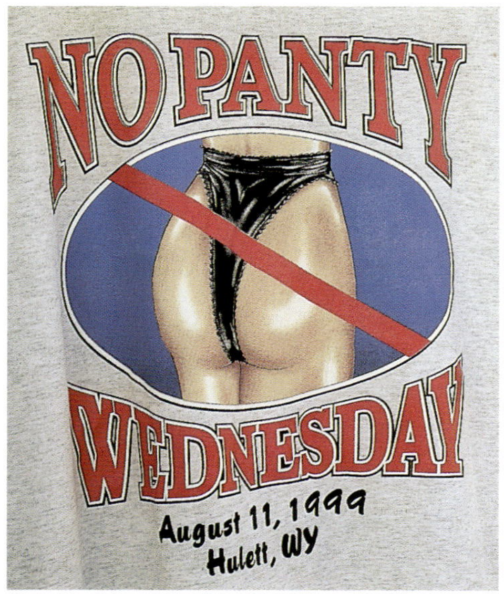

Just like every other big event, there's a great deal of variety at Sturgis. This line-up of custom bikes is presented by Mike la Fores and friends, while others prefer to take it easy on standard bikes. However, it's best not to inquire too deeply into 'No Panty Wednesday'.

OPPOSITE: Sturgis nightlife.

COURTESY PARKING
OASIS
GOLD PAN PIZZA
GOLD PAN PIZZA
SWA
Big BaD BeRT
STOP

CUSTOMIZING

Motorcycling is very much an individual affair and for some a solitary pastime. Whatever it is, most riders usually have their own strong opinions and are not afraid to voice them or stand out from the crowd. While for some, the standard model Harley in the same trim as it left the factory is sufficient to satisfy their needs, others regard this as a starting point. These are the customizers who just cannot resist altering their bike's appearance or performance.

What does 'customizing' actually mean? The precise definition is to personalize, which is achieved by changing various components to suit the individual's taste. This may simply be a case of a new paint scheme, but for others it may be more radical, involving totally altering components, reducing the bike's weight, or making it fit them personally in an effort to make it more comfortable to ride. And for some reason, Harleys seem to be the most popular bikes for this purpose.

There are, of course, very practical reasons why this is so. Harleys, unlike many modern machines, can be worked on with ease. They are not as mechanically complicated as others and require fewer special tools for the job. Also, because they are built and designed with no end life, every part is replaceable

So parts are easily obtainable, and the cost of making changes, often a huge factor, can be spread over a long period of time. In other words, one can do small amounts at a time without having to strip the whole bike from top to bottom. Lastly, of course, there is the major fact that the engine is a magnificent lump of metal and the sound of the V-twin motor makes the whole effect complete. With most custom bikes, the top speed is immaterial, unless it is designed as a race bike; but even then there are sufficient tuning parts available to make it a viable part of the rebuild as well.

Custom bikes, from the excesses of the 1970s to today's exotic and varied creations, have become a permanent feature of the motorcycle scene. But the modern custom movement didn't get going until after the Second World War. As already discussed, the end of the war saw many GIs returning to America with the desire to ride a bike. Ex-army bikes were plentiful and cheap, but they were heavy and cumbersome. The cheapest and easiest way to make them faster was to cut the

ABOVE:
'Shovel Trouble' was built in the UK.

OPPOSITE: A high-tech chopper. On the face of it, this bike is simply repeating a 30-year-old style. But look more closely, and innumerable details mark it out as a modern custom.

weight, so certain parts were removed and consigned to the trash can.

It was this practical weight-saving exercise that led to the verb 'to chop' entering the biker's vocabulary – hence the 'chopper', used to refer to bikes with long raked out front ends, that were built out of machines and parts that had been chopped. Some maintain the name comes from chopping the frame to alter the steering geometry, but the end result is the same.

But like most things, chopping isn't as simple as it seems, as removing one part can make a bike appear lopsided or unbalanced; so owners began to build bikes from scratch rather than modifying a piece

ABOVE:
'Has Advantages', another modern chopper, with belt primary drive and twin front disc brakes, though the low seat/high bars look is an enduring one.

OPPOSITE: A mid-1970s Ironhead Sportster chopper with prism-shaped tank.

here or there as the fancy took them. This means *carte blanche* with surfaces, and ever more elaborate paintwork, murals and chrome began to make an appearance. Neither was it just cosmetic, with raked forks and high bars getting ever more radical year on year. Peter Fonda's classic chopper in *Easy Rider* inspired a generation of riders to do the same to their bikes, of whatever make, but during the 1970s there was a suspicion that it had all gone too far. Specialized custom shows encouraged people to build bikes that were designed for showing only, rather than riding: some custom bikes were becoming less and less practical for the road – more modern art than motorcycle – and some even had no internal engine.

While these initial efforts were nothing more than attempts to outdo the next guy, shrewd machinists and sprayers realized that a business opportunity had presented itself to make parts for others less skilled than themselves. As a result, a multimillion-dollar industry has resulted, allowing people to alter parts here and there by replacing components using the existing mountings.

Bolt-on customizing had arrived, and it is still an important part of the movement today, and of which Harley-Davidson takes full advantage. Early on, Harley realized it was its own bikes that were receiving all the attention. So it too began marketing its own performance and tuning parts as well as the more normal accessories like bags and screens. Now, the Harley-Davidson range of accessories and Screamin' Eagle tuning parts is vast, to rival any of the after-market manufacturers.

However, for the committed customizer, merely buying the part and bolting it on is not an option. They have to make as much of the bike as possible, in order to end up with something that is unique.

Another alternative, albeit an expensive one, is offered by the firms that will build a custom bike from scratch, to the buyer's own specification, when he chooses from various components in a catalogue. Some of these firms work on the sound principle that people do not want to buy a new bike, then discard half of it and spend thousands on new parts.

What they like to do instead is build something different from the start. Companies such as Titan, Big Dog Motorcycles, American Eagle, and American Legend build up the bike from scratch. Most use the S&S motor and frames built by various manufacturers, but put their own names on the tank. These tend to be classed as clones, as a quick uneducated glance would mark all of them down as Harley-Davidsons. In reality, not a single part will bear the Harley stamp.

ABOVE: A Ron Simms paint job, with skull and flames a favourite theme.

OPPOSITE: The ever-innovative Arlen Ness (left) with 'Convertible', one of the new breed of 'bodywork bikes'.

ACCEL
ACCEL
ACCEL
ACCEL
HARLEY-DAVIDSON
NAVY
USA

Not the real things, but replicas of the bikes Dennis Hopper and Peter Fonda rode in *Easy Rider*.

'Jesse James' is the name of this bike, the flames and staggered pipes giving a vivid impression of movement even when it's standing still.

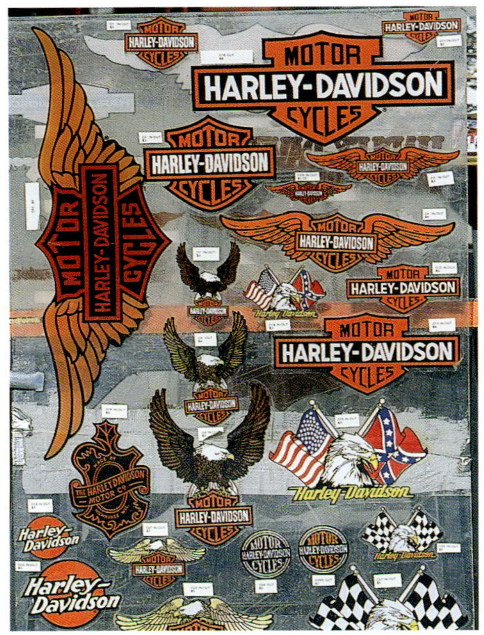

Arlen Ness is undoubtedly the doyen of commercial customizing, having been involved in painting, modifying and building Harleys since the late 1960s. He now owns a multimillion-dollar business by providing ready-made components, building bikes and advertising them, in the process becoming the acknowledged guru of the custom scene. Based in California, his influence has spread the world over, his catalogue of parts and accessories being almost a wish list for those intending to personalize a bike, and he has provided the inspiration for many who have gone on to create their own masterpieces of motorcycle art.

Custom Sub-Groups

If one thing has distinguished the custom movement in recent years, it's the sheer diversity of bikes now being built. Given that the idea is to build unique, personalized machines, you might think that all of them are different. But within this diversity, it is still possible to find general categories of custom bike, in which the builder may or may not have intended to place it. What follows is a brief guide to the major types.

Choppers - The classic custom bike, typified by long raked front ends as exemplified by Peter Fonda's Captain

ABOVE LEFT: Indian Larry's Custom Shop.

ABOVE: Arlen Ness' 'Ferrari Bike' was inspired by the Italian sports car.

LEFT: If your budget won't run to a professionally-built bike, buy a sticker instead.

OPPOSITE: Arlen Ness' son Cory has followed him into the custom business. This is his 'Y2K'.

Y2K

America. The headstock of the frame is angled at 45 degrees and it normally has long extended forks or, if these are short, high handlebars. Most of these are hardtail frames, without rear suspension. For the well-informed these can be broken down into many sub-categories of chopper, depending on the style.

Full Dressers - These are the large touring bikes that have had just about every part of their exterior covered in extra lights and chrome, not to mention a few extra novelties. The name stems from the 1930s, when motorcycle tourists chose to 'dress' their Harleys or Indians with a screen and panniers, or maybe spotlights. Where paintwork can still be seen, a mural is usually depicted, which will often be lit by multi-coloured lights in the dark. You'll often find a stereo and sometimes a CB radio on the fairing, and nowadays even a personal phone or a satellite navigation system is likely to be there. Their forte is to parade during the hours of darkness, when their presence can hardly be ignored!

ABOVE: A hardtail Evo chopper, with Weber carburettor and neat pipe arrangement.

RIGHT: You need to be dedicated to ride in the classic apehanger position.

OPPOSITE: Something different. A mildly customized dresser: but the closer you look, the more you see. There is the paint job, the aftermarket front discs, as well as billet handlebar grips.

DANGER
600 VOLTS
READY TO RIDE

Rat Bikes - These are different from other custom bikes (which are usually kept painstakingly clean and spotless, as if they had never travelled the road) and are at the opposite extreme, dirty workhorses that make a point of being just that and sport patches, bodges (replaced parts from non-motorcycle sources), and more grease and grime than paint and chrome. The typical rat bike shows signs of having covered thousand of miles with a hard-riding owner, who paradoxically may love his bike as much as anyone else.

But like any other customizers, rat bikers can go to extremes. Some owners now set out to encourage the process of decay, deliberately causing the bikes to rust, creating oil leaks and manufacturing the bodges before leaving the bike out in extremes of weather to create the desired effect and in their eyes making them mobile works of art in their own right.

Add a few dead critters, along with some faded ex-military luggage, and hey presto! you have a bike capable of winning a class at a custom show.

OPPOSITE: A gloriously tatty rat bike: one wonders how it manages to move at all!

RIGHT: A superbly decayed rat bike in all its flea-bitten glory. Is that a trailer behind?

Custom Bikes - This covers just about anything, but there are various classes within the category.

The term Lowrider speaks for itself: a bike that is long and low. This can be further sub-divided into a Taildragger, where the rear fender covers the rear wheel down to ground level. The Streamliner is where most of the bike is enclosed, save maybe the engine. The Classic, is where it is based on classic styling themes from the past, such as the Indian Chief, for example, or the Harley Knucklehead. A Street Dragster, is a long and low machine with a performance engine.

The Café Racer or Warbird - This is a bodywork or styling kit which is fitted to give the impression of a fighter plane. It is made by seat makers Corbin.

Multi-Engine speaks for itself. At least two or more engines are slotted into an extended frame, usually inspired by multi-engine drag-racing bikes – not always useable on the road! The Art Class is where the bike is made to look like a well-known object or represent an advertising campaign. Trikes are again a self-explanatory term, together with the Sidecar classes.

ABOVE LEFT: On some rat bikes, the less you can see of the bike, the prouder the owner will be.

LEFT: A radical Knucklehead chop from Milwaukee Iron.

OPPOSITE
ABOVE BOTH: The Corbin Warbird dramatically transforms the standard Sportster. This one has the classic Harley racing colours of black and orange.

BELOW: The Sportster as modern Café Racer – just to prove it can be done.

The Rat's Hole Custom Chopper Show

Having built a custom bike, most serious builders want to show it off to an admiring audience and pit it against other builders in organized competitions. There are many shows worldwide, but the best known and most prestigious takes place at Daytona each year during Bike Week. For some, the Rat's Hole is the only one worthwhile.

Karl Smith, a local Daytona businessman, known as Big Daddy Rat, started it in 1972. Smith acquired the name Rat when playing football at university in Ohio and the Big comes from his size. In fact he used to sell T-shirts.

The inaugural meeting was held on the parking lot of Johns Family Market. There was so much interest, that it is repeated each year on the last Saturday of Bike Week, when a modest entrance fee is charged. Its popularity grew so quickly that it was moved to the boardwalk in front of the ocean and that year 15,000 people came to view the bikes, which were vying for victory in 44 classes. Nowadays, over 400 bikes come from all over the world to compete, Big Daddy Rat still being one of the main judges.

Since 1986 it has also run at Sturgis and has been linked to various other custom shows throughout the world. The free newspaper, *Ratty Rag*, always gives visitors plenty of information on what is happening in the area and the low-down on the various bikes that scoop the prizes.

Despite being copied the world over, the Rat's Hole Custom Chopper Show is still the number one and one of the few to have Big Daddy's own custom trike, the 'Cheese Hauler' on display.

ROCHFORD
POP. 25
ELEV. 5305

The One Percenters
Perhaps the strongest identification of all, they are the people who love being different, still dress in the style of the 1960s rider and feel they should act in the same way as the people responsible for Hollister.

Having said that, most of them keep themselves to themselves and fraternize only with their own kind. They ride heavily modified bikes and do not conform to society in general.

Stereotypes
Closely allied to the one percenters is the stereotype biker or, as some would call them, the cult bikers. They dress all in black, wear heavy Indian jewellery, have tattoos and behave as though they were real outlaws. In truth, most are hardworking family members who like to pose and escape at weekends. They ride pretty much standard bikes.

The Normal Rider
These form the bulk of Harley riders. They are often family people who work in normal jobs and run a Harley as a means of escaping at weekends. Their bikes are not always new, but they don't care as long as it is a Harley, to which they are unswervingly loyal. Not as heavily into the tattoo/jewellery image as the 'stereotypes', they are just as committed to riding Harley-Davidsons.

RUBs or Rubbies
RUB is short for Rich Urban Biker. This sector comes from the high-earning faction, many of whom have come recently to motorcycling and see it as a leisure activity, like hang-gliding or mountain-biking. They require a bike to match the new car each year and are prepared to spend whatever it takes. Money is not a problem and they ride for relaxation rather than escapism. Because of their spending power, this group, far more than any other, is now Harley's target audience.

Although some profess disdain for the Rubbies, their practice of changing bikes almost every year ensures a constant supply of low-mileage secondhand Harleys onto the market. They keep the public profile of Harley-Davidson high, not to mention making a substantial contribution to company profits.

The Rich and Famous
Harley-Davidson has had a long relationship with celebrity riders who have been keen to associate themselves with the image. It is product placement of the highest order, and a marketing man's dream. In reality, celebrities only ride spasmodically, having to fit riding time into busy schedules. For Harley-Davidson though, they remain one of the best advertisements it can have.

So there you have it, a short resumé of the Harley lifestyle with all its associations and events. It will go some way to explaining what makes these people tick, and why they buy and ride Harley-Davidsons. At the end of the day, however, every rider has his or her own personal agenda, and what matters is that it works for them. Ride Free!

RIGHT: Ten per cent of Harley riders are female – you want to argue?

OPPOSITE: Taildraggers: modern customs with plenty of nods to the past.

L315
EHO

MILITARY HARLEY-DAVIDSON

It was in April 1917 that America actually entered the First World War, although long before this it had been supporting the Allies with supplies and vehicles. Like all the major US bike manufacturers, Harley-Davidson had been fulfilling military contracts that came its way since 1915, so the mechanism was already in place to supply in volume the distinctive drab olive green machines to the armed forces.

The company had also learned valuable lessons concerning military requirements when its bikes, in both solo and sidecar trim, had been part of a home- based military operation. This took place in 1916 after the Mexican government had allowed the US to enter its territory to deal with the problems that it was encountering with bandits, who had been crossing the border and attacking and sometimes killing Americans in adjacent townships.

The force of 20,000 soldiers, under the command of General John 'Black Jack' Pershing, failed to capture the freedom fighter Pancho Villa and his men, but they did gain valuable experience in the use of motorized units, known as 'Mounted Infantry', or 'Motor Mobile Infantry', and they were used where heavier vehicles were unable to go because of the mountainous terrain. Pancho Villa himself was a motorcyclist, owning an Indian, but despite the failure of Pershing's campaign, motorcycles had proved their worth on the battlefield. There was a roughly fifty-fifty split between Harleys and Indians in the unit, with just a few Excelsior machines thrown in for good measure.

On the strength of this, and the huge orders which were beginning to come through from Europe at the outset of the First World War, Indian nearly bankrupted itself and its dealers by turning over virtually all its production to the military, and starving civilian showrooms. It also sold the military machines at almost cost, which may have been laudable for the war effort, but didn't do its cash flow any good.

Harley was rather more cautious, and it was not until the middle of 1917, after America had entered the war, that it actually turned 50 per cent of its production over to military machines. Many of these bikes were shipped abroad, either to the front lines or to be used at bases as supply vehicles, or as part of intelligence-gathering operations.

ABOVE: This 1917 military motorcycle with box sidecar was exported to Russia.

OPPOSITE LEFT: *The Enthusiast*, Harley-Davidson's in-house magazine, was happy to emphasize its part in the war effort. This 1942 cover shows a US Army-specification WLA.

RIGHT: This rare machine, a flathead big twin, was supplied to the British military (hence its sidecar position). Most of the Harleys bought by Britain during WWII were 45-ci (750-cc) WLAs, brought in when its own motorcycle factories were destroyed by the Luftwaffe.

The Enthusiast
A MAGAZINE FOR MOTORCYCLISTS
DECEMBER
1942

ABOVE: Another view of the British outfit, with spare wheel, large solo saddle and hand gearchange.

ABOVE RIGHT: The WLA, with its rifle scabbard to the right of the front wheel, documents box on the left, and siren mounted on the handlbars. The 'L' plate is genuine, for many servicemen learned to drive during the war.

RIGHT: The speedometer may have promised a full 120-mph (193km/h), but a fully laden WLA could barely manage half of that. How would Steve McQueen have fared if he had used this in *The Great Escape?*

A fully-equipped WLA. The changes from the civilian bike were extensive, unlike earlier military machines.

One problem that was readily apparent was the lack of training, not only of the riders, but also of those expected to maintain the bikes and keep them running, often in difficult conditions and sometimes with incorrect equipment. Training schools were eventually established by Thomas Butler, a government representative appointed to solve this problem, who liased with the various manufacturers meeting military contracts and formulated a common strategy. These 'Training and Servicing Facilities', as they were known, were most successful, and paved the way for bike and vehicle manufacturers to found their own service schools after the war. The Harley-Davidson school was set up by Joseph Ryan, and eventually trained about 300 people during the war, going on to train dealers afterwards.

In fact, Harley-Davidson did well out of the First World War. Quite apart from making sure to keep the home market supplied, as well as the military contracts, it was able to increase exports while most of the competition was overseas and out of the picture due to war. It even managed to bag a few military contracts from foreign forces long before it really supplied the US military in a big way. One of these was the Dutch army and the other, the Russian, which took delivery of gun-carrying outfits along with ambulance units. The latter carried the stretcher at saddle height, the patient benefiting from the soft ride provided by the decent sidecar springs.

Apart from special outfits like these, there were few differences between the specifications of civilian and military machines, and in fact both benefited at the time from the general improvements in Harleys. The V-twin Model 8 had progressed in leaps and bounds with better handling as a result of the rider being repositioned near the centre of the bike. Rider comfort had also improved, due to better springing, a lower seat height, and a more acceptable riding position that made it less tiring to ride

The motors themselves had been re-engineered, which resulted in more power: the capacity was also upped to 61ci (1000cc) to bring it in line with the opposition. Then there was a hub clutch, step starter – the list went on. In 1915 the latest version of the Model 11J appeared with 98 improvements. Now with a three-speed gearbox and dry multiplate clutch that could easily be adjusted, it had a proper gate on the gear lever. There were numerous engine improvements and it now gave 15bhp: this, coupled with the improved gearbox, made it popular as a sidecar hauler. With this in mind, the frame was strengthened around the head-stock and it was generally beefed up to cope. Perhaps more importantly, it was the first bike to be offered with an electric headlight as a standard fitment.

OPPOSITE: Some people collect racers, others scooters, but many riders are keen on military machines. These three riders from Holland have a trio of similar WLAs, originally used by military police. Note the rudimentary screen on the bike on the right: protection obviously came before asthetics!

BELOW: The power unit for the WLA (though this is actually a WLC, the Canadian-specification machine) was a detuned version of the long-running 45-ci (750-cc) side-valve V-twin. Lower compression, aluminium cylinder heads and a bash plate were the main changes from the civilian specification. From its role as the workhorse of Main Street, whether powering a bike or Servicar trike, this motor found its moment of glory in wartime.

In 1916 the pedals as a means of starting disappeared totally, replaced by the kick-starter, and new forks appeared on the Model 16J, which had various styling alterations to make it better-looking as well as functional. Bigger fillers on the petrol tanks went with petrol taps below, while filters were fitted inside to help stop blockages in the carburettor due to poor filtration. These benefits were further enhanced a year later on the 17J with the introduction of a generator and battery system. This was just in time to make it the model that would take Harley to war, albeit with a reduction in bright nickel plating.

Some Harleys had already seen action, having gone to war with the Canadians, who declared war shortly after Great Britain. The riders, who were often the owners as well, soon found the weight to be a problem, compared to the smaller British-built machines in service alongside them. This was particularly noticeable when bikes needed to be pushed through the inevitable mud of battlegrounds. As a result, most were retained and used for long runs on better-surfaced roads, where reliability and comfort were more important than off-road capability.

The army made certain demands when it placed those first contracts to ensure that bikes became as universal as possible. This was to make rider training, such as it was, easier, and to facilitate the stocking and supply of spare parts. All bikes had to have a right-hand twistgrip, the gearshift lever on the left of the fuel

OPPOSITE: Matching military bikes, with civilian cousins.

ABOVE: The Canadian specification WLC had left-hand clutch and brake lever. After the war (in fact for about another ten years) it was still possible to buy surplus WLAs that had never actually been used and spares weren't a problem either. The spotlight is an extra.

RIGHT: The rifle scabbard was standard, with the gun stock poking out, ready for use. In the event of suddenly coming under enemy fire, riders were instructed to leap straight off the bike and grab the gun: there was to be no nonsense about stopping first!

in solo trim also came with a repeat order in 1936 for a few more. A year on, the WL had its first mention in the order books, but for only 50 machines.

Although in military terms business was slow, with just a few overseas contracts keeping things going, Harley was doing quite well. What with police machines and sales in general it was certainly doing better than Indian, while Excelsior, the third member of the Big Three, had long since gone to the wall.

But the situation was soon to change: even before war was again declared in Europe, the US had been convinced that it would be involved. But as had happened 20 years before, Harley-Davidson was supplying military contracts well before war was declared by the White House. Five thousand solo machines were shipped to

England, for example, after the Luftwaffe destroyed the Triumph factory at Coventry. Not to be caught out this time, Harley re-opened its service school, training mechanics and at the same time submitting bikes for the army's extensive testing programme. Harley itself had two test tracks at Fort Knox and Louisiana, where bikes were subjected to merciless endurance and engine fatigue tests.

Finally, after a great deal of thought, Harley produced what it considered to be the best possible machine for military use, the WLA, the A standing for Army. This was a militarized version of the 45-ci (750-cc) side-valve twin that had been selling well in the late 1930s in the normal civilian market. Its compression had been lowered and it now sported aluminium cylinder heads to prevent overheating problems caused by escort work. Larger bearings, a better oil pump, and a stronger gearbox and clutch were also part of the improvements to deal with low-speed convoy work. It was tuned to give power at lower revs at the expense of top-speed performance, which was not seen as important. Fitted with a special large air filter for the Linkert carburettor, this would enable it to deal with hostile running conditions. The tubular forks were lengthened by 2.75in (7cm) to help increase ground clearance, and comfort for the rider was ensured by a large pan saddle supported on a sprung pillar. There were crash bars to help protect it from damage and a bash plate to look after the underside of the engine and frame. In addition, there were various attachment points for different weapons. The standard package also included a fork-mounted scabbard for a gun and a rack for carrying items at the rear. This was capable of carrying a 40-lb (18-kg) field radio and two saddlebags full of other items, without breaking down too often.

ABOVE: There is a good set of lights on this bike, but those stationed in Britain had to comply with blackout regulations. Headlights had to be heavily shielded to prevent being spotted by night bombers. This, and the lack of street lighting, led to a high accident rate during World War II.

OPPOSITE
TOP LEFT: A neat personalized plate for this WLA.

TOP CENTRE: The extra-large air cleaner was a standard fitment.

TOP RIGHT: No lack of storage space.

BELOW: One estimate numbers 88,000 WLAs like this rolling out of Milwaukee, though the true figure is a mystery. An order for 11,000 was cancelled in 1945, when victory seemed just over the horizon.

There was also a WLC model, the C indicating Canada. This model had the clutch lever and brake on the left with the gearchange on the right as specified by requisition orders from the Canadian military. It also had a box on the front fender to carry miscellaneous items. The WSR also appeared and was a sidecar version for the Russian forces, which other military outfits also bought, finding it suitable for their needs.

Unlike the previous conflict, civilian sales virtually ceased as the war progressed, though bikes were built for the home market. Home-based forces got grey-painted bikes, as opposed to olive green, with SP painted on them to denote their function as Shore Patrol. Apart from these general-purpose machines, records show several 74-ci (1200-cc) UA models along with some 61-ci (1000-cc) E and ELC OHV bikes bought for military staff use.

But the WLA was Harley-Davidson's standard as well as most often produced military machine. The US Amy ordered an initial 421 of them in 1940, followed by two further orders totalling 2,451 bikes. In 1942 a slightly modified version was produced, remaining the stock army bike until 1945. It has been suggested that over 88,000 bikes were produced for the forces at Milwaukee during the war, but this is always going to be difficult to confirm, especially with the various versions on offer. In 1945 Harley had an order for over 11,000 bikes cancelled when the government sensed victory, which confuses matters even more.

Certainly, when the hostilities ceased, there were several thousand brand-new bikes left with enough parts to build an estimated 30,000 more. This is confirmed by the fact that the bikes were still being sold into the mid-1950s at rock-bottom prices, complete with all military fittings! The price was set at $450, for which you

could have a ten-year-old bike that had never been used.

But while the WLA was well known as America's ubiquitous military machine, it was almost eclipsed by something very different – the XA. This was a horizontally-opposed twin, and an almost exact copy of the equivalent BMW. It was Harley's second departure from the V-twin and again came as a result of its military involvement: despite some reluctance from Harley-Davidson, the top brass was convinced that a shaft-drive flat-twin would make the perfect military motorcycle.

The company's answer, after purchasing a BMW for evaluation, was the XA, which produced 23bhp at 4,600 rpm and had a low 5.7:1 compression ratio to allow it to function in desert conditions. With its cylinders sticking out into the air for additional cooling, it was hoped that it would fare better than a V-twin or single in hot weather. It also featured a wet sump to tackle the heat problem and was estimated to run nearly 100 degrees cooler than a V-twin motor.

Other notable design features on this

The XA. Built at the US military's insistence, and heavily based on a pre-war BMW, the XA is the only Harley-Davidson (to date) with shaft drive. It was more advanced than the WLA, but more expensive as well. Harley-Davidson did produce a civilianized prototype in 1946, but nothing came of it.

ABOVE: The XA used the standard springer fork.

RIGHT: The power unit was a mildly tuned side-valve flat twin of 23bhp. It ran cooler than the WLA, due to those well-aired cylinders. One thousand XAs were produced, a tiny number by wartime mass-production standards.

model included a hand clutch, foot gearchange, with a down-for-down and up-for-up shift pattern on the lever. Some were fitted with sidecars and Harley experimented with a drive to the sidecar wheel like the BMWs, although this was never put into production.

During testing, a few problems emerged, but not enough to stall the project: at one time there was talk of an order for 25,000 XAs, and a reduction of WLA orders. In the end, the army took 1,000 in 1943 and then no more. None of them actually saw active service, remaining in the US at military bases. Despite all the development work that had been put into the XA and the lessons learned about such things as shaft drive,

LEFT and ABOVE: The Military Police-specification WLA, with shrouded light sitting alongside the main (legal) headlight, and the siren beneath it. It was enough to strike terror into the bravest AWOL recruit, unless he happened to be riding a Knucklehead.

OPPOSITE: This is not a weapon, contrary to rumour, but a standard-issue baseball bat for MPs! Look closely and you'll see the gloves, confirming that it is a recreational implement.

MP
Liberator
MP

Harley never used any of this knowledge or engineering in post-war bikes, almost to the extent of pretending the XA had never happened.

Other prototypes and experiments tried during the Second World War included a sidecar with a driven wheel like the BMW outfits. This was a totally separate project to the XA and was called the Model XAS. It was tried at the insistence of the US government in 1942, but the right-hand mounted sidecar machine failed to get further than the prototype. There was a separate design study for a machine using a 45ci (750cc) engine with special forks, frame and sidecar, which again came to nothing.

Slightly more suitable were the various trike projects (see A–Z chapter) one being a Servicar with a shaft-drive Knucklehead motor and which rolled along on solid disc wheels. Designated the TA, it suffered from handling problems, lots of vibration, and leaked oil in copious amounts. These problems were rectified, and despite being approved by the army in 1941, it went no further and certainly not into production. Another undeveloped prototype was a Canadian mini-tank, powered by two Knucklehead motors mated together.

But while experiments were in process, the solo motorcycle continued to have a role in the war effort. While radios were more reliable than ever before, messages could still be intercepted and there was still no substitute for hand delivery, for which a motorcycle was invaribly the fastest means. Dispatch riders were also the best people to keep intelligence up to date in the constantly changing scene of the battleground. However, the role was still basically one of support: leading convoys was probably the closest the rider got to the front line along with DR work. This non-

OPPOSITE: These are the items no WLA rider could afford to be without: gas mask, waterproof boots, radio, canvas bucket, tyre pump, full toolkit, spares, bed roll, mess kit and instruction book. The radio alone weighed 40lb (18kg).

ABOVE: Not a hint of chrome anywhere, the WLA was functional rather than pretty.

combat role did not prevent a fairly high casualty rate; but this was caused by accidents rather than enemy gunfire.

It should also be remembered that it was not only in Europe that Harley WLAs performed their duty. North Africa was another, and they appeared throughout the Commonwealth as countries began to use them as security or shore patrols like the US. In fact the US Navy also had a fair number of them travelling on ships to provide transport when docking at the various ports.

It should also be remembered that the Japanese were still producing Harley copies under the Rikuo, or 'King of the Road' brand name, having taken over total control of the Japanese factory from Harley in the late 1930s. These machines could also be found as part of the war machine in both solo and three-wheeler trim. There was also a much rarer Kuro Hagare, or 'Black Iron' machine in use, which was also Harley-based.

Meanwhile, Harley-Davidson was given the US Army/Navy 'E' award for excellence in 1943 and 1945, in recognition of its significant contribution to the war effort. This was all very well, but it seems that the government had failed to honour the total cost of spares produced by the company, and Harley is reported to have only made a 10 per cent profit on each bike it sold to the military. But as before, there were spin-offs: many overseas riders saw Harleys for the first time, and many GIs learned to ride at Uncle Sam's expense, with the result that when they returned after the war, they were keen to have their own bikes.

Throughout the 1950s, military production was almost non-existent. In 1963, Harley gained an order from the military police and shore patrols, totalling 418 bikes. These were XLAs, basically Sportsters fitted with fibreglass panniers and a windshield, and were more use on base rather than out in the operations field. In fact, the XLA was the last true Harley-Davidson supplied for military use. No new contracts came out of the Vietnam war, and in any case the army still had plenty of old WLAs in stock.

But Harley's military story doesn't end here, as the company did land another contract as late as the mid-1990s. Ten years earlier it had purchased the world manufacturing rights to the Armstrong MT500. Armstrong was a British company which was making a trail-style military bike powered by a 30-ci (500-cc) Austrian Rotax engine.

This successful machine had followed on from the Can Am Bombardier 15-ci (250-cc) single that had been acquired by many armed forces throughout the world, between 1977 and 1985.

It had revolutionized the military market with its lightness and responsiveness and was capable of withstanding both off-road as well as on-road situations without a weight penalty. Needless to say, it was popular with riders due to its all-round ability.

The Armstrong machine provided more power, being double the size, and had taken over quite nicely from the Can Am and was selling well. Harley was to re-badge these as its own, so when in 1995 the British Ministry of Defence awarded a new contract, it was to Harley-Davidson.

A total of 1,450 bikes were ordered and all were delivered on time; but three years later the contract was cancelled, possibly because the use of bikes for anything other than convoy work was diminishing.

On the road, the Japanese Honda Pan Europeans, military variants of police bikes, provided the escort bikes and other Japanese manufacturers now sell their trail bikes and Quads to military buyers who only need the odd off-road machine, rather than standardized military bikes as in the past. So it appears, as Harley moves towards its 100th birthday, that it will no longer be selling military bikes in any quantity: hopefully, there will be no need for it to do so again.

FAR LEFT: So that there was no confusion, gas and oil tanks were clearly labelled

LEFT: The metal plate below the speedometer listed essential information like oil type, spark plug gap, and recommended cruising speed.

BELOW: Crash bars were useful if you needed to dive for cover and leave the bike to its own devices.

OPPOSITE: The WLC model intended for Canada.

CAL .30M1
AMMUNITION BOX
Liberator

POLICE BIKES

Today, when Harley-Davidson is one of the foremost suppliers of motorcycles to the police, it's hard to believe that, not so long ago, it lost much lucrative business in that area. Everyone remembers *CHiPs*, the famous TV series based on the California Highway Patrol, that achieved a huge following the world over. But who remembers what they rode? They were not in fact Harleys, but Kawasaki KXp bikes. Actually, *CHiPs* was shown after Harley-Davidson had lost much of its traditional police business in the home market, as a result of poor quality control. It took several years of hard work, building up quality and holding down costs before the company was able to persuade US police forces to buy its motorcycles again.

But years before this, Harley had dominated the police market in America, from the early days right up until the late 1960s. Even when it was receiving stiff opposition from Indian and Excelsior-Henderson (with their super-smooth four-cylinder bikes that the patrolmen liked so much), Harley still had the lion's share of the market and always had since bikes were first used for the purpose.

Although for many years it was thought that the first bikes had been in service in Pittsburgh in 1909, it now emerges through detailed research, that Detroit had motorcycle cops in 1907. These were Harley-mounted in a town dominated by Ford and General Motors and demonstrates that they were far more use in congested streets than their four-wheeled counterparts. At the end of the century, some forces began to show renewed interest in two-wheelers, and for exactly the same reason!

In 1911, San Francisco and Los Angeles are recorded as having Harley motorcycle patrols and in 1913 Cincinnati followed suit with a bike squad. These early bikes are likely to have been the Model 5, a belt drive, pedal-assisted, single-speed machine of 30ci (500cc). Top speed was in the region of 45mph (72km/h), which was quite an achievement over basic rough roads more used to horse traffic. It is, of course, very easy to draw the analogy between the horse-mounted sheriff and the motorcycle-mounted officer who always worked alone. The difference was that the lawman was now more often chasing speeding cars than cattle rustlers. Tickets were handed out with a set lecture

ABOVE: Through the 1950s and '60s, Harley-Davidson was *the* US police motorcycle. The company was able to rely on police business, not only through patriotic loyalty, but also because no other bike of the time suited American conditions so well.

OPPOSITE: A magnificent 1938 UH side-valve sidecar outfit. Flashing lights and paint scheme are the only visible changes to the machine, though some details were different as well.

BP
HARLEY-DAVIDSON
POLICE SPECIAL

designed to make a person feel suitably chastened. The legend of the bike-mounted traffic policeman arose from the very outset of motorized transport, when he grew to be feared, even hated at times; but people also developed a sneaking regard for him as well, and of all the other police jobs, perhaps the motorcycle cop has retained an element of glamour to this day.

The work was also dangerous, and not only due to criminals: roads were poor and protective clothing minimal. But the cops were a highly visible deterrent to bad driving, and did a good job in reducing fatal accidents, a point Harley was quick to use as positive publicity. The publicity angle has been an important one ever since, and explains the cut-throat competition between American bike manufacturers to secure police business. Quite apart from the lucrative sales, bikes chosen by the police were seen as tough, dependable and speedy, and always on the side of the good guys.

However, it would be some years before Harley-Davidson set up a department dealing specifically with police contracts. It would be longer still before the bikes and the options available were any different from those on offer to the general public. However, sidecars were useful for police business, and after the First World War, the ex-military 61-ci (1000-cc) Model J (in both solo and sidecar trim) proved ideal for traffic enforcement and was sold on to the police to use up stocks no longer required by the armed forces. But in prohibition days, America was suffering a crime wave in which bootleggers on more powerful 74-ci (1200-cc) Harleys and Indians could outrun any lawman on a 61: the police had no option but to follow suit.

It was not until 1926 that Harley recognized the need to have a separate department to deal with police sales and

OPPOSITE: By the late 1950s, police bikes were more specialized. This bike started out as a touring specification Duo Glide with screen, spotlights and large rigid panniers. A police radio, fire extinguisher, flashing lights, siren and crash bars have been added. There were now painted wheel rims (not chrome), to save public money.

ABOVE: The standard Panhead V-twin, low-revving and reliable.

ABOVE RIGHT: A rider's eye view: the screen was essential for all-year-round use.

opened one up to liase with the 1,400 police departments which by then were using its bikes. It also launched an advertising campaign to capitalize on this huge success, 'Harley-Davidson will curb this traffic slaughter', referring to its huge take-up by police, compared to opposition like Indian. It worked and by the end of the 1920s 3,000 police departments were Harley-mounted!

Police riders themselves received no formal training, but were only required to prove that they could ride before joining up, and the'official' equipment was still almost non-existent. They were expected to work at all hours of the day and night and in all weathers. They had to be mechanics, paramedics and crime busters and be able to survive on their own in remote areas with little outside help. Radios had begun to appear in police cars in the late 1920s, but it took another ten years for some early unreliable units to appear on bikes. Until that time, the risks to the individual were high and the rider had to use public phone boxes to ring in every hour to give and receive relevant information concerning accidents and crimes.

Likewise, the relationship between police and dealers, who were expected to service these contracts, was not all that rosy. Although initial publicity was good, dealing thereafter with service contracts on tight budgets proved tough work, despite the bikes being almost identical to the standard civilian machines they were also selling. Most police bikes, however, had at least a tank-mounted speedometer driven off the rear wheel for enforcement use. But items like red lights, spotlights and klaxon horns were fitted by the riders themselves, often at their own expense. Even the bags containing report forms were often supplied by riders themselves and it was not until the R-model that the bikes came ready equipped for police.

Although the early 1930s was a generally bad time for Harley, police business continued to expand, while in 1932 the new Servicar provided an additional outlet. It quickly found favour with police departments, especially as a means of enforcing parking, thus helping sales figures which it would continue to do until the 1970s.

Harley realized its dependence on the increasing importance of police work at a time when civilian sales were slumping, and some specific pieces of police equipment were introduced: a police speedometer, a front-wheel siren, first aid kit and fire extinguisher. Then there was the Burgess muffler (which actually appeared in 1931) to direct exhaust gases downward, vital in the heavy traffic in which police riders often found themselves. Later safety guards and pursuit lamps were added as options.

By 1935 these options had been rationalized into two groups from which the police department could chose. The basic package included a safety guard, jiffy stand, 100-mile (161-km) maximum lighted speedometer and rear-wheel siren. The de luxe option added a steering damper, ride control, stop light, luggage carrier, first aid kit, fire extinguisher and pursuit lights.

There was also a 'Safety Silver' colour scheme to make the bike more visible at night, though it is fair to say that there was no standardization of police colours and markings at this time, each force doing its own thing as it does now; but the take-up on this option was not high. Uniforms varied from state to state, though most consisted of boots, breeches, jacket and caps.

What of the bikes themselves? Many departments had moved across to the JD-model for its extra speed and famous reliability, the California Highway Patrol being just one of note. It was, in fact, one of the first to switch to the twin headlight, side-valve Model D in 1930, and three years later designated the Harley as its official police bike throughout the state.

However, the later side-valve 74-ci (1200-cc) VL, with its early vibration problems, lost police sales.

In 1935, apart from new colour schemes, Harley was introducing a lighter feel to its bikes, achieved by relocating the tool box from the front fork to the main frame. Better brakes were also emphasized in advertising, achieved by fitting harder linings and using carburized brake drums.

Lights also received a makeover with a 'beehive'-style rear lamp, making it visible from either side as well as from the back. At the front was a pre-focused headlight, which concentrated the beam on the road ahead and allowed higher speeds at night, helping to cut accidents.

The factory was also offering a service that allowed older bikes to be returned and updated, effectively giving the forces a later model which was a cheaper option than buying new. (In the last few years, restorers have seen this as something of a problem when they realize they have incorrect or later parts on bikes they have purchased from a particular year.)

OPPOSITE: Caught in the act: 'May I see your driver's licence, sir?'

RIGHT: Police departments were long-time users of the Servicar, finding it useful for parking enforcement. One method was for an officer to ride past a line of parked cars with a piece of chalk on a long stick, when he would mark the cars illegally parked, leaving it to the meter maids to slap on the tickets. A Servicar could also carry a good amount of equipment in its box (or two or three hefty officers) but was still smaller and handier than a full-size squad car. It was not a pursuit vehicle, however.

In Milwaukee, the police department has been testing police-specification Buells, called Defenders. Based on the civilian Thunderbolt sports tourer, these have police equipment built into the panniers.

Now there *is* a pursuit vehicle. In town, the Buell Cyclone is fast and nimble enough to act as chaser and get to situations quickly. Note the blue lights peeking out from the headlamp fairing.

ETOWAH
FOOTBALL
FOR SALE

OPPOSITE: Like most police vehicles, the Servicars were sold off once past their prime. However, this has been given a new lease of life, having been converted into an ice cream van.

BELOW and RIGHT: The police were responsible for keeping the Servicar in production long after civilian orders dried up, and it continued to do so up to 1973. There were various updates over the years, such as three-wheel brakes with electric start later on. The Servicar was also fitted with a radio before the bikes, because it had available space (you can see the handset on this late model). It still sported a hand gearchange, but when you had three wheels to keep you upright, that hardly mattered.

The Knucklehead of 1936, though initially seen as a sports bike, soon found favour with the police, who were having to catch criminals using ever faster automobiles. That was also the first year the factory offered radios as a police option. The radio receiver cost $75 and the purchaser could state which radio frequency or wavelength they required, the whole unit being mounted on the luggage rack. Two years later, the price had risen to $135, and it was now manufactured by RCA, rather than Harley-Davidson itself.

The Servicar was still selling well, helped by updates such as a constant mesh gearbox, a strengthened rear axle, and brakes on all three wheels. It was also restyled, and by 1939 the box at the rear came in a variety of sizes. But despite all this, sales had been dwindling as some police departments sold off bikes in favour of cars. The reasons were many, but some felt that the car carried more equipment and was not that much slower; neither was there the massive price differential there once had been. Nevertheless, during the early 1940s, 3,500 police departments were using Harleys, with Indian on the slippery slope. It looked as though Harley-Davidson would have what police business was left all to itself.

During the war, Harley-Davidson cut

back its production for the civilian market to concentrate on police and military needs. So although the civilian market was virtually moribund, police sales and service were the means of Harley dealers staying in business. In 1942 and '43, production peaked at 29,000 bikes, the vast majority going to police and military contracts. Meanwhile, police work had undergone a change during wartime, with many forces carrying out duties that can best be described as paramilitary. Patrols of locations deemed to be sensitive were carried out by police officers, sometimes with Thompson sub-machine guns as part of their equipment, though never listed as a factory option, it has to be said.

Some unusual problems did surface. In 1942 all the American speed limits were reduced to conserve fuel, but bikes were wearing out cylinders in 6 –800 miles, despite the lower speeds. It turned out that the recently introduced radios required the engine to be kept almost constantly running, to maintain the battery charge and power it. Also, the nickel-alloy Harley-Davidson had used for plating cylinders, was now in short supply, and it had adopted a silicon substitute that was not up to the task, hence the high rate of wear. It tried chrome-plating the bores, which did the trick, and cylinder life increased to a rather more acceptable 20,000 miles or so.

This was fortunate, as the workload had increased, with many additional officers being employed for all the extra duty and patrol work. The bikes remained largely unchanged throughout this period, and Harley reduced the amount of chrome and brightwork by painting over these areas instead.

Around 1947, Harley resumed full-scale production of its bikes and began to enjoy something of a post-war boom. Thousands of returning servicemen wanted to buy bikes, having ridden them during their military service. They had enjoyed the experience, and wished to continue riding for pleasure. But the technical advances which had been achieved now benefited police riders as much as the emergent civilians; the Hydra-Glide fork, rubber-mounted handlebars, sprung seat post and 5 x 16-in (127 x 406-mm) tyres, all gave a more comfortable ride to officers facing a long hard day in the saddle.

Also welcome was the first of the two-way radios developed by RCA. They were first used on the Servicar, which had all the necessary space to house the hardware required. By 1948 the units had been trimmed to a compact 27lb (12kg) and were offered for solo use. They were fitted to the carrier at the back of the bike, taking up space once occupied by saddlebags. On the

ABOVE: On the look-out at Daytona. The FXRP was Harley's police bike for the 1990s, faster and more nimble than the traditional Electra Glide-based machines.

OPPOSITE: On duty at Sturgis. The police contingent at premier bike events, such as Sturgis and Daytona, are well aware that Harley-mounted officers will receive more co-operation and respect from the public.

POLICE
POLICE
STURGIS
POLICE
STURGIS
POLICE
61
60

SHERIFF
SHERIFF
SHERIFF
223
SHERIFF

OPPOSITE: Deterrents don't get more visible than this. This sheriff is based in Daytona.

ABOVE: Something to show the guys back home – speeding tickets, courtesy of the Daytona police.

RIGHT: A late-model Electra Glide: some forces still opted for the traditional police mount.

left was the receiver and on the right the transmitter. An aerial completed the rear equipment, but at the front was a speaker box mounted on the handlebars. Adjustable for volume and squelch, it sat right in front of the rider. The microphone came from the rear units and clipped onto the back end of the gas tank, so was easily accessible while on the move.

The problem was that while the officer now had two-way communication, the radios took all the available electrical power as well as storage space. The bike still needed to be kept at high revs to power it, which of course meant that it was virtually inaudible thanks to the engine noise drowning out the primitive speaker. It would not be until 12-volt systems came into being that the problem would really be solved, but at least the engines could now take the abuse, Harley having reverted to decent peacetime materials. Other police radios, including one from Motorola, soon followed, though the additional expense meant that many forces preferred their riders to carry on calling in from public call boxes. It wasn't until the 1950s that the introduction of transistors made radios lighter, more reliable, cheaper and less power-hungry.

By the 1950s Harley was enjoying something of a post-war boom and they were to be good years for the Milwaukee concern. In 1953 Harley lost its major competitor in the police market when Indian finally went to the wall. This left Harley as the only American motorcycle manufacturer, and certainly the only police supplier. A few forces did manage to keep their Indians going for another ten years before finally switching to Harleys, but most adopted them straight away as they required the back-up service.

But while there was now no home-based competition, smaller, lighter and cheaper Japanese bikes had begun to make an appearance on the general market and were of course later to have some bearing and effect on supply to the police.

Meanwhile, Harley-Davidson at long last offered a foot gearchange option, though oddly enough many police riders still specified the hand change, still available until the mid-1970s, which left hands free for traffic regulation and signalling. There were also new options on offer for the 74-ci (1200-cc) Panhead and these included high- and low-compression motors to cope with town use or the open road. The latter came with a lower ratio gearbox. Other options available were chromed safety bars, wheel rims and muffler, which all cost extra, as did the screen, which came in three colour options. Sirens were still driven off the rear wheel by a friction device, which occasionally failed to deal with new tyres that expanded as a result of heat. This caused a few of the units to self-destruct and many riders found it safer and quieter to have them behind rather than in front of them.

Surprisingly, the radio also remained on the options lists and was not standard, so in some US states tooled pannier bags were still seen attached to the rear carrier. Motorola had in fact introduced a smaller more compact unit specifically designed for a bike in 1957, instead of one adapted from a car. This was called the Dispatcher Radiophone and took very little power to keep going with a clarity not previously experienced. Designed to mount on the handlebars, it moved rearward in 1958 when the Duo Glide was introduced as a police motorcycle. This was of particular interest to police as it offered comfort levels as yet unexperienced, as well as increased safety features like hydraulic rear brakes. In reality, it was the venerable FLH-model, fitted with a rear suspension system. This was in response to claims of the rear carriers splitting due to vibration,

ABOVE: Close to home: This FXRP was spotted outside the Juneau Avenue works in Milwaukee.

OPPOSITE: Electra Glides of the Milwaukee Police Department – hardly a force to buy Hondas!

POLICE
POLIC

Harley's 95th-anniversary celebrations in Milwaukee saw a parade of police from all over the USA.

ABOVE: On patrol in Sturgis.

ABOVE RIGHT: No force is better known than CHiP, the California Highway Patrol, with its strong association with Harley-Davidson. It has been a loyal customer since 1933, so it was a big shock when it decided to switch to Kawasakis in 1975, a move which influenced many other forces to consider doing the same. Oddly enough, *CHiPs*, the hit TV series, featured patrolmen riding Kawasakis, though the public perception has long been that Californian patrolmen ride Harleys. This has turned out to be true since Harley-Davidson won back this prestigious contract.

especially when the weight of a radio was added. The lack of springing also caused some of these to fail on a regular basis, despite the new radio technology. There was also the fact that the FLH had a stronger transmission and clutch, along with an optional larger oil tank, all helping to keep it on the road longer and reduce mechanical breakdown. The weight had risen by 50lb (23kg) as result, but it was a small price to pay compared to all the other advantages.

Along with increased bike safety features, rider safety was now regarded as paramount: riders now got fibre helmets to replace the garrison-style caps that they had used before, and waterproofs were also beginning to appear in a uniform style. Rider training had also been taken on board with many forces organizing proper programmes to equip their riders to deal with all the likely hazards they would encounter on the road, like railroad tracks, cobblestones and errant drivers. Instruction was given on both road and dirt-riding techniques, especially in country areas. Low-speed as well as high-speed riding was also part of the equation, riders needing to be able to control their bikes in heavy traffic. In addition, of course, riders were required to complete daily safety checks of their bikes and maintain them in a clean condition.

Out of this grew many display teams, which allowed riders to demonstrate their prowess in front of large audiences. While they were representing their various forces in the often very impressive displays, they were at the same time providing much valuable publicity for Harley-Davidson.

But the 1960s were to have as much impact on the police motorcycle and ordinary markets as they did on society in general. 'Times they are a-changing', sang Bob Dylan and so it was with Harley, especially in 1969 when it was taken over by American Machine and Foundry (AMF). By then, police Harleys, like their civilian counterparts, had already gained 12-volt electrics and electric starting, both big leaps forward for long-suffering

STURGIS
THE
KEY CITY OF
THE
BLACK
HILLS
SD
POLICE
DEPARTMENT
SPD

OPPOSITE: Relaxed and in control: an officer of the Sturgis Police Department.

ABOVE: It's not only city police who ride Harleys: this FXRP is run by the San Francisco Airport Police.

motorcycle cops, who no longer had to jump on a kickstarter several times a day, or worry about a marginal electrical system keeping up with the demands of the radio.

The electric start had first appeared a year earlier on the Servicar, which was still going strong, in spite of the police being the only customers for the little workhorse. The change meant that a new primary casing had to be designed to deal with the starter motor drive, which in turn required the oil tank to move to the other side of the solo machine, which required a new frame in which to house it all. As a result, the weight of the bike rose yet again and new uprated brakes were therefore needed. Changes cost money and the price of the bike rose to the point that it was close to being double the cost of a car, which did not endear it to the purchasing departments.

Despite a power increase for the Shovelhead engine, police Harleys were finding themselves increasingly outclassed on the road by faster cars and bikes. In fact, the classic chase scenario grew less common as there was increased reliance on radio and radar to catch miscreants and discourage speeders. Harley-Davidson no longer fitted radios at the factory, leaving the market open for Motorola and others to meet the demand.

The Servicar received the12-volt alternator system as well, though there were still many early bikes in service. As before, a conversion kit was offered so that departments could update their Harley-Davidsons rather than buy new. The Servicar got its final update in 1967, when the box was made of fibreglass by the Tomahawk Company and incorporated the fenders into the moulding. In this form, the machine remained in production until 1973 when virtually all the orders for it dried up and it was no longer viable to make the hybrid solely for the police.

Meanwhile, as well as proper training and improved equipment, riders were now given extra pay in the form of a hazard allowance. Some claimed this was more for fixing the bikes, as the standard of quality control had declined and the machines were being stretched beyond their capabilities in the modern traffic conditions.

Certainly, Harley seemed to be increasingly indifferent to the needs of its police customer base, which had kept it going through some lean times in the past. This was a serious mistake and would take it a long time to repair the damage caused by lost sales. The Japanese were moving in on Harley's traditional market, building bigger bikes, while other manufacturers, who had a share of their own home police markets, were looking to expand there as well.

By this time, the California Highway Patrol had established itself as the most thorough tester and evaluator of police vehicles, so many other forces and agencies took a lead from them. If CHiP thought it was OK, it was good enough for them and it saved costs on doing their own trials.

So when they began looking at the Moto Guzzi 850 and Honda 750 police bikes in 1972, other departments sat up and took notice. But it was Kawasaki that secured the CHiP contract in 1975, when the department ordered 130 of its 55-ci (900-cc) police specials. The reason was very simple: Kawasaki had asked the officers who rode the bikes exactly what they expected of an official bike, as opposed to one for pleasure, and noted the replies; it then built a bike to correspond with those needs. It was faster, smaller and lighter than the equivalent Harley, easier to ride and faster to accelerate. The four-cylinder engine was smooth, and was mated to a five-speed gearbox. It ran longer between maintenance checks, broke down less often and spent less time off the road waiting for parts. So effective was the Kawasaki that it stole what few US police sales the Moto Guzzi had achieved. To show how serious it was about the police market, Kawasaki even set up a factory in Lincoln, Nebraska to manufacture the KZp 1000, as it later became. Harley-Davidson couldn't even accuse the Kawasaki-mounted officers of being unpatriotic!

It should also be borne in mind that, since 1933, CHiP had continued to have Harley as its official police motorcycle and its defection served to open the floodgates to other manufacturers. The official tenders now specified items that at that time deliberately excluded Harley bikes, so disenchanted were some authorities. Towards the end of the 1970s, Harley didn't even bother to offer bikes for the annual police evaluation tests, as some forces had stopped even looking at them as early as 1975.

But it would be wrong to infer that Harley did absolutely nothing for the police market during this period. In 1971 it had introduced the FLH Electra Glide with a 74-ci (1200-cc) high-compression engine and 10-inch (25.4-cm) front disc brakes, although the latter were found to be seriously lacking when under pressure. Electric sirens began to replace the wheel-driven mechanical items, and in 1978 an 80-ci (1340-cc) V-twin replaced the existing engine. This, however, still had an unacceptable level of vibration and the weight had increased even more. The suspension was not good and riders found the ergonomics totally unacceptable. Rider fatigue, poor handlebars, switches, operating of the clutch and gears were items adversely mentioned by the test riders. In the stability category they criticized it because it grounded out too easily and the suspension didn't cope too well. The engine was quoted as lacking in speed and power, to mention just a few of the complaints.

Not only that, but some forces were again having second thoughts about using motorcycles of any description: employment legislation was making them twitchy concerning the risks to which riders were being exposed. The effectiveness of the bike always relied on the skill of the rider, and not everyone had these skills. And patrol cars were cheaper than ever, forcing motorcycles into an ever more specialized role. Something had to change.

Change it did, for in the early 1980s, a management buyout caused a new attitude to emerge from Milwaukee, and one of the things that changed was the attitude to police business. Once again, the company made a real effort to obtain police sales, but it promised to be an uphill struggle. Whenever a Harley was tested against the big Kawasaki, the 1340-cc Harley came off worse in almost every category. Harley's police sales continued to decline.

It was probably the introduction of the Evolution engine in 1984 that was really the turnaround for Harley. After seven years at the drawing board, the rubber-mounted, belt drive, five-speed powerplant was radical indeed by Harley standards, but it was just what was needed to win back police business. A year later, Harley launched the FXRP and began to actively court police buyers again, which was to be a hard struggle with many having abandoned bikes altogether. There was also an east/west divide, with the Western police forces favouring Kawasakis; but Harley was at least making some progress in recapturing a little of its market share.

This was helped somewhat by the increasing amount of congestion on the roads, especially in towns and cities, which hampered response times to calls at a time when accountability was becoming a real issue. The only real solution was to revert to motorcycles.

Harley was able to climb back onto the evaluation lists, although it still came in for some serious criticism. But this at least pressurized it into dealing with complaints. Gradually, forces recapitulated and the familar sight of a traffic cop on a Harley reappeared on the scene. To achieve this, the company had had to offer deals that no doubt lost money. But it was necessary to regain a toehold in the market, and convince police buyers that the bad old days were finally over.

In 1991 the FXRP was substantially updated with new frame, forks, engine-mounting system, instruments, carburettor, seat, and starter: in fact the list ran to over 300 changes. Sitting behind the familiar fibreglass fairing was yet another level of sophistication with regard to law enforcement items. Now, a radar gun was part of the kit, as well as a PA system and improved emergency lights.

Three years later the bike was discontinued and the FLHTP appeared, with adjustable air suspension at both ends, complete with anti-dive on the front forks. Rider comfort was paramount and Harley was aware of this, realizing that if it wanted to continue winning back contracts it would have to meet this need and get the riders on its side. That is why items like the footboards became adjustable, the control layout became easier to read and everything was more accessible. The range of the bikes was improved with bigger gas tanks and the electrics were uprated. In short, Harley had listened to the needs of the professional rider, just as Kawasaki had nearly 20 years earlier. It had been a hard lesson, but Milwaukee had profited from its mistakes.

As the century drew to a close, Harley-Davidson introduced its first new engine for 15 years, the Twin Cam 88. Now increased in size to 88ci (1450cc), but still the distinctive 45-degree-twin design, it offered more power, was smoother-running and met all the relevant and increasingly stringent noise and pollution regulations. It was to revolutionize general market sales, and was attractive to the police market that was gradually improving. The Road King FLHPI and the Electra Glide FLHTPI were offered in fuel injection form.

In 1999, recognizing the almost cult status of police bikes among some civilian riders, Harley-Davidson offered a civilian 'Peace Officer Special' to enable officers to ride off-duty what they rode at work. In fact, it wasn't available to just anyone. It was a clever marketing strategy, but the FLHTCUI, in two-tone blue, was available only to those producing proof of employment with law enforcement agencies, to make sure it remained something special. As an end-of-century statement, it underlined the status of the police-specification Harley to the public at large. Although Harley-Davidson would never again have a monopoly on police business (BMW, for one, made inroads in the 1990s), its firm place in the market place seems at last assured.

OPPOSITE: Cruising by the Boothill Saloon in perfect formation, just to keep an eye on things.

RIGHT: Can you honestly imagine this officer happy on a Honda scooter?

THE BUELL STORY

On 19 February 1998, Erik Buell announced to the world that he had sold almost all his remaining interest in the Buell Motorcycle Company to his business partner, Harley-Davidson Inc, the latter having held a 49 per cent stake in the company since 1993. Buell was named chairman and chief technical officer, but his boss was Harley-Davidson, and he was effectively back to how he started out.

In one way or another, Harley-Davidson had long been associated with Buell the company: it had supplied engines from 1987, and in 1993 suggested to Erik that it might take a stake in his company. By the early 1990s, Harley had become a successful producer of cruisers, only one sector – albeit a large one – of the motorcycle market, and Buell was involved in something quite different: muscle bikes. From Erik's point of view, it was a chance for Buell to benefit from Harley-Davidson's huge resources and secure its future at a time that there was difficulty paying the monthly wage bill. So in early 1994 the new Buell Motorcycle Company was born, with 49 per cent of it owned by Harley-Davidson. For Buell, the association has worked well, with production rocketing from a few hundred a year to over 6,000.

ABOVE: Erik Buell, the man behind the bikes.

OPPOSITE: The Buell RR1000 Battletwin.

The Early Days

But none of this would have happened without Erik Buell himself, his hard work, and his dream. He had ridden his first motorcycle at the age of 12, and thereafter had been well and truly hooked. To pay for night school, he decided to take a job at a bike dealer's and worked his way up from trainee mechanic to service manager in a short space of time. At the same time, he began road racing at amateur level, and his dedication and talent soon established him in the realms of the professional. Racing a TZ750 in the Formula One and a Ducati 900SS in Superbike events he was soon up with the front runners and in 1978 recorded the fastest qualifying time for a newcomer in the Daytona 200. This will give some idea of his talent.

Despite his racing exploits, however, Erik still found time to study his textbooks and win a degree in engineering in 1979. He had worked for this with one intention in mind: to stay in America and to continue to work with motorcycles. This was at a time that there was only one major manufacturer: he went to work for Harley-Davidson as a chassis engineer.

It looked like the start of a successful career, and the still-young Buell managed to achieve several motorcycle-related component patents in his few years working at Milwaukee. But the dream of building his own race bike was still there, and in 1982 he left to do just that, forming the Buell Motorcycle Company in the process.

The first motorcycle that he designed and built was the RW750 racer in 1983. This was a 45-ci (750-cc) two-stroke, 'square-four', rotary-valve machine, designed to compete in the AMA Formula One road-racing class. His inspiration actually came from within Harley-Davidson, as the company had already produced drawings for a two-stroke four-cylinder engine of its own as a feasibility study, but had quickly abandoned the idea. It did, however, plant the seed of an idea in Erik Buell's mind.

The RW's engine actually came from England, being the same Barton-built motor that had powered David Essex's Silver Dream racer in the film of the same name. Buell also bought a frame from Barton, but was not satisfied with it and built his own chassis to house the motor. This was the first bike he built with his own name on the tank.

The prototype first hit the track at Pocono Speedway in the summer of 1983 in

Buell
AMERICAN MOTORCYCLES
Buell
RR 1000
BATTLETWIN
POWERED BY/HARLEY-DAVIDSON

an AMA national race. Gradually, its builder (and rider) resolved what problems there were and by the the autumn the bike clocked 178mph (286.4km/h) on the Talladega track in Alabama. At this time, the Buell racing business was being funded by distributing parts brought in from England: exhaust pipes, Dymag wheels, Interstate leathers, to name but a few. He also sold items made by himself, with a view to keeping the cash coming in to allow development of the race bike to progress.

Buell carried on developing the RW750 throughout 1984 and a production version was released in the autumn of that year. He managed to sell just one RW750 to the American Machinists Racing Team, before the AMA announced that the following year

ABOVE: An RR1200 of 1990 used the 74-ci (1203-cc) Sportster motor, suitably tuned.

RIGHT: Buells rapidly took to the tracks, which was fitting, given that their roots lay in racing.

LEFT: But of course, Harley-Davidson had its own long racing history before Buell came on the scene. These are side-valve WR racers.

would be the last season of Formula One racing. Despite having the parts to build more RW750s, the project folded, so did the company and Buell lost everything, including his home.

At a stroke, the governing body of racing had effectively eliminated any future market for the RW750 by giving the premier slot to Superbike. But Buell viewed the rule change philosophically, and although tempted to go back and work for Harley, he changed direction and focused his attentions on building street bikes – machines that would be capable of making the transition to the track if the need arose.

The First Buells

Buell went back to work, this time with the goal of building the first world-class sports bike for the road, designed and built in the USA. Using his knowledge of what worked on the race track and his experience as a Harley-Davidson engineer, Buell went on to design his first candidate for the sports bike market: the RR1000.

This bike was remarkable in many ways, not the least of which was the fact that the project was started in September and the bike was up and running three months later. This was something of a record and one that Buell has yet to equal again. The RR used a Harley-Davidson XR1000 engine, which was mounted in a stiff, lightweight chassis. The frame was designed to allow rubber mounting of the powerplant, a system that has become a patented engineering trademark of Buell sports bikes. In addition, the design also utilized the engine as a planar stressed member of the frame.

Another Buell RR1000 first was the use of a rear suspension unit, working in tension, and mounted horizontally beneath the motor to allow a short wheelbase. This is another feature that can still be seen today on all the latest Buells. A total of 50 RR1000s were produced during the 1987–88 season, but Buell had already seen the new 74-ci (1203-cc) Evolution engine and saw it as a further step forward and another way of improving his bikes. Being mass-produced, it would be cheaper than the XR1000 V-twin, at the same time giving away little in power. He redesigned the chassis to accommodate the new engine and the RR1200 was introduced in 1988, with 65 produced for sale in 1989.

The same year, Buell also introduced the RS1200, a two-seat version of the RR model for riders who demanded both performance and two-up comfort. Just over 100 of these were produced through to 1990. The bikes also featured all-enveloping, streamlined bodywork, a fact which caused something of a stir at the time: never before had a bike powered by Harley-Davidson looked so modern or sporting.

Five-speed transmission was a new feature of the 1203-cc engine in 1991. Buell responded to revised engine-mounting points with further design improvements to the RS chassis. These bikes were the first production motorcycles

OPPOSITE: The limited-edition Buell XI, with 'upside-down' forks and that characteristic headlamp fairing.

BELOW: Buells have always been all-engine, bare-boned, minimalist machines.

RIGHT: RSS1200, a single-seat version of the RS1200.

BELOW RIGHT: Buells were the first Harley-powered bikes on which one could wear full leathers and a full-face helmet and not feel overdressed.

Buell

Buell
N537
CVV

Buell
AMERICAN MOTORCYCLES

OPPOSITE
ABOVE LEFT: The frame wraps around the engine: note the steering damper.

ABOVE RIGHT: Only a brave passenger would use that tiny pillion seat.

BELOW LEFT: The Lightning at rest.

BELOW RIGHT: There is a cutaway fuel tank on this show Buell to demonstrate the induction system.

ABOVE: The Uniplanar rubber mounting system quells vibration.

ABOVE RIGHT: A good view of the horizontally mounted shock absorber.

to use inverted front forks, stainless steel braided brake lines and a six-piston front-brake caliper. Manufacturing capability and quality control were also improved by the installation of a new composite and paint shop within the factory during this year. This allowed greater design flexibility and the capability to make changes quickly as well as aiding design work. A single-seat version of the RS1200 was then introduced for the 1992 model year. Dubbed the RSS1200, it won the enthusiastic approval of the motorcycle press on account of its lean, clean lines. Combined production of RSS and RS models through 1993 totalled 325.

But the company was still very small: Buell was employing just seven people at the time and needed a fresh injection of cash if it was to survive and develop. Hence the involvement of Harley-Davidson and the subsequent takeover. This amalgamation was to bring a new generation of mould-breaking muscle bikes, and facilitated the transition of Buell from a tiny producer of limited editions to a true manufacturer.

The first bike to roll out as part of the new era was the S2 Thunderbolt, a generously proportioned machine which could offer a measure of weather protection to the rider. Although sport-orientated, it was really what the Japanese describe as a sports tourer. There was even a police version of this and it was used by several departments. Then came the 'proper' touring version of this, the S2T, complete with panniers and leg shields.

From this evolved the S1 Lightning, a minimalist bike devoid of all but the barest of necessities and looking very different from anything else on the market. A cheaper, entry-level machine was the M2 Cyclone, with conventional forks, as opposed to the inverted ones, and a lower state of tune.

Further variations on these models has since produced the S3 Thunderbolt, the White Lightning and Lightning Strike and the X1, but they are still based on the same principles and basic themes. All are selling well and attracting a growing band of fans that might not have otherwise considered a bike with a Harley engine.

But all these bikes were still at the enthusiast end of the market: what Harley-Davidson required was a genuine entry-level machine for first-time riders who,

Buell
Blast!

OPPOSITE: Blast! The first small Buell ever – in fact the first small civilian Harley for over 20 years.

ABOVE: Erik Buell, left, being interviewed by *Big Twin* magazine.

having found their motorcycling feet, would go on to buy bigger, more expensive Harleys. Even the Sportster 883 was too large and intimidating for some, but the answer came in the form of the single-cylinder Blast! (the exclamation mark is part of the name) which was unveiled in 1999.

It utilized a new 30-ci (492-cc) single-cylinder motor, though it owed something to the existing Sportster V-twin, that was mounted in a steel backbone frame which rolled along on 16-in (406-mm) wheels to keep the seat height low. However, typical Buell design details could still be seen, such as the silencer under the motor and the large front disc brake, which add confirmation to the name emblazoned on the tank. It is a Buell, but not as people have come to know it.

Buell was quick to acknowledge that the tooling and processes used on the V-twin motor were used in the production of the single, along with a few components like the gearbox. However, it was also stressed that it was much more than an existing V-twin powerplant, chopped in half. The overall philosophy was simplicity and ease of riding, seen as most important in attracting the learner rider. The small stature of the bike, and the total lack of complication to enable the rider to concentrate on learning to ride, were also paramount in the execution of the design.

All this shows that Buell is not content to stand still and will continue to move forward, providing Harley with what it wants, and is an innovative arm that will try out new markets. This leaves Harley in a safe position and precludes it from compromising its important heritage.

Buell's Philosophy

Erik Buell has always been acknowledged as a deep thinker and a person who can recognize a problem and formulate its solution. There are two words, however, which encapsulate his philosophy of engineering: innovation and simplicity. When one looks at both the V-twins and the single, one can see that these principles are uppermost in the design and manufacturing of the Buell machines.

Buell also believes that the objective is to 'create motorcycles for enthusiasts who believe the fun of the actual riding experience is every bit as important as the destination itself'. Every Buell motorcycle appears to have been designed, engineered and manufactured with this in mind.

Buell insists that some bikes are becoming over-complicated and designed with top-speed as the ultimate goal. This, he argues, is only relevant on the race track or on unrestricted autobahns: for most riders, speed is irrelevant. So he provides a bike which can be ridden and enjoyed in a real-world scenario. Likewise, the style and form is designed with function in mind rather than with a view to pleasing design committees. This has led to the bikes' very distinctive style, one that is unmatched by other manufacturers.

Buell
AMERICAN MOTORCYCLES

OPPOSITE: Buells can be as much about looks as performance. This one has many aftermarket parts.

ABOVE: Buells and Harleys don't appear to be on speaking terms!

ABOVE RIGHT: Thunderbolt has a bigger seat and more bodywork than the Lightning.

RIGHT: Buells are more slimline, far less heavy machines than Harleys.

Apart from this, all the bikes have been built bearing in mind Erik Buell's three fundamental design concepts. These are: Mass Centralization, Frame Rigidity and Low Unsprung Weight.

Mass centralization is the practice of concentrating the greatest amount of mass as close to the centre of the motorcycle as possible. By positioning the engine, exhaust and suspension components of the motorcycle near the centre of the bike, frame stress is reduced and there is a lower polar moment of inertia. The horizontal rear shock-absorber not only helps to keep the mass of the bike centred, but also reduces weight.

Frame rigidity results in a machine that offers the rider greater control. To this end, the light 29-lb (13-kg), but immensely strong chrome-moly geodesic perimeter-frame design has many triangulated sections for maximum frame stiffness.

Minimizing the weight of those components not supported by the motorcycle's suspension springs is also critical to control and the feel of the ride. Low unsprung weight allows the tyres to maintain contact with the road, irrespective of the surface conditions. The six-piston, single-disc front brake on all Buell models is one way in which unsprung weight has been reduced, saving around 4–6lb (2–3kg) compared with conventional twin disc set-ups. Whilst saving weight, the 13.38-in (340-mm) disc and six-piston caliper also offers superb braking power.

Maybe it is the pursuit of these goals that have given Buells their quite unique look. One of the most distinctive features is the large-volume Helmholtz Resonator

HARLEY-DAVIDSON

OPPOSITE: Lightning: Buell has become the definitive muscle bike.

RIGHT: This Victory cruiser has inspired Erik Buell to make best use of a big torquey V-twin engine, but in a sporting chassis.

airbox, which has been the subject of much criticim and has been seen as spoiling the bike's looks as well as restricting leg room. But there is a reason for its size. It is not an empty box: inside, tuned ducts and chambers allow free air-flow for instant throttle response and peak performance, while cutting down induction noise to meet European noise regulations. The X1 features the same large-volume airbox, but the design is now integrated within the form of the fuel tank cover for improved styling and as an answer to the styling critics.

Buell's Thunderstorm V-twin engine is built by Harley-Davidson, of course, but it is not the same motor you will find in the Sportster. There is more torque and power compared to the stock item, obtained by using different cams, lighter flywheels and high-compression cylinder heads with bigger valves and revised ports. The motor is fuelled by Buell's Dynamic Digital Fuel Injection (DDFI) system. Its closed loop design monitors conditions inside and outside the engine, constantly checking fuel, air mixture, ambient air temperature and humidity, engine temperature, barometric pressure and exhaust output to ensure optimum performance.

The stainless-steel exhaust pipes that offer a tortuous route into the tuned large-volume muffler, running under the bike, have often been the subject of comment, like the air box. The reason for the design is that it enables exhaust gases to flow more freely. Likewise, the large muffler has internals that are scientifically tuned to cancel out problem noise frequencies and allow the bike to pass noise tests. The patented Uniplanar engine vibration isolation system uses rubber mounts to isolate vibration from the engine, to which the swing-arm assembly is bolted. Rose-jointed tie-rods restrict the movement allowed by the rubber mounts to a vertical plane to ensure constant wheel alignment.

Allied to this, Buell, like Harley, uses a strong and durable toothed belt for the final drive. The belt is clean, quiet and smooth in operation, with low maintenance requirements, and its lightness contributes to the Buell's low unsprung weight.

As one might expect from his background, Erik Buell still uses the race track to test and validate technological developments and to improve the production motorcycles. They are now the important issue, but it is well known that, 'racing improves the breed', and allows bikes to be pushed hard and tested to their limits. Buells compete in the AMA Pro Thunder Series, and the experience gained is fed directly back to street-bike development.

Riding the Brand

Riding a Buell is different from riding any other Harley. While the expression, 'laid-back', may be appropriate to most of the Harley-Davidson range, it cannot be applied to any of the Buells and that includes the Blast, civilized though it may be. To give an idea of what Buell has created, here are a few examples.

S1 Lightning

The S1 Lightning adheres to the Buell philosophy in every respect. It is a sports bike, but an unconventional one. Visually the bike is different: it is devoid of the acres of plastic that one normally associates with

OPPOSITE: Lightning at speed: that large black box beneath the motor is a silencer.

RIGHT: Buells have proved their worth on the track.

a sports machine. Its looks are nonetheless striking, due to its abbreviated bodywork, large circular halogen headlight, and triangulated chrome-moly frame. But the critics are right: that large airbox on the offside, not to mention the huge silencer beneath the engine, are eyesores, disturbing the overall appearance. Buell trademarks they may be, but they are far from attractive.

At the time of its introduction, the first S1 was the lightest bike ever produced by Buell, weighing in at 425lb (193kg).

The mass is centralized and as low as possible to keep the centre of gravity low, so aiding stability and responsiveness according to Buell's basic design concepts. Allied to this is a short and stubby 55-inch (140-cm) wheelbase, which adds to its desirably responsive handling characteristics.

Right in the centre of the bike is the Harley 74-ci (1203-cc) engine, but in a very different state of tune to the one that powers the Thunderbolt. Buell utilized the skill of Harley engine tuner Gary Stippich to provide an engine that, as strange as it may seem, began life in the same factory. Most of the work was done on the top end of the engine, although 3lb (1.4kg) of metal was shaved from the flywheels.

Completely new cylinder heads were cast, and these now combine the combustion chamber shape of the smaller 54-ci (883-cc) motor, with the larger valves of the bigger engine. While they were at it,

91
SUPERMONO
DESMOQUATTRO
buell
AMERICAN MOTORCYCLES
91

the compression ratio was also raised from 9:1 to 10:1. Some parts were replaced by Screamin' Eagle performance components, such as a cam with a longer lift duration.

As one might expect while the work was in progress, the inlet tracts were modified and the carb re-jetted to cope with change in demand. This also necessitated alterations either side to deal with changes in volumetric efficiency. Independent specialists were brought in to redesign the airbox and exhaust systems so that the bike would meet all the relevant noise and emission tests, not only in the US, but also in other world markets. These changes boosted power to a claimed 91bhp, being 15 more than a Thunderbolt engine and a massive 50 per cent more than a standard Harley.

The modifications have not affected the rideability of the bike. It is all usable power, allowing the S1 to pull all the way from tickover to the red line, with no fluffing or flat spots: there is raw power all the way through. Open the twistgrip and the large airbox gulps air in as the exhaust note rises and the horizon approaches. Exit a bend or corner with an open road ahead, feed the power in hard and the bike gallops towards the next.

Reduce speed for the approaching corner (normally backing off the throttle is sufficient, due to the masses of engine-braking available), take an appropriate gear, and drop the bike in, with the knowledge that it will hold its line without complaint. The inverted multi-adjustable WP forks at the front seem to have no problems dealing with any road irregularities, and only the worst of these cause a little twitch from time to time. At the rear and below, the single shock in tension does a similarly reassuring job keeping the back wheel in check.

The reworked motor can be spun hard towards the 6,300rpm red line with no worry and if this is done the bike will go as fast as its name suggests. one can feel the engine working, but the vibes are not intrusive. The five-speed box is always precise when shifting through the various well-chosen ratios, but changes cannot be hurried if they are not to result in a clashing of cogs. Certainly, there is also no noticeable transmission snatch from the Kevlar belt drive, or for that matter delay in responsiveness.

The chassis too has no flex or ever seems to get out of shape, even when the bike is pushed hard, which is what you would expect from a design by an ex-racer, while, overall, the handling is always predictable when the bike is being ridden

ABOVE: Buells handle well due to a stiff frame and relatively low weight.

OPPOSITE: Battle of the Twins racing has allowed Buells to race competitively against Ducatis and other twins.

through the twists and turns of meandering roads. At the other end of the scale, relaxed riding is possible, but not necessarily recommended. The heat getting through to the rider can be uncomfortable, especially in heavy or slow-moving traffic, though of course this was never intended to be Lightning's environment.

On the freeway, legal speeds are fine, but the upright riding position, due to the straight bars, becomes tiring once legal limits are exceeded. Head towards the 130-mph (209-km) top-speed and you will need to hang on grimly, which puts strain onto wrists, arms and shoulders. This is not the machine for 'green' riders: if you head for the red line, or cruise at high speeds, it is a real gas-guzzler. Painful at any speed is the seat, which is too sparse and minimalist, as is the leg-room for those on the tall side. Pillion riders also get a raw deal, and should not be contemplated for any length of journey, unless their rear end is on the small side and you are good friends!

Time spent cruising at steady speeds on an straight and open highway gives one time to appreciate the quality of the bike, the deep lustre of the paint on the petrol tank, cowling and mudguard glowing expensively in the sun. One begins to appreciate the neat alloy housing for the clocks and sensible well-made switchgear and controls that are very un-Harley. In fact, one will be aware of all the attention to detail that makes the bike what it is: a hybrid of mass-production and the hand-built. Operation of all the controls and operating systems, from the wet clutch through to the brakes, is smooth and gives plenty of feedback. The massive front brake is excellent, the six pistons gripping the 13.4-in (340-mm) disc with enough power to slow from whatever speed one likes and deal with whatever weight of rider may be on board at the time. At the back end, a single cast-iron disc takes control, gripped by a Brembo caliper, which is sufficient when considering the front's effectiveness.

Top quality components are the name of the game in all departments, like the 17-in (432-mm) Marchesini cast-alloy rims, shod with Dunlop Sportmax tyres. Both are well suited to one another and the bike in general, and allow the handling limits to be explored to the full.

Erik Buell said that it was his intention to create a bike for riders who believe that the actual riding experience is more important than arriving at the destination. The Lightning fulfills this promise: it goes like a bat out of hell, handles well and more importantly stops when required. It has its faults, of course, but more than anything else, it is the embodiment of the Buell philosophy: enjoyment.

ABOVE: WR-series Harley racers from a different era.

OPPOSITE: The 30.5-ci (500-cc) Buell Blast offers ease of riding and a low seat height for novices.

The Blast!

At the other end of the scale is the Blast. Criticized for its name, that suggests a level of performance and handling it clearly does not possess, it is nevertheless an important model in the range and demonstrates the level of proficiency within the company and the fact that it is now looking to diversification.

The signs are that the Blast is already selling in large numbers; no doubt some view it as a cheap option or as a ride-to-work machine. Others will want it for its light weight and compact dimensions, not to mention low running costs. And of course, it is a genuine American, which to many is an important factor. But the Blast's main target is the novice rider, and is designed to give them a taste of motorcycling before they move on to something bigger. Consequently, the small, compact, single-cylinder machine is uncomplicated and user-friendly in all departments.

The plastic bodywork is virtually indestructible and will resist damage if the bike is knocked, so reducing repair bills and making the learner more relaxed; it has a clean, almost maintenance-free belt drive, together with ease of use. In theory, all the novice has to do is acquire the on-road skills. To help remove anxiety regarding its perceived complexity, the Blast has an automatic choke, like a car, easy-to-use controls, minimal instrumentation and nothing more to do apart from filling the tank with fuel and putting air in the tyres. It is significant that it is not only private individuals that are buying Blasts, but also bike-training schools.

The Blast is surprisingly plush and comfortable, and despite its small size can be ridden by a 6-footer with no trouble at all. The bike's main frame is composed of a wide steel beam which forms a backbone and an attachment point to which the other frame parts are mounted. These consist of a box section and formed flat plates, rather than the round section tube we are more accustomed to seeing. The main section is also an integral oil reservoir to make use of the space and reduce weight. The engine is mounted using the Uniplanar system with which Buell riders are now familiar.

The 30-ci (492-cc) motor is relatively low-tech in some respects, having just two valves operated by hydraulic adjusters, which are self-regulating. Cooling is by air and the unit-construction powerplant has just five gears contained within its case. There are no balance shafts to smooth out vibration, unlike Japanese singles; instead, it relies on the unique engine-mounting system, which actually allows the factory to dial in a few vibrations to get riders accustomed to them, so that when they move up to a full Harley engine they will be unfazed by their presence.

An automatic choke on the 1.5-in (40-mm) Keihin carb makes starting a one-touch affair (no kick-start) to get the single spinning, and it soon settles down to an even if lumpy tickover. It is then a case of engaging the first gear, that has the inevitable clunk from rest, and moving out into the traffic to experience the full Blast sensation. The exhaust note has been tuned to make the bike sound right and in fairness it does sound pleasant as one pulls away from rest. Around town, or using the available power to go through a series of bends, it also sounds good, spot-on, in fact. However, at cruising speeds, the wind cancels this out and the exhaust is almost inaudible above the rush of the wind.

At low speeds, the built-in vibes are noticeable. However, like most American motors, once out on the open road, everything smoothes out and the bike will happily cruise at all legal speeds with enough in reserve to make it less tiring than when it is at full stretch. In traffic, it is nimble enough to allow gaps in traffic to be manoeuvred and the low-down punch of the engine means the gearbox does not have to be worked too hard: there is good throttle response as well.

Controls and layout are functional and easy to use. There are no frills – just the basic legal and safety requirements. The electronic speedometer with its LED odometer, is easy to read at a glance, while still giving sufficient information to the rider. Petrol consumption is quoted as an average of 70-mpg of mixed riding, which is quite useful considering that the tank capacity is only 2.8gal (12.7 litres), and means that there could be around 200 miles (322km) between fill-ups.

The bike copes as well with two up as it does with one and the small five-spoke 16-in (406-mm) wheels, shod with Dunlop tyres, allow the bike to track well through

Buell
Blast!

sweeping, flowing bends. It is just as efficient on tight ones, due to the good overall balance and weight distribution. At the front, the 1.5-in (37-mm) conventional Showa telescopic forks also cope well, considering that there are no means of adjustment present. While this is not uncommon on the front, no adjustment on the rear suspension, not even pre-load, seems a strange omission. The idea, of course, is to reduce the complexity of riding the bike: it must be remembered that there are experienced riders who have no idea how to set up the suspension system for optimum handling and comfort. However, the Showa gas-charged shock seems up to the task and the bike is unlikely to get out of control or ran out of ground clearance. The traditionally large (for a Buell) single front disc brake has no trouble slowing the Blast's relatively light weight.

There are a host of factory extras, such as tank bags, panniers and tail racks to give the rider a chance to use the bike for touring or for carrying shopping. For those wishing to be different, there is a rather tacky decal kit available as a factory option to allow one to personalize the bike.

All in all, when looking at the Blast in relation to the market for which it is intended, there is little to criticise. Experienced riders may want more, but in reality, provided one is willing to work the bike hard, it provides an acceptable balance between function and fun.

Initial production has had to be stepped up to meet demand, so it looks as though this latest extension of the Buell philosophy has succeeded. Inextricably part of the Harley-Davidson story, it looks as though Buell will remain there for some time to come.

OPPOSITE: Harley-Davidson hopes the Blast will take it into a whole new market.

LEFT: Five-speed gearbox and single-cylinder engine have much in common with Harley components.

RIGHT: Blast is styled to look like the bigger Buells.

THE RACING GAME

It would be impossible to catalogue all of Harley-Davidson's racing exploits. Over the years, the bikes from Milwaukee have competed on board tracks, dirt tracks and tarmac. They've climbed cliffs, hills and mountains, raced on the road and off it, indoors and out and with short circuits and gruelling endurance runs, they've done the lot. It is all the more surprising when one learns that the four Founders, William Harley and the Davidson brothers, were no great fans of competition, seeing it as an expensive distraction from the main business of making and selling bikes for the road: in fact, it would take over ten years for them to be persuaded to officially enter a race. This was the 300-mile (500-km) event at Dodge City in 1914, and even then Harley failed to win! But from this inauspicious start, Harley-Davidson was to go on to be a major force in almost every aspect of motorcycle sport. This is so large a subject that it would fill a complete book, so we shall concentrate instead on circuit racing in all its forms.

In fact, the company had officially taken part in competition before that Dodge City race, but these were endurance and reliability runs, not races, the trick being not to finish first, but simply to finish, which with pioneer bikes on rudimentary roads, was an achievement in itself.

A board-track racer and FLXI 'flexy' sidecar.

Walter Davidson himself had entered and ridden in these events, notably the two-day endurance run to Kokomo in Indiana in 1907, and back to Chicago. He was one of only three riders to complete the 400-mile (650-km) event with a perfect score, which was no mean feat considering that roads were little more than horse trails in places. A year later he did the same on a belt-drive single-cylinder Model 4, from Catskill, New York to New York City, completing the run with a perfect score, which was good publicity for Harley-Davidson. Both events revealed the ruggedness and reliability of these early Harleys, something the company was keen to accentuate, even in those very early days. If it were to survive in competition, it needed to demonstrate the extreme reliability of its bikes, rather than their speed, and endurance runs did just that.

Racing, on the other hand, involved bikes being stripped and tuned, often over-tuned. They could blow up, crash or break down, giving the impression that they were unreliable. This, in other words, was the

Not all surviving board-track racers are museum pieces: these bikes are part of a demonstration race at Davenport, Iowa. None of the board tracks themselves have survived but they were relatively cheap to erect and allowed high-speed racing in almost every county. But the crashes could be horrific, and they were superseded by the dirt-track ovals which took precedence in America over European-style twisty tarmac circuits.

antithesis of the Founders' philosophy.

But this didn't mean that Harley-Davidson had no involvement in racing. There were some very good privateer riders around, keen to ride Harleys, so the solution was to quietly build race bikes for certain favoured riders, which Harley did, as early as 1910. If they subsequently won, all well and good. If they didn't, then their entry could be regarded as a purely private affair, and nothing to do with the factory!

Around this time, the 'Speed Bowls' or 'Thunderdromes' were also popular, with bikes being raced on 60-degree wooden slopes. Speeds of up to 90mph (145km/h) on rotting wooden boards were achieved by bikes with no brakes and cut-off exhaust pipes and often meant that riders were badly hurt, splinters being the main cause of injury. This led to the nickname 'Arena of Death' and initially, at least, Harley had no wish to be associated with them, especially after the 1912 catastrophe at Newark, when a bike spun into the crowd, killing spectators.

Arch-rival Indian, along with many other marques, had a different attitude to racing, and was never slow to capitalize on achievements in the sporting arena and extract the maximum media coverage available. It was also taking part in events outside America. When it took the hat-trick of wins at the 1911 TT races on the Isle of

ABOVE: Early singles raced the board tracks as well.

OPPOSITE: This is Steven Wright's restored eight-valve racer of 1915/16, which was Harley-Davidson's first serious stab at a purpose-built racing bike.

HARLEY-
DAVIDSON

Man, the publicity was immense and Indian bikes sold like there was no tomorrow. Harley-Davidson noted this and began to advertise its privateer wins; but to a certain extent it was still sitting on the fence, as was seen in a 1913 desert race at Phoenix, Arizona. The Harleys failed to do well, and the factory covered its shame by hastily pointing out that the entries were private, not factory-sanctioned!

In fact, that same year it issued a statement to the effect that 'race victories are just a needless desire to impress and something a serious manufacturer has no need to do'. But behind the scenes it had decided to adopt an official racing policy, though more out of commercial necessity than any genuine love of the sport.

RIGHT: The 1916 F-head motor wasn't as powerful as the eight-valve, but was often more reliable.

OPPOSİTE: The F-head V-twin formed the backbone of Harley-Davidson's early racing effort, though special racers from other manufacturers soon persuaded it that something more exotic was required.

HARLEY-
DAVIDSON
10

LEFT: An eight-valve racer being fettled at a modern race meeting.

OPPOSITE
ABOVE LEFT: A 1916 board-track racer. The white tyres, wheels and frame are period items.

ABOVE RIGHT: A 1922 74-ci (1200-cc) JD board-track racer with exposed chain drive and extreme riding position.

BELOW LEFT: A 1920 61-ci (1000-cc) racer, as ridden by Dewey Sims.

BELOW RIGHT: An eight-valve board-track racer. Note the exposed pushrods for the four valves per cylinder and friction-type damper for the leading-link front forks.

The Ottaway Era

William 'Bill' Ottaway had been poached from Thor, another fledgling manufacturer in the US at this time, and one which was doing well in competition. Ottaway was largely responsible for this racing success and was given the task of setting up Harley-Davidson's race department. As a result, he was to be central to the company's early golden era of racing victories.

Harley-Davidson's reasons for this dramatic U-turn in policy may have occurred when the Founders realized that they had already achieved their initial goal, without compromising the principles on which the company had been based. The bikes were solid and reliable, capable of lasting the distance; now they needed to make sure they were the first to cross the finishing line, which would be Ottaway's job. Competition among manufacturers was growing more intense, with smaller players falling by the wayside: racing seemed to be just another important weapon in the growing sales war.

HARLEY-
DAVIDSON

HARLEY-
DAVIDSON

Bill Ottaway certainly came up with the goods in the form of an 8-valve V-twin (four valves per cylinder), based only loosely on the existing 61-ci (1000-cc) engine, which cost over $25,000 to develop. Part of these costs went to Englishman Harry Ricardo, the engine breathing specialist. He solved the intake and exhaust problems that were initially experienced with the engines, and when he had finished, they were putting out a very respectable 50bhp.

Well-heeled private racers could in theory buy one of these new 8-valvers from Ottaway's new racing department at Milwaukee, but at $1,500 each there weren't many takers. There was a rumour that the high price was to discourage privateers, enabling the best bikes to be kept for the new factory race team. For the first well-publicized race on dirt, which coincidentally was on 4 July, Harley entered five brand-new type-11K 8-valve machines, all personally prepared by Ottaway himself. It was not a fairytale debut, and there was strong competition from eight-valve Indians: the Harleys were not placed, although all finished the course.

The factory was not impressed, and expected better things considering the money it had invested in the new racing project. However, matters were to improve and by the end of that year it had claimed a championship win at Alabama. The next year, Harley was featuring a little more prominently in the results as its knowledge and expertise in the world of racing increased. In the road-racing series Otto Walker took a couple of prestigious wins and the opposition began to realize that Harley was now serious in its intentions towards racing.

When the 1916 racing season began, the famous Wrecking Crew was formed. The name was coined by journalists of the time, but was descriptive in describing the

ABOVE: A 1926 61ci (1000-cc) board-track racer V-twin. There was nothing to these early racers, which were little more than bicycles with overpowered engines bolted in.

RIGHT: Rear view of Steven Wright's eight-valve board-track racer.

OPPOSITE: A museum reconstruction of a board-track Motordrome scene, courtesy of the Otis Chandler Collection.

HARLEY-DAVIDSON
HARLEY-DAVIDSON
MILWAUKEE

HARLEY-DAVIDSON

OPPOSITE
LEFT: A single board-tracker: how simple can a racing engine be?

RIGHT ABOVE and BELOW: A 1922 74-ci (1200-cc) JD board-track racer. But a year before, Harley had withdrawn official support from the racing programme.

RIGHT: An original factory racer, the only known complete and original pre-1916 bike, with original race 'M' motor.

way Harley wrecked other teams' chances of success. Brought together by Ottaway, this hard core of riders changed over time, but some names remained in the picture and included Jim Davis, Ralph Hepburn, Fred Ludlow, Otto Walker, Maldwyn Jones, and Ray Weishaar.

Harley not only employed the best riders and kept them trained, it also provided the best back-up and logistical support. It is rumoured that Ottaway trained them with military-like precision, for pit stops and tactics, pit signals and so on. Whatever the truth, the results gave credence to this, with 15 National Championships coming its way in 1916. On the subject of tactics: the 8-valve machines were fast but still not as reliable as the production-based F-head bikes. So a favourite ploy was to enter teams consisting of both. In the early stages of a race, the 8-valvers would storm away to the front, daring other riders to keep up. When they and their rivals were sidelined, the slower F-heads were sitting near the back of the field, ready to take the lead.

At the end of 1916, top-level motorcycle sport in the US ceased after there was an agreement between all the major factories, who were gearing up for military contracts, and it would be three years before racing again began in earnest. The line-up of the Wrecking Crew had by now altered slightly, but their lust for victory had increased and one could see that they meant business straight away. New pit crew and extensive practice drills meant that a bike could be refuelled, given oil and a new rear tyre in 38 seconds, which is quite something even by today's standards and gave them a big advantage over the opposition.

One record after another began to fall to Harley and the team won the most important race of the year in Indiana, taking second and third place for good measure.

Meanwhile, the endurance runs had not been shelved, and Harley-Davidson dealer Hap Scherer set several new records on the new Sport Twin, which was designed more for reliability than outright speed. He rode it from New York to Chicago, covering the 1,012-mile (1630-km) route in 31 hours and 24 minutes. He also completed the famous 'Three Flags Run', from Canada to Mexico on a Harley. Incidentally, one of these bikes

HARLEY-
DAVIDSON
HARLEY-DAVIDSON
OF SUFFOLK
BARON
2

was also the first motorized vehicle of any type to climb 'Old Baldy' or Mount Antonio in California, whose 10,114-ft (3083-m) summit was accessed over a 7.5-mile (12-km) burro trail and demonstrated that the lightweight machine was as robust as it was light. Many a specialized vehicle had tried before and failed, so it was no mean achievement.

While 1919 may have been a good year in terms of results from the reformed race department, the following year was even better, with five national championships won that season. At the beginning of the year, Red Parkhurst competed at Daytona, setting several short-distance records, and using an 8-valve to take a combination record. Meanwhile, records were also being broken in the Dodge City race, re-run for the first time since 1916, and making up for Harley-Davidson's poor start at its debut event. In the same year, the origin of 'Hog' as the quintessential Harley nickname was established. Ray Weishaar had adopted a pig as his mascot at race meetings, and when he won the Marion in Indiana, setting a new record time in the process, he took it for a victory ride around the track. The pig, wearing a Harley vest, was named 'Hog'.

But despite success on the track, motorcycle sales were gradually decreasing and dealers and manufacturers alike felt that the bike had become more of a luxurious toy than a means of transportation. Cars were getting cheaper all the while, wresting the family transport role from the bike and sidecar. Walter Davidson was also increasingly frustrated that race success was failing to encourage more direct sales, and when the local police department at Dodge City still persisted in buying Indians, in spite of Harley's wins, it was the final straw. The Founders decided to pull out.

So although Harley entered the 1921 season with full works' support, it was to be its last for a few years, though the team itself did not know it at the time, neither did Bill Ottaway. Ironically, it had been its best season to date, the Wrecking Crew having won just about everything. By the end of the year Harley had won the 1, 5, 10, 25, 50, 100, 200 and 300-mile championships – in other words all of them – winning one race with an average speed of over 100mph (161km/h), the first time that this had ever been achieved.

Harley had spent $200,000 on racing that season and it is rumoured that certain riders had been paid as much as $20,000. But sales were failing to recoup this expenditure and the plug was pulled without the riders being aware of it until it was too late. (The cynical would describe this as a ploy to keep the riders committed until the end of the season.) The end was acrimonious, many of the Wrecking Crew having been left stranded in Arizona at that last meeting, with no money to get home. They were reduced to borrowing their train fares from a local Harley dealer and when they did get home, Walter Davidson refused to take their calls. It was a sad end to a glorious era.

OPPOSITE: The Flexy was quite unlike modern racing sidecars in that it leaned around corners, just like the bike.

ABOVE: A board-track racing sculpture captures the essence of this close and dangerous sport.

LEFT: A 1927 21.35-ci (350-cc) OHV factory Peashooter. The little overhead-valve single was derived from a road model, and proved highly successful in Class A racing.

OPPOSITE: A later-model Peashooter with Art Deco-style tank finish. There were still no brakes, however.

Petrali & the Peashooters

Harley-Davidson may have pulled out, but its bikes were still being raced; even some of the Wrecking Crew continued on a private basis. There was also success overseas and Harley received some favourable publicity in 1923, when Englishman Freddie Dixon, riding a Harley, took the 61-ci (1,000-cc) World Championship at the Brooklands circuit in Surrey, England. In fact, two years earlier, a Harley had been the first-ever machine to lap the circuit at over 100mph (161km/h), ridden by Douglas Davidson (no relation).

With racing still so popular and with many more events taking place, the American Motorcycle Association (AMA) was formed in 1924 to organize the world of competition and act as an official regulatory body. One of its first problems was that the performance of the big V-twin racers was now outstripping the capabilities of their chassis. The number of fatalities and accidents was increasing, and the obvious solution was to slow race bikes down by reducing the engine size permitted. Harley-Davidson was ready to take advantage of this, having developed 21 and 31-ci (350 and 500-cc) single-cylinder OHV machines for the export market. Speedway was particularly popular in England and Australia and the bikes were designed with this in mind. Bill Ottaway, Arthur Constantine and Hank Syvertson developed the motors so that they were competitive from the start, but they also formed the basis of one of the

most famous racing Harleys ever, the short, cobby little Peashooter. This lightweight 350 single was to feature strongly in various race disciplines right through the 1920s.

In 1925 Harley made something of a comeback to racing when Joe Petrali beat all the established top-name riders at Pennsylvania on a twin-cam Harley. Relatively unknown until then, he had been riding Indian singles until the age of 21, mainly on the board tracks that still proliferated around the USA. He turned up at the meeting without a bike, but managed to talk his way onto a works Harley, replacing the injured Ralph Hepburn. He began the race as an unknown and ended it as a legend, having averaged 100mph (161km/h) for the 100-mile race and beaten all the established stars along the way. This was the start of a long racing career in which he was to maintain a virtual monopoly of the chequered flag for the next ten years.

To give an idea of Joe's record: he won the National Hill Climb Championship every year from 1932 to 1937, and speedway races in record times. From 1933 to 1936 his wins included the 200-mile (322-km) Expert Championship, the miniature TT championship, the 200-mile Speedway Championship and all 13 Dirt Track Championships. He also set four AMA records along the way. In 1935, for instance, he took first place in every single National Dirt Track race of the season. At one event, he managed five wins in a day at a New York meeting. Not content with actual track racing, in 1937 he took a streamlined Knucklehead Harley to a world record speed of 136.183mph (219.15km/h) at Daytona Beach.

One might imagine that a rider of this calibre would have received handsome remuneration from Harley for his trouble, and given every conceivable help there was; but this was not the case and would eventually lead to a parting of the ways. Even that new Knucklehead record at Daytona failed to net him a bonus, because he had failed to reach 150mph (241km/h)!

Meanwhile, with the country in the grip of the Depression, and the manufacturers more inclined to concentrate on survival, racing was in the doldrums. In an effort to give it a boost, the AMA introduced Class C racing in 1934. This was designed to be cheaper and more accessible, attracting more amateur riders without destroying the existing Class A and B professional factory-backed events. The bikes in Class A (21-ci/350-cc ohv singles and 45-ci/750-cc side-valve twins) and B (45-ci twins or 80-ci/1300-cc fours) were pure racers, but in Class C, the bikes (limited to 30.5-ci/500-cc ohv or 45-ci side-valves) had to be production-based.

As it happened, both Harley and Indian both made machines ideally suited to the 750-cc end of the class, which was hardly surprising, as both were highly influential with the AMA, having rescued it from financial ruin. Because of this, the AMA was obviously not totally independent at that time, both companies having a commercial interest in the rules, something that would later change. But for the moment, the rules undoubtedly favoured home-produced bikes over the imported ohv singles.

The initial rules also stated that bikes

had to be ridden to meetings, so Harley sold its WLDR fully equipped with all road-legal equipment such as lights. Moreover, the bikes had to be part of a production run of at least 25, to prevent the use of one-off specials, and even had to be offered for sale after winning a race for a maximum of $1,000. A sound marketing move, perhaps, but Harley ended up finishing behind Indian in most events. So in 1941, Harley marketed two bikes solely for racing, the WR and WRTT. The W indicated the 45 engine and the R racing and it was the sort of bike that the dealer and private owner would end up with after stripping down the WLDR.

This appeared at a time that the rules had slackened and because many of the bikes were now being towed on trailers to meetings. These new models already came stripped of heavyweight road-legal equipment. Tuning was becoming *de rigueur* and Harley was listing many tuning items in its parts catalogue.

The WR racer is still ridden in anger, on the tarmac circuit at Daytona (below) and on dirt track (opposite).

6P

However, the rules had the effect of creating a level playing field and if a rider did not win, it was now down to him rather than the bike. By 1938 the AMA had actually suspended the A-class races as there was now little interest, due to the success of the C-class having captured people's imagination.

Of course, racing was also taking place on the dirt ovals and the sight of 130-mph (210-km/h) V-twins broadside with no brakes made the European Speedway scene look rather tame. It was at these events that riders learned to control 45-ci (750-cc) machines on the slide and taught riders how to totally master the bikes so that, on returning to tarmac, they cleaned up there as well. It was this dirt-based apprenticeship that made so many American riders adept at high-speed slides, and was one reason why they went on to dominate world-class racing in the 1970s and '80s.

ABOVE: WR racers still relied on hand gearchange. There is a Hummer racer in the background.

OPPOSITE: The WR was a bitter rival of the Indian Scout.

Indian
Indian
HARLEY
DAVIDSON
1
46x

The Enthusiast
A MAGAZINE FOR MOTORCYCLISTS
MARCH
1940

HARLEY
DAVIDSON
STEVE COE
213
HARLEY DAVIDSON
213
213

213
213
213

OPPOSITE
LEFT: March 1940 was a time of war in Europe, while racing in the USA continued, as this cover of Harley-Davidson's *The Enthusiast* magazine shows. Races at Daytona really were run on the beach.

ABOVE RIGHT: Steve Coe, a well known WR racer, with his son at Daytona.

BELOW RIGHT: WR racers aren't yet valuable museum pieces, or at least, not enough to prevent amateur racers like Steve Coe using them for their intended purpose.

RIGHT: A WR racer in action at Daytona. The classic races are an integral part of Daytona Bike Week.

LEFT: These two late-model WRs, both owned by Max Middlebosch in Holland, appear to be fitted out for the road, with moderately high bars and a silencer.

OPPOSITE

ABOVE LEFT and RIGHT: KR dirt trackers like the WR can still be seen racing. On the left is Rusty Lowry.

BELOW LEFT: The KR marked a departure from the WR, with unit-construction four-speed foot-change gearbox and hand clutch.

BELOW RIGHT: The engine, however, remained a side-valve, seen here in twin carburettor guise.

Post-War Challenges

In 1941, shortly after Harley had launched the WR production-race machines, all motor sport ceased due to the Second World War. It would not be until 1948 that things got back to normal, when riders on Harley-Davidsons took 19 of the 23 American Championships. They were sweet victories, as the year before Harley had been trounced at Daytona by an Indian ridden by Johnny Spiegelhoff.

But the real challenge came not from Indian, but from imports. Politically-motivated, many of the import tariffs had been dropped in the US, and there was a flood of British bikes in particular taking advantage of the fact and that overall they were considerably cheaper than Harleys in terms of cost. While the racing rules did not favour them, their nimble handling and foot gearchanges meant they were still able to win races. It also encouraged the British factories to speed up development and they were soon challenging Harley on the track as well as in the showrooms. The rule bias meant that the engineers worked harder to extract the necessary horsepower and top speed from the smaller engines, so in effect their progress was accelerated by the restrictive rule book.

By 1952, the ageing WRTT was struggling and was now well off-pace. Fortunately for Harley-Davidson, its replacement, the KRTT, was ready and

94
BORE TECH
94

33

98
98

waiting to become the most successful racing Harley ever. This side-valve 45ci (750cc) was the most radical innovation ever made to the famous V- twin, and gave the company its first-ever Daytona win in 1953. Ridden by Paul Goldsmith, the 37-bhp KR managed to stay ahead of the opposition to take the flag. This was really the first full season for the race bike, having been launched late in 1952, and with the inevitable teething problems that needed to be resolved.

The KR had taken many of its new features from the British bikes that were now doing so well. There was swinging-arm rear suspension, controlled by twin shock-absorbers, and a four-speed transmission with foot change was the order of the day, with a hand-operated clutch immersed in oil. The telescopics and newly-sprung rear ends may have made for a good ride on the street, but the track revealed faults and they were difficult to ride. Harley eventually resolved the problems and factory rider Joe Leonard ended the 1954 season as the winner of the AMA No 1 plate.

LEFT: 24 hours, 1,366 miles (2200km), and an average 57mph (92km/h) – enough for a WR45 to win this marathon.

ABOVE: With a swinging-arm rear frame, the KR was used in road racing as well.

OPPOSITE: But as with the WR, its real forte lay in dirt-track racing – the heart of American competition riding. This is the 1/2-mile circuit at Davenport, Iowa.

DPC Racing
20
DPC
Racing
HOGWASH
20

302
Harley Davidson
784

ABOVE: The 883 racing class has allowed lightly modified Sportster 883s to race as well.

OPPOSITE: A 54ci (883cc) at Montlhéry. Pipes, low-set handlebars and rear-set footrests are among the modifications.

extract even more power from the ageing engine, enough to enable Rayborn to win his race at Daytona with an average speed of over 100mph (161km/h). In doing so, he also lapped all but the second- and third-place finishers.

In 1969 only eight bikes were built, and these remained with the official factory team. Featuring Italian Ceriani forks and a massive Fontana four-leading-shoe front brake it was sufficient to keep the Harley team up front, allowing Mert Lanwill to take the overall championship. But change was to come when the AMF takeover revealed that racing was not part of its great master plan: it was to be a lean time for the Harley-Davidson racing department.

It was, however, to spawn the legendary XR750, initially designed by Dutchman Pieter Zylstra. That first bike made its debut in February 1970 at the Houston Astrodome and was claimed to be an all-new bike. In reality it was little more than a revamped collection of parts, the chassis being a hybrid of KR parts and the motor coming from the XL Sportster. It retained the iron barrels and this was to be its initial downfall and meant that a revamped motor would appear two years later with aluminium components instead.

Those first bikes were fast enough, but they overheated and often failed to finish, earning the nickname 'Waffle Iron', although they fared better on the flat-track circuit. All the XR750s retired at Daytona in what proved to be Harley's worst-ever race season; its highest placed rider was Lanwill in sixth place in the overall championship. During the winter months, O'Brien and his team worked hard, but 1972 saw the bike beginning to suffer from gearbox breakages, although there were noticeable improvements to the engines.

The first sign that the XR750 was finally working out came at Easter in England when Cal Rayborn took a bike prepared by Walt Faulk to compete in the Anglo-American race series called the 'Transatlantic', and achieved success. He managed three wins out of six at three different race tracks and showed not only what a superb rider he was, but that the bike did indeed have potential, provided the problems could be cured. Rayborn also managed to give Harley some more positive publicity with an XLR-engined streamliner, by taking it to a land speed record of 265mph (426.5km/h) on the Bonneville Salt Flats.

Mark Brelesford was able to take the new alloy-engined XR to the AMA plate in 1972 and Rayborn returned to England and gave the legendary Jarno Saarinen, riding a works Yamaha, a run for his money until the Harley magneto failed. But this was to prove nothing more than a slight upturn of fortune and Harley entered the 1973 season with the factory as a whole suffering poor sales. On the dirt tracks, the thundering V-twins were losing out to the likes of Kenny Roberts on a Yamaha 650 twin. He was dominating the dirt scene and racing in general, taking the AMA plate before moving on to the world of GP racing.

ABOVE LEFT: Cal Rayborn's XR750 run by Team Obsolete.

ABOVE: American racing legend Cal Rayborn.

LEFT: Even the little 7.6-ci (125-cc) Hummer has been raced.

OPPOSITE
LEFT: Although the XR750 was designed as a dirt-track racer, it has taken to the circuits as well. This is George Roeder's bike.

RIGHT ABOVE and BELOW: Late-model Aermacchi Sprint racers with full fairings but still with drum, rather than disc, front brakes.

George Roeder

MOTUL
906

65

Daytona was also to pave the way to a poor season on the black stuff. Brelesford suffered an accident on lap 11, in which his bike was engulfed with flames after a crash. He unfortunately broke both legs and a hand, which was his racing demise. In the same race, Rayborn also crashed after his engine seized, resulting in a broken collarbone. Being a hard man, he was still able to captain the American squad in the Transatlantic shortly after, but he returned too soon and was not able to give of his best. He was also frustrated by the lack of Harley performance against the Japanese bikes and left at the end of the season.

This was not only unfortunate for Harley, but was also a loss to the world of racing, as the 33-year-old was killed when riding a Suzuki the following year. Many people believe that Rayborn, although never crowned champion, was one of the best riders the US had ever produced and consider him on a par with that other Harley legend, Joe Petrali.

RIGHT: Built for Daytona: George Roeder's XR on the infield.

FAR RIGHT: The XR750 in its element in a classic dirt-track.event. Scott Parker leads the field.

1
31
56

LEFT: Preparation is all: Chuck Davis' XR being prepared for an outing at Daytona.

OPPOSITE
ABOVE LEFT: A dirt-track XR, very different from the road racer, though it now also uses a swinging-arm frame.

ABOVE RIGHT: The XR750 engine is now a distant relative of the road-going V-twin, with a short stroke and twin carburettors.

BELOW RIGHT: There are high-level pipes on all XRs, whatever the discipline.

Revival

But if the early 1970s was a bad time for Harley racing, 1974 was to be its finest hour, albeit that the bikes were basically Aermacchi racers now wearing the AMF/Harley-Davidson logo on their tanks. Based at Varese, Italy, the team was led by Walter Villa, one of two brilliant rider-designer brothers who had evolved a brand-new water-cooled two-stroke twin. Villa began the season by taking the lap record and winning his home Italian GP at Monza. This was the start of numerous wins that saw him crowned 250 World Champion at the end of the season, which was a feat he repeated in 1975, giving Harley-Davidson its second World Championship in a row. It was also looking good for 1976 with a 31-ci (500-cc) having been produced, although it failed to live up to expectations and was given to the Bimota squad to race, eventually coming to nothing.

Even if the engine did not live up to expectations, Villa did, and took the 250-cc (15-ci) title for a third year on the trot with a victory at the Czechoslovakian GP in August of that year. A week later he added

the 350-cc (21-ci) World Championship at the West German GP, giving him and Harley a world double and their last win in the world racing arena. The bikes were at that time unbeatable, but the factory could not afford to race the following year and it was a privately funded outfit that ran in the championships. Villa still managed a third place in the 250-cc championship, with team-mate Franco Uncini taking second, but came nowhere in the 350-cc class at all. The following year the Varese plant was declared bankrupt and was sold on to Cagiva.

Back in the US, things were not going as well and the name Harley-Davidson was gradually slipping from rostrum places on tarmac. On the dirt ovals, things were better and the big V-twins could still be seen taking the top places and entertaining the crowds.

It was here that they managed to give their riders enough points to take the No 1 plate, among them Jay 'Springer' Springsteen, who took it for several years in succession. He was fast becoming another Harley legend and a name that would be associated with the factory into the next century.

At the beginning of the 1980s it looked as though Harley and racing would be relegated to just words in the history books, unless it was classic events, where the old war-horses still showed their mettle. Then came the 'Battle of the Twins' (BOTT) class for twin-cylinder machines, which proved the saviour of Harley's racing heritage.

The AMA may not have liked it, with a full racing programme already established at most race meetings, but when the inaugural event took place at Daytona in 1981, 88 riders turned up to race. The names on the bikes they rode

read like a list of the world's motorcycle manufacturers. Four different classes were run, the displacement limit was set at 61ci (1000cc), and an eight-race series began.

Harley may have been there with a half-hearted effort based on the XLCR Café Racer, but Ducati took the first honours. After the 1982 season, when the series was well-established, Harley then decided to have a serious crack at winning, with 'Lucifer's Hammer'. Built by Dick O'Brien, it was to feature in the history books along with other great race bikes. Ridden by Jay Springsteen, the Hammer stole the win in the BOTT race in 1983, winning by 24 seconds from the following machine. This was ten years after the last factory bike appeared in the US, ridden by Gary Scott at Laguna; Harley was now back with a vengeance.

Basically, the Hammer was constructed from a road-race chassis taken from the early 1970s and fitted with the XR1000 engine taken from Harley's then street bike on sale to the public. The bike achieved 104bhp and was to go on and win the BOTT titles in the Grand Prix category for the 1984–86 seasons. However, Springsteen was not in the hot seat for, although he rode it a few more time in 1983, his contract was with the dirt-track events, which always took precedence. Instead, Gene Church took the ride and the championship and the bike went on to be sponsored by the Harley Owners Group (HOG).

The bike was actually run by dealer Don Tilley, who managed to keep the bike competitive until 1989 when it was replaced by a Buell-based machine. Although having a relatively short career, the Hammer had proved its worth and paved the way for Harley to make a comeback to full-time racing.

In 1998 Harley itself launched a one-make racing series for 883 Sportsters. Run on a semi-professional basis, the slightly modified street bike-series quickly attracted the public's attention. Strictly policed, the racing was tight and close as a result. That inaugural year's championship was actually won by a British rider by the name of Nigel Gale who had won 'Rookie of the Year' in 1986 and went on to ride for Bartels, the Harley dealer who sponsored race bikes.

LEFT: With Cal Rayborn, Jay Springsteen is one of the most accomplished XR750 riders of all time. 'Springer' became the youngest ever winner of the AMA National Championship in 1976 when he was 19, and if there was still any doubt, he quashed it by winning in 1977 and '78 as well, all three times as an official Harley rider. Despite being diagnosed with a rare type of diabetes, he was to accumulate 40 race wins on short track, TT, $^1/_2$-mile and 1-mile dirt tracks, proving an able road user too.

OPPOSITE: Springsteen in full sliding flight on an XR750.

42
SPRINGSTEEN
9
BIG TWIN
42

Japanese to the Rescue!

By now, Harley had recognized the need to be back in racing and was once more in control of its own destiny, feeling that it was able to embark on a new project aimed purely at racing, the VR1000. This was to be a costly endeavour, not to mention an historic one, as for the first time ever the race bike would not be based on a standard production machine. It did make a road-legal copy of the race bike in limited numbers and it was on offer to the public for nearly $50,000, making it somewhat exclusive: a set of race parts was also included in the price for those who wanted to race them privately.

Harley announced that it would campaign the VR1000 in AMA Superbike and Grand Prix racing, but the first time the bike appeared was at Daytona in 1994, after five years' development. Once again it was a repetition of other bikes in Harley's racing past – no fairytale debut, but failure on its maiden run. The rider was Miguel Du Hamel, a top competitor and son of racing legend Yvon, who qualified nearly 8 seconds slower than the pole position Ducati. Cynics were beginning to

ABOVE: The VR1000 (seen here at Daytona) has so far failed to match the XR750's success on the tarmac circuits.

ABOVE RIGHT: Scott Zampach is one rider who has popularized a VR1000.

crow that in spite of Harley's massive resources, it could no longer build a race bike and compete with the world's best on the race track. When the bike lasted just 22 of the 57 race laps, due to the failure of a weld on the balance shaft, they seemed to have been proved right.

Two weeks later the bike failed to finish again, at Phoenix, Arizona, when a defective fuel line was responsible for starving the engine. It was to be a similar story the entire season, despite some creditable performances by the rider. Was it to be the early XR750 all over again?

The design team at that time consisted of a few well known names connected with the Harley brand: Jerry Branch, best known for his Harley cylinder-head work, Erik Buell, designer and bike builder, Dick O'Brien the former race boss, and last but not least, Don Tilley, the tuner of Lucifer's Hammer. The man in charge was Mark Millar, who actually got the whole project up to prototype stage.

The fairing was designed by none other than Willie G. Davidson, and it would later be his son who managed the race team in the late 1990s. Rousch Racing was used for getting the cylinder-head work done, thanks to its Nascar expertise, as a result of Harley's desire to keep the whole project as American as possible. Steve Scheibe did much of the initial work and he was eventually hired to take over the project.

The new engine retained the familiar Harley V-twin format, although it was moved to a 60-degree angle between the cylinders. The bore and stroke were 3.85 and 2.6in (98 and 66mm) and the cylinders nikasil-coated aluminium. The crank had two titanium plain bearing rods running on ball-bearing big ends, while the bike ran with a dry sump. Double overhead cams were used to operate four valves per cylinder and were chain driven, while a counterbalancer was driven off the crank to smooth things out. Water-cooling was used for the first time on a Harley motor, as was fuel injection.

Primary drive to the 'dry' clutch was by gears, the gearbox itself being five-speed. That first motor used at Daytona pumped out 140bhp at 10,400rpm with the rev limiter set at 11,000rpm. Weight-wise they were at a disadvantage, as well as having only test data to set the bike up, weighing in at 355lb (161kg), 20lb more than the minimum allowed weight under AMA rules.

While they could have used magnesium for crankcases, shaving 8lb (3.6kg) in one hit, this would have meant that all the road bikes produced would also have to use this expensive commodity, making them more expensive than they already were.

Eric Buell initially designed the frame, although his tubular design was later rejected along with a British design that was tried. Scheibe then used computer modelling to design Harley's own version, made by a company called Anodizing Incorporated which specialized in aluminium extrusion. The chassis finally ended up as a conventional aluminium beam with a massive head area for strength around the forks.

Despite attempting to remain totally US-based and initially using Penske suspension front and rear, Harley found that the Indy car team could not match Ohlins for handling. It therefore used the Dutch firm's 1.8-in (46-mm) inverted front forks. The brakes were taken from the car-racing world and Wilwood supplied the calipers, which featured pistons of different sizes for better load distribution across the pads.

Du Hamel was frustrated by the bike's failures and, unaccustomed to being off the pace, left at the end of the year, leaving Chris Carr (synonymous with Harley in dirt-track circles) and GP veteran, Doug Chandler, to get the bike onto the rostrum. Despite their considerable expertise in setting things up, progress was slow and it was not until 1998 that the bike began to be a serious contender.

With Canadian Pascal Picotte having moved into the hot seat, the Harley team

ABOVE: The VR1000 was a real departure for Harley-Davidson, with a water-cooled 60-degree V-twin and 11,000rpm rev limit. An aluminium beam frame and top-notch Ohlins suspension complete the specification.

OPPOSITE: Earl Small touches down a knee slider, trying hard on the VR.

Arai
vanson
RACING
EARL SMALL'S
HARLEY-DAVIDSON
EBC

Scott Zampach riding a VR1000.

Chris Carr on a VR1000.

once again lost the opportunity to stand on the Daytona rostrum through poor pit-stop technique, something that would not have happened under Bill Ottaway's leadership. But it was not all bad news, for at that 1998 Daytona meeting, a Harley became the fastest-ever pushrod machine to lap the Daytona banking.

However, a Harley did win the Battle of the Twins F1 that year, with the all-black 'Daytona Weapon', which was making its sixth appearance in the Battle of the Twins F1. Constructed in 1992 by Japanese Harley guru Takehiko Shibazaki, it represented a totally different approach to the one the factory had been taking. Instead of a new high-tech engine, it featured the standard 45-degree V-twin, tuned and worked on by the man himself, who had previously built his own version of Lucifer's Hammer in the late 1980s. Shibazaki had himself given up racing to wield the spanner, and built the futuristic Weapon himself. He inserted his own tuned motor into a Japanese Over frame and using top components like Ohlins suspension and Brembo brakes to complement it, attempted to achieve reliability.

That first bike, like the factory VR, did not win on its initial outing, but the team kept coming back with improved versions. In 1997 another bike appeared called 'Golden Balls', so named because of its colour scheme. This was unfaired and the bike on which Jay Springsteen won the F2 race at Daytona.

In 1998 'Weapon 2' was brought along as well, complete with a new frame made by the British firm, Saxon. But this, like its forebears, did not manage to finish its first race, sitting it out to watch its older brother achieve that historic win. The rider was 'Mr Daytona', Scott Russell, taking his fifth win at the banked tri-oval circuit.

Harley was to go on to sign him to ride for the 1999 season, confident that the bike was now competitive enough to win in the hands of the right rider.

Before the race, however, a brawl in a Daytona bar prevented him from taking part and Pascal Picotte crashed out, leaving the team bereft of that much-coveted win. Harley was to be disappointed in 2000 as well, Russell failing to produce a win. So while the official Harley race team continues to persevere, it seems that its legions of loyal fans will have to wait before a new Wrecking Crew is formed and Harley is once again seen as a racing force to be feared.

OPPOSITE: Chris Carr on a VR at Daytona. At the time of writing, after five years of development and six years of racing, that coveted Daytona win has yet to happen.

RIGHT: Meanwhile, the Sportster 883 series provides an affordable means of racing for many amateur riders around the world.

Milestones: the Eight-Valve Racer

Designed in 1916 after Bill Ottaway had joined Harley from Thor, the eight-valve racers started life using already available basic components. Many engine parts came from the existing road-going F-head engine and the standard bore was kept, which gave the maximum 61ci (1000cc) allowable for race meetings.

However, the major change was to the cylinder heads, which gained two inlet and two exhaust ports per cylinder. A single camshaft located in the timing case operated the valves by long pushrods running up the outside of the barrels. The adoption of this design was for the normally accepted engine design principles of volumetric efficiency and better breathing. The design team used Harry Ricardo from England to make the four-valve system work to the best possible advantage after initially experiencing problems.

Fuel was supplied by one carburettor located between the V of the barrels and a Bosch Magneto provided ignition. There was still total-loss lubrication, although both a hand and automatic pump were provided to deal with racing conditions. (Englishman Freddie Dixon did run with a twin carb at Brooklands, but this was not a factory modification.) Three versions were produced with different rocker arms and cam formations. There was a one-shaft four-lobe camshaft, or one with two-lobe camshafts. Some engines had short exhaust pipes, others had none, relying on open ports to allow the burnt gases to escape very close to the rider's legs.

It would seem that no two engines were alike, due to the fact that only parts available at that time were used to build the engines. Power is thought to be around 50bhp, but varies due to the different compression ratios used to allow the bike to run on alcohol.

The motor was housed in a keystone frame, so called because it was a diamond shape dissected into fore-and-aft triangles. Plates at the bottom, which carried the bottom of the motor providing the main mounting points, joined the two frame down tubes, making for a rigid structure.

The rear of the frame was rigid with no suspension, while Harley leading-link forks were located in the front wheel with the occasional use of Flying Merkel forks. There were no brakes supplied, although a rear was listed as an optional extra for road races. Likewise there was no clutch, no

The legendary Jay Springsteen is pictured before the start of a dirt-track race (ABOVE) and at Daytona (OPPOSITE).

23
49
BELL
DUNLOP
JAY

OPPOSITE: An evocative sight: a string of XR750s chase one another towards the flag in the Del Mar races, California.

ABOVE: An Aermacchi Sprint, in road-race trim.

gearbox or starter mechanism, although some were listed for road-racing options, like the brakes. The bike was started by means of the compression release being dropped when sufficient speed had been obtained by a tow start, due to the high-compression ratios used.

The basic bikes were priced at $1,500 for the eight-valve and $1,400 for the four-valve single-cylinder version, which was the same bike with only the front cylinder fitted. This made it eligible for the professional classes run at the smaller tracks. Indian at that time priced its similar-specification bikes at $350 and $300! The number built and the specification is somewhat sketchy, estimates suggesting that somewhere between 20 and 60 bikes were made, the vast majority going overseas. Suffice to say that today they are rare – and still expensive.

The Peashooter

In 1925, when Indian brought out its budget-priced Prince with a 21-ci (350-cc) side-valve motor, Harley thought that it should respond. That response was in the form of two 350-cc side-valve machines giving 8bhp in both an A and B form costing $210 and $235 respectively. In smaller numbers, it also introduced the OHV AA with magneto ignition and the BA with a battery ignition system, both costing $250 and $275.

They were aimed primarily at the overseas markets, as the American buyer in the main still wanted the big V-twins. However, the sport of Speedway was growing fast abroad and the OHV AA-model quickly found favour with competitors. Tuned up and slimmed down to 215lb (97.5kg), the bike could easily top 80mph (129km/h). In the US it also found favour, and riders like Joe Petrali were almost unbeatable on it, often winning when competing with 30-ci (500-cc) machines. It also did well in hill-climb events as well as on the track.

Harley brought out a competition 'S'-version, which quickly earned the name Peashooter because of the characteristic dry popping sound made by its short exhaust pipe. This bike had a Sager front fork and 27-in (686-mm) wheels and cost $300. In 1926 it won 14 international titles and a year later all the US titles in the 350-cc class.

There were three basic versions: the prototype had an altered frame and special gas tank. The engine was fitted with a single port head and exhaust pipe, along with a Deluxe Schebler carburettor. This had no gearbox, just a countershaft with two different sprockets.There were flat-track versions with three-speed gearboxes and in 1927 there was another modified frame and a manual oil pump, a new carburettor and underslung handlebars.

One year on, the Peashooter was offered with a twin port head, two exhaust pipes and an Indianapolis carb. The engine also had a beefed-up con-rod for greater reliability. The export TT and road-racer had the three-speed box fitted along with the twin port head.

Another variant intended for A-class racing had a high compression ratio of 12:1 and was intended to be run on alcohol. There were no gears or brakes, the frame was very stiff with little suspension movement and the forks were set at a steep angle for quick, responsive turning. The bikes did well until the beginning of the 1930s when they began to lose out to machines fitted with the 30-ci (500-cc) British JAP engine that had been developed for Speedway use.

ABOVE: The classic foot-sliding technique demonstrated on an XR750 dirt-track racer.

OPPOSITE: An XR750, shown at the Sammy Miller Museum, Hampshire, England.

XR
750 HARLEY-DAVIDSON
38

ABOVE: The WR was Harley's backbone racer for 15 years.

OPPOSITE: The RR250, on which Walter Villa won three World Championships for Harley-Davidson. This is the 1975 factory machine.

The WR Racers

When Class C racing was introduced, Harley had to rely on the high-compression DL and DLD sports versions of the side-valve D-series on which riders could compete. In 1937 the new W-series succeeded it, with its own racing version, the WLDR, which developed 7bhp more than a standard WLD and cost $380. The WLDD that followed was a pure race machine formed by paying $109 for a conversion kit to transform the WLDR. The kit consisted of cylinder heads, cam, valve lifter and air intake.

However, the bikes always benefited from tuning and one well-known exponent of the model was Tom Sifton, a dealer from San Jose. His bikes, ridden by Sam Arena, took many race wins and Harley itself eventually incorporated many of his modifications into production models.

In 1942 the factory itself produced the WR and WRTT. The former was a standard production racer and was based on the type of bikes people were producing when they stripped the WLDR and did things made possible by a relaxation of the AMA rules. The WR weighed in at 300lb (136kg) and could run at a genuine 100mph (161km/h). The bike came with a vertically-mounted Edison-Splitdorf magneto, this later being replaced by a Wico unit. The motor itself had larger polished inlet ports and altered combustion chambers and stronger valve springs.

The engine came with boxes of spares and charts to allow it to be altered for the type of racing for which it was intended. Different gearing, wheel hubs and a choice of 18 or 19-in (457 or 483-mm) wheels were part of the package. The frame was strengthened for the rigours of competition and there were small or large gas tanks. In fact most of the bike was alterable making the rider and tuner's life that much easier.

There was also a WRTT special for road racing and hill climbing only and this had cut-down fenders along with clutch and brake pedals. This version had brakes compared to other disciplines where they were not allowed.

In essence, the WR race bikes brought Harley-Davidson's racing effort to a level playing field and allowed races to be decided on rider skill, rather than the amount of tuning they could afford or sponsorship they had received.

Cal Rayborn
TEAM OBSOLETE
Domi Racer
KOSMAN
HARLEY-DAVIDSON
17
AVON TYRES
AirTECH
TEAM OBSOLETE
TEAM OBSOLETE
MEGACYCLE

managing over 245mph (394.3km/h) on an 89-ci (1460-cc) Sportster five years later.

Perhaps it was too late for Harley to dominate drag racing, but Marion Owens and a few others fought a rearguard action against multi-cylinder Japanese bikes in 1979 and 1980, Owens running a fearsome twin-engined machine. But even before then, Harleys were still prominent in US drag racing, with private riders campaigning successfully. Ron Fringer, for example, posted a 10.09-second 1/4-mile time, with a 133.72mph (215.2km/h) terminal speed at Beech Bend Raceway in Kentucky, in 1971. Four years earlier, Les Payne had ducked below the 10-second barrier (9.71 seconds) with 152.38mph (245.2km/h) at Connecticut Raceway. Both used modified Sportster motors.

But even today, Harleys continue to do well in drag racing. This is due largely to the power characteristics of the big V-twin, which offers excellent torque from low revs, unlike the more 'peaky' four-cylinder opposition. UK specialist RMD of Reading, built a Harley-powered drag bike, and was delighted when it out-dragged a tuned Kawasaki ZZR1100. Although the Kawasaki posted a significantly higher terminal speed, and was quieter in the final stages of the 1/4-mile, it was slower in those crucial first 200 yards (180m) from a standing start. It

Drag Racing

Drag racing has long had an association with Harley-Davidson, right from the very first unofficial Saturday night races outside town, to the present day. This form of sport rapidly gained popularity in the US in the 1950s, taking over from hill climbs, although these still exist as a thriving branch of the the sport. But there was no official involvement until the late 1970s when, according to author David Wright, some of the proceeds from the sale of Aermacchi went into a factory drag-racing effort. Before that, Milwaukee seemed more keen on investing in outright speed records at the Bonneville Salt Flats with George Roeder piloting a streamlined Aermacchi Sprint 250 to 177.255mph (285.3km/h) in 1962 and Cal Rayborn

LEFT: Only in America: skid racing involves removal of the entire front end and allowing the bike to rest on a skid plate. The rider places his weight as far back as possible, to minimize friction.

OPPOSITE: This stock-appearance bike and purpose-built dragster, both Harley-powered, make for an interesting comparison.

M 19

LEFT: A fearsome single-engined bike. Note the size of the exhaust pipes and carburettor intake!

OPPOSITE: Buells have been quickly accepted into drag racing. Wheelie bars, to prevent the bike flipping over backwards, are a mandatory fitment.

is this sort of performance that should keep Harleys well to the forefront of modern drag racing.

In fact, so popular have Harleys remained, that the American Drag Bike Association (www.americandragbike.com) runs its own drag series for Harley-engined bikes only. There are eight separate classes as follows:

MODIFIED – Largely standard Sportsters and Big Twins. Engines can be enlarged to 76ci/1245cc (Sportsters) or 84ci/1375cc (Big Twins) but no four-valve or twin carburettors are allowed. Engines must run on standard pump gas, and the lights and changing systems must be operational. The current class record is 9.355 seconds, 138.90mph (223.5km/h).

SUPER MODIFIED – Larger engines allowed, 76–120ci/1245–1965cc (Sportsters) or 84–120ci (Big Twins) but no four-valve or twin carburettor heads. Any aftermarket part is permitted as long as it is deemed safe by the ABDA. The transmission must be Harley-Davidson or a 'like same' replacement, gear ratios can be altered, but auto, rpm or computer-controlled changers are not allowed. An aftermarket frame is permitted, but only the sit-up type. The current class record is 8.522 seconds, 155.96mph (251km/h).

TOP GAS – Highly modified engines. These must be pushrod 45-degree V-twins, petrol-powered and naturally-aspirated, but

AHDRA
SEP 12

LEFT: Harley drag bikes are LOUD!

OPPOSITE: Underneath all the pipes and wires is a much-modified Evolution V-twin. The twin small-bore pipes are to operate the airshifter transmission, while the huge ones are for fuel.

any size induction and fuelling system is allowed. Any transmission is allowed, including rpm or computer shifting, and an aftermarket frame of any wheelbase. The current class record is 8.275 seconds, 157.53mph (253.5km/h).

PRO STOCK – Stock appearance with carburettors, XL, FX, or VR-based. Maximum 160ci (2620cc) on gas, but four-valve and twin carburettor heads are allowed. VRs must have factory cases and heads. Any transmission (including auto), 3–6-speeds, airshifters permitted. Racing slick tyres allowed, and maximum 70-in (178-cm) wheelbase. The current class record is 7.706 seconds, 166.13mph (267.3km/h).

PRO MODIFIED – Petrol or alcohol fuelling, naturally aspirated up to 151ci (2475cc), turbo or supercharged up to 122ci (2000cc). Any transmission (including auto), 68-in (173-cm) maximum wheelbase, any aftermarket frame (sit- or lay-down type) and V-rated or drag-racing tyres. The current class record is 7.915 seconds, 167.97mph (270.3km/h).

PRO DRAGSTER – Nitro-fuelled, carburettor, single engine up to 122ci (2000cc), pushrods, four-valve heads. No transmission, any clutch, any frame, 68-in (173-cm) maximum wheelbase. The current class record is 7.297 seconds, 174.75mph (281.2km/h).

J. BENK BROWN TRUCK
Chassis By...
LANSDALE AMUSEMENT COMPANY

LEFT Drag bikes run over 1/8- or 1/4-mile, and although they run in pairs, the chief competitor is the clock. Start times are controlled by a 'Christmas Tree' of seven coloured lights (five amber, one red, one green), but the clock only starts and stops when the electronic timing beam is broken.

OPPOSITE: Smoking the rear tyre looks spectacular, but it's not for show, being the inevitable result of wheelspin.

PRO FUEL – Nitro-fuelled bikes designed specifically for drag racing. Maximum 151ci/2475cc (carburettors) or 132ci/2165cc (fuel injection), pushrods, four-valve heads. One- or two-speed transmission, any frame, maximum 72-in (183-cm) wheelbase, unlimited tyre size with single-speed transmission. The current class record is 6.921 seconds.

TOP FUEL – Nitro-fuelled with super or turbocharging. Single or double engines up to 175ci/2870cc combined. Must still use pushrods, but four-valve heads, any transmission or clutch allowed. 78-in (198-cm) maximum wheelbase and any tyres specified for drag racing. The current class record is 6.601 seconds, 213.52mph (343.6km/h).

OPPOSITE: A typical specification of a serious drag bike like this would read: 140-ci (2295-cc) motor, 3-inch (76.2-mm) belt drive, custom five-speed transmission, pneumatic shifting, chrome-moly tube frame, 10-in (254-mm) wide rear tyre on 15-in (381-mm) rim. The rider will be experiencing three times his own weight in accelerating 0–100 in less than 2 seconds.

RIGHT: This bike might look relatively stock, but underneath it's highly modified.

OPPOSITE: Notice that the whites of this rider's eyes are showing as he blasts off the line.

RIGHT: Blast-off! An ultimate nitro-fuelled drag bike.

CLES
eiser
AHDRA
C 9

OPPOSITE: Just to prove they're not all exotic and expensive, here is a stock-like 883 Sportster powering modestly away.

RIGHT: The very spirit of drag racing.

GUIDE TO MODEL NAMES

Model 0 – 1904, 24.74-ci (410-cc) F-head single
1 – 1905, 24.74-ci (410-cc) F-head single
2 – 1906, 26.8-ci (440-cc) F-head single
3 – 1907, 26.8-ci (440-cc) F-head single
4 – 1908, 26.8-ci (440-cc) F-head single
5/A/B/C – 1909, 30.5-ci (500-cc) F-head single
5D – 1909, 50-ci (820-cc) F-head V-twin, magneto
6/A/B/C/E – 1910, 30.5-ci (500-cc) F-head single
7/A/B/C – 1911, 30.5-ci (500-cc) F-head single
7D – 1911, 50-ci (820-cc) F-head V-twin
8/A/X – 1912, 30.5-ci (500-cc) F-head single
8D – 1912, 50-ci (820-cc) F-head V-twin
8E – 1912, 61-ci (1000-cc) F-head V-twin
9A/B – 1913, 35-ci (575-cc) F-head single
9E/F – 1913, 61-ci (1000-cc) F-head V-twin, chain drive
9G – 1913, 61-ci (1000-cc) F-head V-twin with front-mounted delivery box
10A/B/C – 1914, 35-ci (575-cc) F-head single
10E/F – 1914, 61-ci (1000-cc) F-head V-twin, chain drive
10G – 1914, 61-ci (1000-cc) F-head V-twin, Delivery Van, two-speed
11B/C – 1915, 35-ci (575-cc) F-head single
11E/F/H – 1915, 61-ci (1000-cc) F-head V-twin
11G – 1915, 61-ci (1000-cc) F-head V-twin Delivery Van
16B/C – 1916, 35-ci (575-cc) F-head single
16E/F – 1916, 61-ci (1000-cc) F-head V-twin
17B/C – 1917, 35-ci (575-cc) F-head single
17E/F – 61-ci (1000-cc) F-head V-twin

A – 1926–27, 21-ci (350-cc) side-valve single, magneto
A Topper scooter – 1960, 10-ci (165-cc) two-stroke single, automatic
AA – 1926–27 21-ci (350-cc) OHV single, magneto
AH Topper scooter – 1961–65, 10-ci (165-cc) two-stroke single, automatic
AU Topper scooter – 1960–64, 10-ci (165-cc) two-stroke single, automatic

B – 1918, 35-ci (375-cc) F-head single
B – 1926–34, 21-ci (350-cc) flat-head single, electrics
B – 1955–59, 7-ci (125-cc) two-stroke single
BA – 1926–29, 21-ci (350-cc) OHV single, electrics
BT – 1960–61, Super 10, 10-ci (165-cc) two-stroke single
BT Pacer – 1962–64, 11-ci (175-cc) two-stroke single
BTF Ranger – 1962, 10-ci (165-cc) two-stroke single
BTH Scat – 1962–65, 11-ci (175-cc) two-stroke single
BTH Bobcat – 1966, 11-ci (175-cc) two-stroke single
BTU Super 10 – 1960–61, 10-ci (165-cc) two-stroke single, restricted
BTU Pacer – 1962–64, 10-ci (165-cc) two-stroke single, restricted

C – 1918, 35-ci (575-cc) F-head single, magneto
C – 1929–34, 30.5-ci (500-cc) side-valve single, electrics
C Sprint – 1961–66, 15-ci (246-cc) OHV single
CB – 1933–34, 30.5-ci (500-cc) side-valve single
CD Commercial – 1921–22, 37-ci (606-cc) F-head single

D – 1929–31, 45-ci (750-cc) side-valve V-twin
DL – 1929–31, 45-ci (750-cc) flathead V-twin, high-compression
DLD – 1930–31, 45-ci (750-cc) side-valve V-twin, high-compression
DS – 1930–31, 45-ci (750-cc) side-valve V-twin, sidecar model

E – 1918, 61-ci (1000-cc) F-head V-twin
E – 1936–38 & 1944-50, 61-ci (1000-cc) OHV V-twin, medium-compression
EL – 1936–52, 61-ci (1000-cc) OHV V-twin, high-compression
ELF – 1952, 61-ci (1000-cc) OHV V-twin, foot-change
ELP – 1949, 61-ci (1000-cc) OHV V-twin, medium-compression, spring fork

ELS – 1951–52, 61-ci (1000-cc) OHV V-twin, medium-compression, sidecar gearing
EP – 1949, 61-ci (1000-cc) OHV V-twin, medium-compression, spring fork, sidecar gearing
ERS Sprint Scrambler – 1969–70, Sprint Scrambler, 15-ci (246-cc) single
ES – 1936–41, 1945–50, 61-ci (1000-cc) OHV V-twin, sidecar gearing

F – 1918–23, 61-ci (1000-cc) F-head V-twin, three-speed
F – 1942–49, 74-ci (1200-cc) OHV V-twin, medium-compression
F – 1950, 74-ci (1200-cc) OHV V-twin, medium-compression
FD – 1921–24, 74-ci (1200-cc) F-head V-twin, magneto
FDCA – 1924, 74-ci (1200-cc) F-head V-twin, alloy pistons, magneto
FDCB – 1925, 74-ci (1200-cc) F-head V-twin, iron-alloy pistons, magneto
FDCBS – 1925, 74-ci (1200-cc) F-head V-twin, iron-alloy pistons, magneto, sidecar gearing
FDS – 1921–24, 74-ci (1200-cc) F-head V-twin, magneto, sidecar gearing
FDSCA – 1924, 74-ci (1200-cc) F-head V-twin, alloy pistons, magneto, sidecar gearing
FE – 1924–25, 61-ci (1000-cc) F-head V-twin, alloy pistons, magneto
FES – 1924–25, 61-ci (1000-cc)F-head V-twin, alloy pistons, magneto, sidecar gearing
FL – 1941–64, 74-ci (1200-cc) OHV V-twin, high-compression
FL – 1973–75, 74-ci (1200-cc) OHV V-twin
FLB – 1965–69, 74-ci (1200-cc) OHV V-twin, hand-change
FLE – 1953–56, 74-ci (1200-cc) OHV V-twin, combination
FLEF – 1953–56, 74-ci (1200-cc) OHV V-twin, foot-change, combination
FLF – 1952–64, 74-ci (1200-cc) OHV V-twin, foot-change
FLFB – 1965–67, 74-ci (1200-cc) OHV V-twin, foot-change
FLH – 1955–64, 74-ci (1200-cc) OHV V-twin, high-compression
FLH – 1970–73, 74-ci (1200-cc) OHV V-twin, high-compression, hand-change
FLH – 1974–80, 74-ci (1200-cc) OHV V-twin
FLH-80 Electra Glide – 1978–84, 82-ci (1340-cc) OHV V-twin
FLH Electra Glide Heritage – 1981, 82-ci (1340-cc) OHV V-twin
FLHB Electra Glide – 1965–69, 74-ci (1200-cc) OHV V-twin, high-compression, hand-change
FLHC Electra Glide Classic – 1979–81, 82-ci (1340-cc) OHV V-twin
FLHF Hydra Glide – 1955–64, 74-ci (1200-cc) OHV V-twin, high-compression, foot-change
FLHF – 1970–73, 74-ci (1200-cc) OHV V-twin, high-compression, foot-change
FLHFB Electra Glide – 1965–69, 74-ci (1200-cc) OHV V-twin, high-compression, foot-change
FLHP Road King Police – 1998, 82-ci (1340-cc) Evo V-twin
FLHPI Road King Police – 1998, 82-ci (1340-cc) Evo V-twin, e.f.i.
FLHR Electra Glide Road King – 1994–98, 82-ci (1340-cc) Evo V-twin
FLHR Road King – 1999–2001, 88-ci (1450-cc) Twin Cam V-twin
FLHRCI Road King Classic – 1998, 82-ci (1340-cc) Evo V-twin e.f.i.
FLHRCI Road King Classic – 1999–2001, 88-ci (1450-cc) Twin Cam V-twin e.f.i.
FLHRI Electra Glide Road King – 1996–98, 82-ci (1340-cc) Evo V-twin e.f.i.
FLHS Electra Glide – 1977, 74-ci (1200-cc) OHV V-twin
FLHS Electra Glide – 1980–81, 82-ci (1340-cc) OHV V-twin
FLHS Electra Glide Sport – 1983–93, 82-ci (1340-cc) OHV V-twin
FLHT Electra Glide – 1983, 82-ci (1340-cc) OHV V-twin
FLHT Electra Glide Standard – 1995–98, 82-ci (1340-cc) Evo V-twin
FLHT Electra Glide Standard – 1999, 88-ci (1450-cc) Twin Cam V-twin
FLHT Electra Glide Standard – 2001 on, 88-ci (1450-cc) Twin Cam V-twin
FLHTC Electra Glide Classic – 1983–97, 82-ci (1340-cc) OHV V-twin
FLHTC Electra Glide Classic – 1998, 82-ci (1340-cc) Evo V-twin
FLHTCI Electra Glide Classic – 1996–

98, 82-ci (1340-cc) Evo V-twin e.f.i.
FLHTCU Electra Glide Classic Ultra – 1989–97, 82-ci (1340-cc) Evo V-twin
FLHTCUI Electra Glide Classic Ultra – 1995–98, 82-ci (1340-cc) Evo V-twin e.f.i.
FLHTCUI Electra Glide Classic Ultra – 2000 on, 88-ci (1450-cc) Twin Cam V-twin e.f.i.
FLHTPI Electra Glide Police – 1998, 82-ci (1340-cc) Evo V-twin e.f.i.
FLHX Electra Glide – 1984, 82-ci (1340-cc) OHV V-twin
FLP – 1949, 74-ci (1200-cc) OHV V-twin, high-compression, spring fork
FLP – 1970–72, 74-ci (1200-cc) OHV V-twin, hand-change
FLPF– 1970–72, 74-ci (1200-cc) OHV V-twin, foot-change
FLS Hydra-Glide – 1951–52, 74-ci (1200-cc) OHV V-twin, medium-compression, sidecar model
FLST Heritage Softail – 1986–90 82-ci (1340-cc) V-twin
FLSTC Heritage Softail Special – 1987–98, 82-ci (1340-cc) Evo V-twin
FLSTC Heritage Softail Classic – 2000 on, 88-ci (1450-cc) Twin Cam 88B V-twin
FLSTCI Heritage Softail Classic – 2001 on, 88-ci (1450-cc) Twin Cam 88B V-twin e.f.i.
FLSTF Fat Boy – 1990–99, 82-ci (1340-cc) Evo V-twin
FLSTF Fat Boy – 2000 on, 88-ci (1450-cc) Twin Cam 88B V-twin
FLSTFI Fat Boy – 2001 on, 88-ci (1450-cc) Twin Cam 88B V-twin e.f.i.
FLSTN Heritage Softail Nostalgia – 1993–95, 82-ci (1340-cc) Evo V-twin
FLSTN Heritage Softail Special – 1996, 82-ci (1340-cc) Evo V-twin
FLSTS Heritage Springer – 1997–98, 82-ci (1340-cc) Evo V-twin
FLSTS Heritage Springer – 2001 on, 88-ci (1450-cc) Twin Cam 88B V-twin
FLT Tour Glide – 1980–82, 82-ci (1340-cc) OHV V-twin
FLTC Tour Glide Classic– 1981–94, 82-ci (1340-cc) OHV V-twin
FLTCU Tour Glide Classic Ultra – 1989–95, 82-ci (1340-cc) Evo V-twin
FLTCUI Tour Glide Classic Ultra – 1996–97, 82-ci (1340-cc) Evo V-twin e.f.i.
FLTR Road Glide – 1998, 82-ci (1340-cc) Evo V-twin
FLTRI Road Glide – 1998, 82-ci (1340-cc) Evo V-twin e.f.i.
FLTRI Road Glide – 1999–2001, 88-ci (1450-cc) Twin Cam V-twin e.f.i.
FP – 1949, 74-ci (1200-cc) OHV V-twin, medium-compression, spring fork, sidecar gearing
FS – 1919–23, 61-ci (1000-cc) F-head V-twin sidecar motor
FS – 1941, 1945–50, 74-ci (1200-cc) OHV V-twin, sidecar gearing
FUS – 1918, 61-ci (1000-cc) F-head V-twin, Presto-lite
FX Super Glide – 1971–78, 74-ci (1200-cc) OHV V-twin
FXB Sturgis – 1980–82, 82-ci (1340-cc) OHV V-twin
FXD Dyna Super Glide – 1995–98, 82-ci (1340-cc) Evo V-twin
FXD Dyna Super Glide – 1999–2001, 88-ci (1450-cc) Twin Cam V-twin
FXDB Dyna Glide Sturgis – 1991, 82-ci (1340-cc) Evo V-twin
FXDB Dyna Glide Daytona – 1992, 82-ci (1340-cc) Evo V-twin
FXDC Dyna Glide Custom – 1992, 82-ci (1340-cc) Evo V-twin
FXDL Dyna Low Rider – 1994–98, 82-ci (1340-cc) Evo V-twin
FXDL Dyna Low Rider – 1999–2001, 88-ci (1450-cc) Twin Cam V-twin
FXDG Disc Glide – 1983, 82-ci (1340-cc) OHV V-twin
FXDWG Dyna Wide Glide – 1993–98, 82-ci (1340-cc) Evo V-twin
FXDWG Dyna Wide Glide – 1999–2001, 88-ci (1450-cc) Twin Cam V-twin
FXDX Dyna Super Glide Sport – 1999–2001, 88-ci (1450-cc) Twin Cam V-twin
FXDXT Dyna Super Glide T-Sport – 2001 on, 88-ci (1450-cc) Twin Cam V-twin
FXWG Wide Glide – 1980–86, 82-ci (1340-cc) OHV V-twin
FXE Super Glide – 1974–80, 74-ci (1200-cc) OHV V-twin, electric start
FXE Super Glide – 1981–84, 82-ci (1340-cc) OHV V-twin, electric start
FXEF-1200 Fat Bob – 1979–80, 74-ci (1200-cc) OHV V-twin
FXEF-80 Fat Bob – 1979–81, 82-ci (1340-cc) OHV V-twin
FXEF Fat Bob – 1985, 82-ci (1340-cc) V-twin
FXLR Low Rider Custom – 1987–94, 82-ci (1340-cc) Evo V-twin
FXR Super Glide II – 1982–83, 82-ci (1340-cc) OHV V-twin
FXR Super Glide – 1986–94, 82-ci (1340-cc) Evo V-twin
FXRC Low Glide Custom – 1985, 82-ci (1340-cc) V-twin
FXRD Sport Glide Grand Touring – 1986, 82-ci (1340-cc) V-twin
FXRDG Disc Glide – 1984, 82-ci (1340-cc) OHV V-twin
FXRP Police – 1984, 82-ci (1340-cc) V-twin
FXRS Super Glide II – 1982, 82-ci (1340-cc) OHV V-twin
FXRS Low Glide – 1984–88, 82-ci (1340-cc) Evolution V-twin
FXRS Low Rider – 1990–91, 82-ci (1340-cc) Evo V-twin

FXRS-CONV Low Rider Convertible – 1989–98, 82-ci (1340-cc) Evo V-twin
FXRS-SP Low Rider Sport – 1987–93, 82-ci (1340-cc) Evo V-twin
FXRT Sport Glide – 1983–92, 82-ci (1340-cc) OHV V-twin
FXS Low Rider – 1977–80, 74-ci (1200-cc) OHV V-twin
FXS-80 Low Rider – 1979–82, 82-ci (1340-cc) OHV V-twin
FXSB Low Rider – 1983–85, 82-ci (1340-cc) OHV V-twin
FXST Softail – 1984–90, 82-ci (1340-cc) Evo V-twin
FXST Softail Standard – 1999–2001, 88-ci (1450-cc) Twin Cam V-twin
FXSTB Night Train – 1998–99, 82-ci (1340-cc) Evo V-twin
FXSTB Night Train – 2000 on, 88-ci (1450-cc) Twin Cam 88B V-twin
FXSTC Softail Custom – 1986–98, 82-ci (1340-cc) Evo V-twin
FXSTD Softail Deuce – 2000 on, 88-ci (1450-cc) Twin Cam V-twin
FXSTDI Softail Deuce – 2001 on, 88-ci (1450-cc) Twin Cam 88B V-twin, e.f.i.
FXSTS Springer Softail – 1988–99 82-ci (1340-cc) Evo V-twin
FXSTS Springer Softail – 2001 on, 88-ci (1450-cc) Twin Cam 88B
FXSTSB Bad Boy – 1995–97, 82-ci (1340-cc) Evo V-twin

G Servicar – 1932–63, 45-ci (750-cc) side-valve V-twin
GD Servicar – 1932–41, 45-ci (750-cc) side-valve V-twin, large compartment
GE Servicar – 1933–37, 45-ci (750-cc) side-valve V-twin, air tank
GE Servicar – 1964–73, 45-ci (750-cc) side-valve V-twin, electric start

H Sprint H - 1962-68, 246cc OHV single, high compression

J – 1915–29, 61-ci (1000-cc) F-head V-twin, three-speed, electrics
JD – 1921–29, 74-ci (1200-cc) F-head V-twin, electrics
JDCA – 1924, 74-ci (1200-cc) F-head V-twin, alloy pistons, electrics
JDCB – 1925, 74-ci (1200-cc) F-head V-twin, iron-alloy pistons, electrics
JDCBS – 1925, 74-ci (1200-cc) F-head V-twin, iron-alloy pistons, electrics, sidecar gearing
JDH Two Cam – 1928–29, 74-ci (1200-cc) F-head V-twin, Dow-metal pistons, electrics
JDS – 1921–29, 74-ci (1200-cc) F-head V-twin, three-speed, electrics, sidecar gearing
JDSCA – 1924, 74-ci (1200-cc) F-head V-twin, alloy pistons, electrics, sidecar gearing
JDX– 1928, 74-ci (1200-cc) F-head V-twin, electrics
JDXL – 1928, 74-ci (1200-cc) F-head V-twin, Dow-metal pistons, electrics
JE – 1924–25, 61-ci (1000-cc) F-head V-twin, alloy pistons, electrics
JES – 1924–25, 61-ci (1000-cc) F-head V-twin, alloy pistons, electrics, sidecar gearing
JH Two Cam – 1928–29, 61-ci (1000-cc) F-head V-twin, Dow-metal pistons, electrics
JS – 1919–29, 61-ci (1000-cc) F-head V-twin sidecar motor, electrics
JX – 1928, 61-ci (1000-cc) F-head V-twin, three-speed, electrics
JXL – 1928, 61-ci (1000-cc) F-head V-twin, Dow-metal pistons, electrics

K – 1952–53, 45-ci (750-cc) side-valve V-twin
KH – 1954–56, 55-ci (900-cc) side-valve V-twin
KHK –1955–56, 55-ci (900-cc) side-valve V-twin, tuned
KK – 1953, 45-ci (750-cc) side-valve V-twin, tuned

M50 –1965–67, 3-ci (49-cc) two-stroke single
M65 – 1967–69, 4-ci (65-cc) two-stroke single
MC65 Shortster – 1972, 4-ci (65-cc) two-stroke single
M125 Rapido – 1968–72, 7.5-ci (124-cc), two-stroke single
MS50 – 1966–68, 3-ci (49-cc) two-stroke single
MS65 – 1967–72, 4-ci (65-cc) two-stroke single
MSR Baja – 1970–72, 6-ci (98-cc) two-stroke single

R – 1932–36, 45-ci (750-cc) side-valve V-twin, low-compression
RC125 – 1974, 7.5-ci (124-cc) two-stroke single
RL – 1932–36, 45-ci (750-cc) side-valve V-twin
RLD – 1932–36, 45-ci (750-cc) side-valve V-twin, high-compression
RS – 1932–36, 45-ci (750-cc) side-valve V-twin, sidecar model

S – 1926–27, 21-ci (350-cc) OHV single racer, magneto
S – 1948–52, 7.5ci (125-cc) two-stroke single
SR100 – 1973–74, 6-ci (98-cc) two-stroke single
SS Sprint SS – 1967–71, 15-ci (246-cc) OHV single
SS Sprint SS – 1973–74, 21-ci (346-cc) OHV single
SS125 – 1976–77, 7-ci (124-cc) two-stroke single

OPPOSITE: Riding into the millennium.

SS175 – 1976–77, 10.5-ci (174-cc) two-stroke single
SS250 – 1975–77, 14.8-ci (242-cc) two-stroke single
ST – 1953–59, 10-ci (165-cc) two-stroke single
STU – 1954–59, 10-ci (165-cc) two-stroke single, restricted
SX Sprint SX – 1971–74, 21-ci (346-cc) OHV single
SX 125 – 1974–75, 7.5-ci (124-cc) two-stroke single
SX 175 – 1974–75, 10.5-ci (174-cc) two-stroke single
SX 250 – 1975–78, 14.8-ci (242-cc) two-stroke single
SXT 125 – 1976–77, 7.5-ci (124-cc) two-stroke single

TX 125 – 1973, 7.5-ci (124-cc) two-stroke single

U – 1937–48, 74-ci (1200-cc) side-valve V-twin, low-compression
UC – 1945, 74-ci (1200-cc) side-valve V-twin, medium-compression
UH – 1937–41, 80-ci (1300-cc) side-valve V-twin, medium-compression
UHS – 1937–41, 80-ci (1300-cc) side-valve V-twin, sidecar gearing
UL – 1937–48, 74-ci (1200-cc) side-valve V-twin, high-compression
ULH – 1937–41, 80-ci (1300-cc) side-valve V-twin, high compression
US – 1937–41, 1945–48, 74-ci (1200-cc) side-valve V-twin, sidecar gearing

V – 1930–33, 74-ci (1200-cc) side-valve V-twin, medium-compression
VC – 1930–33, 74-ci (1200-cc) side-valve V-twin, commercial model
VD – 1934–36, 74-ci (1200-cc) side-valve V-twin, low-compression
VDDS – 1935, 80-ci (1300-cc) side-valve V-twin, sidecar model
VDS – 1934–36, 74-ci (1200-cc) side-valve V-twin, low-compression, sidecar gearing
VHS – 1936, 80-ci (1300-cc) side-valve V-twin, sidecar gearing
VL – 1930–33, 74-ci (1200-cc) side-valve V-twin, high-compression
VLD – 1933–35, 74-ci (1200-cc) side-valve V-twin, high compression
VLDD – 1935, 80-ci (1300-cc) side-valve V-twin, high compression
VLH – 1936, 80-ci (1300-cc) side-valve V-twin, high compression
VS – 1930–33, 74-ci (1200-cc) side-valve V-twin, sidecar model

W – 1937, 45-ci (750-cc) side-valve V-twin, low-compression
WF – 1919–23, 35-ci (575-cc) ho twin, three-speed, magneto
WJ – 1919–23, 35-ci (575-cc) ho twin, three-speed, electrics
WL – 1937–42, 1945–51, 45-ci (750-cc) side-valve V-twin, high-compression
WLA – 1941–45, 45-ci (750-cc) side-valve V-twin, US Army
WLC – 1941–44, 45-ci (750-cc) side-valve V-twin, Canadian Army
WLD – 1937–42, 45-ci (750-cc) side-valve V-twin, extra-high compression
WLR – 1945, 45-ci (750-cc) side-valve V-twin, Russian Army
WS – 1937, 45-ci (750-cc) side-valve V-twin, sidecar gearing

X90 – 1973–75, 5.5-ci (90-cc) two-stroke single
XA – 1942–43, 45-ci (750-cc) flathead ho twin, US Army
XL Sportster – 1957–59, 54-ci (883-cc) OHV V-twin, medium-compression
XL53C Sportster Custom 53 – 1998–2001, 54-ci (883-cc) Evo V-twin
XL883 Sportster - 2000 on, 54-ci (883-cc) Evo V-twin
XL883 Sportster Hugger – 2000 on, 54-ci (883-cc) Evo V-twin
XL1200C Sportster Custom – 1996–2001, 73-ci (1203-cc) Evo V-twin
XL1200S Sportster Sport – 1996–2001, 73-ci (1203-cc) Evo V-twin
XLC Sportster – 1958–59, 54-ci (883-cc) OHV V-twin, high-compression
XLH Sportster – 1958–71, 54-ci (883-cc) OHV V-twin, high-compression
XLH Sportster – 1972–85, 61-ci (997-cc) OHV V-twin
XLH Sportster – 1986–98, 54-ci (883-cc) Evo V-twin
XLH Sportster – 1988–99, 73-ci (1203-cc) Evo V-twin
XLH Sportster Hugger – 1988–99, 54-ci (883-cc) Evo V-twin
XLH Sportster Deluxe – 1988–95, 54-ci (883-cc) Evo V-twin
XLH1100 Sportster – 1986–87, 67-ci (1100-cc) V-twin
XLCH Sportster – 1958–71, 54-ci (883-cc) OHV V-twin, high-compression
XLCH Sportster – 1972–79, 997cc OHV V-twin
XLCR Cafe Racer -1977-78, 61-ci (997-cc) OHV V-twin
XLS Sportster/Roadster – 1979–85, 61-ci (997-cc) OHV V-twin
XLT Sportster – 1977–78, 61-ci (997-cc) OHV V-twin
XLX Sportster – 1983–85, 61-ci (997-cc) OHV V-twin
XR1000 Sportster – 1983–84, 61-ci (1000-cc) OHV V-twin

Z90 – 1973–75, 5.5-ci (90-cc) two-stroke single

PARKING
TEXACO

INTO THE MILLENNIUM

ABOVE and OPPOSITE: For 2000, a genuine factory custom limited edition by Black Bear of the UK.

HARLEY-DAVIDSON
BLACK BEAR
HARLEY-DAVIDSON

2000 Range

RIGHT: A Buell SI Lightning at Daytona.

OPPOSITE: A Buell SI Lightning.

VANCE & HINES

BELOW: A Sportster XL 883.

OPPOSITE: A Sportster XL1200S

Sportster
Harley-Davidson
1200

RIGHT and OPPOSITE: Softail Standard.

Softail with custom accessories.

Softail with custom add-ons.

RIGHT and BELOW: Dyna Super Glide.

OPPOSITE: A Dyna Wide Glide at the Daytona 2000.

PIER
WELCOME TO DAYTONA BEACH

ABOVE and OPPOSITE: Dyna Super Glide Sport.

HARLEY-DAVIDSON

RIGHT: Road King.

OPPOSITE: Road King Classic.

HARLEY-DAVIDSON

RIGHT and OPPOSITE: Ultra Classic Electra Glide.

SUPER UNLEADED
Coors LIGHT
WELCOME
Bud
Bud
Bud

HARLEY-DAVIDSON
Touring

OPPOSITE: Ultra Classic Electra Glide.

RIGHT: Heritage Softail.

Three Road Glides at Sturgis.

ABOVE: Road Glide.

N ROCHFORD
S ROCHFORD
HARLEY-DAVIDSON

OPPOSITE and BELOW: Electra Glide Classics.

Screamin' Eagle Road Glide.

Limited edition Screamin' Eagle Road Glide.

Cory Ness

OPPOSITE and THIS PAGE
Y2K Khaos Millennium bike, customized by
Cory Ness.

2001 Range

Sportster Range
Sportster 883 (54ci/883cc)
Sportster 883 Hugger (54ci/883cc)
Sportster 53C Custom (54ci/883cc)
Sportster 1200 Custom (74ci/1199cc)
Sportster 1200 Sport (74ci/1199cc)

Dyna Glide Range
Super Glide (88ci/1449cc)
Low Rider (88ci/1449cc)
Wide Glide (88ci/1449cc)
Super Glide Sport (88ci/1449cc)
Super Glide T-Sport (88ci/1449cc)

Softail Range
Softail Standard (88ci/1449cc)
Night Train (88ci/1449cc)
Fat Boy (88ci/1449cc)
Fat Boy EFI (88ci/1449cc)
Heritage Classic (88ci/1449cc)
Heritage Classic EFI (88ci/1449cc)
Softail Deuce (88ci/1449cc)
Softail Deuce EFI (88ci/1449cc)

Touring Range
Road King (88ci/1449cc)
Road King Classic (88ci/1449cc)
Ultra Classic Electra Glide (88ci/1449cc)

RIGHT and OPPOSITE: XL Sportster 883.

Sportster

BELOW: Dyna Super Glide Sport.
OPPOSITE: Dyna Super Glide T-Sport.

HARLEY
DAVIDSON

Harley-Davidson

OPPOSITE: Softail Deuce.

RIGHT: Fat Boy Softail.

RIGHT and OPPOSITE: Road King Classic.

Service
LAMALOU
Tel : 04.6795.26.96
CESSENON
Tel :
MURVIEL
04.67.3796.22
St PONS
Tel : 04.67.97.01.89

RIGHT and OPPOSITE: Ultra Classic Electra Glide.

2002
Harley-Davidson V-Rod: An American Revolution

Every 20 years or so, Harley-Davidson comes up with a new engine, or at least a new development of what has gone before. Consequently, the overhead-valve Knucklehead of 1936 owed much to previous side-valve Harley V-twins; the Panhead (1947) and Shovelhead (1966) which succeeded it were all developments of the same thing. Even the Evolution of 1983 – a great leap forward for Harley – was based on the same old bottom end, albeit with lots of new parts. But in the summer of 2001 came something completely new – the V-Rod.

At the dawn of the 21st century, Harley-Davidson was selling more bikes and making bigger profits than ever before. But it couldn't afford to rest on its laurels. Quite apart from ever more sophisticated (and authentic-looking) Japanese cruisers, there was homegrown opposition from the Victory. The V-Rod was the answer – the first all-new Harley-Davidson for over half a century. All previous Harley road twins had been based on the same air-cooled 45-degree format, but the V-Rod was completely different. Not for nothing was the new engine named 'Revolution'.

The 60-degree angle between the cylinders was a huge break with tradition, as was the use of water-cooling and four valves per cylinder. It is tempting to add fuel injection and a balance shaft to that list (the V-Rod has both), but Harley-Davidson had already applied these to the faithful Evolution V-twin. By Harley standards, at 1130cc the new engine was suprisingly small compared to the 1450-cc (88-ci) Twin Cam. But it was more powerful than any road engine Harley-Davidson had ever put its name to, with 115bhp at 8,000rpm. This was the sort of power needed to counter the latest generation of sports cruisers, and the V-Rod delivered.

But the new engine wasn't quite as fresh as it seemed, being a much changed, sanitized version of the VR1000 race motor. As a racer, that was a flop (it secured a single second place and no wins in seven years of trying), but as the basis for a new generation of road bikes it looks more promising.

Harley-Davidson recruited sports car maker Porsche to transform the VR1000 engine into a 21st-century V-twin, and build the first prototypes. Harley finalized the R&D on the engine, as well as designing the chassis that would house it. However, the new engine is a product of a joint venture between Porsche and Harley, though built in Harley-Davidson's own Kansas City factory.

In comparison with this revolutionary engine (by Harley standards), the chassis was decidedly conservative: tubular steel (not alloy) frame, with short-travel rear shocks adjustable only for pre-load; the front forks weren't adjustable at all, and the cornering clearance was a mere 32 degrees; like most Harleys, even this one would soon ground out if the rider got too enthusiastic on twisty roads!

However, most people agree that here at last is a half-modern-looking Harley, power-packed and muscular, all bright silver and polished alloy. And who led the styling team? Willie G. Davidson. Respected motorcycle journalist Alan Cathcart summed it all up after riding the V-Rod in August 2001: 'This is one of the most important motorcycles of the past decade – Harley-Davidson has reinvented itself.'

SPECIFICATIONS
Harley-Davidson VRSC V-Rod

Engine	Liquid-cooled 60-degree V-twin, DOHC, 4 valves/cylinder
Bore x stroke	100 x 72mm
Capacity	1130cc (69ci)
Compression ratio	11.3:1
Power	115bhp @ 8,000rpm
Torque	65lb ft @ 6,300rpm
Gearbox	Five-speed
Frame Type	Tubular steel
Rake/Trail	34 degrees/99mm
Wheelbase	1713mm
Suspension	F: 49mm forks, non-adj, 100mm travel R: Twin shocks, pre-load, adj, 60mm travel
Brakes	F: Twin discs, 292mm R: Disc, 292mm
Tyres	F: 120/70 19 R: 180/55 18
Seat Height	689mm
Fuel Capacity	15 litres (3.3)
Weight (dry)	270kg (595lb)

HARLEY-DAVIDSON
V-ROD

Page numbers in italics refer to illustrations

ACKNOWLEDGEMENTS

A book as comprehensive as this is never possible without the help of a great many people. In particular, the authors would like to thank Garry Stuart for his excellent photographs, also Dr Martin Rosenblum for permission to reproduce his Harley-Davidson poems.

All the photographs in this book are courtesy of Gary Stuart, with the exception of those on pages 6, 7, 9, 14, 24 right, 191, 192, and 430–439, which are courtesy of Harley-Davidson UK: page 70 above left which is by John Carroll: pages 300 right, 301, 306, 311, 313–319 which are by Ian Kerr.

Garry Stuart wishes to thank the following people and organizations for their help and co-operation:
Glenn Bator, The Otis Chandler Museum, Oxnard, California; Lawayne Mathies, Mystery Design, Dallas, Texas; Dale and Martha of Greeley, Colorado; Martin Henderson, Boothill Motorcycles, England; Lonnie Isam Competition Motorcycles, Pearland, Texas; Mr Cole, Cole Motorcycle Museum, Sturgis, South Dakota;
Harley-Davidson UK Press PR, Brackley, UK; John and Helen Stuart, Holyhead, UK; Advantage Photo Services, Swindon, UK; and Harley-Davidson riders everywhere.

In memory of Dr 'Maz' Harris, HAMC, Kent, UK.

Eadweard Muybridge

Philip Brookman

with contributions by Marta Braun, Andy Grundberg, Corey Keller, Rebecca Solnit

Tate Publishing

15
21 22 23 24 25 26 27

Contents

The Terra Foundation for American Art is pleased to sponsor *Eadweard Muybridge*, an important reassessment of Muybridge's career and the first exhibition in the United Kingdom to address the full range of his work. We offer our congratulations to Tate for their long-standing commitment to presenting scholarly and challenging exhibitions and look forward to continuing to collaborate with them on exhibitions of American art. Through grants and partnerships, the Terra Foundation for American Art has made it a priority to bring works of American art, scholarly resources, and educational materials to audiences worldwide in order to foster global perspectives on the visual arts of the United States. By engaging audiences through exhibitions such as *Eadweard Muybridge*, we hope to enhance, and thereby strengthen, the understanding and enjoyment of historical American art throughout the world. Implicit in such activities is the belief that art has the potential both to distinguish cultures and to unite them.

Elizabeth Glassman
President and Chief Executive Officer
Terra Foundation for American Art

TERRA
FOUNDATION FOR AMERICAN ART

Foreword

In 1901, the seventy-one year old Eadweard Muybridge published his last, and arguably his most influential, volume, *The Human Figure in Motion.* Still in print today, this well-known selection of Muybridge's photographs of the human form has been regularly used by photographers and artists. Best known among these was Francis Bacon, whose studio contained perhaps as many as five copies of the book, as well as hundreds of individually annotated sheets of Muybridge's images. Their influence can be traced repeatedly in Bacon's paintings, as was apparent in the exhibition at Tate Britain in 2008.

Many people are unaware, however, that these familiar images are drawn from Muybridge's much larger project, *Animal Locomotion: An Electro-Photographic Investigation of Consecutive Phases of Animal Movement* (1887). Originally issued in eleven volumes, containing 781 plates, the total project comprised 19,347 individual shots. Although this immense labour was presented as a scientific endeavour, in recent years scholars such as Marta Braun have detected Muybridge's intervention in the series much more than had previously been realised. In the last two decades this line of enquiry has been enriched as a result of conservation work at the National Museum of American History in Washington, which effectively reclaimed Muybridge's neglected cyanotype proofs. Originally presented to the museum in 1927, these vital clues to the way Muybridge designed the published versions of his images remained unpacked until 1985. During the following ten years, each batch was diligently conserved and rehoused, enabling new research to get under way towards the end of the 1990s. Previously unseen outside the United States, these beautiful images provide many unexpected insights into the ways in which Muybridge staged and crafted his work.

The enduring fascination of Muybridge's attempts to record movement, both in *Animal Locomotion* and in the earlier *The Attitudes of Animals in Motion* albums (1881), has tended to obscure his very real achievements as a landscape and documentary photographer. Born in Kingston-upon-Thames, near London, in 1830, Muybridge had emigrated to America by 1852. Eventually settling in California, he embarked on his career as a photographer in 1867, though where he trained remains a mystery. His astounding views of the Yosemite Valley equal those of his contemporary Carleton Watkins, and anticipate the celebrated views taken by Ansel Adams sixty years later. Constantly challenging the perceived limits of photography, Muybridge also astonished his contemporaries with his huge 360-degree panoramas of his adopted city San Francisco. These achievements are part of a life story so full of truly sensational incident that it would stretch credulity had it not been so scrupulously documented by recent biographers.

The need for a more integrated picture of Muybridge's attainments—particularly in the creative field—inspired Philip Brookman, chief curator at the Corcoran Gallery of Art, to conceive this publication. The exhibition which it accompanies has already been seen in Washington, and will travel on to San Francisco after its showing at Tate Britain. The images gathered here derive from an intensive survey of photographic collections across the United States and Great Britain, both public and private, and reveal many unknown aspects of Muybridge's production.

During his later years Muybridge travelled extensively, in America, Europe and in Britain, presenting his images of animals and humans in motion in hugely popular lectures. The last of these was given to the Artists' Society in St Ives, Cornwall, in 1897. Around that time, having spent more than forty years mostly in the States, Muybridge returned for good to England, choosing to settle in his native Kingston. Outside North America, Muybridge's work is best represented by the extensive body of material that he bequeathed on his death in 1904 to Kingston. In partnership with Kingston University, the Kingston Museum has now developed a new website portal to facilitate future Muybridge studies (www.eadweardmuybridge.co.uk).

An exhibition on this scale, which effectively straddles the Atlantic in both its subject and its organisation, could not have been achieved without the

Fig. 3. *Edgington—Fast (Phase 16)*, 1878

active support of so many Muybridge enthusiasts. We are grateful to everyone who has enabled the loan of works that will be shown at Tate Britain, and a full list of lenders appears on p. 360. In addition to thanking lenders, we should also like to offer special thanks to the Terra Foundation for American Art for their vital support in staging this important exhibition in London. We are extremely grateful to Elizabeth Glassman, President and Chief Executive Officer of the Terra Foundation, and her fellow Trustees, for supporting not only this exhibition, but also Tate's wider ambitions in the field of American art.

At Tate Britain, *Eadweard Muybridge* has been curated by Ian Warrell and Carolyn Kerr, assisted by Sofia Karamani and Emma Baker. The exhibition helps to develop our commitment to photography, in the form both of exhibitions and of acquisitions for the collection. Though Muybridge's practice is rooted in the nineteenth century, there are few photographers of that period whose images have cast such a spell on modern and contemporary artists, in Britain and abroad. And though his greatest work was produced in America, one might think of this exhibition as a suitable home-coming for an artist who retained his British roots.

Penelope Curtis
Director, Tate Britain

Paul Greenhalgh
Director and President, Corcoran Gallery of Art
and Corcoran College of Art and Design

Introduction
Andy Grundberg

> Four generations of artists, of the most diverse persuasions, have acknowledged the fascination of [Eadweard Muybridge's] work, and it is obvious that many have learned from it, if only at second or third hand: that alone justifies our curiosity about the genesis of his sensibility.[1]
>
> Hollis Frampton

Eadweard Muybridge's career took place on the cusp of what we now call modern vision. His medium, photography, the most influential form of visual representation to be born in the nineteenth century, played a paradoxical role in this crucial phenomenological shift. On the one hand, it enshrined the single-point pictorial perspective that had defined painting since the Renaissance—a not-surprising development considering that the vast majority of photographic images have been produced by the same kinds of lenses that artists used as drawing aids for hundreds of years. (Lenses, not incidentally, were often called "objectives.") On the other hand, photography could produce more subjective, contingent experiences such as those generated by the stereograph, through which the viewer's eyes and brain conspire to produce an illusion of three-dimensional space, and the panorama, which flattens a full 360-degree view so one can in effect see the world behind one's head.

Photography's ability to produce subject-centered experiences was just one facet of the nineteenth century's fascination with individualized vision. The phenakistoscope, zoetrope, and other spinning devices used as parlor entertainments fashioned a wondrous, semiperfect illusion of continuous motion for an intimate audience, albeit using images drawn by hand. Their popularity declined only with the advent of more sophisticated cinematic devices that projected photographs. More central to the prevailing narrative of art history is the turn of painting toward its own form of subjective vision. By most accounts, what we now call modern art began in the second half of the nineteenth century with the rise of impressionism, which made the case that what mattered in a painting was the way an artist saw light, not objects. In short, for modern artists and modern viewers alike, the retina became the primary location of the real, as Jonathan Crary has cogently argued in his 1990 book *Techniques of the Observer.*

Eadweard Muybridge is an essential figure for art today in part because his work straddles these two compelling and, to an extent, competing ideologies of representation. In his early career, he produced large-format, wet-plate landscape pictures that adhere in most important respects to a classical, premodern model of what a landscape should be, as derived from Claude Lorraine's example centuries earlier: a scene viewed from an overarching vantage point, with a foreground often populated by a human figure or figures who look out over a prospect into a distant, pristine natural world. At the same time, however, Muybridge also was making smaller, side-by-side images that could be put in a stereo viewer and enjoyed in three dimensions from the comfort of one's armchair. He also produced what is one of the most famous panoramas in the history of photography, an 1878 view of San Francisco from atop Nob Hill, assembled from thirteen mammoth glass plates.

Then, from that year onward, he devoted himself almost entirely to devising ways to depict instantaneous, sequential views of motion that, when assembled together, could create what appear to us now as rudimentary movies. Aside from the impressive technical innovations involved in going from the assembly of individual negatives into a long but coherent single cityscape to the consolidation of split-second exposures into cinematic narratives, the conceptual leap is vast: from suggesting a third dimension in space to disassembling a fourth dimension, time. As Phillip Prodger has written, "In the new world Muybridge created, the depiction of moving things was freed from the limitations of memory or preconception."[2] It was the novelty and inventiveness of this preoccupation with dissecting time that made Muybridge a phenomenon in his lifetime and a hero to artists of the modern era. Marcel Duchamp, for one, credited Muybridge as an inspiration, although his *Nude Descending a Staircase (No. 2)* and the allied works of the Italian Futurists seem to have taken their cues most directly from the contemporaneous photographs of the French scientist Étienne-Jules Marey. (Then again, Marey's own corpus is directly indebted to Muybridge, so the lineage is not in dispute.) Much has been said about how the English painter Francis Bacon was inspired by Muybridge to make multiple, motion-based portraits in the 1950s and 1960s. However, Muybridge's prominence in the contemporary art world stems mainly from the great relevance his work held for American and European minimalist and conceptual artists of the late 1960s and early 1970s, as well as for structuralist filmmakers of the same period.

Sol LeWitt, Mel Bochner, Jan Dibbets, and Vito Acconci were among a generation of artists whose work emphasized process, information, systems, and what Bochner has called "serial order."[3] Among other media, these artists used photography as a way of recording procedures and systems that occurred over time or that existed outside of the gallery and museum walls. Their quasi-scientific work often resembled Muybridge's in taking the shape of grids (LeWitt, Bochner) and panoramas (Dibbets), and in adopting his experimental mode of investigation (Acconci). Most important, they saw in Muybridge an example of an artist whose passions were contained within and expressed by a rigor no less strict for being poetic.

Writing recently in the *New Yorker*, Peter Schjeldahl called these artists "last-gasp Romantics … with time on their hands and on their minds, in the wilds of geographic and imaginary space."[4] Muybridge's influence on their work was direct, not speculative. In a 1967 essay published in *Artforum*, "The Serial Attitude," Bochner cited Muybridge as a primary antecedent for what was, for Bochner, the most vanguard art of his day.[5] LeWitt once related how during his early days in New York City, he would browse through sheets of prints from a broken-up copy of *Animal Locomotion* at the Weyhe Book Store and Gallery on Lexington Avenue, occasionally buying them and sending them on to friends like Dibbets, who lived in Holland.[6]

Michael Snow, Ken Jacobs, and Hollis Frampton were independent filmmakers of the same period, seeking to break film down into its essential elements, of which time, and the splitting of time, was the sine qua non. All saw Muybridge's work as a kind of Rosetta stone of cinematic possibility. Frampton, who also was a still photographer, wrote an influential essay about Muybridge and his work, "Eadweard Muybridge: Fragments of a Tesseract," for *Artforum* in 1973.[7] Appearing just as the art world's growing interest in photographs collided with photography's emerging presence in museums and galleries, Frampton's essay was both a recuperative form of art history and a confirmation of how much sway the fourth dimension of time held over contemporary artists. Two years later, Frampton and his wife, Marion Faller, embarked on an homage to Muybridge, a comical series of photographs called *Sixteen Studies from Vegetable Locomotion*.

A similar interest in Muybridge's life and work spurred the musician Philip Glass to write a three-act "music-drama" called *The Photographer*, which had its premiere in Amsterdam in 1982 and appeared at the Brooklyn Academy of Music's Next Wave Festival in 1983. The composer of *Einstein on the Beach* no doubt felt a kinship between the rhythmic pulses of his musical style and the intervals that define Muybridge's motion studies, and *The Photographer* is propelled by repetitive, short notes performed by a violin and other chamber instruments. Assembled together in time, the notes make up a kind of postmodern melody. The theater piece also featured lyrics written by David Byrne, a troupe of dancers, and projections of Muybridge's photographs onto the stage backdrop. By the end of the 1980s, the photographer's influence on all the arts was obvious and inescapable.

Independent of his status within the vanguard art scene in New York, Muybridge gained renewed appreciation among photographers starting in the 1970s, but based on his ostensibly objective views of the western United States rather than on his more contingent motion studies. The rise of what might be called post–Ansel Adams landscape photography, codified by the 1975 exhibition *New Topographics: Photographs of a Man-Altered Landscape*, at the George Eastman House in Rochester, New York, led to a reinvestigation of nineteenth-century photographs of the American West by contemporary photographers and collectors.[8] Another exhibition that same year, *Era of Exploration: The Rise of Landscape Photography in the American West, 1860–1885*, organized by the Albright-Knox Art Gallery in Buffalo and the Metropolitan Museum of Art in New York, focused specifically on five photographers: Carleton E. Watkins, Timothy O'Sullivan, Andrew Joseph Russell, William Henry Jackson, and Muybridge. The curators singled out Muybridge's Yosemite views of 1867 for possessing aesthetic qualities unsurpassed by any of his compatriots.[9]

Many of the photographers included in *New Topographics*—Robert Adams, Lewis Baltz, Joe Deal, and Stephen Shore, among them—were highly skeptical of landscapes that romanticized the condition of the American West, having seen the highways and telephone poles that belie the Edenic renditions of Ansel Adams and his followers. Thus they were more attuned in sensibility to Timothy O'Sullivan's brutish sublime than to Muybridge's grand scenic views, but they still had to admire the ambition and enterprise of Muybridge's early work documenting California. Other photographers of the time sympathetic to the *New Topographics* aesthetic, prominent among them Mark Klett, took a page from Muybridge's career and embarked on a series of latter-day surveys of the West. Klett's enterprise was called the Rephotographic Survey Project, and he and his colleagues ventured out to find the exact locations where Muybridge, O'Sullivan, and other frontier photographers had taken their pictures some one hundred years earlier. At each spot, they created a duplicate view that showed a century's span of time.

Today, some thirty years later, Muybridge seems more central, not less, to the concerns and methodologies of contemporary art. This is true, I would argue, not because of the precinematic, time-stopping, interval-based nature of his motion studies, which is what appealed so much to the conceptual artists, nor because of the spectacular but fact-based beauty of his best landscapes, which the *New Topographics* photographers admired. Instead, it is a matter of the massive archive he created in assembling *Animal Locomotion*—the sheer volume of the volume itself.[10] Somewhere between a catalog and a library, the archive of motion studies that Muybridge created lies along a central axis of contemporary art that finds meaning in an accumulation of images rather than in the exceptionality of a chosen few.

To fully understand the import of the motion studies, we need to remember that Muybridge was part of a nineteenth-century generation intent on knowing the world by recording and assembling its constituent parts. Charles Darwin is probably the most well-known exemplar, attempting through his own observations to create a comprehensive map of the evolution of the species through time. In photography, examples range from French neurologist Duchenne de Boulogne's portraits of human expressions, achieved using electrical probes on the face, to Alphonse Bertillon's mug-shot images of Parisian criminals, which he hoped would lead to the identification of criminal "types." In the United States, Mathew Brady assembled a more high-minded catalog consisting of his studio's portraits of what he called illustrious Americans. We might also see Walt Whitman's *Leaves of Grass*, for example, as a poetic archive of his momentary feelings and observations that, when taken together, constitute a portrait of both Whitman and his age.

Examples of this kind of exhaustive urge to collect abound in contemporary art, going back at least as far as Bernd and Hilla Becker's gridded photographs of water towers and other industrial structures begun in the early 1970s. Just as Muybridge's archive of animal locomotion by no means exhausted the category, the Bechers did not attempt to photograph every water tower in the world; instead, they were content to demonstrate the incredible variability of form that had accreted around a single function. Hiroshi Sujimoto's long exposures of the insides of movie theaters illuminated by every frame of the film on view, and later of expanses of ocean water, follow the same strategy: they surround their subject with the multiplicity of its representations.

It is easy to carry this tradition forward: Sol LeWitt's 1980 cataloging of the contents of his New York apartment, consisting of hundreds of black-and-white prints; Gerhard Richter's *Atlas*, a forty-years-and-counting recording of what interests him visually; Thomas Ruff's formally identical portraits of his fellow Düsseldorf art students, taken at the start of the 1990s; Roni Horn's mid-1990s *You Are the Weather*, a series of close-up color photographs of an Icelandic

18
23 24 25 26 27 28 29

woman's face, and 1999 *Still Water*, focused on the flowing water of the river Thames; and Mike and Doug Starn's twenty-first-century pictures of individual snowflakes. One could even group together in this enterprise such disparate collections as Lee Friedlander's *American Monuments* series, published in 1976, and Cindy Sherman's *Untitled (film stills)*, first exhibited a few years later.

What Hollis Frampton called the "encyclopaedic enormity"[11] of Muybridge's undertaking exists somewhere between science and poetry, objectivity and personal whim, and consequently it is far from a complete record of the range of all possible human and animal motions. There are pictures of naked men wrestling (plates 147, 169) and a naked woman spanking a baby, for example, but few scenes of ordinary labor. This is, perhaps, why it appeals so strongly as an antecedent for contemporary artists; far from laboring under the illusion that they might ever encompass the entire range of a human experience, they seek to suggest its richness and complexity without comprehensiveness or closure. And as with Muybridge, the intentionality of their work focuses our attention, ultimately, on the creative mysteries of the artistic process.

Of course, no archive is ever complete, which is what makes such collections so fascinating for artists, curators, and researchers. Their interpretation requires us to temporarily suspend judgment on the quality of individual pictures while allowing image after image to spill out in front of us, like the pieces of a large jigsaw puzzle that have no guarantee of ever assembling into one coherent shape.

How we interpret archives of photographs is one of today's most interesting aesthetic questions, and one that bears on the fact that we now live in a world filled with the accumulation of 170-odd years of picture taking. It is, one might say, photography's unique contribution to the art of the last two centuries, and it makes Muybridge a central figure in any history of that art.

[1] Hollis Frampton, "Eadweard Muybridge: Fragments of a Tesseract," *Artforum* 11 (March 1973), reprinted in Hollis Frampton, *Circles of Confusion: Film, Photography, Video, Texts, 1968–1980* (Rochester, NY: Visual Studies Workshop Press, 1983), 71.

[2] Phillip Prodger, *Time Stands Still: Muybridge and the Instantaneous Photography Movement* (New York: Iris & B. Gerald Cantor Center for Visual Arts at Stanford University in association with Oxford University Press, 2003), 4.

[3] Mel Bochner, "The Serial Attitude," *Artforum* 6, no. 4 (December 1967), reprinted in *Looking Critically: Twenty-one Years of Artforum Magazine*, ed. Amy Baker Sandback (Ann Arbor, MI: UMI Research Press, 1984), 73.

[4] Peter Schjeldahl, "Conceptual Motion," *New Yorker*, August 3, 2009, 77.

[5] Bochner, "The Serial Attitude."

[6] LeWitt in conversation with the author, May 2006. The plates may not have been the 1887 originals but from the Dover reissues titled *Human Locomotion* and *Animal Locomotion* of the mid-1950s, for which Weyhe supplied the original plates for reproduction.

[7] Frampton, "Eadweard Muybridge: Fragments of a Tesseract," 69–80.

[8] William Jenkins, *New Topographics: Photographs of a Man-Altered Landscape* (Rochester, NY: International Museum of Photography at George Eastman House, 1975).

[9] Weston J. Naef, James N. Wood, and Therese Thau Heyman, *Era of Exploration: The Rise of Landscape Photography in the American West, 1860–1885* (Buffalo, NY: Albright-Knox Art Gallery; Boston: New York Graphic Society, 1975). The authors discuss the 1867 Yosemite views on pages 170–73.

[10] Eadweard Muybridge, *Animal Locomotion: An Electro-Photographic Investigation of Consecutive Phases of Animal Movements*, the most ambitious publication of Muybridge's work, appeared in 1887 and included a total of 781 plates.

[11] Frampton, "Eadweard Muybridge: Fragments of a Tesseract," 79.

19
25 26 27 28 29 30 31

The Astonished Woodchopper

1. *The Astonished Woodchopper*, 1867

2. *The Flying Studio, Photographer's Equipment in the Field (114)*, 1867

3. *Charon at the Ferry (21)*, 1867

4. *Base of First Fall of the Yosemite, from the Spirit's Voice Cave (40)*, 1867

5. *Views in Yosemite Valley*, page 14, 1867

Fig. 6. *Contemplation Rock, Glacier Point (1385)*, 1872

Helios: Eadweard Muybridge in a Time of Change

Philip Brookman

Eadweard Muybridge was above all an innovator who played a pioneering role in the creative transformation of late nineteenth-century American and European culture. He pushed the limits of existing photographic technologies during a time of rapid social and political change, which encompassed the world between 1850 and the dawn of the twentieth century. Inspired by a complex set of goals that were artistic, scientific, commercial, and especially personal in nature, Muybridge reinterpreted the grand landscape traditions of American and British photography, surveyed the construction of railroads and the growth of San Francisco and other areas of the Pacific Coast, documented the Modoc War and Central American landscape, devised the photographic facility to freeze sequences describing human and animal locomotion, and reanimated these studies in some of the first projected motion pictures. He helped change our understanding of the world, yet the complexity and totality of his art are little known today, in part because the landmark locomotion studies completed late in his career have since overshadowed the breadth of his vision (fig. 6).

In the wake of Muybridge's myriad experiments of the 1870s and 1880s, the United States evolved from a remote and provincial nation, pushing hard at its own frontier, into the hyper-connected, globalized society that today straddles a shifting border between terra firma and cyberspace. During the later half of the nineteenth century—Muybridge's time—the seeds of this change were planted by vast and rapid shifts in science, technology, economics, and art. The nature of the landscape itself took on new meaning as railroads, steamships, and advanced forms of speedy communication connected the world, weaving the networks we still use today and redefining the ways people experienced time and space. Muybridge worked at the heart of these social and cultural transformations. His groundbreaking depictions—he created innovative views of both motion and the landscape—changed how we see and relate to the world, and his wide influence is manifested in the histories of visual art, photography, graphics, communications, and cinema. Today, the impact of Muybridge's art can be found in many places, from Thomas Edison's pioneering kinetoscope viewer and Marcel Duchamp's *Nude Descending a Staircase* to the poignant paintings of Francis Bacon, the blockbuster film *The Matrix*, and U2's video for their hit song "Lemon." It is interesting now to imagine Muybridge's world, little more than a century ago, when our sense of time was turned upside down for good.

I – Lights

> Photography appears on the scene as though someone had found a way to freeze the water of passing time; appearances that were once as fluid as water running through one's fingers became solid objects.[1]
>
> Rebecca Solnit

As an artist working between 1867 and 1893, Eadweard Muybridge embraced the difference between the solid, or geologic, view of world and one that was fluid, characterized by the unfixed motion of water, people walking, or a horse running at full gallop. Artists of his time often danced around the philosophical division between reality and illusion, yet Muybridge understood the camera could address both with equal objectivity (fig. 7). In photography circles, the dichotomy between science and art was also a major point of discussion from the first years of the medium, which was launched in earnest by 1839.[2] The notion that photography would help prove scientific conjectures and also generate significant aesthetic debate was the principal fulcrum on which this evocative new "pencil of nature" was balanced.[3] A photograph represented at the time an altogether new kind of image, created by fixing the light reflected by or emitted from an

Fig. 7. *Athletes. Twisting Summersault. Plate 106*, 1879, from *The Attitudes of Animals in Motion*, 1881

object. It could unveil information not seen or understood by the eye, or help make sense of our uneasy place in the world. From its 1820s origins in experimental optics and chemical technology, photography grew during the second half of the nineteenth century into an enormously popular new window on the world, carrying with it new ways to see and communicate across the traditional border between perception and cognition. After all, a moment recorded by a camera is already past once a picture is made—it is no longer "real"—yet its connection to the authenticity of that moment and to time passing gives it a representational authority that fuses its function as evidence with its aesthetic character as art. This intersection—through photography—of the physical and the fluid, of matter and mind, eventually led Muybridge to a redefinition of both science and art.

Practiced primarily by amateurs before 1850, both science and photography were ripe for rapid development in the second half of the nineteenth century. For the scientist, photography (itself an invention born of both scientific and artistic achievement) was primarily a tool for recording "truthful" data used to systematically analyze visual information. Because photographs were presumed to provide factual evidence, they added to the authority of scientific research. For visual artists and writers, photographs introduced new ways to explore the physical nature of the world, its people, and their relationship to the landscape.

Photography quickly grew into a new artistic medium, rooted in the visual and literary traditions of painting and art history, as well as the narrative customs of language. Even though the rapidly evolving technology grew organically from both science and art, its application to both disciplines was frequently disconnected. Like the introduction of the microscope and telescope, the institutionalization of the camera as a research tool provided scientists with unprecedented methodologies based on direct observation. The emerging disciplines of biology, chemistry, geology, and astronomy, for example, were each rooted in the analysis and interpretation of observed facts. What had been invisible to the eye became more evident with photography, and what was made visible by the camera could be recorded, categorized, and structured for analytic study within a topological framework. Artists, on the other hand, often created photographic images as much from the imagination as from what could be seen. Photography—born of a technological revolution, which itself was transformed by its precocious offspring, the camera—brought new aesthetic challenges by foregrounding issues of time, memory, realism, abstraction, and mechanical reproduction.

Following the public declaration of Louis-Jacques-Mandé Daguerre's daguerreotype and William Henry Fox Talbot's paper photography process, both in January 1839, photography became the focus of significant public interest in France and England. The eminent British astronomer and mathematician Sir John Herschel, who conducted many experiments in photographic chemistry and optics, had met Talbot in 1824 and had corresponded with him for some time. On January 29, 1839, Talbot wrote to Herschel, "The Editors of the 2 London principal scientific newspapers, 'the Athenæum' & 'the Literary Gazette' have both taken up with zeal (much more than I could have expected) the subject of my discovery of Photogenic Drawing."[4] Just weeks behind

Daguerre, Talbot was intent on announcing, with a poetic flourish, his own photographic research, in which he found the apparent "character of the marvelous, almost as much as any fact which physical investigation has yet brought to our knowledge." In a written manuscript addressed to London's Royal Society, he concluded, "The most transitory of things, a shadow, the emblem of all that is fleeting and momentary, may be fettered by the spells of our 'natural magic,' and may be fixed forever in the position which it seemed only destined for a single instant to occupy."[5] Rather than reveal all the technical details of his process, which enabled him for the first time to permanently fix an image on paper, on January 31, Talbot spoke in concrete though eloquent terms about the phenomenological connections between light, shadow, and the image: "Such is the fact, that we may receive on paper the fleeting shadow, arrest it there, and in the space of a single minute fix it there so firmly as to be no more capable of change, even if thrown back into the sunbeam from which it derived its origin."[6] Herschel soon argued in his own comments to the Royal Society, on March 14, that photographs could be used to defend scientific experiments. He described his own numerous attempts to fix an image so it would not disappear and his discovery of the relative sensitivity of silver emulsions to different colors of light. He also considered how photography could be used to copy works of art, and argued that his own pictures created during the experimental process would "suffice to justify considerable expectation of the complete success in this new branch of art."[7] And Talbot, who was in rigorous competition with Daguerre, began to seek new uses for his discoveries, capturing through a microscope details of things that could not be seen by the unaided eye. In a short time, Talbot—in frequent correspondence with others—improved on his techniques. With his introduction of the calotype, or Talbotype, process in 1841, he could create a paper negative by exposing a sensitized sheet of fine, translucent stationery in the camera.[8] Since this paper negative process allowed photographers to make numerous positive prints from a single negative, photography became a more accessible medium. Whereas Herschel's photographs were always experiments in process, Talbot's were subject-driven—he worked to create images that reflected and interpreted his surroundings: his home at Lacock Abbey, his family, the gardens, windows, casts, and sculptures of his time and place. Here the science and art of photography came together in a form that would inspire many of the photographers who followed.

Edward James Muggeridge, Kingston upon Thames

Eadweard Muybridge is a name the illustrious photographer assumed much later in his life. He was born Edward James Muggeridge on April 9, 1830, in Kingston upon Thames, a historic market and trading town that flourished where a bridge crossed the river about fifteen miles southwest of London. For several hundred years, the bridge at Kingston, dating to the twelfth century, was the only other crossing of the Thames aside from the London Bridge. Kingston is also celebrated as the site where seven Saxon kings were crowned between 900 and 1016, including two bearing the name Edward. According to legend, a Coronation Stone was used in the ceremonies. On September 19, 1850, the stone, which had survived numerous moves throughout the centuries, was inaugurated in a Masonic ceremony to celebrate its new position of dignity in the yard of the Courts. The names of the kings are engraved on the stone, and Edward the Elder and Edward the Martyr are both spelled Eadweard, as they would have been when it was cut. The stone now stands beside the Guildhall on High Street, across from the medieval Clattern Bridge, all within site of Muybridge's childhood home.

Edward was the second of four sons born to John Muggeridge (1797–1843) and Susannah Smith Muggeridge (1808–1874). The Smith family was from Hampton Wick, just across the river from Kingston, near the royal Hampton Court Palace. Muybridge's father was a grain and coal merchant. He bought and sold corn, coal, and lumber that was shipped up the river from London or down from outlying farms. Their home was near the center of business on West-by-Thames, now High Street. Its yard backed up to the river, where docks allowed barges to stop to load and unload their freight. The Smith family was in the barge business, transporting goods along the Thames before railroads supplanted the barges. One story handed down in a memoir by Muybridge's cousin Maybanke Susannah Anderson refers to Muybridge's maternal grandfather and his commerce: "His business is almost unknown in these days of steam. He owned barges on the Thames, and many men, as well as horses, were needed for the work. When he drove his gig to London, to buy wheat or coal, he took under the seat of his gig a carrier pigeon, and in his pocket a quill or two, and when he bought a cargo, he wrote on a small piece of paper the number of barges he needed, put the paper in the quill, tied it under the wing of the pigeon and set it free. Someone watching for the bird's arrival unfastened the quill, took the message to the barges, and they started."[9]

The advent of the railroad, the invention of the telegraph, and the rise of capitalism and free trade changed their world forever. The 1840s—the years just following photography's launch—were a politically tumultuous time in Europe. As industrial and technological growth began to displace the dominant agrarian economies, and as machines increasingly replaced the livelihood of agricultural workers, people in Britain began to move from the country into the cities or to emigrate. Class structures were also changing as land ownership gave way to new industrially derived wealth in determining social status. Capitalism became an overriding economic force, driving opportunities to expand investment in railroads, shipping, mills, and manufacturing. As the demand for cheap labor grew to fill the ranks of mechanized manufacturing, and poverty grew in the cities, riots broke out in England and France—workers began to demand living wages and better working conditions. In Britain, growing poverty and famines, the Chartist Movement (which campaigned for voting rights), and disputes over the Corn Laws (tariffs that protected grain prices to the benefit of British land owners), each contributed to political turmoil brought on, in part, by agitating workers during the second half of the 1840s. After widespread protests, the Importation Act of 1846 was enacted to repeal the Corn Laws, opening the door to free trade. The second half of the nineteenth century opened with a wider view toward the economic potential of the British Empire and its relationships with the rest of the world. "At present, Englishmen may purchase their grain wherever they please, without paying duty; and they receive raw materials from every part of the globe without taxes or privilege of flag," wrote publisher and cartographer John Tallis in 1851. "Assuredly never was economical reform more radical than this. It attacked at one fell swoop landed property in its income, the state in its financial resources, and the national susceptibility in its most ticklish points."[10]

Fig. 8. Unknown photographer, *The earliest known photograph of Eadweard Muybridge*, ca. 1850

The gradual opening of international markets and new manufacturing methods meant more competition and reduced prices for goods and raw materials and more business for the merchant class. Kingston, as a crossroads of trade, never suffered the rampant poverty of the larger cities. When Muybridge's father died in 1843, his mother took over the family business and ran it successfully into the 1850s. His older brother, John, was studying to be a doctor, but died in 1847. Muybridge became the eldest son in a matriarchal family held together by the strength and business acumen of his mother and grandmother (fig. 8). Still a youth in the 1840s, Muybridge was, according to his cousin, "an eccentric boy, rather mischievous, always doing something or saying something unusual, or inventing a new toy, or a fresh trick.... He wanted adventure, the world was wide and he wanted to see it."[11] Growing up beside the Thames, he could swim the river from the docks and wander the pastoral Surrey landscape, watching in wonder the same billowing clouds painted by J. M. W. Turner and John Constable as they crossed the sky, like so many ships setting sail for unknown lands. Coming of age in the late 1840s in a small market town, Muybridge must have dreamed of the opportunities he would later find downriver, first in London and then abroad.

The Great Exhibition of 1851, London

When the Great Exhibition of the Works of Industry of All Nations went on view at the new Crystal Palace in London in May 1851, a wide spectrum of the public had its first chance to see many kinds of photographs. Over six million visitors paid admission to this first international World's Fair of design and industrial production, and it influenced many aspects of modern life at the time (fig. 9). Photographs from six nations were displayed, most in the category titled "Class X: Philosophical Instruments and Processes depending upon their use; Musical, Horological, and Surgical Instruments." Photographs were shown together with optical equipment having to do with light in a subsection titled "Application of Mechanical and Physical Science to Useful Purposes, not included in any of the preceding of subsequent sections." Some photographs—those considered more innovative in process—were also exhibited in the Fine Arts category, Class XXX, although they were judged together with those shown in Class X. Herschel was a member of the jury for the fine arts category, which was led by the eminent physicist Sir David Brewster. Medals were given to the Frenchman Hippolyte Bayard for his photographs of Parisian art and architecture and to the American Mathew Brady for his daguerreotype portraits, among others. (Talbot did not show his photographs in the exhibition because he had assumed that individual artists could not qualify.[12]) A special commendation was entered into the record, honoring Daguerre, who had just died on July 10.

The Great Exhibition showed examples of most of the current techniques, including daguerreotypes, calotypes, and albumen glass negatives. The brand-new wet-plate collodion-on-glass process, developed in London by Frederick Scott Archer, was barely present.[13] Another new technology that captured the imagination of visitors, including Queen Victoria, was stereography, a camera and presentation method first invented by Charles Wheatstone and improved by Brewster, which used two lenses and two paper prints or daguerreotypes mounted side-by-side to give the illusion of three-dimensional depth when looked at through a special viewer. The published comments of the jury focused on the relationship of photography to both art and science. "With art, doubtless, its future destiny will be closely linked; but so far from becoming a rival, it will prove a most useful auxiliary and a means by which the artist of merit may rise higher in reputation and eminence," they wrote. "Let us now view Photography in its applications to science. A process by which transient actions are rendered permanent, and which enables nature to do her own work, or in other words, which causes facts permanently to record themselves, is too well fitted for the purposes of science to be long overlooked."[14] Stressing that photography was still ripe for improvement and experimentation, they concluded, "Its pursuit, as an elegant and most extensively useful art, affords a strong motive for inquiry and experiment in the improvement of its process."[15]

Fig. 9. Claude-Marie Ferrier, *The Great Exhibition of Works of Industry of All Nations: VOL. IV—Reports by the Juries: View of Western Nave*, 1851

Although there is no evidence that Muybridge attended the Great Exhibition, as the son of a merchant family living close to London, he may have been one of the six million visitors. His uncle, Henry Selfe, was an inventor and plumber who exhibited a new high-pressure firehose there. One can imagine Muybridge as an ambitious, excited young man of twenty-one, astounded by the sheer range of mechanical and industrial inventiveness spread throughout the Crystal Palace. He would have been impressed by the determination of the internationalized vision on display, and by the modern presentation inside the steel-and-glass building, which itself conveyed this vast competition for progress, not only between exhibitors but also between nations.[16] He would have observed advances in the use of steam engines to power ships, railroads, and mills, transforming transportation and manufacturing on a global scale. And the optical instruments—the microscopes, telescopes, and cameras, along with the photographs and their connecting threads that wove together science and art—would have equally intrigued him. The technical nature of photography as well as its dreamlike ability to grab hold of a moment—a segment of time—and represent or "fix" it as an image would have appeared to Muybridge then as a vision of possibility for the future.

The aftereffects of the Great Exhibition were resounding, reflecting a growing optimism. Photography in Europe and the United States was poised for new invention and growth in both its scientific and aesthetic dimensions, as the technologies were improved and industrialized. American photography had been singled out by the jury: "Her works, with few exceptions, reject all accessories, present a faithful transcript of the subject, and yield to none in excellence of execution."[17] In the aftermath of the exhibition, journals and clubs were formed to debate both techniques and aesthetics; the public was captivated by such modern pictures made from something ethereal like light and shadow; and photographers began to seriously consider their real subjects—memory, history, time, and motion.[18]

The New World, America

Time and motion are the issues that captivated Muybridge for much of his career, which culminated some four decades after the Great Exhibition. In 1893,

Fig. 10. Unknown photographer, *Zoopraxographical Hall, World's Columbian Exposition, Chicago*, 1893

he attended another major World's Fair, which summarized the vast changes that had occurred since 1851. By then Muybridge was a world-renowned photographer and lecturer, on hand to share his achievements with visitors eager to view and experience the new technologies that were transforming their lives. When the World's Columbian Exposition, organized to commemorate the 400th anniversary of Columbus's landing in the Americas, opened in Chicago in May 1893, photographs were pervasive throughout the displays and proved to be a significant tool in communicating the theme of progress.[19] Indeed, it may have been this cacophony of images that supplanted the unified, encyclopedic view of the world that the fair's organizers had hoped to engender. Photographs were used to frame and contextualize the elaborate staging, questioning the authenticity of the displays, which included not only the requisite halls of manufacturing, transportation, agriculture, and machinery but also ethnographic tableaux that incorporated dioramas and native performers. Laid out next to the Ferris wheel on the "Midway Plaisance," which focused on amusements and entertainment, were a Chinese village, an Indian bazaar, a Javanese settlement, and the fantastically popular Cairo Street concession. Almost hidden in this mix of artifacts and fictional destinations, sandwiched between a Persian theater and the Cairo Street, was Muybridge's Zoopraxographical Hall (fig. 10), where he lectured on the science of "descriptive zoopraxography." Best known at the time for his encyclopedic visual study of human and animal motion—published in 1887 as *Animal Locomotion: An Electro-Photographic Investigation of Consecutive Phases of Animal Movements*—Muybridge was a practiced performer who analyzed physical movements based on countless sequences of photographs he had created since the late 1870s. In his lectures, he described the technical breakthroughs that advanced his photographic science and weighed his own motion studies against the depiction of movement in the history of art. The climax of his talk was the animation of selected images using his zoopraxiscope, an altered lantern-slide projector that could bring his photographs to life. This was, in effect, one of the earliest examples of projected motion pictures. By freezing motion in a sequence of still images using a bank of cameras and then reanimating the photographs to re-create the recorded action, he forever altered the way people would perceive and experience the passage of time.

Eager for popular entertainment, the audience at the Exposition chose Cairo over Muybridge's Zoopraxographical Hall, which closed within a few short months, only to be replaced by a painted panorama of Pompeii.[20] Rather than continue his lectures in the United States, he soon returned overseas to his hometown, from which he had emigrated around the time of the Great Exhibition more than forty years before.

Muybridge, then still named Muggeridge, had left Kingston upon Thames in 1851 or early 1852, in search of a new identity he would create for himself first in London, New York, and San Francisco, working as a sales agent for the London Printing and Publishing Company.[21] As a printer and publisher of books, maps, engravings, and music with offices in Great Britain and the United States, the London Printing and Publishing Company packaged and distributed works by contemporary artists and authors as well as reprints of the classics. They sold some of John Tallis's maps and *Tallis's History and Description of the Crystal Palace and the Exhibition of the World's Industry in 1851* and created numerous engraved portraits of politicians, military heroes, and poets, as well as landscapes, still-lifes, and copies of works of art. Before it was possible to reproduce photographs in mass publications, printers relied on the engraving process for

Fig. 11. Thomas Hollis and John Mayall, *Great Exhibition, Main Avenue, looking west*, 1852

illustrations. By the early 1850s, engravers frequently copied photographs to illustrate people and places. Tallis's book on the Great Exhibition was illustrated profusely with steel engravings of the Crystal Palace and its displays made from daguerreotypes by Richard Beard, John Mayall, and others (fig. 11). The London Printing and Publishing Company also reprinted engravings from old plates by engravers such as William and Edward Finden. To expand their distribution in the United States, the company hired sales agents to create new markets and take orders for works to be imported. When Muybridge was sent to New York as one such agent, his new job demanded an understanding of sales and marketing promotion for British books and illustrations as well as the printing, binding, and distribution process.

Muybridge's assignment in New York was to import unbound books and oversee their binding, sales, and distribution. He also traveled with shipments, and in a 1902 letter to the editor of *The Standard* he later recalled, "My engagements frequently required visits to the levee [in New Orleans] where steamboats from upriver discharged their cargoes of cotton and other merchandise."[22] One of his first friends in the United States was Silas Wright Selleck, a photographer close to Muybridge's age and skilled as a portrait daguerreotypist. Selleck worked in the New York photography studio of Mathew Brady, and other members of his family were printers and bookbinders in the city. Following the westward course taken by many fortune seekers at the time, by autumn 1852 Selleck was working as an operator in the Sacramento daguerreotype gallery of George Howard Johnson, best known for his meticulous portraits and outdoor images of the gold mines. Selleck's arrival was accompanied by the gallery's claim that he had taken the prize-winning portraits by Brady that were exhibited at the Great Exhibition.[23] The following year, Selleck joined Johnson in business in San Francisco and soon opened his own studio there. Muybridge continued to work from New York until 1855, when he too traveled to San Francisco, where he created a new outlet for the London Printing and Publishing Company.[24] One of the first published records of his arrival in California is an advertisement in the *San Francisco Daily Evening Bulletin* on April 28, 1856, soliciting "A gentleman well qualified to obtain subscribers for a new illustrated standard work."[25] It was signed E. J. Muggridge [*sic*] at 113 Montgomery Street. He may have already changed his name to Muygridge (the first alteration of his last name), leaving behind that part of his past. He opened a bookstore on the ground floor and an office of the London Printing and Publishing Company on the second. One of his first projects was to issue a book from the popular Encyclopaedia Britannica series titled *Entomology*. Dated 1855, it carried the imprint Edinburgh: A. and C. Black; San Francisco: E. J. Muygridge.[26]

In 1848, the United States had taken the southwestern territories from Mexico in the aftermath of the Mexican-American War. The 1849 discovery of gold at Sutter's Mill in the Sierra foothills precipitated a huge increase in California's population during the 1850s, and the young territory exploded as a bookend to the nation's colonization. California became the thirty-first state on September 9, 1850. San Francisco was the principal port of supply for the gold rush, and between 1847 and 1860, the city's population grew from about four hundred and sixty to fifty-six thousand people.[27] Photography played a major role in describing and documenting these substantial changes, and many photographers came west from Boston, New York, and Philadelphia. In 1851, lured by the gold rush, Carleton E. Watkins traveled from Oneonta, New York, to California with his friend Collis P. Huntington, a shopkeeper and hardware supplier who later helped finance construction of the Central Pacific Railroad. Watkins first worked for Huntington as a store clerk in Sacramento, and soon moved to San Francisco, probably to assist the city's best-known photographer, Robert H. Vance (who had arrived from South America in 1850), in his studio and outdoor work. Watkins became a skilled daguerreotypist and taught himself the wet-plate collodion process. By 1856, Watkins had established his own studio in the city. In reviewing the state of photographic art in California, critic Robert A. Carden wrote in 1857, "In San Francisco, the Art is of course more generally practiced than in any other city or town in the state. Mr. R. H. Vance, corner of Montgomery and Sacramento streets, has long been established there, and has now one of the most extensive galleries in the United States.... The Art in California bids fair to rival the cities of the Eastern States, and it may be said, with credit, that there are many who honor the profession in this once far off land."[28]

Fig. 12. Robert H. Vance, *James King of William*, ca. 1854

During the 1850s, San Francisco became a melting pot of immigrants and refugees from across the country and around the world, who had arrived by steamer, stage, horseback, or on foot. It was a place of redemption and liberation, where lives were remade and names changed to make a fresh start. It was also a place of lawlessness and fierce competition. The assassination, on May 14, 1856, of reform-minded publisher James King of William precipitated one of the more momentous periods in the city's history, pitting Protestants against Irish Catholic immigrants. With the formation of the San Francisco Vigilance Committee of 1856, residents witnessed an armed paramilitary insurrection that took power from the police and state judiciary.[29] Muybridge immediately connected with the dramatic Wild West culture of his new home. Following King's murder, Muybridge placed an advertisement proposing "to issue, on or about September 1st, a large and beautifully executed [lithograph] portrait of the late Mr. King."[30] For this project, Charles Fenderich—renowned for his realistic renderings of politicians—portrayed King from a daguerreotype by Vance, whose studio at 115 Montgomery was next door to Muybridge's shop (fig. 12). By the mid-1850s, Vance was considered one of the most important photographers in the West and had built an impressive business documenting the gold rush and the growing mercantile class of Northern California. His 1851 exhibit of daguerreotypes in New York had helped define for easterners the rough-and-tumble look of the forty-niners and their mining camps. Because of his aesthetic interest in the landscape, his prescient documentation of the gold fields, his expressive consideration of emotions in his portraiture, and his technical mastery of the daguerreotype medium, Vance is remembered as a pioneer of nineteenth-century American photography. Another significant portrait photographer, William Shew, also moved into Muybridge's building in the fall of 1856 and became a direct competitor of Vance.[31] With engraver and lithographer W. H. Oakes, his partner on the Fenderich print, Muybridge also published the sheet music for a melodramatic song memorializing King called "He Fell at His Post Doing Duty."

Muybridge's business grew throughout the 1850s. He continued to distribute books, engravings, lithographs, and sheet music, and showed his books in the 1857 Mechanics Fair in San Francisco. He soon began to sell photographs and photographic copies of paintings, and also began to lend money to supplement his income. His stock came to include the book *Brunel's Mammoth Ship: View, Section, Plan, and Description of the Great Eastern Steam Ship*, about the colossal sailing steamer popularized by Robert Howlett's 1857 construction photographs. Howlett's iconic image of engineer Isambard Kingdom Brunel, the ship's designer, standing in front of its launching chains, came to symbolize the industrialization of Britain in the 1850s. Muybridge also had his imprint on *Appleton's New American Encyclopedia*, *France Illustrated*, *Pictorial World*, and *Finden's Beauties of Moore and Flowers of Loveliness*, a London Printing and Publishing Company reprint of "a series of imaginary portraits for the heroines of the author of *Lalla Rookh*" (fig. 13).[32] Muybridge's growing shop encompassed books on British art and artists, the National Gallery, special editions of Shakespeare and John James Audubon's *Birds of America*, as well as *The History and Topography of the United States of America* edited by John Howard Hinton, containing "a series of views drawn on the spot and engraved on steel expressly for this work," and the *History of the United States from the Earliest Period to the Administration of James Buchanan* by J. A. Spencer. In an advertisement on the back cover of the latter, Muybridge described himself as a "publisher and importer of illustrated and standard works" and offered to buy European books on commission and import them (fig. 14).[33] Striving to

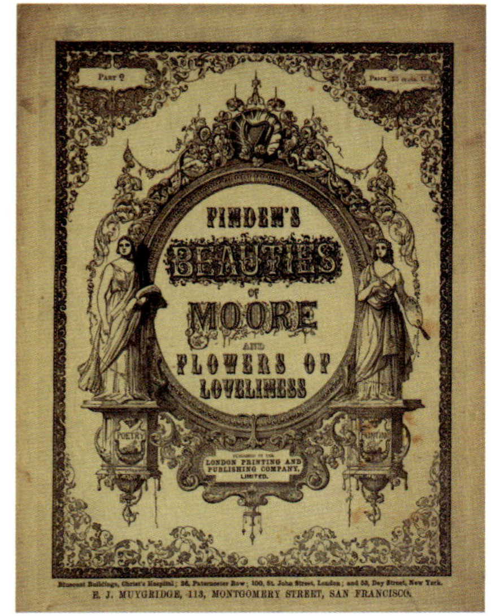

Fig. 13. William and Edward Finden, *Finden's Beauties of Moore and Flowers of Loveliness*, 1857

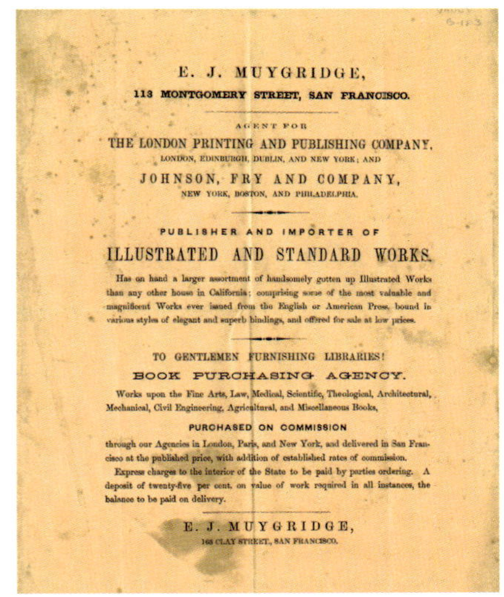

E. J. MUYGRIDGE,
113 MONTGOMERY STREET, SAN FRANCISCO.

AGENT FOR
THE LONDON PRINTING AND PUBLISHING COMPANY,
LONDON, EDINBURGH, DUBLIN, AND NEW YORK; AND
JOHNSON, FRY AND COMPANY,
NEW YORK, BOSTON, AND PHILADELPHIA.

PUBLISHER AND IMPORTER OF
ILLUSTRATED AND STANDARD WORKS.

Has on hand a larger assortment of handsomely gotten up Illustrated Works than any other house in California; comprising some of the most valuable and magnificent Works ever issued from the English or American Press, bound in various styles of elegant and superb bindings, and offered for sale at low prices.

TO GENTLEMEN FURNISHING LIBRARIES!
BOOK PURCHASING AGENCY.

Works upon the Fine Arts, Law, Medical, Scientific, Theological, Architectural, Mechanical, Civil Engineering, Agricultural, and Miscellaneous Books,

PURCHASED ON COMMISSION

through our Agencies in London, Paris, and New York, and delivered in San Francisco at the published price, with addition of established rates of commission.

Express charges to the interior of the State to be paid by parties ordering. A deposit of twenty-five per cent. on value of work required in all instances, the balance to be paid on delivery.

E. J. MUYGRIDGE,
163 CLAY STREET., SAN FRANCISCO.

Fig. 14. *E.J. Muygridge, 113 Montgomery Street, San Francisco*, 1858

maintain his business and connections to the photographic community, in April 1858 Muybridge moved into Silas Selleck's gallery in the Rabe's Building at 163 Clay Street. Toward the end of the year, the *Daily Alta California* published a short review of a new photograph by Watkins. Sandwiched between reports of an assault and a celebrated trotting horse named Princess, the review stated, "We are indebted to Mr. Watkins, photographic artist, of 166 Clay street, for a very beautiful photograph of the Golden Gate, taken from Telegraph Hill. The picture is the first and only one that has ever come under our notice that gives a perfect delineation of the beautiful passage-way that connects San Francisco harbor with the ocean, together with the surrounding landscape on either side.... In fact, the picture is a most perfect one in every respect, and will be valued by all who desire to procure views in California for preservation. It is for sale by Mr. E. J. Muygridge, at his store, on Clay street, a few doors above Montgomery."[34]

By 1860, Muybridge had brought both his younger brothers, George and Thomas, to San Francisco, and on May 15, 1860, announced in the *San Francisco Daily Bulletin*, "I have this day sold to my brother, Thomas S. Muygridge, my entire stock of Books, Engravings, etc., and respectively request a continuance of public patronage in his favor. After my return from the Yosemite I shall, on 5th June, leave for New York, London, Paris, Rome, Berlin, Vienna, etc., and all orders or commissions for the purchase of Works of literature or Art entrusted to me, will be properly attended to."[35] There is no evidence that Muybridge actually traveled to Yosemite in May or June of 1860 (his first known journey there was in 1867), but on July 2, he did board the Butterfield Overland Stage for the long and arduous trip from San Francisco to Saint Louis, where he was to connect by train for New York on his way to Europe.

Deep in the Cross Timbers region of north-central Texas, Muybridge was seriously injured when the stage he was riding broke away from its team and hit a tree. He was thrown out the back and sustained a head injury that left him in a coma for days. When he regained consciousness in Fort Smith, Arkansas, where he had been taken for treatment, he had no memory of the accident. Years later he reminisced, "I found myself lying in a bed at Fort Smith, 180 miles from the place. I found a scar on my head. I had double vision—saw two objects at once; had no sense of smell or taste; also had confused ideas."[36] When able to travel, Muybridge went on to New York for treatment and then to London, where he undertook therapy from the distinguished physician Sir William Gull.

The Lost Years, London

Little is known about Muybridge's activities between 1861 and 1866, a period in his life spent primarily in England that biographer Robert Bartlett Haas refers to as the "Lost Years."[37] Muybridge returned briefly to New York in 1861, a trip made primarily to file a $10,000.00 lawsuit against the Butterfield Overland Mail Company. He won a compromise settlement of $2,500.00, which allowed him to take time to recover and to freely pursue his own interests in London before returning to work in San Francisco. While convalescing, his inventiveness surfaced and he applied for and received two patents in Britain: the first, in 1860, was for a new metal plate-printing device; the second, in 1861, for a washing machine described as a "peculiar construction and arrangement of machinery or apparatus, whereby the operation of washing clothes and other articles of a textile nature is greatly facilitated."[38] Muybridge exhibited prototypes of both his patents at the International Exhibition of 1862 in London, a World's Fair on the scale of the 1851 Great Exhibition (fig. 15). In the *Official Catalogue of the Industrial Department*, his entry number 5310A was described as follows: "Specimens of his plate-printing, the ink being introduced from underneath by perforations."[39] It was exhibited in the section for "paper, stationery, printing, and bookbinding," along with many examples of beautifully printed books, lithography, engraving, binding, and state-of-the-art printing equipment. Inspired by the Exhibition's atmosphere of limitless industrial growth, and fueled by the creativity of invention, Muybridge was clearly looking to expand his business potential beyond publishing and distributing the work of others. At the same time, he had been counseled by his doctor to spend time outdoors, and he may have begun experimenting with photography at this time.

Fig. 15. London Stereoscopic and Photographic Company, *No. 2—The Nave, from the Western Dome*, 1862

Muybridge was just over thirty years old when he arrived back in England from California, having been in the United States for almost a decade. What had been intended as a book-buying trip became a long period of recovery from his injuries, and an opportunity for a fresh start. He presented himself in 1862 as Edward J. Muygridge, an inventor, and he had money to invest from his suit against the Butterfield Overland Mail Company. He had written to his uncle Henry Selfe, who had moved his family to Australia, offering the opportunity to patent his washing-machine design there.[40] Then, in July 1865, he was listed—for the first time—as "Edwd. J. Muybridge (late of California), 4 Brompton-square," among the directors of the Austin Consolidated Silver Mines Company Limited, in an advertisement for a stock offering posted in *The Times*.[41] The new company was established to develop silver mines and crushing mills in Austin, Nevada.[42] In late November 1865, in a series of advertisements that appeared in Glasgow, Dublin, and Edinburgh, he was listed again as "Edward J. Muybridge, Esq. (late Merchant, San Francisco), 4 Brompton Square, London."[43] He appears as one of nine directors representing the Ottoman Company, offering for sale 20,000 shares of stock to capitalize the Bank of Turkey Limited, "for the purpose of extending to the mercantile community of Turkey that legitimate aid, which, in more favored countries, is rendered to commerce by well-conducted banking institutions."[44] His future as a venture capitalist was short-lived, however, as the Bank of Turkey quickly collapsed under the weight of the Panic of 1866 and the subsequent withdrawal of major financing.[45] A financial crisis swept Britain that year, caused by a severe winnowing of credit and capital, and a rise in interest rates following years of excessive economic confidence. Many banks were shuttered. By summer, only months after the collapse of the Bank of Turkey, the Austin Consolidated Silver Mines Company was also dissolved. His investment income now seriously compromised, Muybridge resolved to return to San Francisco, where he had friends in the photography business, to remake his life yet again.

The International Exhibition, London

A self-generated 1881 article in the *San Francisco Examiner* about Muybridge's extraordinary inventiveness as a photographer reported that "Mr. Muybridge came to San Francisco in 1855 and most of the time since and all of the time since 1860, he had been diligently, and at the same time studiously, engaged in photography."[46] He purposefully conveyed the notion that he had been closely connected to photography since leaving San Francisco for Europe, and during his time in England he in fact became an accomplished photographer. In Britain, as in the United States and France, the young art of photography had matured during its formative years into a significant new medium of communication, one that could represent the world and reimagine it at the same time. During Muybridge's sojourn in Britain, between 1861 and 1866, he would have witnessed the rapid growth of photography as an art and a science, and the strengthening of the photographic societies that did much to promote the medium. The Photographic Society of London was formed in 1853 from the scientific and industrial interests of the Royal Society of Arts, which had hosted a major international exhibition of photographs in late 1852. The Photographic Society developed around the ideas and interests of photographers such as

Roger Fenton, Antoine-François-Jean Claudet, and Charles Eastlake. It later became known as the Royal Photographic Society. The *Journal of the Photographic Society* grew into a significant forum for the discussion and promotion of the evolving science and aesthetics of photography, and the growing organization hosted serious annual exhibitions.

The Photographic Society was also a noteworthy participant in the International Exhibition of 1862, and it was there that Muybridge would have seen a wide variety of photographs by Claudet, Fenton, and Talbot, along with those by Francis Bedford, Valentine Blanchard, Charles S. Breese, George Robinson Fardon, Francis Frith, David Octavius Hill and Robert Adamson, William Edward Kilburn, the London Stereoscopic and Photographic Company, John Jabez Edwin Mayall, James Mudd, Oscar Gustave Rejlander, Henry Peach Robinson, William Russell Sedgfield, Benjamin Brecknell Turner, George Washington Wilson, Henry Stuart Wortley, and many others. "Photography, as an art, has taken a high position; and with the facilities afforded by the collodion process, and by the preserving zeal of the lovers of photography, a remarkable degree of excellence has been obtained," wrote Robert Hunt about the photographs exhibited at the fair in his *Handbook to the Industrial Department of the International Exhibition 1862*. "In the application of photography to the investigations of science, much has been effected," he continued. "In another direction, it may be said that scientific knowledge has been brought to bear on photographic art."[47] Several new photographic enlarging and engraving processes were displayed, including a photo-etching technique developed by Talbot, along with cameras, lenses, stereographic equipment, and dark tents needed for outdoor work. In addition to studio portraiture, many of the major themes that Muybridge would later address in his own photographic work were represented, including landscapes, urban surveys, architectural images, instantaneous photographs, travel albums, and photographs of works of art. A number of the British photographers shown in the exhibition also had a hand in book and print publishing, which may have also brought Muybridge, who was exhibiting his patent models in the printing and machinery categories, in proximity to the photographic community.

The ongoing debate about whether photography was an art or whether it was better classified as an optical and technical process was amplified throughout the country. The positioning of photography in the International Exhibition of 1862 was a monumental issue for photographers, even during the fair's organization. As in the Great Exhibition of 1851, photography was not categorized with the fine arts; instead, it was placed in the Industrial Division in Section II, Machinery. Many photographers and journalists argued that photographs should have an aesthetic status independent from the tools used to make them. The Photographic Society accepted a compromise that placed photographs in a separate room from the perfunctory equipment, which elicited even more heated discussion. Photography's diminished status at the fair gave the society less credibility, even as its exhibitions were criticized for their lack of new, inventive ideas.[48] Antoine Claudet, one of the jurors, responded by invoking the work of Francis Bedford and Roger Fenton, among others, discussing them as photographic artists, rather than technicians, and comparing them to other great artists of the time. As historian Steve Edwards has argued, Claudet "sought a way to authorize the machine, a way for it to serve *him*, so that the products of the apparatus would attest to his capacities rather than defining him through those associations with *it*."[49] Photography could be a means for artists to bring into play their own identity in the face of an increasingly industrialized culture. Elevating photographs to the place of art, above mechanical equipment within the rigid taxonomy of the Exhibition, would imply recognition of the individual identity and creativity of the photographer. On January 2, 1863, the *Photographic News*, a British publication that had argued this point from the start, reported that "the advent of the International Exhibition, and the position of photography therein, was a topic of all-absorbing interest.... It is true that in this Exhibition photography for the first time received definite recognition as a distinct and independent art; and it is true that it fully maintained, by the excellence of the various contributions and variety of its applications, its claim to such a position."[50]

While the debate over photography as art was central to the discourse of the International Exhibition of 1862, there was also a significant focus on technical innovations like instantaneous photography, which may have also attracted Muybridge's attention at this time. Instantaneous photography was becoming more relevant as the technology improved, making it possible to freeze motion on a sensitized glass plate given a short, or instantaneous, exposure time and enough light from the sun. And with the stopping of time, photographers could envision the types of aesthetic and scientific discoveries that lay on the horizon. In the 1850s, several factors combined to make instantaneity more possible: the increased sensitivity of photographic emulsions; the adoption of bright, outdoor subjects; improvements in lens technology; and infrequent trials with mechanical lens shutters to establish more control over exposure times. Photographers such as Gustave le Gray, Jean Victor Warnod, Edmond Bacot, and John Dillwyn Llewelyn experimented with instantaneity during this time. The International Exhibition of 1862 featured excellent examples of the British experiments, stressing that technical progress had been an important element in the selection. "Instantaneous photography has excited much attention during the year, and its conditions seem to be better understood as consisting in simplicity rather than complexity. Good chemicals, good lenses, and good light; bromo-iodized collodion, and iron development, are generally known to be the main requisites," recounted the *Photographic News* in the aftermath of the fair.[51] George Washington Wilson had displayed several of his seminal 1859 *Loch of Park* photographs, in which he depicted his family in a rowboat moving across a mirrored lake, all set against a backdrop of the setting sun (fig. 16). He also showed photographs that stopped the motion of urban life, such as *Princes Street, Edinburgh*, 1860, and a series called *Six Instantaneous Views: H.M.S. Cambridge* from 1857 that portrayed the billowing smoke emitted from the ship during gunnery practice, frozen and solid in space like clouds hugging water. Henry Stuart Wortley showed views of Mount Vesuvius during the eruption of 1861, and Charles Breese showed a series titled *Transparent Photographs on Glass. Stereoscopic Views, some instantaneous, and a Statue in one is taken by moonlight.* "Breese's wondrous instantaneous transparencies have excited so much admiration, that he has been induced to appoint London agents for their sale," reported the *Photographic News.* They concluded, "His moonlight pictures, at one time regarded by many as a hoax, are now received as veritable results."[52] In his "Reports by the Juries" on the exhibition, photographer and juror Hugh Welch Diamond contributed to the discussion of Breese by adding that "every drop of foaming water or foaming cataracts is reproduced by an operation perfectly instantaneous and with a transparency beyond comparison."[53]

Fig. 16. George Washington Wilson, *Lock of Park, Aberdeenshire*, 1859

In some ways, it was the expansion of subject matter displayed at the International Exhibition—all made possible by steady technical progress and the increasing acceptance of photography as an aesthetic pursuit—that most intrigued the artists in attendance. The organizers' focus on technique, technology, and the accompanying apparatus seemed to be subsumed by the beauty and meaning of many of the photographs on view and their inherent connection to the history and traditions of fine arts. Though he had given up photography by 1862, and his showing at the International Exhibition was in effect his last, Roger Fenton displayed a broad range of his landscapes, architectural images, and several recent still-life compositions for which he won a Prize Medal. His landscapes and photographs of cathedrals, ruined abbeys, and grand country houses, created between 1854 and 1860, were considered some of the

Fig. 17. Roger Fenton, *Paradise, View down the Hodder, Stonyhurst*, 1859

finest of the time and made him one of the most visible photographers in Europe. His stark, posed documents of the Crimean War had defined the conflict for many viewers in 1855. "He seems to be to photography what Turner was to painting—our great landscape photographer," extolled the *Journal of the Photographic Society* in reviewing his contribution to the 1858 Photographic Society Exhibition.[54] His early training as a painter and his technical background—shaped in proximity to the French experiments with patent-free paper negative and collodion processes in the early 1850s—provided Fenton with a truly aesthetic foundation for his work.[55] His landscape photographs from 1857 to 1860 often focused on the movement of water cascading through the northern river valleys and on the atmospheric effects of nature in the mountains of Scotland, Northern Wales, and the Lake District. These sometimes harbor small figures looking out across the countryside or fishing, singly or in groups, which lend a sense of scale and immensity to the scene (fig. 17). As in the landscape paintings of John Constable or Samuel Palmer in Britain, or Gustave Courbet in France (fig. 18), Fenton's carefully positioned figures invite viewers to enter into his pictures, giving them a way to identify with and connect to the natural world and the place of people within it. Fenton's photographs of ruined abbeys also followed accepted aesthetic precedents in British art, associating the crumbling history of divine architecture with the ageless natural landscape and the temporality of its inhabitants. He also endeavored to make stereographs, often from the same point of view as his larger plates. The preeminence of Fenton's career during this time was an inspiration to others throughout Britain, and Muybridge may have considered it a model as he began to explore his own options in the aftermath of the International Exhibition.

In London, Muybridge had ready exposure to the ongoing dialogues about photography and its bifurcated relationship to art and science. He may have been a member of one of the amateur photography societies, and his continuing interest in the mechanical and reproducible nature of the medium, along with its clear potential as a business, could have pushed him to embrace it as a new vocation. His renewed encounter with British art and photography came at a time when the latter's position was fervently contested and its aesthetic possibilities seemed limitless. Stereography, for example, swept Britain during these years, and millions of stereo images were created and sold. Muybridge was seek-

Fig 18. Gustave Courbet, *Landscape with Fisherman*, 1872

ing an identity he had not yet found and a more secure place within the modern world. It was during these "lost years"—after he had almost died in an accident and explored the possibilities of invention and investment—that he reexamined his life again and settled on photography as his profession. Photography was at the time the visual embodiment of the modern world, a constantly advancing technology that had the potential to revolutionize both art and science. For Muybridge, it made sense to redefine himself as an explorer of art and science in the spirit of the British photographers, but it was to the western frontier of the New World that he chose to return.

II – Camera

> 'Time' is our name for an irreducible condition of our perception of phenomena; therefore, statements which would separate the notion of time from some object of direct perception, are meaningless.[56]
>
> Hollis Frampton

Between 1860 and the turn of the century, photography changed our way of seeing. More than any other art, it proposed new forms of expression drawn from real experience. Photographs exhibited many of the traits of printed text: a linguistic, narrative structure and, in time, a reproducibility that made images pervasive. They represented the solidity of time and space: the way things look and therefore the way they are. At the same time, photography is a malleable process, which easily manipulates how things are represented. It speaks both truth and fiction. Combined with radical transformations in science, transportation, and manufacturing technologies, this more democratic art helped create the world we know today.

With the end of the American Civil War in April 1865, the nation's reconstruction and westward expansion provided many prospects for those seeking new opportunities. The Central Pacific Railroad was already snaking its way eastward from California across the Sierra Nevada Mountains with the promise of connecting the Golden State to the rest of the country by rail. On January 8, 1863, Leland Stanford, California's governor and president of the Central Pacific, had broken ground in Sacramento to inaugurate one of the most complex construction and engineering projects ever attempted: the joining of two railroads to create a transcontinental route. As authorized by the Pacific Railway Act of 1862, the Union Pacific and Central Pacific were given the vast right of way from Omaha to Sacramento, along with additional federal subsidies and enormous land grants along the route. The combined project was intended to unify the country by providing coast-to-coast travel, telegraph services, and the possibility of a military route through territories that were virtually inaccessible. Four Sacramento businessmen—Charles Crocker, Mark Hopkins, Collis P. Huntington, and Stanford—came together to parlay an investment of $15,800.00 into an empire that would earn them $200 million in profits over several decades.[57] About ten thousand Chinese workers were employed by the Central Pacific to blast through the mountains and construct trestles and bridges to complete the project that opened California to the world and changed forever people's perceptions of time and space. When the railroad was finished, one could travel in relative comfort between New York and San Francisco in less than a week. On May 10, 1869, when Stanford hit the final "golden spike" at Promontory Summit, Utah, to commemorate the historic event, three photographers were on hand to record the drama for rest of the world. The meaning of this accomplishment, to the nation and the landscape it occupied, would occupy numerous photographers for years to come.

By the end of the 1860s, American landscape photography was still in its youth; as in Britain, its course was set by strong national traditions in landscape painting and the unique relationships people forged with their physical surroundings. For years, the grand scale and beauty of the North American environment had colored the development of a national consciousness in the United States. During much of the nineteenth century, images of the American landscape were akin to self-definition, helping a growing confederation of immigrants to distinguish themselves from their foreign roots. Artistic portrayals of nature often connoted the psyche of the young country, the sublime embodiment of collective dreams. Frequently inspired by European romanticism, Hudson River School painters created an essentialist view of the land. The lush forests, coursing rivers, and dappled meadows in the paintings of Thomas Cole, Asher B. Durand, Frederic Edwin Church, John Frederick Kensett, and Albert Bierstadt were often expressions of spiritual connections between settlers and the divine. The human spirit was embodied in the natural world as painters created images that placed the civic individual at nature's service. "And, although an enlightened and increasing people have broken in upon the solitude, and with activity and power wrought changes that seem magical, yet the most distinctive, and perhaps the most impressive, characteristic of American scenery is its wildness," wrote Cole in his "Essay on American Scenery."[58] Durand's *Kindred Spirits* of 1849 depicts a fusion of actual, albeit idealized, sites; the composite invites and inspires those who are but visitors within its eternal, invented universe. The "wildness" appears painted by nature, unspoiled by those who trod its paths. By the early 1850s, nature still stood between people and progress, between stasis and the frontier. With its limitless skies, plentiful rivers, and vast mountains, the American landscape was seen as a powerful force, a representation of Eden, and it still exerted a transcendent, spiritual influence.

The enormity and remoteness of wonders such as Niagara Falls and Yosemite Valley pointed to the insignificance of people relative to the scale and power of the natural world. In some ways, Niagara Falls were enigmatic, as their real strength and beauty could rarely be depicted in a work of art.[59] However, photography helped to fashion them as an icon of raw nature. Lacking the grand scale and luminous colors of paintings but uncannily realistic in their topographic portrayals, photographs—even when taken at a remote distance from the subject—appealed to new audiences, who saw them as more factual and accessible. Daguerreotypists Albert Sands Southworth and Josiah Johnson Hawes took pictures of Niagara Falls around 1850. Technical challenges made such outdoor photography a rarity at the time.[60] Shot in the dead of winter, their full-plate daguerreotype *Niagara Falls from the Canadian Side* shows the cascade from an outlook close to the cliff. The foreground and distant shore are frozen over, the usual roar is hushed, and a veil of ice and mist obscures the

Fig. 19. Frederic Edwin Church, *Niagara*, 1857

drama. Here, however, we are close to the scene, peering down over the brink. Our sense of gravity and location is disturbed (an effect Muybridge would also evoke when he first made photographs from the top of Yosemite Falls in 1867). The detail in the frozen branches and the aerial mist that subsumes the horizon cast an unearthliness akin to the fleeting drama sought by the Hudson River School painters.

Frederic Church completed his monumental painting *Niagara* in 1857 (fig. 19). It also places the viewer right on the rim of the gorge. The panoramic composition proved to be a theatrical tour de force, which even today feels like a wide-screen photographic view, cutting off the surrounding landscape and focusing on the plunge of the falls. Illuminated by the rekindled light of a clearing storm, the rushing water engulfs the solid rock below. Church's prismatic rainbow, spawned by the sun on spray, connects the dematerialized, secular ground to the metaphorically sacred sky in a manner that spoke to the American ambition of the time. Composed in

Fig. 20. Thomas Almond Ayres, *The High Falls*, 1855

part from photographs of the scene, *Niagara* quickly became one of the most popular works of art of its day, with one critic praising Church for "the selection of his point of view."[61]

Walt Whitman mentions Niagara Falls in an 1871 version of *Leaves of Grass*, likening the mightiness of nature to the "torrents" of settlers "pouring" into the western cities:

> Not through the mighty woods we go, but through the mightier cities;
> Something for us is pouring now, more than Niagara pouring;
> Torrents of men, (sources and rills of the Northwest, are you indeed inexhaustible?).[62]

Against the layered complexity of Whitman's metaphoric pondering about settlers moving west in the wake of the Civil War, the principal purpose of landscape photography was to survey western topography to map and catalog the natural resources. Yosemite, too, became a subject for artists in the years following the 1851 incursion into the valley by the settlers' militia. Though few had traveled the rough road into Yosemite Valley by stage and horseback, it had by 1860 become an icon of the West for its sheer granite cliffs, magnificent domes, and waterfalls dropping more than 1,000 feet. Like Niagara Falls before it, Yosemite was emblematic of the frontier, an Eden that symbolized the connection between American spiritualism and the unspoiled landscape. The public knew it only through newspaper accounts, literary descriptions, and a few works by adventurous artists like Thomas Ayres, who distributed his images as lithographs (fig. 20). In 1856, James Hutchings began publishing a series of descriptive articles promoting Yosemite as a tourist destination in his *Hutchings' Illustrated California Magazine*. Both Watkins and Charles Leander Weed started to photograph the valley by 1859, and some of their images were made into wood engravings to illustrate the magazine.[63]

Geologist Josiah Dwight Whitney, who organized the California State Geological survey, hired Watkins in 1865 to make photographs of Yosemite and the High Sierra to illustrate his reports; one progress report published in 1866 was accompanied by woodcuts derived from these images. In 1866, Watkins was making photographs for Whitney's *The Yosemite Book*, first published in 1868.[64] Whitney's survey became a model for subsequent government projects such as Clarence King's United States Geological Exploration of the Fortieth Parallel, George Wheeler's United States Geographic Surveys West of the One Hundredth Meridian, Ferdinand Hayden's United States Geological and Geographical Survey of the Territories, and John Wesley Powell's Geographical and Geological Survey of the Rocky Mountain Region. Discussing the importance of photography to such surveys, historian Martha Sandweiss asserts that "they pictured the western landscape as simultaneously awe inspiring in its grandeur and immensity, and as malleable, ready to bend to the iron will and technological prowess of the American people."[65]

In 1868, the same year that Whitney completed *The Yosemite Book*, Muybridge applied to copyright his first photographs of Yosemite, shot in 1867, and Bret Harte published his acclaimed story "The Luck of Roaring Camp" in *Overland Monthly*. Harte's classic parable balanced the concepts of luck and destiny in men's lives with the irreconcilable forces of the natural world. "Nature was his nurse and playfellow," wrote Harte, referring to an orphaned baby boy nicknamed The Luck, taken in by a group of hardscrabble forty-niners who had vowed to raise him. "For him she [nature] would let slip between the leaves golden shafts of sunlight that fell just within his grasp."[66] Despite their belief that the boy would bring them luck, he perished in a flood that swept through the miner's camp. Harte's story suggests a growing tension between the settlers and nature, as their fortune-bound pursuit of a better life ultimately became tied to domestication and destruction of the wilderness.

California, 1867

When Muybridge returned to San Francisco from London in late 1866 or early 1867, he immediately went to work as a photographer. He set himself up in the Cosmopolitan Gallery, run by his old friend Silas Selleck at 415 Montgomery

Fig. 21. Logo for Helios Flying Studio, ca. 1867

Street. Muybridge had brought back with him a new and sophisticated understanding of photography's place in science and art, and a polished picture of the developing history and traditions of the medium. He understood the technical and aesthetic advances British and French photographers had made, and he was now accomplished in the wet-plate collodion process.

To separate his business persona from the artist behind the camera, Muybridge began signing his pictures with the moniker Helios, drawn from the name given to the sun in Greek mythology. The mythic Helios is often pictured as an anthropomorphic god driving a chariot in an arc across the sky, the sun's rays radiating like a halo from his head. The term "sun pictures" had been used for years to refer to photographs printed out from negatives under sunlight, so Muybridge's reference was both to the god and to the natural source of light that made his pictures possible. In the same spirit, he called his business "Helios Flying Studio" and developed a logo showing a camera with wings flying like a chariot across the sun (fig. 21). By attaching this mythic symbolism to his photographs, Muybridge was metaphorically connecting his art to photography's source, its photochemical trigger. The notion of a "flying" studio also connotes a traveling photographer who makes images outdoors rather than under the intricate skylights of a city salon.

Made in mid-1867, Muybridge's first known photographs are landscape views of Yosemite Valley, depicting the mythic waterfalls and mountain peaks that had been described in *Hutchings' Illustrated California Magazine*, and the Calaveras Grove of mammoth sequoias, from which bark had been stripped and exhibited in the Crystal Palace.[67] Muybridge carefully constructed each picture to convey a sense of scale, space, discovery, and wonder, as if his primary interest was in connecting people to the pictorial drama of the landscape. As a group, these images read like a tour of the visual highlights and natural history of Yosemite, as if to re-create the feeling of seeing it for the first time. Muybridge photographed with a half-plate view camera that made negatives 5 1/2 by 8 1/2 inches in size, often creating relatively simple arrangements that used the visual planes established by the towering granite cliffs, falls, and trees to ground his compositions in a geometry that echoes the rectangular shape of the images. Like Watkins and Weed, who preceded him in photographing Yosemite, Muybridge also documented the valley with a stereo camera, sometimes from the same point of view as in his half-plate images. There was a growing market in California, on the East Coast, and abroad for stereo images of western scenery, and over a period of five "productive months," Muybridge made more than two hundred views in Yosemite during this first expedition into the mountains.[68] Like others who worked with stereography, Muybridge often emphasized views with deep, receding perspective to accentuate the three-dimensional effect of the medium.

From early summer into the fall, Muybridge photographed from the valley floor looking up at the peaks and from the heights looking down from the edge in dizzying defiance of traditional landscape forms. In *Summit of the Third Fall of the Yosemite (4035)*, 1867, he perched his camera just over the low water as it glides languorously across the smooth granite precipice (plate 12). Like Church, he has given the viewer no place to stand in the scene, and peering down over the diagonal edge yields a bird's-eye view of the thousand-foot drop. The distant peaks far across the valley, much lighter than the deep, full tones of the foreground, give an extraordinary depth to the image. Even though the print is small, it engenders a sense of the inestimable scale of the mountains, like the

Fig. 22. Frederic Edwin Church, *Niagara Falls, from the American Side*, 1867

aerial perspective of Renaissance painting. Muybridge's strategy is wholly experimental, establishing a dynamic point of view that throws one off balance. *Base of Third Fall of the Yosemite (4038)*, 1867, depicts the same waterfall from below, with the composition divided in two by the glassy white flow of water. With the collodion process, a long exposure was needed to bring out detail in the shadowy tones of the rock, so the moving water appears as a smooth, almost solid form. Muybridge's decision to picture the fall from the side, setting it off against both the dark wall on the left and the bright sky on the right, gives the water an indeterminate quality that emphasizes its structural state as a fluid, between solid and air. The arc of the water, which rushed with tremendous force over the ledge above, is traced on film. The space filled by water appears solid as it moves through time, creating a visual image of time passing set against the timeless backdrop of Yosemite's cliffs. Experimental filmmaker and photographer Hollis Frampton was inspired by this phenomenon in Muybridge's Yosemite photographs, likening his depiction of water to that of a tesseract, a projection of a three-dimensional space into four: "Muybridge, in some of his earliest landscape work, seems positively to seek, of all things, waterfalls; long exposures of which produce images of a strange, ghostly substance that is in fact the tesseract of water: what is to be seen is not water itself, but the virtual volume it occupies during the whole time-interval of the exposure."[69]

In 1867—the same year that Muybridge was photographing Yosemite Falls—Church reprised his celebrated waterfall theme in *Niagara Falls, from the American Side*, a composition that closely mirrors the Southworth and Hawes daguerreotype. Even more forceful than his 1857 rendition, *Niagara Falls, from the American Side* depicts the ferocious cascading water transcended by light and, again, the prismatic connection between earth and spirit (fig. 22). Niagara Falls had come to stand for the passions of manifest destiny and the boundless future potential of the country, and as a result it had become one of the most iconic emblems of the American landscape. The realistic style, deep space, and dizzying viewpoint of Church's new painting immobilized the flow of water as if time itself were momentarily stopped through divine intervention. The viewer peers down into the chasm of the falls as if thrust over the gorge, suspended like the water, without any sense of real time. It evokes a spiritual moment rather than an earthly experience grounded in corporeal materiality. In contrast, Muybridge's prolonged time exposures, evident in his first photographs of Yosemite, blurred the moving water, rendering it as a volume that moves through both space and time. These presented a modern, more prosaic approach to the landscape, while Church's painting connected viewers to a more lyrical, rhythmic representation of the falls that symbolizes our ultimate lack of control over nature—over the inevitable cycle of birth, death, and rebirth.[70] Powerful and beautiful, Church's *Niagara Falls, from the American Side* conveys a fixed, linear narrative, from the calm and eddy of the river above to the formidable torrent, precipitous cascade, and deluge; they climax, like the biblical story of Noah, in a drenching mist and peaceful rainbow, again connecting earth and heaven symbolically. In *Kahchoomah, Wild Cat Fall, 30 feet high, on the main fork of the Merced (4061)*, Muybridge's camera traces the water's path through time and fixes its position, offering evidence of its volume as an integral function of the natural environment rather than as a function of how we perceive or experience it (plate 13). Unlike Church's, Muybridge's narrative is open-ended and seeks to understand

Fig. 23. *Piwyack, (Cataract of Stars;) "Vernal Fall," 450 feet fall (4054)*, 1867

Fig. 24. Carleton Watkins, *River View of Royal Arches, Yosemite*, 1861

nature as a timeless force that has no beginning or end. In Muybridge's view, nature was a universal, impartial force, bound by the laws of science. He was beginning to understand how photography could represent the world in a new, more progressive manner.

Like Roger Fenton and Francis Bedford on their fall excursions to the British countryside, or Watkins and Weed in Yosemite, Muybridge focused on waterfalls as a landscape motif because of their popular, literary, and mythic character. This theme held an important position in nineteenth-century American and British art, and Muybridge's images extol this tradition. *Piwyack, (Cataract of Stars;) "Vernal Fall," 450 feet fall (4054)*, 1867, is a more conventional composition than some of his others, a point of view strikingly similar to photographs made previously by both Watkins and Weed (fig. 23). It is taken from below, and in it the fall appears as a glowing revelation between stands of trees, just as a hiker would discover it en route up the trail from the valley floor. "Upward and onward we climb; and, after passing a bold point, and reading some of the names inscribed on Register Rock, we obtain, suddenly, the first sight of the Pi-wy-ack, or Vernal Fall," wrote James Hutchings of this view in *Scenes of Wonder and Curiosity in California*. "While gazing at its beauties, let us, now and forever, earnestly protest against the perpetuation of any other nomenclature to this wonder, than 'Pi-wy-ack,' the name which is given to it by the Indians, which means 'a shower of sparkling crystals'; ... as the fall itself is one vast sheet of sparkling brightness and snowy whiteness, in which there is not the slightest approximation, even in the tint, to anything 'vernal.'"[71] Muybridge's image of Pi-wy-ack fills the vale with a volume of moving water, classically framed by the forest of trees. A small cataract in the foreground elegantly echoes the much larger fall in the distance, lending a dramatic sense of depth to the image, but the rocks encountered along the bottom edge block the viewer's path, preventing access deeper into the picture. His understanding of the conventions of landscape art is clear, as is his intent on pushing those conventions to create photographs for the market that would surpass any that came before. If the principal subject of Watkins's classic landscape views was space (fig. 24), Muybridge's was time (fig. 25).

Fig. 25. *Tokoya and Hunto. Valley of the Yosemite. (The Basket,) (The Watching Eye,) 3,568 feet high. No. 24*, 1872

Ever the adventurer, Muybridge took his apparatus to precarious heights rarely explored with a camera. The wet-plate collodion process was complex, and it required enormous skill to make quality negatives in the field. Heavy equipment, including cameras, boxes of glass plates, and chemicals, were packed into the wilderness on horses and mules. The glass plates had to be cleaned, coated, and sensitized in a dark tent, which was set up on the site where the photograph was to be made. The camera was secured on a tripod and focused under a black cloth, and then the sensitized plate was placed in a light-tight holder, which was slotted into the back of the camera and exposed by removing the lens cap for a fixed amount of time, which Muybridge came to know intuitively. The plate was developed and fixed in the dark tent before the collodion dried and then washed to remove the chemicals from the glass. Muybridge photographed his outdoor studio that summer in a stereograph called *The Flying Studio, Photographer's Equipment in the Field (114)* (plate 2). It shows his dark tent, bottles of chemicals, and plate boxes scattered among firewood, cups, and bits of clothing drying on a sun-drenched boulder. The wide-brimmed hat he wore on the trail is visible hanging from a log in front of a wooden box labeled "Helios."

That summer, Muybridge was credited with discovering an unusual formation in Eagle Rock. *Fissure in Eagle Rock. 1100 ft. deep*, 1867, was made on the strenuous climb from the valley floor to Glacier Point, Sentinel Dome, and Taft Point, and shows a rending of the smooth mountainside that reveals what Muybridge found as a purely sensual vision (plate 10). Writing about this later, Hutchings recalled, "On our ride down the valley, almost immediately opposite Pompompasus (the 'Three Brothers') ... there is upon the face of the mountain a white irregular spot.... Back of this point, high up toward the top, Mr. E. J. Muybridge ... discovered a remarkable fissure in the wall rock. 'It is,' he says, 'one thousand feet deep, five feet wide at the top and front, and gradually growing narrower as it goes down and back into the mountain. Several stones have fallen into it, and lodged about half way down.'"[72] Hutchings also stressed the ethereal sensation of hiking in the wilderness of Yosemite Valley, where the actuality of nature meshes with the poetic imagination. "Surrounded by such scenes of loveliness and sublimity, we feel a reluctance to break the charm they throw upon us by any speech," he quipped, "when someone is almost sure to cry out—'The Ferry!'"[73] One needed a boat to cross the Merced River, which bisects the valley from east to west, and a ferryman would row passengers across for a fee of 37 1/2 cents per head for men, women, and animals. Muybridge made a stereograph of the ferryman, with his long beard and wide-brimmed hat, that could be a portrait of the artist himself. He stands stiffly at the dock holding in one hand the thick rope that ties the boat, and his features are highlighted in bright light against the dark shadows of the riverbank. Enigmatic cliffs sweep upward, cutting off the distant view. A lone, crooked tree snakes along the left edge of the frame, and the sunlit rails of the ferry dock divide the foreground diagonally, dramatically inviting viewers to board the skiff. Muybridge titled the picture *Charon at the Ferry (21)*, 1867, emblematic of the ferryman in Greek mythology who carries the souls of the dead across the River Styx as they enter Hades (plate 3). The subject of death haunts some of these early landscapes, perhaps in memory of his deceased father and elder brother who never witnessed his coming of age.

Muybridge's first photographs reveal how his intentions were personified in his choice of subjects and developing style. However, they also comment on his search at this juncture for a new identity to replace that of the successful merchant who had left San Francisco almost seven years before. His demanding sense of creativity, his experiments with composition, natural subject matter, and technique, and his curiosity about photography's innate ability to burn from the sun an image about time passing all elevated his interest in exploring the relationships between photography and the land. After all, Muybridge's goal in his new work was to make something permanent of something temporal. When he presented his images publicly for the first time, the *Daily Alta California* called his pictures "a true copy of Nature," concluding, "we are sure that the five months of assiduous labor spent by 'Helios' in that place in securing a collection of views that is unsurpassed by any preceding, will find an abundant return in our appreciative community."[74] He was striving to create a mythic identity—as a landscape artist and explorer—that could, like the Greek myth of Helios, be timeless.

San Francisco and the Pacific Coast

In February 1868, Muybridge published a brochure to promote his first professional photographs, boasting, "I am now preparing for publication twenty views of your world-renowned Yo-Semite Valley, photographed last year by

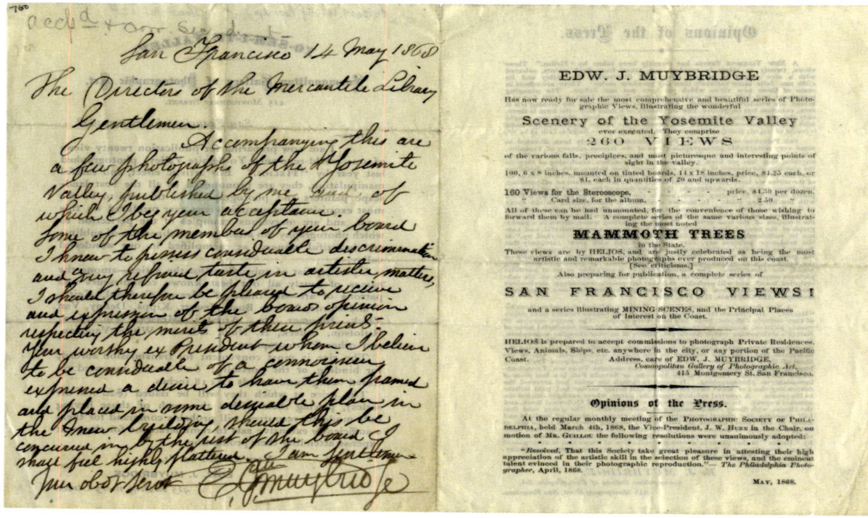

San Francisco 14 May 1868
The Directors of the Mercantile Library
Gentlemen.
Accompanying this are a few photographs of the Yosemite Valley, published by me and of which I beg your acceptance.
Some of the members of your board I know to possess considerable discrimination and very refined taste in artistic matters, I should therefore be pleased to receive an expression of the board's opinion respecting the merits of these prints. Your worthy ex President whom I believe to be considered of a connoisseur expressed a desire to have them framed and placed in some desirable place in the new building, should this be concurred in by the rest of the board I shall feel highly flattered. I am Gentlemen
Your obdt Servt
E J Muybridge

EDW. J. MUYBRIDGE

Has now ready for sale the most comprehensive and beautiful series of Photographic Views, illustrating the wonderful

Scenery of the Yosemite Valley

ever executed. They comprise

260 VIEWS

of the various falls, precipices, and most picturesque and interesting points of sight in the valley.

100, 6 x 8 inches, mounted on tinted boards, 14 x 18 inches, price, $1.25 each, or $1, each in quantities of 20 and upwards.

160 Views for the Stereoscope, ... price, $4.50 per dozen.
" Card size, for the album, ... " 2.50 "

All of these can be had unmounted, for the convenience of those wishing to forward them by mail. A complete series of the same various sizes, illustrating the most noted

MAMMOTH TREES

in the State.

These views are by HELIOS, and are justly celebrated as being the most artistic and remarkable photographs ever produced on this coast. [See criticisms.]

Also preparing for publication, a complete series of

SAN FRANCISCO VIEWS!

and a series illustrating MINING SCENES, and the Principal Places of Interest on the Coast.

HELIOS is prepared to accept commissions to photograph Private Residences, Views, Animals, Ships, etc. anywhere in the city, or any portion of the Pacific Coast. Address, care of EDW. J. MUYBRIDGE, *Cosmopolitan Gallery of Photographic Art,* 415 Montgomery St. San Francisco.

Opinions of the Press.

At the regular monthly meeting of the PHOTOGRAPHIC SOCIETY OF PHILADELPHIA, held March 4th, 1868, the Vice-President, J. W. HURN in the Chair, on motion of MR. GUILLOU the following resolutions were unanimously adopted:

"*Resolved*, That this Society take great pleasure in attesting their high appreciation of the artistic skill in the selection of these views, and the eminent talent evinced in their photographic reproduction."— *The Philadelphia Photographer*, April, 1868.

MAY, 1868.

Fig. 26. *Letter to the Mercantile Society, San Francisco*, 1868

'HELIOS.' For artistic effect, and careful manipulation, they are pronounced by all the best landscape painters and photographers in the city to be the most exquisite photographic views on this coast."[75] They were sold by subscription at $20 for a series of twenty views, and, according to the brochure, subscribers included West Coast painters William Keith, Charles Nahl, Durbin Van Vleck, and Juan Buckingham Wandesforde, along with photographers Alexander Edouart, William Rulofson, Silas Selleck, and William Shew. On February 17, the *San Francisco Call* lauded the pictures, "which have been taken by an artist who conceals his real name under the signature of 'Helios.' The views surpass, in artistic excellence, anything that has yet been published in San Francisco, combining, as they do, the absolute correctness of a good sun picture after nature, with the judicious selection of time, atmospheric conditions, and fortunate points of view."[76] Muybridge advertised a total of "260 views of the various falls, precipices, and most interesting points of sight in the valley," and announced that he was "also preparing for publication, a complete series of SAN FRANCISCO VIEWS! and a series illustrating MINING SCENES and the Principal Places of interest on the Coast"(fig. 26).[77] In 1868, John S. Hittell published twenty original albumen photographs by Muybridge in *Yosemite: Its Wonders and Its Beauties*, one of the first guidebooks to the valley.[78] The photographs, which were reduced to about the size of a *carte-de-visite* image, illustrated written sections about specific sites on one's tour, such as El Capitan or Yosemite Falls. Though Hittell's book relied more heavily on quotes and anecdotes from other writers than had Hutchings's earlier magazine and first edition of *Scenes of Wonder and Curiosity in California*, it was innovative in using actual photographs tipped in to each book rather than engravings. Often drawn from photographs, engravings were far less costly to publish and were therefore standard practice at the time. Hoping for greater national promotion of his Yosemite images, Muybridge sent about 125 pictures to the *Philadelphia Photographer*, the most prominent photographic journal in the country, edited by Edward L. Wilson. The *Philadelphia Photographer* agreed to include one print as its photograph of the month in their November 1869 issue. "'Helios' has outdone all competitors," Wilson wrote in his review. "His views are grand, and, as a photographer, he might vie with the great Wilson of Scotland," he continued, invoking George Washington Wilson, whose instantaneous views of Loch of Park had been so highly praised at the International Exhibition of 1862.[79]

For his own part, Muybridge announced that "HELIOS is prepared to accept commissions to photograph Private Residences, Views, Animals, Ships, etc. anywhere in the city, or any portion of the Pacific Coast."[80] He began to systematically photograph San Francisco, beginning with views of the Golden Gate, the bay, and the city from vantage points such as California Street, Russian, and Rincon Hills, and the docks along the waterfront where ships were loaded for their journeys up and down the coast and across the Pacific to the Orient. His pictures show elegant women pushing baby carriages and picnickers relaxing amidst cattle grazing on grass-covered Telegraph Hill overlooking the water. Muybridge's early photographs of San Francisco depict it as the prosperous port he knew so well from his years as a merchant in the 1850s (fig. 27). Some of these cityscapes reveal the special photographic effects that he had begun to incorporate at this time. He would sometimes point his camera directly into the sun, generating moody, impressionistic images with sunlight reflecting off water

Fig. 27. *Market Street, S.W. from Oriental Square (348)*, 1869

Fig. 28. *Moonlight Effect on the Bay (419)*, 1868

as if broken up by shards of glass; sailing ships dot the backlit bay, silhouetted against the black peak of Mt. Tamalpias north of the city. It was very difficult to get a respectable image this way, and Muybridge's experiments may have been inspired by the dramatic, full-tone landscapes that Wilson had achieved by shooting straight into the sunset. In what he called "moonlight effects," Muybridge would purposefully underexpose some cityscapes and landscapes, paint a tiny moon in the sky on the negative, and then print the image dark as night (fig. 28). Through photographic manipulation, these pictures appeared to be exposed under moonlight, even though they were shot in the full light of day. Muybridge did not shy away from retouching his work or creating effects in the camera or during the printing process, and he also developed new techniques to incorporate dramatic clouds in his photographs. In order to picture clouds, a photographer had to correctly expose the negative so the bright sky would not overexpose the clouds. In the process, however, the darker ground of the image would be underexposed. Furthermore, the collodion emulsion that Muybridge used to coat his plates was much less sensitive to the color blue, which affected its ability to record an image of the sky. Some photographers would simply paint out the sky on their negatives, yielding an even white tone in their print, or they would print clouds into the sky from a second, properly exposed negative. To circumvent these problems, Muybridge developed a "sky shade" he could attach to his camera, which controlled the amount of light reaching the area of the negative that would register the sky. It worked like a spring-operated shutter, allowing him to create proper exposures for both the sky and the ground in a single image.[81] In his later landscape work, he would also master the technique of printing in clouds from a separate negative.

Muybridge was making many stereographs at this time and building an extensive catalogue of city views to supplement his Yosemite images.[82] He traveled throughout the city to document major streets and architectural landmarks such as the Masonic Temple and the Lick House, and he sometimes even included a glimpse of his own process in the pictures. His city photographs are frequently populated with bystanders who have been strategically posed to give a sense of scale and life on the street. In one photograph, titled *Savings and Loan Society, Clay street*, 1869, the horse-drawn cart that served as his outdoor studio is visible at left even as a group of onlookers is positioned on the street (fig. 108). This square wagon covered in a thick canvas tarp both hauled camera equipment, plate boxes, and chemicals and served as a dark tent for coating and processing plates in the field. The back is inscribed "THE FLYING STUDIO" over Muybridge's familiar insignia of a camera with wings. In another example,

Fig. 29. *Steamers Zealous and Scout in Esquimault Harbor (511)*, 1868

Fig. 30. *Group of Indians (489)*, 1868

an uncaptioned photograph taken on a luminous dock looking out toward Goat Island (now called Yerba Buena Island), a box of Muybridge's negatives and a printing frame are lying in the bright sun. Printing his negatives on albumen paper was a tedious process, and Muybridge sometimes would have to travel to find adequate sunlight in the often-foggy metropolis. At the time, negatives were still contact-printed by sandwiching them between glass and a sheet of sensitized photographic paper in a printing frame. After outdoor exposure under direct sunlight, the paper was then processed, fixed, and washed to keep it from fading.

Looking down on San Francisco from the hilltops in 1868, one could observe the grid-like layout of the thriving city, the activity of its people, and the many landmarks of neighborhoods, commerce, worship, and entertainment. Beginning in the early 1850s, several photographers shot multipanel panoramas that encompassed a wider, bird's-eye view of the terrain, and these became very popular. Panoramas synthesized parts of the horizon into one extended image, summarizing the topography beyond the normal limitations of human or mechanical vision. They required technical skill to align various points in each image, along with an unobstructed vantage point, which sometimes meant moving the camera laterally from shot to shot to avoid including unwanted details of buildings or foliage. George Fardon shot a seven-panel panorama of San Francisco in 1855, and it was published the following year in his *San Francisco Album: Photographs of the Most Beautiful Views and Public Buildings of San Francisco.*[83] In the late 1850s, Watkins made a ten-panel stereographic panorama documenting the eastern side of the city from California Street Hill (known today as Nob Hill). In 1864, he created a five-panel mammoth-plate view, peering downtown toward Market Street from above Powell. Watkins's large-format panorama communicated a tremendous amount of detail about the architecture and layout of the streets from Telegraph Hill on the left to Rincon Hill on the right. Its composition mimicked that of a popular engraving by Charles Meryon from the mid-1850s, which was itself derived from unattributed daguerreotype views east and south across the bay.[84] In 1868, Muybridge, too, was surveying the cityscape from many angles and, following Watkins's example, chose to produce his own seven-panel stereographic panorama of San Francisco from a balcony on the north side of Rincon Hill (plates 16, 17). At the time, this was the most exclusive neighborhood south of Market Street by the bay, with a view across the city and the busy port. Muybridge's photographs look north toward California Street Hill, across the residential and financial districts and the lumberyards by the docks, to immortalize in considerable detail a neighborhood in

Fig. 31. *Effect of Earthquake in San Francisco, California St. East of Montgomery St.*, 1868

flux. It is an altogether different view from any of the previous panoramas by others; looking up toward the northern hills in a sweeping series, it was intended to be seen in a special tabletop stereo viewer that could rapidly change one card after another to give the effect of a pan across the landscape.[85]

Because of his growing reputation as an outdoor photographer and his connections with local military personnel, Muybridge received his first major commission: to travel to Alaska with General Henry W. Halleck to document a fact-finding expedition to the new territory. Following his dismissal as general-in-chief during the Civil War, Halleck was appointed to head the Military Division of the Pacific, based in San Francisco. The Alaska Territory was purchased by the United States from Russia in 1867 for $7.2 million. The headquarters of the Military Department of Alaska was established at Sitka, and Halleck's command became responsible for gathering information about the region's geography, resources, and people. Alaska was then a remote outpost understood by few Americans, and secretary of state William H. Seward suggested a photographic survey to help assuage negative public opinion about its acquisition. The British government had already used photography on inspection tours of coastal British Columbia beginning in 1866, and an American expedition led by geographer George Davidson had traveled to Alaska without cameras the previous July to acquire "evidence that could be used to convince the American public in general, and congressmen in particular, that Alaska was worth the purchase price."[86] After Congress finally ratified the purchase in July 1868, Muybridge set off from San Francisco with the Halleck expedition, boarding the steamship *Pacific* on July 29. Between August 4 and August 15, he visited the ports of Esquimalt, Nanaimo, Fort Tongass, and Fort Wrangell, before arriving at the provincial capital of Sitka on August 18. (plates 20, 21, figs. 29, 30). En route, he created atmospheric landscape images of ships in the harbors and moving water from the deck of the *Pacific*.

Fig. 32. Unknown photographer, *Edw. J. Muybridge*, 1869

Once in Alaska, Muybridge documented the coastal landscape and the native Tlingit people, along with local architecture, Russian Orthodox clergy, and the new American military installations that were built to establish control of the territory. In Sitka, he photographed the harbor, the parade ground, the Russian Orthodox church and priests, the Customs House, and military buildings (plate 23). He later wrote, "Until the purchase of this territory … Alaska was scarcely ever heard of, excepting by the fur merchant and geographer, and even now the most erroneous ideas prevail, both with regard to its climate and resources."[87] Muybridge felt that his pictures could present a public face to the unknown region, showing a visual slice of the beauty and "advanced" culture he had experienced there. His images of the Tlingit are some of the earliest photographs of Alaskan natives and helped define for Americans a picture of the indigenous culture. Several stereographs, each titled *Group of Indians*, 1868, depict native Alaskans clustered together looking straight into Muybridge's twin-lens camera, arranged within the strict guidelines of their cultural organization. One shows a large group at Fort Wrangell (spelled Wrangle on most of the cards) staring at the photographer, the individuals posed stiffly in front of a tall totem pole (fig. 30). About a dozen women are huddled together, seated in the foreground, while the men stand in a straight row behind them. The children sit to the side or behind the group. A large animal totem hovers above the sitters—a cultural force signifying the connection of art and nature to everyday life. They all seem unfamiliar with the photographic ritual, cooperative yet uncomfortable in front of the camera. For Tlingit people in 1868, photographic portraiture—the deliberate capture and preservation of a person's image on a sheet of glass—altered their perception of the natural flow of time. A moment from the past was pictured forever within a photographic object. Their understanding of photography evolved from this first encounter with the camera, during which they had witnessed the process of fixing an image cast by the sun's rays. Soon the Tlingit added a word for a photograph to their language—*kaa yahaay'i*—which can be translated as "a person's shadow, departed soul, reflection, or picture."[88] In turn, the American public's first encounter with images of the Tlingit was through Muybridge's photographs, which were shown at Henry William Bradley's National Daguerrean, Photographic, Ambrotype, and Melainotype Depot in San Francisco shortly after Muybridge's return on September 4. After receiving prints from the expedition, General Halleck wrote to Muybridge, acknowledging that "these views besides being beautiful works of art give a more correct idea of Alaska and its scenery and vegetation than can be obtained from any written description of that country."[89]

Fig. 33. Bradley & Rulofson, *Charles Crocker*, no date

Fig. 34. Bradley & Rulofson, *Gov. Leland Stanford*, no date

Muybridge was always adept at promoting his work to a wide audience. After completing his Alaska series, he began to refer to himself as an "official photographer for the U.S. government" and "director of photographic surveys on the Pacific Coast." Seward praised him for the "artistic excellence" of his photographs.[90] On the back of his stereo cards, Muybridge boasted that his were the "most artistic and comprehensive series of Pacific Coast, Yosemite, and Mammoth Tree Views, ever published, of various sizes, adapted for framing, the stereoscope and the album."[91] He was positioning his work in various growing markets, including art galleries, stereo views, and travel albums, and he thought of himself as a "view artist" or survey photographer, rather than as a portrait or studio photographer (fig. 32). In this way, he could sell numerous prints of his most compelling views through a number of commercial outlets across the country, a strategy pioneered by the London Stereoscopic Company during the 1860s.

Muybridge took a photojournalistic approach when he documented the earthquake of October 21, 1868, a serious tremor that struck San Francisco and its eastern suburbs. During the aftermath, he traveled the streets photographing people and damaged buildings in the city and East Bay. He covered the event like a news story, recording images of broken and tilting architecture, sometimes posing the gawking crowds in the corner of the frame to lend scale and a documentary sense of immediacy and civic loss to his photographs. In one *carte-de-visite* view, *Effect of Earthquake in San Francisco, California St. East of Montgomery St.*, 1868, he positioned his camera at a sharp angle to the broken, leaning structures, accentuating the chaotic situation by imposing on the scene a frenzied, undulating, dynamic composition (fig. 31). Muybridge's images that same year of the oceangoing steamship *Colorado* and *The Steamer "Golden City" on the California Dry Dock (4123)* depict the industrial side of the city (plate 19). In

1867, the Pacific Mail Steamship Company had begun the first transpacific service between San Francisco and Japan, and the first permanent dry dock on the Pacific Coast was constructed that year at Hunter's Point to accommodate the extensive shipping traffic that now passed through the port. Muybridge shows the Pacific Mail's *Golden City,* which was used to carry passengers and freight between Panama and San Francisco, head-on in a view of almost perfect symmetry. Using a wide-angle lens to include the horizon and sky, he accentuates the massive, mechanical form of the ship and its cradling environment, emphasizing its huge scale and manufactured form in relation to the small, insignificant figures—both managers and workers—posed around the dock.

Muybridge continued his collaboration with the War Department during this time by photographing San Francisco's defenses, a series he may have started before traveling to Alaska with Halleck. In 1869, he received permission to shoot group portraits of soldiers manning the gun emplacements at strategic locations around the bay, including Alcatraz Island, Fort Point, Point San Jose, and Black Point (plate 35). These photographs depict young recruits relaxing with the huge Rodman guns and stacks of cannonballs layered against picturesque views of the hills and water that sheltered the port of San Francisco, now the West Coast hub of international commerce. The city and its low skyline appear at a distance in these pictures, as fresh and optimistic as the faces of the soldiers posing for his camera. "His photographs gave the public an inspiring impression of gleaming firepower and starched soldiers standing vigilant guard in the heart of San Francisco Bay," wrote historian John Martini about Muybridge's Alcatraz images. "His stereographic view cards of the island fortress were widely reproduced and they turned up in refined Victorian parlors of the era."[92] Muybridge also produced extensive documentation of soldiers in training, as well as the architecture and landscape of the Presidio and Goat Island, the principal army garrisons in the city and headquarters of the military on the West Coast (plate 18). The images of drilling soldiers and artillery practice at the Presidio are carefully posed to accentuate the routines, patterns, discipline, and strict hierarchies that are symbolic of military training. Muybridge's compositions are highly structured, and his rigid groupings of soldiers on parade, assembled for review,

Fig. 35. Andrew J. Russell, *East and West shaking hands at laying last rail,* 1869

or ceremoniously priming the cannon communicate the regimen of martial doctrine. With these photographs, Muybridge continued to earn respect from the army officers, gaining further access to government projects that were off-limits to many photographers.

Muybridge had quit his business relationship with Silas Selleck and the Cosmopolitan Gallery of Photography late in 1868 and began selling his work through Charles G. Ewing's optician's firm in San Francisco, as well as the Montgomery Street gallery of Adrien Gensoul, an importer of prints, photographs, and art supplies and a manufacturer of frames. By April 1869, Muybridge announced that he would begin taking orders through the Nahl Brothers Studio. Charles and Arthur Nahl were well-established painters and photographers whose principal patron was Charles Crocker (fig. 33), one of the Big Four leaders of the Central Pacific Railroad. Muybridge was starting work on his next major project, which would take him away from the city for some time—a series of dramatic photographs to be made along the route of the Central Pacific and Union Pacific Railroads, from Sacramento to Weber Canyon, Utah. It is not known why he took on this considerable task at this stage in his career. It was a huge and costly undertaking that required transporting his camera equipment, assistants, and heavy boxes of glass plates and chemicals over the mountains and deserts for many hundreds of miles. He was clearly working to build up his catalogue of landscape views—there was strong competition in this field from Watkins and others—and completion of the cross-country railroad was but weeks away. It seems possible that Crocker and the Nahl brothers may have encouraged Muybridge to document the course of the Central Pacific with the same "artistic excellence" he had achieved in Yosemite. The biggest news in the country at the time was the impending ceremony that would celebrate the landmark link between east and west, and on May 10, the news was telegraphed across the land from Promontory Summit when Leland Stanford (fig. 34) struck at the final golden spike with a silver hammer, signaling to the world that the monumental task was "done."

Promontory Summit, The Golden Spike

This bridging of the frontiers of space and time via the railroads that day in 1869 inspired the nation in the same way that the first moon landing in 1969 would ignite the world almost exactly a century later. "For a moment the instrument was silent; then the hammer of the magnet tapped the bell," reported the *New York Times*. "'One, two, three,' the signal; another pause of a few seconds, and the lightning came flashing eastward, vibrating over 2,400 miles between the junction of the two roads and Washington, and the blows of the hammer upon the spike were measured instantly in telegraphic accents on the bell here. At 2:47 p.m., Promontory Point gave the signal, 'Done,' and the Continent was spanned with iron."[93] Though Muybridge was not there to record the events, three other photographers—A. J. Russell, Charles Savage, and Alfred A. Hart—created a newsworthy set of images that became instant icons, showing two trains, the Jupiter of the Central Pacific and the Union Pacific's Number 119, standing face to face on a patch of scrubby desert, obscured by a crowd of onlookers.[94] In Russell's classic view *East and West shaking hands at laying last rail*, which was quickly reproduced as an engraving in *Frank Leslie's Illustrated Weekly*, the formal handshake of the railroads' chief construction engineers Samuel Montague and Grenville Dodge is echoed above their heads by the outstretched arms of two locomotive engineers hanging off the front of each engine, clutching bottles to toast their accomplishments (fig. 35). Later that day, the *New York Times* reported that the New York Chamber of Commerce sent a dispatch to their counterpart in San Francisco, extolling how the "new highway thus opened to man will not only develop the resources, extend the commerce, increase the power, exalt the dignity and perpetuate the unity of our Republic, but in its broader relations, as the segment of a world-embracing circle, directly connecting the nations of Europe with those of Asia, will materially facilitate the enlightened and advancing civilization of our age."[95] No mention was made of the uprising that had delayed the ceremony by two days, when hundreds of unpaid tie cutters on the Union Pacific line held hostage the railroad's vice president, Thomas C. Durant, on his way to the ceremony, until compensated. The world became a smaller place that day, as the time it took to travel from coast to coast was cut dramatically and connections were forged to link the United States more closely to Europe and Asia.

Muybridge's 1869 photographs of the Central Pacific Railroad survey the route, scenery, and engineering highlights of the line and show his propensity for arresting and inventive composition. He was following in the footsteps of Hart, who, as the only official photographer for the Central Pacific, had already photographed the railroad for five years. Hart traveled with the engineers and construction workers, documenting their progress, and rode with Stanford en route to Promontory Summit to photograph the ceremony. Crocker admired his work and became his major supporter at the railroad, granting him full access. Hart also thought of himself as an artist and made many impressive landscape views along the route in California, Nevada, and Utah.[96] In his 1873 cat-

Fig. 36. *Shoshone Indians at Corinne (754)*, 1869

alogue, Muybridge described the terrain around the summit of the Sierra Nevada Mountains, where the railroad began its descent into Nevada, as holding "some of the finest mountain views in the world," emphasizing how "views of surpassing grandeur abound."[97] By mid-1869, he was focused on creating a series of dramatic landscapes looking out from the track beds at valleys, lakes, and peaks, furthering the landscape work he had completed in Yosemite. To signify the reengineering of treacherous terrain the railroad had to cross, Muybridge pictured the snow sheds, tunnels, and cuts through hard granite barriers, using these human feats of industrial engineering as the basis for unconventional, modern compositions. Shooting in stereo, he was aware that the illusion of deep space was important to convey a sense of real dimensionality. Consequently, as Hart had done, he composed some of his views looking through tunnels from the inside or up from underneath the long, curving trestles that spanned gorges and canyons obstructing the route (plate 25). He was also interested in the visually complex patterns formed by the matrix of wooden beams that secured the trestles and bridges. Though Hart's images were often more documentary in nature because he was working for the railroad, both Hart and Muybridge found unconventional ways to describe the landscape by elevating their cameras above the tracks or hiking to precarious perches to get to otherwise obstructed vistas. In *Bloomers Cut—85 feet deep, 800 feet long—looking East*, Muybridge divided the square frame into three interlocking triangles, conveying the scale of the towering rock walls through the deep perspective of the tracks vanishing in the distance. The white sky bears down on the rock, jutting into the scene from above like the tip of a spear. In a rare portrait reminiscent of his Alaska images, Muybridge pictured a group of Shoshone natives at Corinne, Utah, posed stereotypically by the tracks in skins and blankets, menacing the photographer with cold stares and a bow and arrow (fig. 36). He traveled beyond Promontory Summit, as far east as Weber Canyon and the 1,000-mile tree marker in Wilhelmina Pass along the route of the Union Pacific, where he photographed a wandering man in the desert, holding his staff like an image of Moses, on the spot where track layers had hung a sign marking their progress as they traveled west from Omaha.

Fig. 37. *Great American Overland Route—Yachting on San Francisco Bay, near San Rafael (380)*, 1873

Muybridge began selling his new stereographs in two series of views depicting the Central Pacific Railroad and the Union Pacific Railroad, for which there was a ready market. The transcontinental railroad soon became known as the Great American Overland Route, and the photographs that Hart, Muybridge, and others made were used to promote it (fig. 37). "When we stood for the first time on the iron-bound shores of the Pacific a generation ago and looked upon their desolate mountains, after a voyage of more than half a year, we thought in our hearts that the last tie that bound us to our native land was broken," wrote Stanford's friend Dr. J. D. B. Stillman, who had accompanied him on the trip to Promontory Summit. "We did not dream that the tie that was to reunite us, and make this our native land forever, was then flourishing as a green bay tree in our woods," referring to the symbolic "last tie" of polished California laurel used that day to link the final rails.[98]

The railroads and telegraphs helped make the clocks run on time, as schedules had to be synchronized across the country and around the world.[99] Clocks were no longer regulated by the sun overhead but by the mathematical calculations of scientists who could plot the position and rotation of the earth and stars with new accuracy. Industrialization of the landscape imposed an unbending conception of time and space on workers—time is money, the saying goes—and as trade increased, even the shipping lanes along the West Coast needed improved management and safety accommodations to keep track of the flow of goods and people traveling in and out of California. By 1870, San Francisco had become the tenth largest city in the country, and one of the wealthiest. The Panic of 1873 and an ensuing six-year depression, brought about in part by rampant speculation in land and the railroads, helped redefine the nation politically. Despite an increasingly dismal economic outlook in both the United States and Europe—as political strife, labor issues, increased poverty, and fiscal problems began to engender destabilizing economic issues that called into question the human cost of rapid industrialization—the West felt the impact to a lesser extent. Because the economy and population of California—especially in San Francisco—had grown so rapidly in the late 1860s and early 1870s, the Golden State was still considered a promised land, full of potential as both an economic and a cultural Eden. This growth was spurred on by completion of the transcontinental railroad in 1869, the ready availability of fertile land and natural resources such as minerals and timber, and the rapid importation of labor from Asia, Latin America, and Europe.

Construction of a new federal Mint began in San Francisco in 1869, and Muybridge created an extended sequence of photographs documenting the project's progress through 1872 (plates 29–31). This was the first of a series of architectural projects for which he would photograph the raising of a building over an extended period, signaling his early interest in depicting how things change through time. He continued his photography of Bay Area military installations and traveled to Great Geyser Springs on Pluton Creek in Sonoma County to produce a series on the hot springs and adjacent hotel that became a weekend retreat for San Francisco's growing middle class. These images were intended to help promote travel to the rural resort. To make his stereographs even more attractive to visitors (and potential buyers), Muybridge employed many of his trademark special effects for this series, including artificial moonlight and the elaborate staging of figures in the steamy, rugged landscape (plate 32). Similarly, he traveled to Calistoga Springs to photograph another resort that had been built at the end the California Pacific rail line, which ran from Vallejo to Calistoga to ferry vacationing tourists from the city.

Fig. 38. *Chang Woo Gow. Celebrated Chinese giant. Virginia, Nevada,* August 31, 1870

Muybridge also began an extensive new series of views of Woodward's Gardens, San Francisco's largest amusement park and a popular destination for the new leisure class. Founded by hotel entrepreneur Robert B. Woodward on the grounds of his former home, the attraction was a formal garden filled with rare plants, a zoo, animals roaming the grounds, and taxidermy specimens. There were also museums and an art gallery, as well as performances and "curiosities" that included Japanese acrobats and Chang Woo Gow, the fabled Chinese Giant

Fig. 39. *Art Gallery at Woodward's Gardens*, 1870

(fig. 38). Like the World's Fairs of the time, Woodward's Gardens imported a wholly artificial view of international ethnic cultures, placing "exotic" stereotypes into an entertainment context where ethnicity became objectified for display in a consumer environment. Seeking to expand his catalogue of popular commercial imagery, Muybridge photographed the gardens, stuffed animals, museum displays, and rides, as well as Hawaiian, Japanese, Indonesian, and Chinese performers. His staged portraits of Chang Woo Gow surrounded by gawking crowds, and of his wife standing alone in her traditional clothing (fig. 109), seem artificial renderings when compared to the ethnographic survey views of Alaskan natives he had made in the field.[100] Muybridge also pictured himself in several of his photographs, seated in the gallery (fig. 39) and reflected in a garden orb (plate 37). He even opened his own retail shop on the grounds as yet another way to create new markets for his work, and he publicly released the pictures by September 1870. He had already been advertising his services and several new series, including coastal views, for some time; in August he posted the following notice in the *San Francisco Evening Daily Bulletin*: "Private Residence, Horses, Monuments, Ships, etc, are photographed in the best manner by HELIOS' Flying Studio."[101]

The Pacific Coast, Lighthouses, and the Buena Vista Vineyard

In January 1871, Muybridge accepted a new government commission from the Lighthouse Board, a federal office created in 1852 to oversee construction, operation, and maintenance of all lighthouse facilities in the United States. Beginning in March, this venture would take him up and down the Pacific coast for several months by steamer on a survey project to document lighthouses from Cape Flattery on the northernmost tip of the Olympic Peninsula to Point Loma on the southern coast at San Diego. Muybridge came to the following arrangement with the Lighthouse Board: "In the event of your department providing for my personal expenses, safe transportation for myself and apparatus, and remuneration for my time, chemicals and use of apparatus at a rate of Twenty Dollars per day, for each day or part of a day, during my absence from San Francisco upon any expedition connected with the objects required, I will agree to furnish you with two photographic prints, views of all the Light Houses upon this coast, and such further views as you may direct of other points, without extra charge."[102] He set off in mid-March with the crew of the steamer *Shubrick*, a lighthouse tender named for William Shubrick, a career naval officer who became chairman of the Lighthouse Board in 1852. Leaving the city, Muybridge photographed the Golden Gate from the deck of the ship, looking back off the stern at the nautical, machinelike forms of the boom and rigging that divided his picture geometrically, in stark contrast to the organically composed seascape beyond.

Muybridge's photographs of lighthouses and the surrounding coastal terrain are some of his most luminous and meaningful landscapes. He seems to have been at home around the water, and he used the ebb and flow of tides and time to animate his images. These pictures are about the natural forces of land and sea, and how the two meet in a timeless geologic ballet. The water carves the land as the land holds the water; they embrace and caress each other back and forth along the vast, mutable coastline where they have interacted together for eons. The west coast of the United States sits astride a powerful tectonic fault line—it forms one edge of the Pacific Rim—and Muybridge's photographs are

Fig 40. *First-Order Light-house at Punta de los Reyes, Seacoast of California, 296 Feet Above Sea (4136)*, 1871

permeated by evidence of the seismic activity that helped form the windswept cliffs, islands, points, and capes. He brings forth at once both the moment fixed in the photograph and the eternal nature of time, represented here by the implied restlessness of the water and its flow against the rocks. The ocean often appears in these pictures like a glassy fog; its motion appears on film as a fluid mist. The waves seem to move endlessly—infinite like time, with no beginning or end—which implies what comes before and after the depicted moment. The landscape is always changing, and the lighthouses, strategically placed to alert and guide ships traveling near the coast, illuminate the inherent dangers. In *Light House at Punta de los Reyes, Coast of California (8074)*, 1871, the titular structure sits perched on a sheer cliff three hundred feet above the waves; a vast stretch of foggy ocean subsumes this tiny beacon at Point Reyes, which juts ten miles into the sea along the San Andreas Fault just north of San Francisco (plate 43). Sky and water dominate the image, and the horizon is barely discernable in the fog, implying an endless space that is broken only by the craggy outcrop of land and the miniscule road leading to the light. There is no place to stand in the picture; the viewer is not grounded, and the camera seems to be floating in space, conveying the illusion that the lighthouse and its perch are hovering as well. The structure's symbolic relationship to the natural world—the light is a warning signal to sailors about a physical danger they do not see—is transcended by the vastness of the ocean. The sailor's journey is one, like that of Odysseus, with no ground on which to stand, no horizon, and no safe harbor in sight.

Muybridge was quite serious about obtaining views that would satisfy the documentary requirements of the commission, but that did not prevent him from pushing the conventions of landscape photography. "Point of view about 250 feet above base of Light House," he wrote on the back of a print of *Light House at Punta de los Reyes* he gave to the Lighthouse Board, offering a scientific explanation for his unusual perspective. "A view of a distant object below the stand point at this elevation, in true perspective, fails to render the declivity apparent; and although as in this view, the Light House is 250 feet below the point of view, approached by a descending trail, it is apparently more elevated than the standing point. There is no stand point for a profile view of the rock projection on which the Light House is built."[103]

When the *Shubrick* put in at the Farallon Islands, an isolated, windswept group of craggy rocks populated by seals and sea birds about thirty miles off the coast of San Francisco, Muybridge photographed the lighthouse sitting atop a 360-foot peak and the landing, where a derrick was constructed to offload supplies for maintaining the light; he also photographed views looking out to the clouds from the peak of South Farallon Island. These images are inspired landscapes that balance the enormity of water, earth, and sky with tiny figures and the distant lighthouse, each silhouetted in long exposures against the ethereal instability of nature (plates 38, 39). Known for its rugged beauty, South Farallon Island was the subject of a series of mammoth-plate photographs and stereographs by Watkins, made in 1869, and Muybridge photographed some of the same landmarks, such as Sugar Loaf Island, Murr Bridge, and Fisherman's Bay.[104] Watkins fills the center of his frames with his subjects, striving for a compositional balance that is simpler and more straightforward than Muybridge's expositions, and he appears at ease with the natural world. Muybridge's views are more atmospheric, graphic, and filled with tension. Even his distant observations call attention to special details, building taut relationships between land and sea rather than opening up his images to the sky, as did Watkins. Incorporating the tightly strung visual drama he learned from British landscape painters like Turner, Constable, and Samuel Palmer, Muybridge fills his skies with atmospheric clouds, which are reflected in the gossamer movement of water, helping to unify the space as a coherent whole. In *The Ramparts, Funnel Rock, Hole in the Wall, Pyramid, Sugar Loaf, Oil House, and Landing Cove on Fisherman's Bay, South Farallon Island (4150)*, 1871, Muybridge cleverly balances the temporal, finite world of humanity with the timeless, inestimable realm of nature (plate 39). His elevated viewpoint encompasses Fisherman's Bay and Landing Cove, "a fine estuary, where, with a little improvement, small schooners can enter at any season of the year, and where the oil and other supplies are landed for the lighthouse," as Hutchings described it during his own visit to South Farallon. "Like the other views, it is singular and wild—each eminence covered with birds, each sea-washed rock occupied by seals, and the air almost darkened by the sea gulls skimming backward and forward, like swallows, and by the rapid and apparently difficult flight of the murres."[105] Freight was offloaded using a triangular derrick set up next to a shed used to store incoming oil for the lamp. From this landing, a mule carried the supplies up the hill to the light and the keeper's home. The industrial angularity of the crane and shed at Landing Cove sits in stark contrast to the miasmic view across Fisherman's Bay, which takes in the primeval shapes of small islands called Hole in the Wall, Pyramid, and Sugar Loaf, each the result of natural erosion from water and weather through the ages. Muybridge posed two men, seemingly in dialogue, in the foreground of the scene, opposite the derrick. Their presence instills an earthly quality, pulling the viewer into the space and lending a sense of scale to an otherwise ambiguous image. The composition is a cacophony of triangular shapes: the geometric construction of the crane and shed; the angled spit of rock that juts into the frame from the left; the pyramid-like islands; one man's arm akimbo; and the form drawn across the image as one's eye moves from the shed to the Hole in the Wall to the men. These forms combine to extract clear relationships between nature and the men within it.

Muybridge was in top form as a landscape photographer. He continued to travel for the Lighthouse Board and by mid-August photographed a total of twenty-three lighthouses along the West Coast, allowing him to expand his catalogue of Pacific Coast views and strengthen his relationship with government agencies. He printed many of his stereographs and 7-by-9-inch photographs of the Farallons, describing that landscape as "some of the most rugged, singular and interesting scenery upon the coast."[106] Some of the lights he pictured, like Point Reyes, had just opened, and others, like South Farallon Island, had been in service for years. The last lighthouse he photographed was at Pigeon Point south of San Francisco, which had begun to operate when its five-wick lard-oil lamp was lit on November 15, 1872. Probably photographed in the spring of 1873, Muybridge used a mammoth-plate camera that produced huge, detailed negatives on glass plates approximately 17 by 22-inches in size. *Pigeon Point Lighthouse*, 1873, is a classically composed image depicting a luminous moment in which the fog banks are lifting from the horizon, allowing full illumination from the sun (plate 45). The great resolution Muybridge achieved with such a large negative allowed him to pinpoint tiny figures situated on the shoreline and bluffs and at the top of the lighthouse, giving a clear sense of how vast and open the space really is. The photograph is divided vertically by the skyward thrust of the engineered tower and the meandering, rocky shoreline separating the domain of people from the frothy marine waters offshore. Such images symbolized at the time a new frontier beyond the coast, as American commerce began looking farther out across the Pacific, toward Asia.

Early 1872 was a fecund time for Muybridge. In fall 1871 and early 1872, he photographed the Buena Vista Vineyard in Sonoma for a series called *A Vintage in California*. Muybridge first exhibited some of his Buena Vista views at the San Francisco Art Association in January, where he met painter Albert Bierstadt at the reception.[107] Founded in 1857 by Count Agoston Haraszthy, Buena Vista is one of California's oldest wineries. Haraszthy had brought vine cuttings from Europe and promoted the region as a rival to France in its suitability for grape growing. Muybridge photographed the cycle of wine production in a documentary manner, from the plowing and planting of the fields through the harvesting and crushing of the grapes to the aging and bottling of the finished wine in what is now regarded as "the first photographic essay on vineyard labor in California."[108] Comprising some of the earliest photographs of the state's farmworkers, Muybridge's Buena Vista images reflect his discernment of landscape composition. They show workers and supervisors at a distance, integrating them into the pictures as part of the farming process rather than as individuals (plates 50, 51). He also depicted Chinese workers side-by-side with others, an unusual representation at the time. Novelist Helen Hunt Jackson, who saw the series on a visit to Muybridge's studio, compared the photographs to Jean-François Millet's Barbizon School paintings of peasant farmers. "One of the first pictures in the series, representing the first breaking of the soil for the vineyard, is as perfect a Millet as could be imagined," she mused in *Bits of Travel at Home*. "The soft tender distance, outlined by low mountain ranges; a winding road, losing itself in a wood; a bare and stricken tree on the right of the foreground; and in the centre a solitary man, plowing the ground" (plate 52).[109]

In February 1872, Muybridge recorded the cornerstone ceremony that inaugurated construction of the new San Francisco City Hall and Law Courts (fig. 41). Photography was a process capable of documenting events over a period of time; as he had done with his images of the Mint, Muybridge went on to picture various phases of City Hall's construction in sequence as it progressed over several years (plate 46). In April, he was commissioned to photograph Leland Stanford's palatial Sacramento home, and in May, according to the artist's later recollections, he made his first attempts to stop motion with a camera by photographing Stanford's prized trotting horse Occident at the Agricultural Park racetrack in the state capital.[110] Stanford was at the center of a widespread debate at the time about whether a running horse had all four hooves off the ground at one point in its gait, an idea that could not be proved with the naked eye. Stanford engaged Muybridge with the goal of understanding the detailed physiology of this movement so as to breed and train horses for greater speed. To accomplish this Muybridge had to develop the technical means to photograph Occident while trotting, freezing the action to record the positioning of the horse's hooves while in motion. This was a complex proposal for him since the technology of instantaneous photography had not advanced far enough to make it possible and he would need to develop new techniques and a scientific methodology to meet the challenge, which he would not fully achieve until 1878.

Yosemite Valley, the High Sierra, and the Mariposa Grove

Also in the spring of 1872, Muybridge was planning a significant new series of large-scale views of California and the West, to be produced with a mammoth-plate camera. Even though he had not yet mastered the technically difficult process of producing mammoth-plate negatives—the largest he would make—Muybridge needed bigger images with new and more dramatic compositions to compete in the lucrative landscape market. He announced, in May, from the Nahl Brothers Studio: "At the suggestion of several Artists and Patrons of Art, I propose devoting the approaching season to the production of a series of large-size photographic negatives, illustrating the Yosemite and other grand picturesque portions of our Coast."[111] His list of subscribers already included the Central and Union Pacific Railroads, the Pacific Mail Steamship Company, Central Pacific legal advisor Edwin B. Crocker (the older brother of Charles Crocker), photographers and agents Alexander Edouart, Henry William Bradley, William Herman Rulofson, and Isaiah West Taber, and painters Bierstadt and Charles Nahl.

In June, while ascending a trail out of Yosemite, Helen Hunt Jackson passed Muybridge as he was packing his equipment down into the valley while working on his new series. "As we slowly climbed the trail, a long line of pack-mules met us," she wrote. "We drew aside to let them pass. They were loaded with a photographer's apparatus, lenses, plates, camera, carefully packed boxes of chemicals.... Their owner, Mr. Muybridge, has just established himself in the valley for the purpose of taking a series of views, larger and more perfect than any heretofore attempted."[112] His work had always been compared favorably with that of Watkins, and Jackson went one step further, commenting that "the scenery of California is known to Eastern people chiefly through the big but inartistic pictures of Watkins. When it is known through the pictures which Mr. Muybridge is now engaged in taking, it will be seen in its true beauty and true proportions. Everything depends on stand-point; very few photographs of landscapes really render them."[113] It was as though Muybridge himself were speaking to promote his work through Jackson.

Fig. 41. *Grand Masonic Ceremony Laying the Cornerstone of the City Hall and Law Courts,* 1872

Fig. 42. *Base of Yowiye Fall (1318),* 1872

Given the difficulty of mammoth-plate photography, this would be Muybridge's most complex project to date. In the span of about six months, between June and November, he made approximately five hundred photographs of Yosemite, the Sierra Nevada Mountains above the valley, and the Mariposa Grove of sequoias, including mammoth, full-plate (5 1/2-by-8 1/2-inch), and stereographic views. Beginning on the valley floor and along the tranquil Merced River, and then moving upward to the sheer cliffs and peaks more than half a mile above, he focused much of his energy on realizing the fifty-one large-scale photographs that would define his work in the coming years.

In his 1867 Yosemite landscapes, Muybridge had already displayed a keen interest in the temporal effects of nature. In 1872, his images grew more complex, depicting sunlight playing on the surface of water and rocks, clouds advancing in dramatic forms across the sky, a stenciled moon playing peek-a-boo above dark cliffs, and water, glassy smooth or broken into a sparkling spray, that resembled stars spilling from an atmospheric mist (fig. 42). His 1872 mammoth plates represented a technical tour de force; his camera was gigantic, and each negative measured about 17 by 22-inches. In *Valley of the Yosemite. From Mosquito Camp. No. 22*, an image that engenders both real and illusory views, Muybridge looks into and across the river that both serves as a barrier to and reflects the distant cliffs (plate 64). One mirrors the other, just as a bleached white branch lying in the foreground mirrors the dead white tree emerging from the water smack in the middle of the image. A small axe lying on the ground near the river symbolizes the photographer's presence, and a nearly invisible man can be discerned in the trees across the water. The level of detail achieved—in which every leaf on the close-in trees has as much presence as the far-off cliffs—helps to convey the vastness of the space contained within the frame. Muybridge correlates the importance of the microscopic with the macroscopic. *Pi-Wi-Ack. Valley of the Yosemite. (Shower of Stars) "Vernal Fall." 400 Feet Fall. No. 29* shows Vernal Falls emerging from a sheer cliff face (plate 66). There is no visible horizon and, like Timothy O'Sullivan's renowned survey photograph *Ancient Ruins in the Cañon De Chelle, N.M.*, 1873, the pictorial space is quite flat and the image compositionally advanced (fig. 43). Because of the long exposure time, the waterfall appears as a light-filled volume, a representation of dimensional space occupied through time. In contrast to the rock-solid cliffs, water is portrayed here within a different aesthetic framework; its flow is not so much documented as virtually conceived.

Muybridge's pictures were now significantly distinct from those of Watkins or any other Yosemite photographer to date, as he shot from dizzying heights and precipitous angles, looking over cliffs and waterfalls, and focused on the effects of light, clouds, and moving water rather than on the solidity of things. He was painting the temporality of nature with the visual tools of light, atmosphere, and composition. It was later reported in the *Daily Alta California* that "he has cut down trees by the score that interfered with the cameras from the best point of sight; he had himself lowered by ropes down precipices to establish his instruments in places where the full beauty of the object to be photographed could be transferred to the negative; he has gone to points where his packers refused to follow."[114] The larger scale of his negatives also helped Muybridge achieve the softer, more atmospheric effects he sought, in which the relationships of scale between foreground details and distant peaks and clouds became manifest. And it accentuated the feeling of disorientation found in these pictures made from precipitous ledges and heights, which seem to fling the camera out into space in gravity-defying leaps of confidence. Muybridge was purposefully destabilizing traditional nineteenth-century landscape values, based on balance and order, to create more dynamic images that disorient rather than orient the viewer. In other photographs, diagonal swaths of textured rocks and trees break the images into zones of light and dark, and vales of deep space punctuate the flat surface, yielding new visual structures that are more dramatic and emotionally driven than Watkins's views. In *Tenaya Canyon. Valley of the Yosemite. From Union Point. No. 35*, the scene is divided into a grid of inverted triangles, organized in receding planes from dark to light, accentuating the remarkable depth of the valley and the granite textures that map its geological formation (plate 70). The subject of geologic time, signified by eons of glacial formation visible in the cliffs, is balanced by the transient grandeur of ever-changing light and foliage. The intricate relationship between the vastly different measures of human and geologic time, and their delicate equilibrium, is inherent in the structure of this photograph; it establishes an environmental assertion that was distinctly modern. This was a land worth visiting and preserving since its very history was discernible.

Fig. 43. Timothy O'Sullivan, *Ancient Ruins in the Cañon De Chelle, N.M., In a niche 50 feet above present Cañon bed*, 1873

Jackson, who insisted on using the indigenous word *Ah-wah-ne* to name Yosemite in her writing, praised Muybridge's landscapes for their aesthetic use of clouds. "The skies are always most exquisitely rendered," she wrote. "His cloud photographs alone fill a volume; and many of them remind one vividly of Turner's studies of skies."[115] Muybridge had already produced numerous stereographs of clouds, and in Yosemite he also shot large, new negatives to use in overlaying the dramatic, billowing forms onto his skies. In printing some landscapes, such as *Valley of the Yosemite from Union Point. No. 33*, he remained true to what the camera saw, using the contrast of sky and ground to etch a sharp linear edge across the surface of the image (plate 69); in others, such as *Cottonwood Bend. Valley of the Yo-semite. No. 15*, he judiciously left the sky white while showing clouds reflected in the Merced River, adding texture to the foreground while allowing the distant sky to recede (plate 60). In still other views, such as *Glacier Channels. Valley of the Yosemite. From Panorama Rock. No. 41*, he produced the full effect of weather spilling across the distant mountain ridge by printing clouds across the top of the photograph (plate 72). These were aesthetic decisions designed to bring into play the full range of visual and symbolic effects then available to him; he emulated the range of emblematic manipulation also found in nineteenth-century landscape painting. As art historian John Ruskin posited in his 1884 lecture "The Storm-Cloud of the Nineteenth Century," clouds indicate a special relationship between our view of the sun and the terrestrial environment. Critiquing rampant industrialization, which blackened the sky, and the Victorian predilection for science, which he connected to increasing political violence in Europe, Ruskin stated, "You have your *sun*-measure, and can tell exactly at any moment how strong, or how weak, or how wanting, the sun is. But the sun-measurer can't tell you whether the rays are stopped by a dense, *shallow* cloud, or a thin, *deep* one. In healthy weather, the sun is hidden behind a cloud, as it is behind a tree; and, when the cloud is past, it comes out again, as bright as before."[116] Muybridge introduced his picturesque clouds in part to subvert the scientific, analytical nature of photography in favor of its aesthetic, malleable capabilities. Helios, the artist, could manipulate the sun's rays in this visionary side of his work, just as he could create moonlight in the darkroom. He represented clouds in his art to change the landscape he experienced into a more transcendent view, one free of the limitations of technology. Simultaneously, works such as *Glacier Channels* also documented the visible striation in the granite, providing scientists with data about Yosemite Valley's formation. If scientists could show that the valley was formed by the slow movement of glaciers over millions of years, rather than by a cataclysmic rending of the earth as many believed, the evidence would help date its creation. If that were not enough to substantiate the role of the camera in debates over evolution and geology, Muybridge also photographed an assistant posed before giant boulders that had been deposited in place by moving sheets of ice (fig. 44). His landscape sensibility shifted back and forth between science and art.

While working in Yosemite in 1872, Muybridge also pictured ancestors of the valley's enduring inhabitants, a group of Mono Piute natives found relaxing by the Merced River and camped around their traditional lodges. They had moved to the valley years after the indigenous Ahwaneechees were run out in 1851 by the Mariposa Battalion, a band of armed forty-niners authorized by the California governor to clear the land for white settlement. Central and Southern Sierra Miwoks were also there seasonally, and by the 1870s, commerce and the growing tourist culture had irrevocably altered Native American subsistence and customs. The name "Yosemite" was derived by white settlers from the Miwok word *Yosemeti* or *Yohemeti*, which means "they are killers," referencing another native group living nearby.[117]

Fig. 44. *Ancient Glacier Boulders at Lake Tenaya (1545)*, 1872

Like Muybridge's Alaska images, these presented a decidedly ethnographic look at the indigenous people whose language and stories were used to identify many of the valley's landmarks but who had been written out of much of Yosemite's history. *An Octogenarian (1580)* depicts an old woman seated gracefully on a log in front of a conical dwelling made of branches. A child joins her on the log, uniting two generations to signify established roots and a possible future for native people in Yosemite (plate 53). Albert Bierstadt, who was also in the valley that summer making sketches for paintings, had asked Muybridge to document the Indian camp, thinking that his photographs could be used as a foundation for later works. Muybridge photographed Bierstadt at work in his outdoor "studio," sketching the camp and some of the Paiute men who also appear in Muybridge's pictures (fig. 45). Bierstadt's painting of *Indians in Council, California*, 1872, was modeled directly from one of these photographs.[118] Muybridge documented the encampment in a sequence of images that show the village from a distance and from the inside. He made group portraits of men and women; showed their lodges, baskets, and sifting trays; and pictured two girls making bread. Like his photographs of the Buena Vista Vineyard, these provide a sequential, ethnographic view over time. As documents, these pictures portray the Paiute villagers as exotic outsiders by typing them as "belles," "bucks," and "medicine men," without examining individual personality or character. Bradley and Rulofson later published them as a series of stereographs called "The Indians of California," about whom Muybridge wrote that "the majority prefer the nomadic life of their forefathers, and are fast dwindling away before the advancing tide of civilization."[119]

In early November 1872, the *Daily Alta California* reported on various artists still working in Yosemite late in the season: "The scenery is not without its attractions in Winter," the paper emphasized. "Bierstadt went in before the snow had disappeared last Spring; Keith is now there—studying the atmosphere and foliage of Autumn; and when artists, so skillful in the disposition of effect and color, are not contented with one visit to such a place, but go again and again, better pleased with every trip, there must be extraordinary natural beauties.... Notwithstanding the multitude of photographs that have been taken in previous seasons, a man employed by Mr. Houseworth has been at work there for months. Mr. Muybridge is now there, and Mr. Watkins will start in a few days; and a multitude of views, finely executed, from points well selected, will be the result."[120] On his way out of Yosemite that fall, Muybridge visited Mariposa Grove, a stand of giant sequoias on a protected ridge south of the valley. There he photographed a man holding an axe in front of Galen Clark's log cabin. Clark found the Mariposa Grove of big trees in 1853 and had already been portrayed by Watkins in 1861 at the base of the Grizzly Giant, gripping his hunting rifle like an iconic western explorer. Clark's environmental awareness and his public image as guardian of the trees helped secure the Yosemite Grant, a federal deed signed in 1864 by President Lincoln that protected the land around the valley and the Mariposa Grove "for all time." Muybridge's photograph *Wm. H. Seward. 85 feet in Circumference, 268 feet high. Mariposa Grove of Mammoth Trees. No. 51* is focused on the base of an enormous tree named for the former secretary of state; the massive trunk dominates the scene as it shelters Clark's diminutive log home behind it (plate 73). The sheer scale and texture of the forest environment envelop the diminutive figure, which appears in perfect harmony with the outsized trees. Seward, who had encouraged Muybridge for his pictures of Alaska, had died just weeks before the negative was exposed.

Fig. 45. *Albert Bierstadt's Studio (1586)*, 1872

Muybridge himself was photographed in Mariposa Grove, appearing tired and indefinite, propped up on a plate box and cradled by the bark of the General Grant, another giant sequoia in the upper grove near Clark's cabin (plate 74). With a circumference of 67 feet and a height of 268 feet, the trunk towers over the photographer, who appears reticent in the cool November air, posing for the mammoth-plate portrait in the guise of a weary outdoorsman, whose tattered cuffs, tightly buttoned wool vest, and well-worn boots belie the image of a prominent artist from the city. It is not known who made the portrait of Muybridge—it could be a self-portrait he staged himself, or he may have set it up with his one known assistant in Yosemite that year, William Henry Towne.[121] Towne was a photographer previously affiliated with Robert Vance and Alexander Edouart. In 1872, Towne was a business partner of George Daniels Morse, who became Muybridge's agent in 1877. The plate (or equipment) box on which Muybridge is seated is inscribed "California Views, Thom Houseworth & Co, San Francisco," implying that a photographer affiliated with Houseworth was present at the time. The "man employed by Mr. Houseworth" reported by the *Daily Alta California* could have made the portrait; Charles

Leander Weed was then affiliated with Houseworth on other jobs, and Weed—who had already produced many large plate views of Yosemite as early as 1859—may have been there assisting Muybridge or working on Houseworth's already extensive catalogue of Yosemite views.[122] As Peter Palmquist speculates, "It would … have been an ideal arrangement, since Weed had experience producing large plates in Yosemite. Even more important … is Weed's long record of working as a team member and of eschewing personal glorification."[123] Muybridge may also have been subsidized by Houseworth and working with Houseworth's equipment, in exchange for the opportunity to publish some of his Yosemite views. George Fiske, who later lived in Yosemite Valley and made an extensive photographic survey of the region, was also working outdoors as a photographer for Houseworth at times between 1869 and 1874.[124] It is possible, too, that Watkins, who was in Yosemite in fall 1872, photographed his old friend and former sales agent in the Mariposa Grove.[125] The portrait of Muybridge was later published in different formats on a Houseworth and Company mount, and it appeared in several albums and books, side-by-side with other photographs by Watkins, Weed, and Muybridge. The best clue to its attribution is the image itself, and on close examination, it seems likely that Muybridge had a hand in its construction. The cynosure of the photograph—the weary artist returned to earth from the mountaintop—rests by the mythic tree, dwarfed by the presence of nature. In a sophisticated manner, the portrait concludes the odyssey Muybridge began in 1867 as *The Astonished Woodchopper*, his allegorical stereoview of a small, dumbfounded figure holding out an axe in awe before the overwhelming symbol of nature's tenacity, the giant sequoia (plate 1). Muybridge's woodchopper recalls Thomas Cole's painting of *The Woodchopper, Lake Featherstonhaugh*, 1826, in which the structured beauty of trees and sky subsumes an industrious figure clearing land by hand with an axe.[126] Here the woodchopper represents the duality between the sublime nature of the wilderness and its destruction, thought necessary to develop the full potential of civilization. This became an important, emerging theme in American art through much of the nineteenth century. In the 1872 portrait, Muybridge is sheltered, embraced, and sustained by the grandeur of nature, having now transcended the reckless abandon of his youthful woodchopper. He seems to harbor a better understanding of the complex relationship between civilization and the natural world that sustains it, a sensitivity that informed much of his work that followed.

In the early 1870s, there was a good deal of competition among the major galleries—including Watkins's Yosemite Art Gallery, Bradley and Rulofson Palace of Photographic Art, and Thomas Houseworth and Company—for the lucrative photographic view and stereograph market (fig. 46). Some of Muybridge's striking mammoth-plate views had initially appeared on Houseworth mounts. However, Bradley and Rulofson, which was fast becoming the premiere photographic establishment in California, sought and eventually won Muybridge's endorsement to publish his 1872 Yosemite photographs, promising lavish production values and fanfare. "To most persons in California the name of this artist is as familiar as those of the majestic scenes he illustrates," they crowed.[127] As Peter Palmquist noted, "Muybridge's star was on the rise at this time and whichever publisher acquired his Yosemite mammoth-plate series would have great advantage over all others. Though Muybridge announced his plan from Nahl's gallery, it appears that he went to Yosemite Valley with his commission from Houseworth. Meanwhile, Bradley and Rulofson were secretly (and successfully) courting Muybridge with offers of the finest production and distribution arrangement they could put together."[128] By mid-1873, Muybridge had contracted with Bradley and Rulofson to publish his entire "catalogue of photographic views," including his new Yosemite pictures and most of the work he had made since 1867. This precipitated a nasty public debate carried out in the *San Francisco Daily Evening Bulletin*, in which Bradley and Rulofson accused Houseworth of placing "an old soiled print from a condemned negative of Muybridge's" in their shop window. Houseworth responded that it was "one of a set of forty furnished to a subscriber by Bradley & Rulofson … and it is now in the same condition as when received by the original purchaser." Muybridge had the final say when he replied to Houseworth, "Silence and contempt, says Aesop, are the best acknowledgements for the insults of those whom we despise."[129] Eventually, a few of Muybridge's pictures were given over to Houseworth for publication in exchange for their previous assistance, including some alternate, mammoth-plate images of Yosemite. In 1874, Houseworth collaborated with the Chicago publisher Coyne and Relyea to produce a book titled *Sun Pictures of the Yo-Semite Valley, Cal.*, 1872, that included some of these.[130] It presented forty-four photographs by Weed and Muybridge, reduced in scale from their original mammoth-plate size, including the Mariposa Grove portrait of Muybridge. *The Domes* (fig. 47) was shot from an almost identical viewpoint as another Muybridge view, *The Domes. Valley of the Yosemite. No. 20*, 1872.[131] Watkins had also made comparable images (fig. 49). Interestingly, five of the Muybridge photographs in the Houseworth book display exactly the same clouds in the sky—all were printed in later from a negative that was used in other prints produced by Houseworth (fig. 48).

Fig. 46. Thomas Houseworth & Co., *Thomas Houseworth's store at No. 9 Montgomery Street, San Francisco*, ca. 1873

In September 1873, the *Philadelphia Photographer* praised Muybridge's dramatic use of clouds in his Yosemite pictures, which were on display at the Vienna International Exhibition in Austria. Launched on May 1, this new World's Fair was organized to call attention to the economic and industrial progress of the Habsburg monarchy, and to the significance of Austria-Hungary as one center of European art and culture. A new structure, the Rotunda, was engineered for the fair to rival memories of the Crystal Palace and enhance the architectural profile of the city. However, just as the fair opened, the stock market crashed, plunging much of Europe into an economic depression. Hermann Wilhelm Vogel, a German photographer and chemist on the jury of the exhibition, wrote that "Muybridge, above all others, is distinguished by superb cloud effects."[132] Then Vogel announced that Muybridge, Houseworth, and Watkins had all won medals at the exhibition for "taste with acknowledgement of improvement," praising Muybridge's Yosemite images in particular for going beyond what had been done before: "The large pictures taken at the Yosemite Valley received also the medal for progress, because, considering the locality and compared with older pictures, they show an improvement."[133] Muybridge and his new representatives, Bradley and Rulofson, took full advantage of the honor by advertising his "award-winning" new pictures and reproducing the medal on their cards (fig. 50).

Muybridge had been busy printing his photographs and preparing them for publication following his return from Yosemite in November 1872. In April, 1873, the *Daily Alta California* praised his Yosemite photographs, and on the same page announced publicly for the first time the instantaneous photographs of Occident trotting he had taken in Sacramento the previous year. "The space of time was so small that the spokes of the wheels of the sulky were caught as if they were not in motion," it was reported. "This is considered a great triumph as a curiosity in photography—a horse's picture taken while going thirty eight feet in a second!"[134] On May 2, the *New York Times* picked up on the story—"A San Francisco photographer is declared to have obtained a perfect likeness of the horse Occident going at full speed"—and soon articles about Muybridge's achievement began appearing in papers around the world.[135]

Fig. 47. *The Domes*, 1872

Fig. 48. *Looking Down Yo-Semite Valley*, 1872

The Modoc War

In April 1873, Muybridge was commissioned by the U.S. Army to photograph the Lava Beds at Tule Lake, where war had erupted the previous winter. Fought across the desolate northeastern border of California, the Modoc War was a symbolic turning point in relations between West Coast Indians and the government, which had by 1873 moved many Native Americans onto reservations. The Modocs' homeland, set in the shadow of Mount Shasta, straddled the Lost River and Tule Lake region between California and Oregon; it was their territory for as long as they could remember and offered abundant wildlife, water, and shelter in an area of volcanic caves and warm springs. In 1864, the Modocs were removed from their aboriginal homes and sent to live on the Klamath Indian reservation in Oregon to the north. Cut off from their land, mistreated by the Klamaths, and unable to hunt or grow crops, they went hungry; then some Modocs, led by Kintpuash (also known as Captain Jack), left the reservation and went to live again at Lost River until the end of 1869, when the government negotiated their return back to Oregon. In 1870, a larger group again went back to Tule Lake. They were able to move back and forth and began to discuss creating their own reservation there. When negotiations deteriorated, the U.S. Army arrived to arrest Captain Jack, and a number of people on both sides were killed in escalating violence between the soldiers, the Modocs, and white vigilantes who also entered the fray. The Modocs escaped to the Lava

Fig. 49. Carleton Watkins (attributed), *The Domes from the Merced River, Yosemite*, ca. 1872

Beds, a warren of volcanic caves known only to the natives, and by the end of 1872, war had broken out near the shores of Tule Lake. "For five months in 1873, fifty-three Modoc warriors, ensconced with their families in a northern California lunar landscape known as the Lava Beds, held off twelve times their number of Army regulars," writes historian Peter Cozzens. "Before they were rounded up, the Modocs had killed forty soldiers (including Edward R. S. Canby, the only general officer to die in an Indian war) and wounded sixty-eight, at a loss of half a dozen Modoc fighting men killed or wounded."[136] Facing dwindling supplies and army scouts from the neighboring Warm Springs tribe, the Modocs eventually quit their stronghold and moved south. By June, they were either captured or had surrendered, after almost defeating the U.S. Army in one of the more memorable fights of the Indian Wars.

The government proposed using photography to document the inhospitable terrain and the participants for their official reports. Muybridge received the commission and set off in late April, through Sacramento and Yreka, for the army encampment at Tule Lake. "It is understood here that General Gillem has been directed by the War Department to have prepared photographic views of the different approaches to the Lava Beds and of Captain Jack's famous cave and fortification," the *San Francisco Chronicle* reported on April 29. "E. J. Muybridge, the well-known photographic artist of this city, who has been charged with the execution of this plan, left for the seat of war yesterday with a full battery of chemicals."[137] In about ten days, Muybridge photographed details of the recently abandoned Lava Beds and made many group portraits of the soldiers and Warm Springs scouts. To achieve the level of documentary realism and portability the assignment required, he used a stereo camera and created two significant panoramas of the landscape. His twelve-panel panorama *The Lava Beds* is a rough, highly textured, and multifaceted sequence of images showing the dark volcanic crags of lava against the sky, with the comparatively insignificant soldiers concealed in the rocks (plates 79–90). It details the intimacies and intricacies of the hidden warrens and reveals Muybridge's uncanny ability to explore the coarse, abstract beauty of light and shadow, while simultaneously documenting details of the military topography. This series, along with *Captain Jack's Cave in the Lava Beds*, was intended to assist the War Department in mapping the inhospitable terrain for current reporting and future reference, and both the aesthetic and scientific possibilities of photography are visi-

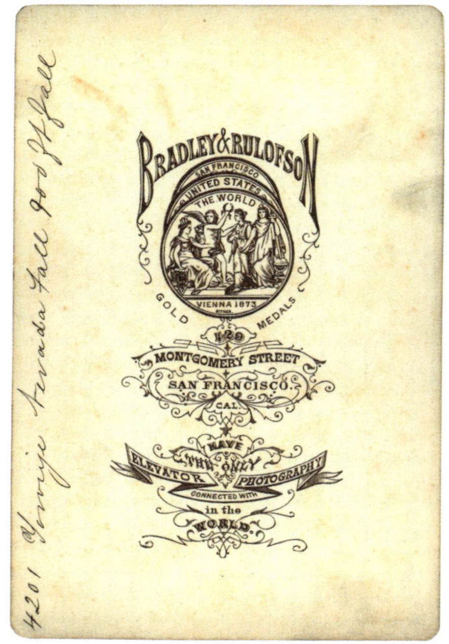

Fig. 50. *Yowiye, Nevada Fall, 600 feet fall (4201)* (verso), 1872

Fig. 51. *Captain Jack's Cave in the Lava Beds (1602)*, 1873

Fig. 52. *Panorama of Lava Beds from Signal Station at Tule Lake Camp, South (1604)*, 1873

ble in the work (fig. 51). Another more traditional five-panel panorama was shot near a signal station from a bluff high above the army camp, looking down at an alignment of tents by the lakeshore. It surveys the army hospital, General Gillem's camp and officers' quarters, soldiers' tents, interpreters' camps, a restaurant, and a store established for the troops (fig. 52, plates 94–96). Muybridge's posed portraits such as *Toby, (the squaw who warned General Canby of his impending fate), and four old Modoc squaws (1624)* and *Donald Mc Ky, the celebrated Warm Springs Indian Scout and his chief men (1623)* pictured with dignity some of the key native players in the unfolding drama (plate 91). Other group portraits are staged like school pictures, segregating the officers, soldiers, and scouts by type and lining them up to create souvenir records of their experience. These were some of the first photographs created by the military to record field operations during a war, and the Bureau of Ethnology also collected them at the time as a document of disappearing Native American cultures.

By the time Muybridge arrived at the army camp on May 2, the Modocs had already fled their hiding place in the Lava Beds, which he was then able to photograph readily. He attempted to convey a visual impression of the battle outside the encampment, even if that meant fictionalizing an image. Though he got no closer to the front lines than General Jefferson C. Davis's Tule Lake headquarters and the abandoned Lava Beds, Muybridge photographed the Warm Springs military scout Loa-kum Ar-nuk on picket duty, posed for the camera while aiming an army carbine over a pile of rocks used for cover. This picture was published on June 21 in *Harper's Weekly: A Journal of Civilization* as an engraving over the caption "Modoc Brave Lying in Wait for a Shot," along with several other engravings from Muybridge's pictures of soldiers and Warm Springs scouts (fig. 54). Bradley and Rulofson later published it as a stereograph titled *A Modoc Brave on the War Path*, 1873 (plate 78).[138] Muybridge's conscious deception—staged with the implicit cooperation of the military command—calls into question the veracity of his entire survey. His images of the Modoc War had a dual meaning; they were at once documents of the landscape and publicity pictures of the military operations, used to provide visual cues to an unpopular war that was reported like a daily serial in the news media. As the media and public began to question the competence of army leadership and the expense of pitting thousands of soldiers against a handful of Modoc insurgents, Muybridge's photographs were used by the War Department to present an image of seasoned progress in the news.

Fig. 53. *One-Eyed Dixie and Modoc Squaws (1625)*, 1873

Muybridge's project produced an archive of images that documents and maps space, resources, and people. It was not a survey, and it was hardly a journalistic representation of the war; in some ways, the Modoc War was a turning point for him, as it was for the changing culture he pictured. He diagrammed both the topography and the topology of history, a study that linked the austere homeland of the Modocs with geopolitical events happening in real time. Muybridge professed that "the wide spread and accurate knowledge of the topography of the memorable Lava Beds and the country round-about … is due chiefly to the innumerable and valuable photographs" he had taken.[139] Some of these pictures were sold commercially in the stereoscopic market, feeding the demand for timely illustrations of frontline news. Others were intended only for military use. Yet they represented for him a new way of working, and their construction counters that of his mammoth-plate pictures completed just months before. Muybridge's large views of Yosemite had been produced for the art market and function as works of art intended for exhibition. In scale and composition, they assimilate the characteristics of modern painting, compressing the illusion of deep space and perspective to focus on arrangements of light and shadow across the surface of his prints. Consequently, his Yosemite images

MODOC BRAVE LYING IN WAIT FOR A SHOT.

ON THE LOOK-OUT FOR AN ATTACK ON A PICKET STATION.

the cause of the uproar, and at once the whole camp was in commotion. Down the level plain north of the house came a cavalcade of horsemen. The steeds rushed forward at a furious rate, and soon neared the groups of spectators scattered about the premises.—"Captain JACK is captured!" shouted a sturdy sergeant. Again the valley echoed with cheers and yells, as the form of the dreaded chief was recognized among the prisoners. They were soon grouped in a field near the camp, and surrounded by a strong guard. Captain JACK is described as apparently about forty years old, with a large and well-shaped face, full of character and individuality. Even in his worn and ragged dress he looked every inch a chief, and bore himself with a proud and disdainful indifference toward his captors. He paid no attention to those who crowded about him with eager curiosity, and maintained a dogged silence when spoken to. He was mute as a statue.

Before night Captain JACK and SCHONSHIN were ironed together, and, with the other warriors, placed in a small building under guard. Neither Captain JACK nor SCHONSHIN spoke a word or manifested the slightest feeling. SCAR-FACED CHARLEY, however, protested against the indignity in behalf of his fellow-captives, and said that none of them intended to escape, even if the opportunity offered. He obtained but little satisfaction, and retired in disgust.

What to do with the captured Modocs is now an embarrassing question. It is held by the military authorities at Washington that the Modocs could not surrender as "prisoners of war" in the sense known to nations where war is declared in accordance with constituted forms. Not having been so received, they are not entitled to consideration as prisoners of war. The orders issued to

MODOC SQUAWS.

CAPTAIN JACK'S LATE STRONGHOLD.

WARM SPRING INDIANS SCOUTING.

PICTURES FROM THE LAVA BEDS.—[FROM PHOTOGRAPHS BY MUYBRIDGE, FURNISHED BY THE COURTESY OF BRADLEY & RULOFSON, SAN FRANCISCO.]

Fig. 54. *Pictures from the lava beds*, from "The Last of the Modocs," *Harper's Weekly*, June 21, 1873

mimic the flatness of the walls on which they were made to be shown, in contrast to his documentary views of the Modoc War. Art historian Rosalind Krauss wrote that, "given its function as the physical vehicle of exhibition, the gallery wall became the signifier of inclusion ... a representation of *exhibitionality*, of that which was developing as the crucial medium of exchange between patrons and artists within the changing structure of art in the nineteenth century."[140] Krauss distinguishes between the aesthetic and archival discourses of nineteenth-century photography, arguing for the importance of considering a work's function and subject in order to understand its meaning. Muybridge was clearly trained as a landscape artist, but his temperament was that of a documentarian. After 1873, he never produced another mammoth-plate view.

Flora Muybridge

Muybridge achieved a new level of fame during this time, stemming from the April 1873 announcement of his photograph of Occident trotting at full speed, the release of his Modoc War images in May, the publication of his *Catalogue of Photographic Views* by Bradley and Rulofson in June, and the medal he received at the Vienna International Exhibition that fall for his mammoth Yosemite views. At the age of forty-three, he had become one of San Francisco's most prominent artists. Two years earlier, in May 1871, Muybridge had married Flora Downs, an attractive young woman of twenty-one, originally from Alabama, who had worked with him at the Nahl Brothers Studio as a retoucher, and later as a retoucher and set-dresser at Bradley and Rulofson. C. H. Nash, a stage driver in Calistoga, remembered them together: "Mr. and Mrs. Muybridge came to the Geysers about 1873.... His whole thought seemed to be of photography. He was very nice to his wife and apparently devoted to her. M. had the appearance of a man who would stand no trifling."[141] Muybridge, who traveled constantly for his work, was protective of their relationship and grew suspicious as she developed her own interests and began to spend time with others, including a theater critic and raconteur named Harry Larkyns.

Early in 1874, Muybridge photographed Flora, who was very pregnant with their only child, in the Bradley and Rulofson studio; she stands awkwardly between an enormous bough of pears and an animal skin, which taken together symbolized the fertility and fruitfulness of California's bountiful agriculture (fig. 55).[142] Flora appears in profile wearing a white pinafore, reaching out pensively to the outsized pears, her rounded abdomen echoing the shape of the hanging fruit, which appears larger when shown next to her petite form. In mid-April their son, Floredo Helios, was born, and tragically, a few days later, Muybridge's mother, Susannah Smith Muggeridge, died. Then, in July, an exquisite portrait of Flora from a series taken by William Rulofson appeared in the *Philadelphia Photographer*, showing her poised, elegant profile sensuously illuminated against a plain dark background. The Bradley and Rulofson studio was awarded a gold medal for the series by the magazine, a pinnacle for American portraiture for the year.[143] Another portrait of Flora from the same series appears in the *Brandenburg Album of Bradley & Rulofson "Celebrities" and Muybridge Photographs*, an astounding 138-page collage book that was found in the early 1950s in a Sebastapol thrift shop. It contains celebrity portraiture, studio studies, and period chromolithographs, as well as a wide variety of more than 300 photographs by Muybridge made between 1867 and 1873. No one knows the album's origins or how it was made, and, judging from its eclectic contents and construction, it may have been the work of several people. Some speculate that Flora created it as a sort of scrapbook of her days with Muybridge, using leftover, damaged, or cut up pictures she brought home from Bradley and Rulofson. Muybridge and his wife could have worked on it together as a record of their new affiliation with the studio. The album contains images from most of Muybridge's projects over the years, from the first photographs he took in Yosemite to his more recent Pacific Coast landscapes, and it remains one of the best records of his early work. The pages are often collages of disparate images, sometimes combined with one or more studio portraits of actors and performers associated with the theater. These are regularly centered on the page, surrounded by Muybridge's photographs. One page reveals Stanford's horses at a Sacramento track (plate 75). Another shows a stereo view of Pi-Wy-Ack (Vernal

Falls) paired with a steamship docked on the San Francisco waterfront. Pasted down in a tight grid on a now brittle sheet is another startling group of pictures, juxtaposing two different stereo images of wilderness mountain lakes with two portraits of elegant urban women dressed in crinoline and fancy lace (fig. 56). One is the portrait of Flora, and the other an actress from Bradley and Rulofson's catalogue of celebrities. The bottom stereo image is titled *The Deer Slayer of Tuolomne* and shows a frontiersman, possibly one of Muybridge's assistants, posed in the foreground staring out at a lake, transfixed by the glassy view reflected in the water. He is holding a rifle, and the carcass of a deer hangs from his broad shoulders. Like a view out a window, in which the eye moves from foreground to middle to deep space at the horizon, this landscape is classically composed to function as a stereographic illusion, illustrating a literary character from the frontier fiction of James Fenimore Cooper. The mountain man may also be read as a stand-in for Muybridge himself. The top stereo pair is titled *Mount Stanford, from Lake Flora, near Summit Valley, Sierra Nevada Mountains*, an allusion to both Muybridge's patron and his wife, each personified in these wilderness landmarks. The crystalline lake surrounded by mountains, a fitting namesake for Flora Muybridge, could also be drawn from Cooper's literary image of an unspoiled American landscape. Flora appears below left in a print from the series of award-winning studio portraits of her. As a view into San Francisco's social sphere, this image positions Flora as an archetype of sophistication and refinement–a sharp contrast with the wilderness sensibility of Muybridge's mountain landscapes. Her face is shown in profile, and the light illuminates her features to carve a powerful, solid presence. A row of buttons on her dress divides the bottom third of the portrait vertically, culminating in a crucifix hanging like a weight around her neck. The ornamental severity of the small cross is broken by the gossamer decoration of white flowers pinned to her bodice. The combination of photographs on this page from the *Brandenburg Album* emphasizes the vociferous clash between the increasingly elegant culture of urban theater society and the rustic lifestyle epitomized by Cooper's frontiersman Natty Bumppo in his 1841 novel *The Deerslayer*.

Fig. 55. *Flora Muybridge*, 1874

After months of suspecting that his wife was having an affair with Harry Larkyns, on October 17, 1874, Muybridge confronted the evidence he found at the home of Floredo's nurse, Susan Smith, where he discovered a photograph he had not seen before depicting his infant son: "What is this on the back of this picture in my wife's handwriting–'Little Harry!'" he was reported to have cried.[144] Finally faced with a sign of his wife's infidelity, in a tortured rage Muybridge sought out Larkyns, her tall, striking lover, who had frequently escorted Flora around town during her husband's long absences. Muybridge ran for the four o'clock ferry from San Francisco to Vallejo, from there caught the train to Calistoga, and then hired a driver to carry him the last stretch of winding road up Mount Saint Helena under a dark new moon to the Yellow Jacket Mine, where Larkyns was then employed. "I found out where he was stopping and went to the door of the house, knocked and asked to see him," Muybridge later recounted to George Smith, a reporter for the *San Francisco Chronicle*, recalling, "As soon as he came to the door I said to him, 'My name is Muybridge; I have received a message about my wife–.'"[145] He later wrote to the reporter, confessing, "What regret I may feel, my heart alone knows."[146]

> The camera was a tool of perfect perspective, but the photographer was powerless to compose his picture. He could only, in the popular phrase, take it. Even in the studio the photographer began not with the comfortable plane of his picture but with the intractably three-dimensional stuff of the world.[147]
>
> Peter Galassi

Fig. 56. *The Brandenburg Album of Bradley & Rulofson "Celebrities" and Muybridge Photographs*, page 98, 1874

"Larkyns came into the hallway and advanced towards the front door, and when within a few feet of it, Muybridge stepped into the full light and said, 'I have brought a message from my wife, take it,' and at the same time fired. The wounds proved fatal almost instantly."[148] Accounts differ as to the details of what happened at the Yellow Jacket Mine on October 17, 1874. Based on Muybridge's pretrial interviews, trial transcripts, and post-trial news reporting, what is agreed is that Muybridge, emotionally distraught over his wife's alleged affair, murdered Larkyns and got away with it. On February 6, 1875, following a lengthy and spectacular trial, reported daily in the news, Muybridge was acquitted of all charges. One of his lawyers, William Wirt Pendegast, delivered an inspired summary, elevating the trial from one of law to one of emotional justice, demanding that the jury of "mostly old men, gray headed and hard fisted" place themselves in the shoes of the avenger.[149] Abandoning his initial plea of insanity, based in part on the injuries Muybridge received in his 1860 stagecoach accident, Pendegast appealed directly to the jury members' hearts. "Across the arch of his fireplace where once was written the words 'Home, Wife, Child, Content and Peace,' now there appears as a substitute for all, in black letters, placed there by the destroyer, the single, awful word DESOLATION," he argued. "But I do ask you to send him forth free–let him take up the thread of his broken life, and resume that profession upon which his genius has shed so much luster–the profession which is now his only love."[150] Flora sued Muybridge twice for a divorce and alimony payments–first in December 1874 (while he was jailed) and again in March 1875. The first petition was denied, but on May 1, the *San Francisco Daily Examiner* reported that "the application of Flora Muybridge for alimony from E. J. Muybridge was granted, yesterday, by Judge Wheeler to the amount of $50 per month, pending the result of the action for divorce."[151] Then, on July 18, while Muybridge was in Guatemala, Flora died tragically of a sudden illness, at the age of twenty-four, leaving Floredo to be cared for in a Catholic orphanage. In September 1876, Muybridge moved his son to the Protestant Orphan Asylum in San Francisco and was still paying his board there in mid-1878.[152]

The first photograph Muybridge took in San Francisco following his acquittal—the second picture in a protracted series he had begun in June 1874—depicts the ongoing construction of the United States Appraisers' Building. Standing in almost the exact spot as he had for the June picture, on February 10, 1875, Muybridge photographed the new government building from Jackson Street, with three well-dressed men standing stiffly in place atop the prepared foundation to provide scale and a look of official supervision (fig. 100). They are echoed by a giant illustration of a similarly dressed man in a long coat and top hat, visible on a poster plastered to the construction fence facing the street. Like his earlier architectural photographs of the U.S. Mint and City Hall, Muybridge's new project, lasting through the course of the building's fabrication, consists of an extended sequence that records material changes to the structure during several years. These images underline his interests in topographic thinking and in using photography as a component of archival documentation.

With various government surveys working throughout the West to document and map terrain, resources, and people during this time, the idea that photographers could help to build a detailed visual archive of the geographic and cultural environment was very much in the air. And some photographers brought their own sense of aesthetic experimentation into the survey process. While working in New Mexico for Lt. George Montague Wheeler's Geographical and Geological Explorations and Surveys West of the 100th Meridian in 1873, Timothy O'Sullivan, intent on documenting the evidence of an expedition through the wilderness that preceded him by almost 150 years, set up his wet-plate camera in front of a historical marker carved by Spanish colonists into the

Fig. 57. Timothy O'Sullivan, *Historic Spanish Record of the Conquest, South Side of Inscription Rock, N.M. No. 3*, 1873

sandstone of Inscription Rock at El Morro. As the leader of a military project, Wheeler was indifferent to the methods of civilian scientists such as King, Hayden, Powell, or Whitney, who led the other western surveys. He insisted on recording information that would be useful in understanding new routes for railroads, the development of economic resources, and the Indian affairs that could affect their progress. He wanted O'Sullivan's pictures to record both geological and geographical data, but he also gave O'Sullivan authority to share the experiences of the survey team and to show the western frontier as a historic and habitable territory that was already colonized.

When O'Sullivan composed his photograph *Historic Spanish Record of the Conquest, South Side of Inscription Rock, N.M. No. 3*, he centered the lines of names and the date carved into the flat, textured stone above the branching leaves of an agave. In the interest of anthropological science—and to mark the scene with his own inscription—he balanced his ruler across the bottom, establishing a grid within the frame to define space, scale, and measurement (fig. 57). It seemed a dispassionate effort, with little pictorial significance at the time beyond its evidentiary role, yet in that gesture O'Sullivan seemed to easily bridge the photographic boundaries between archives, science, and art. He took control and updated the method of inscription by injecting a modern yardstick beneath the words of the Spanish explorers who had preceded him, superimposing a rational, survey-based language that is numerical rather than linguistic. What had commemorated a moment in time—February 18, 1726, as inscribed on the stone—became, in O'Sullivan's view, an image about spatial measurement that transcended time. The picture was included in Wheeler's publication of original photographs by O'Sullivan and William Bell that surveyed his 1871, 1872, and 1873 expeditions.[153] "Makers of survey pictures had to satisfy vague and contradictory needs," writes historian Robin Kelsey. "The forces shaping new approaches came both from above, in the directives and filtering of supervisors, and from below, in the ingenuity and recalcitrance of employees. Because the pictures that resulted bore traces of conflict and accommodation, they operated as rhetorical prisms."[154] Muybridge also struggled with the disparity between the practical needs of his employers and his own creativity, and with the innate differences between his varying roles as a view artist, a documentary or survey photographer, and a scientist or inventor. Throughout the 1870s, his work began to address these issues directly.

Central America

In mid-1874, Muybridge began planning a trip sponsored by the Pacific Mail Steamship Company to photograph in Central America, a project that was long-delayed by his incarceration that fall. The completion of the transcontinental railroad changed the course of travel across the nation. Until 1869, one of the principal routes had been by coastal steamer, with a rail route crossing over the Isthmus of Panama. Once the railroad connected San Francisco and New York more directly, American corporations like Pacific Mail had to seek new economic and trade possibilities in Central America because of declining passenger and freight business. Photographic surveys would help the American public understand the foreign landscape of Panama that was inextricably tied to their own economic growth. O'Sullivan had accompanied naval commander Thomas O. Selfridge's 1870 Darién Survey Expedition to make landscape photographs during their protracted search for a canal route across the Isthmus, coming away with hundreds of glass-plate negatives and stereographs. Now the Pacific Mail Steamship Company wanted to advance their role in Central American trade and tourism to help redefine their business, which also included the transpacific route from San Francisco to Yokohama and Hong Kong.[155] Muybridge's project to photograph the major colonial cities in Panama and coffee production in Guatemala was intended to promote these new efforts.

Just weeks after his acquittal, Muybridge set off from San Francisco for Central America by steamship around February 16, 1875, traveling along the west coast of Mexico, with stops in Mazatlán and Acapulco. He apparently was brought aboard the ship under cover of darkness, secreted away as it steamed toward the Golden Gate to avoid a confrontation related to his pending divorce.[156] Calling himself Eduardo Santiago Muybridge, he arrived in Panama City in early March and quickly began photographing the architecture, landscape, and ruins of colonial churches. He also photographed the coastal region around Colón, known as Aspinwall, after William Henry Aspinwall, founder of the Pacific Mail Steamship Company. The Atlantic terminus of their shipping route from New York was located there, as was the Panama Railroad, which carried passengers across to the Pacific Coast docks at Rio Grande. Muybridge's images of crumbling churches overtaken by tropical decay and jungle growth reveal the slow effects of time passing on the colonial history of Panama. Some of these photographs have an elegiac quality, as though Muybridge hoped that by peering into the deep past he would be able to transcend his own recent suffering. *Ruins of the Church of San Domingo, Panama*, 1875, shows the white sky through the church's missing roof, framed by the giant arch of this once elegant structure (plate 104). Along the bottom of the frame are a number of peasants who stare straight at the camera. They receive no shelter from the church's decrepit, peeling edifice, whose scale nonetheless dwarfs them.

On May 1, Muybridge left Panama for Guatemala on the steamer *Honduras*, photographing briefly en route as the ship called in Costa Rica, Nicaragua, Honduras, and El Salvador. He pictured similar subjects in Guatemala City and Antigua—churches, theaters, and government buildings—before traveling into the mountains to record sweeping views of Quezaltenango from the volcanic peaks. He also made an eleven-panel panorama of Guatemala City from the top of El Cerro del Càrmen overlooking the city. The Roman Catholic Church had lost much of its political influence in 1871, when rebels overthrew the conservative government. Now the liberals who came to power assisted Muybridge in his work. In *Reception of the Artist by the Jefe Politico, Mazatenango*, 1875, Muybridge assembled the troops and military band from the garrison to construct an elaborate self-portrait in which he is welcomed by the local political leader and army officers at a ceremony in the town square. He appears in knee-high boots and broad-brim hat like a propitious Victorian explorer surveying a distant colonial landscape. The camera is positioned above ground on a balcony, looking down on the diagonal lines laid out by the group formations, recalling Muybridge's photographs of San Francisco's Presidio from the late 1860s. Working with local authorities in Quezaltenango, Guatemala City, and Panama City, Muybridge created more diverse images titled *Reception of the Artist*, all with elaborately staged assemblies of soldiers and politicians (plates 100, 110). These uncommon photographs make clear that his Central American survey was meant to represent the point of view of those in power in the colonial states, from the local politicians and military command to the Pacific Mail Steamship Company and plantation owners, whose enterprise Muybridge was visiting and promoting.

Guatemala was experiencing fast-paced political and economic changes in 1875. "Progress was the most sacred word in the Liberal vocabulary," writes historian E. Bradford Burns. "It denoted an irrepressible urge to adopt—rarely to adapt—the latest styles, ideas, inventions, and ways of life of Europe and the United States."[157] Representing the face of American investment, Muybridge ventured farther into the highlands to begin his documentary study of emerging coffee plantations. The photographs were intended to help encourage potential investment in new Guatemalan agribusiness. Cochineal, a natural component of red dye, had dominated the nation's exports for years, but the development of synthetic dyes hastened its demise. Around 1873, reform-minded president Justo Rufino Barrios—himself a coffee grower and exporter—along with other newly empowered plantation owners, proposed strengthening the coffee industry to reinvigorate and diversify the nation's international export trade. When Muybridge visited the Antigua, San Isidro, San Filipe, and Las Nubes plantations, he recorded the complete cycle of coffee production, "in all its stages, from the time the forest is cleared away to receive the seed, all along until the crop is ready to be embarked," just as he had done with wine production at the Buena Vista Vineyard.[158] These images showcase for the first time the emerging conversion of Guatemala to an international market economy based on trade, and they provide one of the earliest documentary photo-essays on such a subject. Muybridge depicted workers clearing fields and transplanting young coffee trees into a vast landscape shadowed by mist-covered volcanoes. The coffee pickers, who earned about one shilling a day, are shown climbing ladders, carrying baskets of beans on their heads, and producing traditional weavings (fig. 58). As art historian Katherine Manthorne has noted, Muybridge's Guatemalan photographs, in their portrayal of "successive stages of work," foreshadow the way his "famous later photographs of animal and human locomotion would break the movements in constituent parts."[159] Together, the coffee plantation pictures form a progressive study of agricultural production taking place over time. "The violence and social devastation underlying this all-over distribution of human beings within the lush vegetation ... is the consequence of the same imperatives of modernization which impelled Muybridge's Palo Alto experiments, which would initiate a ceaseless analysis of the ways the body and its motions could be rendered productive," concludes art historian Jonathan Crary.[160] Also apparent in these images is the social and cultural gulf between native workers and European producers.[161] *Machinery for Pulping Coffee, Las Nubes*, 1875, shows two figures separated by a maze of timbers and machinery; one indigenous worker is barefoot, almost hidden in the shad-

Fig. 58. *Coffee Harvesting at San Isidro, elevation 1000 feet, Guatemala*, 1875

ows, while a foreman stands in the light, centered and visible, overseeing the pulping process (plate 109). The image is divided geometrically, and the viewer's eye follows the diagonal bars that dominate the composition to the center. This reinforces the hierarchy of the workforce and businesslike operation of the plantation to assuage the fears of potential investors in the business. Muybridge was most interested in conveying the image of profitability in a land of promise and tranquility, a stand that upheld the colonial character of his patronage. To promote the sale of his new pictures back in Guatemala City, he announced, "I have formed a collection which includes the most notable public buildings, the most picturesque localities, and the most natural scenes of cultivation of agricultural products. All this is due to the Pacific Steamship Company which has assumed the costs of this venture for the purpose of making the riches and beauty of this interesting land known abroad."[162]

When Muybridge arrived back in San Francisco on November 27, 1875, he began printing his Central American photographs and assembled several copies of an album titled *The Pacific Coast of Central America and Mexico; The Isthmus of Panama; Guatemala; and the Cultivation and Shipment of Coffee*. This book of original prints connects his strong interest in the pictorial landscape with his skill at conveying social narratives in book form. Its title page is a photomontage of numerous Panama and Guatemala pictures spread out across the frame, introducing many of his principal subjects: landscapes, city views, churches, coffee production, and indigenous workers. It is a photograph of photographs, their corners curled and bent in a trompe-l'oeil style that includes the album's title printed on a sheet centered among the pictures, a signed studio portrait of the artist, his calling card, and another card announcing the name of his new dealer, Morse's Gallery, spelled out across an artist's palette (fig. 59). This carefully constructed introduction to the album is an image about the creative process; it presents Muybridge's pictures in a controlled jumble, before they were edited, mounted, and sequenced to create a specific narrative structure. As we turn the pages of his book, he then takes us along on a guided tour of foreign lands. But the introduction makes it clear that this journey has been manipulated and that the point of view is artificial, created by a photographer in the service of commercial interests.[163]

In May 1876, Muybridge's mammoth-plate views of Yosemite were exhibited in the Photographic Hall of the Philadelphia Centennial Exposition, along with those of Watkins, Weed, and John J. Reilly, who owned a commercial photography studio in the valley. That summer, he also exhibited in the Eleventh Industrial Exhibition at the Mechanics Institute Fair in San Francisco, presenting prints from his Central America series. They won a gold medal, with the judges commenting, "These last productions of his camera surpass all his previous efforts, and their examination renders it difficult to believe, that with our present knowledge and taste, photography can make much further progress toward absolute perfection."[164] On May 25, 1876, Muybridge wrote to the widow of Pendegast, his trial lawyer, who had died unexpectedly while he was away. "I am entirely undeserving of the thanks you so kindly proffer me," he said. "It is I who should and do thank you for permitting me to offer you so slight an acknowledgement of my lasting appreciation of the noble and disinterested generosity shown me by your late husband when I bowed down by grief and crushed with broken pride so sadly needed the support and friendship I received from him."[165] In December, he inscribed to her an artist's copy of *The Pacific Coast of Central America and Mexico.*

Fig. 59. Title page from *The Pacific Coast of Central America and Mexico; the Isthmus of Panama; Guatemala; and the Cultivation and Shipment of Coffee, illustrated by Muybridge*, 1876

The Urban Panorama, San Francisco

Muybridge's prescient series of more than ten images documenting the construction of the United States Appraisers' Building in downtown San Francisco culminated in mid-September 1877. He had photographed this building over three years, from the beginning of its construction in 1874 through its completion. Seen together, these images are about the transformation of the city through time, as they document a noteworthy alteration in the urban environment (figs. 99–106). As a historic record, commissioned by the government, they serve an archival purpose; as a sequence, they describe significant growth and change on a human scale. Muybridge engaged and expanded an analogous notion in his architectural photographs made for the Stanford family in 1878. When Leland Stanford chose to move from Sacramento to San Francisco in 1874, he began to construct a palatial new home on what is now called Nob Hill, which was becoming the most prestigious address for the city's wealthy families (fig. 60). "Leland Stanford is about to erect a large and handsome mansion on the east half of the vacant block bounded by California, Pine, Powell, and Mason streets," announced the *San Francisco Chronicle* early in 1875. "The site is one of the finest in the city, commanding a view northward past Alcatraz and the Two Brothers, eastward to the Coast Range, southward to San Jose and beyond, and southwestward to the Twin Peaks. From this point the whole city seems to be in a green basin at the feet of the observer. The depots and offices of the Central Pacific Railroad are in full view from the south windows."[166] Muybridge's photographs—probably shot in 1877 and bound into several albums made for the Stanfords the following year—depict the interior of what was then considered the largest private residence in the state, offering a virtual tour.[167] From the street, viewers are brought up a flight of stone stairs and pass through various public and private grand rooms, stopping for a moment to consider the abundant Greco-Italian frescos by scenic painter G. G. Gariboldi before heading up more stairs to take in the family's quarters with its magnificent panoramic view of the city (plates 113–116).

The narrative, picture-story quality of the *Leland Stanford's Residence Album* reveals Muybridge's intention to document the palatial dwelling from the point of view of its master. Looking down on San Francisco from near its highest point, Stanford would see the city unfold before him like a glittering testament to his own brand of capitalism. In the album, we see it as he did—a burgeoning metropolis built up over thirty years from an influx of immigrants, a rising merchant class, and the financial spoils of westward expansion fed by the gold rush. Muybridge stood in Stanford's shoes, high above the streets, looking through his eyes to create a hagiographic, seven-part partial panorama of San Francisco from three different locations on the roof and upper floors of the home. Laid out in sequence, these cityscapes constitute the climax of the family album (fig. 61). Stanford's unique view of the Central Pacific depots and the ships entering the harbor—documented here by Muybridge—allowed him to watch over, and therefore manage, the transportation networks that fueled the city's growth. In 1880, Muybridge made similar photographs of the San Francisco home and Menlo Park estate of Robert C. and Kate Birdsall Johnson, who were eccentric socialites, art collectors, philanthropists, and Stanford's friends.[168] Like the 1878

Fig. 60. *Panorama of San Francisco from California Street Hill*, 1877

Fig. 61. *View Looking South and West from the Windows of Leland Stanford's House*, 1877

album documenting Stanford's home, the *Kate and Robert Johnson Photo Album* reveals Muybridge's skill at photographing architecture and interior spaces in available light (fig. 62). This album also contains rare examples of Muybridge's sensitive portraiture, as well as some unusual spirit photographs (likely requested by Kate) made while posing family members within the lush environment of the Johnson homes (plates 119–124).

The culmination of Muybridge's interest in urban architecture and topographic documentation came in his astonishing 360-degree panoramic photographs of San Francisco, made in 1877 and 1878. The first, *Panorama of San Francisco from California Street Hill*, was shot in late June or early July 1877 from the almost completed tower capping the home of Mark Hopkins, a director, with Stanford, of the Central Pacific Railroad. Hopkins built his residence at the highest point of the city, next to and just above Stanford's house at California and Mason Streets. Panning his camera in a circle, Muybridge made eleven exposures over the course of about five hours to reveal a complete and detailed view of the city, its bustling harbor, and surrounding islands and hills (plate 117). His work began in the late morning and progressed clockwise until all eleven segments had been exposed. When he finished, the afternoon sun had moved about 90 degrees in the sky, and his first shot (panel 10 in the 1877 panorama) showed starkly different lighting effects than his last (panel 9). Like a sundial, the slow passage of time is visible as shadows move across the city. Muybridge's prints were then mounted to paper, and the eleven individual segments adhered in sequence to a single sheet of linen. This was bound, accordion style, into a quarto folio album. Hollis Frampton insightfully describes how Muybridge, in the 1877 panorama, "condenses an entire rotation of the seeing eye around the horizon (an action that must take place in time) into a simultaneity that is at once completely plausible and perfectly impossible; it is as if a work of sculpture were to be seen turned inside out, by some prodigy of topology."[169] The panorama was published by Morse's Gallery and sold with an accompanying photographic "key" that labeled and listed a number of important buildings and sites. One version of the album shows clouds printed into the sky, and another presents the sky as cloudless. Another iteration of the 1877 panorama

Fig. 62. *Robert C. Johnson in his study*, from *Kate and Robert Johnson Photograph Album*, 1880

was photographed at much the same time and mounted to a set of eleven individual boudoir cards. Marketed together in a box, these cards are more horizontal in format and share a few of the same negatives as the bound version.

The city's business leaders had situated themselves where they could look down on "the great whirling, driving, bustling city of San Francisco," as the *San Francisco Daily Evening Bulletin* called it in an August 1878 review of Muybridge's later 360-degree vista.[170] He created his panoramas in the context of previous hillside views by Fardon, Vance, Watkins, and Weed, but the new standpoint on Hopkins's tower allowed for the first time a circular scan of San Francisco that encompassed the full horizon.[171] "Created as commercial objects, they quite deliberately presented San Francisco in ways intended to flatter contemporary tastes," writes historian David Harris, speaking generally of nineteenth-century California panoramas. "As objects of civic pride, they celebrated at one and the same time the city (its hotels, public and commercial buildings, and remarkable private houses), its unique topography, and its astonishing growth."[172] Whereas previous photographers had surveyed the streets and landscape, equating topography with populace and catering to a commercial market, Muybridge was invested in his growing alliance with those who created the capital markets. He had already solved many of the substantial technical challenges of producing a 360-degree panorama in Guatemala City, producing an image about a historic, colonial capital. His San Francisco panoramas had more to do with the relative speed of people, horses, and the transport of commerce moving into and out of this growing international hub. During construction of his new home, Stanford had told the *San Francisco Chronicle*, "I shall hope to live to sit upon yonder balcony and look down upon a city embracing in itself and its suburbs a million people.... I shall look out though the Golden Gate and I shall see fleets of ocean steamers bearing the trade of India, the commerce of Asia, the traffic of the islands of the ocean."[173] Muybridge set out to survey San Francisco from the viewpoint of its dominant business leaders, placing their personal domain at the center of things, in relation to the rest of the world.

In summer 1878, Muybridge created a second, much larger panorama from the same location. Mounted to linen and bound as an album of mammoth-plate images, *Panorama of San Francisco from California Street Hill, San Francisco, California*, 1878, is one of Muybridge's greatest accomplishments (plate 118). Because of its scale and unprecedented detail, it remains an unparalleled document of San Francisco before the 1906 earthquake changed its face forever. To resolve some of the practical issues he had encountered the previous year, Muybridge this time shot thirteen vertical mammoth-plate glass negatives, again encircling the 360-degree view of the horizon from Hopkins's tower. The focal length of his lens and the size of the 23-by-16-inch mammoth plates allowed him to divide the entire scene into thirteen equal parts. The vertical format also served to flatten and abstract the horizontal space; the viewer becomes disoriented as the horizon and visual perspective are less integrated within the whole long image. It is difficult to take in the entire panorama at once. Just as one would move around to view the actual scene, this panorama requires physical movement of the body to decode its complexity, especially when completely unfolded to its full length of more than seventeen feet.

In both size and technique, this 1878 panorama exquisitely details the urban landscape over an extended period; again, Muybridge exposed his plates in about five hours under mid-summer light.[174] When viewing these images, the eye moves from the residential architecture of the foreground, across city streets to the commercial buildings of the downtown, then out across San Francisco Bay to the hills that ring the city on three sides. Because of the long exposures required, however, little movement is seen on the streets; much of the activity is rendered invisible as people and horses cross the focal plane of Muybridge's giant camera without being fixed to film (fig. 63). Even though the images are static and highly detailed, one senses an unfolding narrative when viewing the complete panorama, as if there is a beginning and an end that can be read like an open-ended story. As the viewer physically moves across the vast horizon of the panorama, it is easy to become lost or disconnected from the streets, as if rambling at will through the great conurbation of San Francisco without a destination in mind. In this way, Muybridge touched upon the themes of movement and time passing as the viewer wanders through the picture—from street to street and frame to frame—across the entire urban and rural expanse he documented. "The panorama's realism hinged on the notion that, to capture life, a display had to reproduce it as bodily, and not merely as visual experience," writes historian Vanessa Schwartz.[175]

By connecting the visual links between the space and historic circumstances of Muybridge's ocular view, the 1878 panorama became an influential experiment in sequencing, narrative construction, and motion. As a photographic sequence depicting a three-dimensional environment presented as a flat, linear object, its beginning tacitly connects to its end, compiling the circular panorama conceptually rather than spatially. By transcending the traditional presentation of landscape space as a static, illusionistic view—governed by the rules of perspective and framed to be hung on a wall—Muybridge embraced the notion that time itself could be the subject of art. He understood that motion could be halted, taken apart, analyzed, and reassembled with photography, a concept that would lead him directly from his masterful landscape practice into new and experimental territory as he returned to Stanford's challenge of picturing movement as a matrix of scientific data.

Fig. 63. *Panorama of San Francisco from California Street Hill, San Francisco, California* (detail), 1878

The Horse in Motion, Palo Alto

Muybridge was working on the heels of other photographers who had slowly developed the potential of photography to stop movement and animate it in sequences of pictures. This was referred to as instantaneous photography.[176] The greatest barrier photographers faced from the 1850s through the 1870s was the long exposure time needed to obtain a detailed image on a collodion emulsion. In the 1850s, André-Adolphe-Eugène Disdéri produced images of people and animals interacting with each other while moving, and Antoine Claudet created a series of sequential self-portraits showing the illusion of action through time. In 1853, Claudet applied to patent a stereo viewer for "pleasing and novel optical illusions..., which are made moveable so as to impart to the picture the appearance of moving figures."[177] During the 1860s, Nadar (Gaspard-Félix Tournachon) shot and printed portraits in series, and American photographer, publisher, and equipment manufacturer Edward Anthony published numerous instantaneous photographs that stopped motion on city streets; his journal, *Anthony's Photographic Bulletin*, was a significant source of information about emerging improvements in photography, cameras, and chemistry. Sir Charles Wheatstone developed a stereoscope around 1870 that rotated stereographs of posed sequences around a drum so they passed quickly in front of the twin lenses, approximating movement in three dimensions. Oscar Gustave Rejlander photographed a variety of individual emotions and expressions of people for Charles Darwin's book *The Expression of the Emotions in Man and Animals*, published in 1872. The faces of children, frozen in anger or fear, were presented in grids of photographs showing how expressions changed based on feelings. Models were portrayed sneering or engaged in acts of defiance. In 1873, Rejlander published an essay titled "On Photographing Horses," in which he proposed using "a battery of cameras and 'quick-acting lenses' ready charged and loaded" to create a sequence of images depicting motion.[178] The same year, Francis Galton experimented by superimposing portraits of criminals on one plate to form composite images intended to distill information about shared characteristics among human types. Astronomers Pierre Jules César Janssen and Achille Cazin photographed successive images of the planet Venus crossing the path of the sun in 1874, and the same year, an English translation of Étienne-Jules Marey's *La machine animale: Locomotion terrestre et aérienne* appeared as *Animal Mechanism*. Marey introduced the public to the practical importance of locomotion in the science of movement and the study of flight. Though he had not yet begun to use photography in his experiments, Marey proposed that "the contemplation of the laws of nature has been a great and noble source of enjoyment to those who have discovered them."[179]

Muybridge appears to have recommenced his experiments in photographing horses sometime in 1876 (fig. 64). The illustrator Thomas Kirby Van Zandt produced a sketch dated September 16, 1876, that seems to be drawn from a Muybridge photograph of Stanford's trotter Abe Edgington racing with all of its feet elevated from the ground. In February 1877, Van Zandt painted a more finished canvas from the sketch. "Both of these works reveal the difficulty Muybridge had producing publishable works during this period," writes historian Phillip Prodger. "Although he was able to obtain results, he struggled to make images suitable for public consumption."[180] It was an accepted practice then to reproduce photographs in alternative graphic media such as engravings, lithographs, chromolithographs, and paintings. Currier and Ives had reproduced Muybridge's 1873 image of Occident; this lithograph, titled *The California Wonder Occident. Owned by Gov. L. Stanford*, is the only surviving version of that picture (fig. 65).[181] A second Currier and Ives lithograph of Occident, presumably drawn from a newer Muybridge photograph, was produced in 1876. On November 23 that year, Muybridge paid thirty dollars to J. Mansur for "a lot of lenses and other photographic material sold him this day."[182] He seemed to be preparing for a major project that would use multiple cameras at the same time. Then on May 21, 1877, Muybridge wrote to Stanford's assistant Alfred Poett, explaining, "I saw Governor Stanford a few days ago, and he told me, you could arrange all about photographing the horses."[183] Working either in Sacramento or in San Francisco in July, Muybridge again photographed Occident trotting a two-and-a-half-minute mile over the track. One eyewitness account, given years later to Stanford University historian Walter Miles by businessman Sherman Blake, goes as follows: "I was the boy employed by the late George D. Morse ... who carried part of the equipment and assisted Mr. Muybridge in photographing Governor Stanford's horses at the old Bay District Track.... It was necessary to improvise a temporary darkroom ... and we took along with us in an express wagon, a heavy orange and red cloth tent, and an improvised ruby light containing a candle."[184] The Bay District Track, located

Fig. 64. Unknown photographer (possibly Eadweard Muybridge), *Harness Racing*, ca. 1876

Fig. 65. Currier & Ives, *The California Wonder Occident. Owned by Gov. L. Stanford*, 1873

then just north of Golden Gate Park in San Francisco, may have been an interim location for Muybridge's project since Stanford was still completing his move from Sacramento. Muybridge's best attempt to stop motion during the July 1877 session, however, was so indistinct that it again had to be copied, this time in watercolor and gouache by John Koch, a retoucher working in George Morse's studio. Koch's painting was then recopied with a camera to produce the image that became Muybridge's first published photograph in his celebrated *Horse in Motion* series (fig. 67). The only element left from his original photograph is the face of the sulky driver, James Tenant, which was cut out and pasted collagelike onto the painting in his one nod to maintaining the verisimilitude of the original (fig. 66). Muybridge had regularly used his camera to copy paintings, graphics, and documents as part of his business. That he went to such lengths, however, to construct an image approximating the photographic record of Occident racing reflects not only the limited technology of the mid-1870s and his inability to fix a clear, reproducible image of the horse on film, but also the prevailing attitudes about photographic reproduction at the time. The public regularly accepted significant retouching and manipulation of negatives as integral to the creation of photographs; they could be reproduced in the form of engravings or paintings, and printing clouds and objects into otherwise realistic views was customary for some photographers in the United States and Europe. To date, many attempts to stop motion had been posed. Cinema historian Tom Gunning comments that such a blatant manipulation of an indistinct image by Muybridge conveyed "a different attitude towards the means of making photographs public, particularly in an era when mass reproduction of photographs through the halftone process of printing was not yet available."[185]

At the same time that he was completing publication of his small panoramas, Muybridge began to promote his new work with Stanford as a breakthrough in the development of photographic technology. He had in fact labored for years to develop the chemistry and new shutters that finally allowed him to freeze the horse in motion. During his long voyage home from Panama in 1875, working

Fig. 66. John Koch, *Occident Trotting*, 1877

Fig. 67. *"Occident," Owned by Leland Stanford; trotting at a 2:30 gait over the Sacramento track, in July*, 1877

to decrease the exposure time of his photographs to a fraction of a second, he had tested new chemical formulas and shutters to produce plates that could render an image with less light coming through the lens. "These experiments resulted in the construction of an apparatus and the preparation of chemicals so as to permit the photographing in outline of a rapidly moving body," he explained to a reporter for the *San Francisco Bulletin* on August 3, 1877, just as he released his new picture of Occident to the press.[186] "Mr. Muybridge sends us a copy of an instantaneous photograph of 'Occident,' taken when he was trotting at a speed of 36 feet per second, or a mile in 2 minutes and 27 seconds," announced the *Daily Alta California* on August 3. "The negative was exposed to the light less than one-thousandth part of a second, so brief a time that the horse did not move a quarter of an inch."[187] Muybridge wrote to the paper's publisher Fred McCrellish that "the length of exposure can be pretty accurately determined by the fact that the whip in the driver's hand did not move the distance of its diameter."[188] He submitted the mounted photograph of Occident to the Library of Congress for copyright on August 11, and wrote on the back of the card, "The exposure of the negative was less than one-thousandth part of a second, and made while the horse was passing about forty feet distant, at a rate of thirty-six feet in a second. The details have been retouched by Koch, for the purpose of giving them a better effect (as is usual with all first class photographs): in every other respect, even to the whip of the driver, and the dust flying from the foot of the horse, the photograph is precisely as it was made in the camera."[189] In November, Muybridge was again awarded a medal at the Twelfth Industrial Exhibition of the Mechanics Institute, this time for his "instantaneous photograph of the race-horse 'Occident.'"[190] The widespread recognition he received provided enough of a boost to encourage a more systematic study of the horse's gait, which was Stanford's announced goal.

Stanford was interested in learning what made one trotting horse run faster than another. He also understood from his advisor, San Francisco stable owner

Harris Covey, that breeding his own horses as thoroughbreds could allow him to study the effect of lineage on their success at racing, and consequently to improve their sport. To accomplish this, Stanford began to develop a giant stock farm on 8,000 acres he purchased at Palo Alto, about thirty-five miles south of San Francisco, where his horses were treated with unusual respect. Muybridge and Stanford began to focus their plans to design a track and studio to enable a systematic photographic study of the horses' movements. The next step was to photograph short sequences of locomotion, which would first require twelve cameras; fast, state-of-the-art lenses; and a new system of tripping the shutters to record the subjects as they moved across the track. With Stanford's unwavering support, Muybridge acquired new Scovill view cameras from New York and Dallmeyer stereo lenses from London. Stanford enlisted help from his engineers in the Oakland office of the Central Pacific Railroad, where the superintendent of bridges and buildings, Arthur Brown, began working with Muybridge in early August 1877 to improve his shutters so they would be capable of shooting in sequence and could be easily triggered by a running horse. The arrangement consisted of two spring-loaded wooden slats that moved rapidly across each other in a frame positioned in front of the twin lenses, allowing light to pass between the slats for a fraction of a second. Brown assigned the young engineer John Isaacs to work on a triggering device for the shutter, and he quickly realized that electric circuitry would be fast, strong, and simple enough to connect a camera to a moving subject. Isaacs recalled how he "thought of putting a magnet on the outside of the apparatus, connecting that magnet with a battery on one side and with a wire which should reach over the track with the other; also connecting the battery with another wire which should run over the track and lie parallel with the first wire."[191] In September 1877, Muybridge again corresponded with Poett, stating, "I have written to the Governor at Palo Alto, suggesting that we make the series of photos illustrating the stride of a trotting horse, at Palo Alto, instead of at Sacramento.... If the Governor decides to have me come to Palo Alto, I would like to see you before you lay out a track, as there are many things to be considered necessary for perfect success."[192] On October 11, Muybridge told Poett, "I would like to see you about the track, etc., and show you the apparatus and electrical machines I am having constructed, and which are very nearly finished."[193] Muybridge was also working with the San Francisco Telegraph Supply Company, which specialized in electrical machinery and signaling equipment, to construct his new shutter mechanisms, and on November 3 he paid $130 for "electrical [work] on photo apparatus."[194] The new Palo Alto studio was completed by May 1878, and Muybridge made his first experiments with sequential photographs at that time.[195]

When reporters were ushered onto the Palo Alto racetrack by Alfred Poett on June 15, they were greeted by Muybridge and given the opportunity to witness an organized demonstration later described as "the successful accomplishment of a feat in photography which has never been excelled, and which marks an era in the art."[196] The track was covered with powdered white lime to reflect as much sunlight as possible back onto the subjects. On one side of the track stood a large wooden screen angled up about 70 degrees and covered with white cotton cloth (fig. 68). Vertical lines were drawn on the screen twenty-one inches apart and numbered from one to twenty across the top. Across the bottom was another low, white, angled screen with horizontal lines drawn four inches apart to measure the height of the horses' hooves as they ran. "In front of this base-board a strip of wood was fastened to the ground, upon the top of which wires were secured at an elevation of about an inch above the ground and extending across the track," explained Stanford's old friend J. D. B. Stillman, who detailed the process in his book *The Horse in Motion.* "The wire was exposed in a groove to one only of the wheels of the sulky, being protected from contact with the horse's feet and the other wheel. Each wire was held in proper tension by a spring on the back of the base-board, so arranged that when the wire crossing the track was depressed by the wheel it should draw upon the spring connected with it, and make contact with a metallic button and complete the electric circuit."[197] Opposite the screen was a low shed with an opening in front that held Muybridge's twelve cameras, fixed with high-quality stereo lenses, placed twenty-one inches apart on a low bench (plate 127, fig. 69).[198] The electric double-slat shutters were placed in frames in front of the lenses (plate 128). Aligned

Fig. 68. *Detail of the Track at Leland Stanford's Horse Farm in Palo Alto*, 1878–79

with the lines drawn on the screen, the wires crossed the track every twenty-one inches and connected to the shutters. Stillman described the arrangement of the slat shutters as "one above and the other below the opening through which light was admitted to the lens, and held by india-rubber springs, constructed in the form of a ring, with a lifting power of one hundred pounds, and secured by latches, to be liberated on the completion of an electric circuit."[199] Situated on a shelf to the side, a bank of batteries powered the whole apparatus. The corresponding shutter was tripped the instant a sulky wheel crossed one of the wires to complete the electric circuit.

As Abe Edgington approached the course, a reporter for the *Morning Call* described his run in the language of a racing authority: "He is surely trotting better than a 2:20 gait and without a 'bobble' he enters on the portion of the track which has been made so white. Every motion is square; no single-footing, hitching, or tendency to pace; it is as firm as a trot can be imagined. The wheel strikes the first wire, or rather, almost continuous in the roll of sound, as the wheel touches other wires, and in a trifle over half a second the twelve pictures are registered." To ensure there would be no question about the authenticity of the photographs he made that day, Muybridge developed his plates in front of the reporters, completing the intricate technological demonstration before their eyes. The following day, the *Morning Call* reporter concluded, "There is a feeling of awe in the mind of the beholder, as he looks at the glass plate which is held before the yellow curtain, and he sees the miniature of the nying [*sic*] horse so perfect that it startles him.... It is a new era in photography and instantaneous is no longer a misnomer" (fig. 72).[200]

For horses running without a sulky to trip the shutter, Muybridge used an array of thin black threads strung across the track at breast height, as he had with the single-camera photographs made the prior year. The threads were broken in sequence to complete the circuits as the horse ran through. These also had a tendency to startle the horses mid-stride, so when the experiments were taken up again the following year, this method was supplemented with a clockwork mechanism that automatically tripped the shutters in sequence.[201] On June 17, 1878, Muybridge applied for a U.S. patent on the new camera shutter and its electro-mechanical release, describing it as a "method and apparatus for photographing objects in motion" (fig. 70).[202] On June 21, he paid the San Francisco Telegraph Supply Company for their labor on electrical wiring of the camera shutters at the Palo Alto track.[203] In August, the journal *Resources of California*

Fig. 69. Unknown photographer, *Interior of the Camera Shed at Stanford's Horse Farm in Palo Alto*, no date

credited the work as a collaborative process by announcing, "The cameras with which this series of elaborate experiments were conducted were made by the Scovill Company, of New York. The shutters by Arthur Brown, of the C. P. R. R. Company. Mr. Tiffany, of the San Francisco Telegraph Supply Company, manufactured the electrical apparatus, and to Messrs. Alfred Poett and Harris Covey were due the admirable arrangements for the reception of the guests, and success of the experiments."[204]

Following the June 15 press demonstration, the response was immediate and pervasive; Muybridge and Stanford were initially showered with praise for revolutionizing the technology of photography, but their work soon engendered a heated debate about the relative merits of Muybridge's mechanized representation of motion. As Muybridge began to capture publishable images of rapid movement on wet-collodion plates, his renderings of Stanford's horses in motion were silhouetted against a bright white background, yielding a distinct outline of animal forms but very little detail. Combined with their innovative decision to depict movement in sequences of still photographs—one following another like the words of a sentence to describe a brief span of time in fragmentary moments—their technique presented a technological breakthrough compared to previous work by Muybridge or others involved with instantaneous photography. Not only did the sequences of photographs dissect rapid motion, they also presented a mechanical, scientific view of action the human eye simply could not decode on its own. The traditional interactions between photographer and subject shifted to include as part of the relationship an electro-mechanical apparatus, which took control of the artist's gaze during the process. Given the mechanical nature of the photographic set-up—Central Pacific engineers had assisted Muybridge in integrating new technologies, especially electrical circuitry—the horses were presented like machines. The framework had become industrial, which lent efficacy to the scientific aspects of Stanford's study but called into question the artistic role of the photographer in the work.

Yet, the project required a complicated mix of art and science, since the problems it addressed were about both representation and the workings of perception. Muybridge's undisputed mastery of photographic seeing and his ongoing ability to adapt technology to solve complex visual problems made him an ideal partner in the study. As an artist, he had already worked with Stanford to convey his powerful position through his previous photographs of Occident and his innovative panoramas from Stanford's home. As a documentarian, he also understood how a systematic archive of startling photographs would construct for the public a transcendent mix of aesthetic and scientific data. Yet it was Stanford's agenda for, and patronage of, this process that established the major characteristics of Muybridge's new series. For the railroad magnate, the implied equation between the mechanics of industry and that of nature linked the speed and power of the locomotive to the gait and beauty of the racehorse. Stanford's success at surmounting nature to engineer a transcontinental route over impossible terrain—in the process transforming public perception of both space and

Fig. 70. *Patent Model for Method and Apparatus for Photographing Objects in Motion*, 1879

time—would translate well into a new photographic symbol of industrialism. And the reimaging of the horse as a machine—the animal force that literally built and defended the nation was then being supplanted by new technologies such as the locomotive—would elicit a powerful metaphor for the escalating transportation markets and accompanying industries that Stanford controlled. As historian John Ott suggests, "Leland Stanford and his advocates publically staged and disseminated these photographs in order to consolidate, promote, and naturalize the developments of industrial capitalism."[205] Even though these experiments were ostensibly proposed by Stanford to study the mechanics of locomotion, they presented abundant opportunities to publicly project an image of benevolent authority and influence. His public relations machinery was running overtime.

For his part, Muybridge was assiduous in promoting his new work. His 1878 sessions resulted in the publication of a series of cabinet cards called *The Horse in Motion*, which included *"Occident" Trotting at a 2:30 Gait*, *"Abe Edgington" Trotting at a 2:24 Gait* (plate 129), *"Mahomet" Cantering at an 8 Minute Gait*, and *"Sallie Gardner" Running at a 1:40 Gait*. As individual grids, the sequences read like comic-book stories describing action taking place within a fragment of time. Seen together, the cards took on the weight of Stanford's greater aspiration to encapsulate the speed of modernity within the conceptual framework of motion itself. The cards, showing sequences of six or twelve images, were quickly sent to newspapers and magazines and were distributed around the country and internationally. Scientific and descriptive details of the horses' gaits were printed on the back of each card (fig. 71).

As news of the pictures spread, Muybridge began to give illustrated lectures, using lantern slides to describe his work, and periodicals such as the *Photographic News* in London began to publish some of the images. The awkward positions of the horses' legs documented in the photographs defied expectations and confirmed that artists had drawn them incorrectly for centuries. Some critics discredited the pictures, calling them impossible or cruel. Muybridge's former dealer William Rulofson wrote to the *Philadelphia Photographer* in August, saying of this new work, "photographically speaking, it is bosh."[206] Others praised the work, calling it a photographic revolution. In October 1878, *Scientific American* published a sequence of detailed drawings from the photographs, writing, "The most careless observers of these figures will not fail to notice that the conventional figure of a trotting horse in motion does not appear in any of them, nor anything like it.... We would suggest that for popular use the photographs should also be mounted on strips for use in the zoetrope."[207] Then in December, the Parisian journal *La Nature* published five of *The Horse in Motion* sequences, which caught the attention of Marey, who had also studied equine locomotion. In 1874, his book *Animal Mechanism* had inspired Stanford to press on with his own research. Through Gaston Tissandier, editor of *La Nature*, Marey wrote to Muybridge, asking for assistance "in the solution of certain physiological problems, so difficult to resolve by other methods. For instance, on the question of the flight of birds, I was dreaming of a kind of photographic gun, to seize the bird in a pose or, even better, in a series of poses marking the successive phases of the movement of its wings.... It is clearly an easy experiment for Mr. Muybridge. And then what beautiful zoetropes he will be able to give us: in them we will be able to see all imaginable animals in their true paces; it will be animated zoology."[208] Muybridge responded through Tissandier, affirming Marey's comments and acknowledging the French scientist's influence: "Would you please assure Professor Marey of my high regard and tell him that the reading of his famous work on animal mechanism inspired Governor Stanford with the first idea of the possibility of resolving the problem of locomotion with the help of photography."[209] Marey's suggestion of animating Muybridge's photographs with a zoetrope would ultimately enrich Muybridge's vision for the use of sequential photographs.

On January 4, 1879, *La Nature* published a diagram and description of a pneumatic clock Muybridge had devised. Work also began around this time on improvements to the Palo Alto studio. Muybridge wrote to Poett, "Do not for-

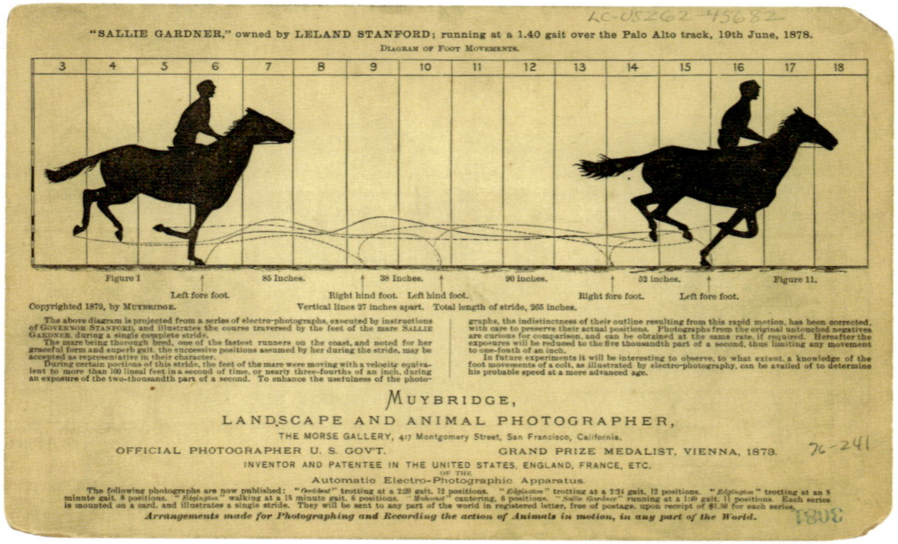

Fig. 71. *The Horse in Motion, illus. by Muybridge. "Sallie Gardner," owned by Leland Stanford, running at a 1:40 gait over the Palo Alto track, 19th June 1878: 2 frames showing diagram of foot movements* (verso), 1878

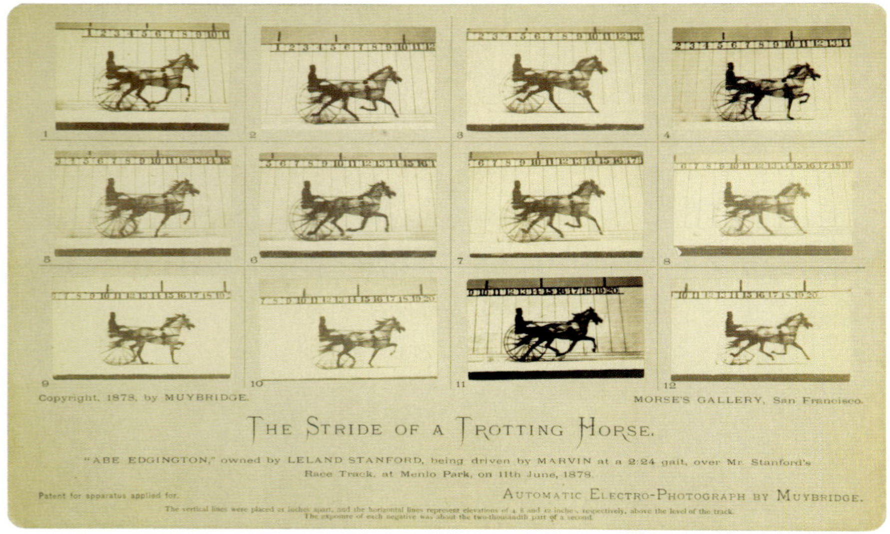

Fig. 72. *The Stride of the Trotting Horse, "Abe Edgington," owned by Leland Stanford; being driven by Marvin, trotting at a 2:24 gait over Mr. Stanford's Race Track at Menlo Park, on 11th June, 1878,* 1878

get to put some of your men during their leisure hours to work upon the new photographical track.... I am very busy now, making some enlargements of rocks with the microscopic polarized light and unpolarized, a very delicate operation.... I have received a copy of 'La Nature', a French Scientific Journal, with reproductions of the horses, in which I am designated 'L eminent Savant Americain.'"[210] Muybridge used the enhanced Palo Alto studio to introduce a more complex visual scheme in his work. The visible numbers that corresponded to each camera and wire were moved from the top of the background to the bottom (fig. 73). By May, he had begun using a battery of up to twenty-four cameras to photograph an array of Stanford's best horses in various gaits, from running, trotting, cantering, walking, and ambling to irregular movements like leaping, pacing, backing, and combinations of actions (fig. 74).[211] He also photographed other animals such as dogs, oxen, cows, deer, goats, and even a boar, using the clockwork mechanism to time the shutters as the subject passed by the camera shed (plate 132). In August, a group of athletes from San Francisco's Olympic Club was invited to perform a variety of physical actions

Fig. 73. *Detail of the Track at Leland Stanford's Horse Farm in Palo Alto*, 1879

for the camera. "Mr. Muybridge, the photographer, had every arrangement made at the race track for carrying out the work," reported the *San Francisco Chronicle* at the time, "and from ten o'clock in the morning until four o'clock in the afternoon boxing, wrestling, fencing, jumping, and tumbling followed in quick succession, and all their intricate movements were instantaneously and exactly pictured" (plate 134).[212] Muybridge even photographed himself nude in a curious progression of sixteen shots, standing against the white screen, playing a sort of allegorical laborer swinging a pickax. Wielding one of the tools used by miners and the men who built the railroad, Muybridge equated his own work with the manual labor used to fuel the industrial development of California during the 1850s and 1860s (plate 136). He also set up banks of cameras outside the shed to document a single moment of time from up to five radically different angles. This series, termed "studies of foreshortenings," depicts a variety of Stanford's horses going through a range of paces such as trotting, running, and leaping (fig. 75). In this case, they were photographed at the exact same time from various camera positions along an arc of about 120-degrees, providing a new set of data, most useful to artists, about how horses appear when viewed from different angles.

Muybridge's 1878–79 experiments were the first photographs to record such specific information about the attitudes of animals and people in motion. With this work, he built upon the allegory of photographic time and space that O'Sullivan had established with his yardstick in 1873. O'Sullivan inscribed the photograph with his own measuring system not only to document scale but also to reference the relationship between scale and the time that had elapsed since the Spanish explorers stood on that same spot to record their mark—the history evinced between the "historic record" and the present moment. He engendered a new construct that symbolically merged the time his photograph was made with the space transcribed within the frame. In his first successful work about motion, Muybridge created an analogous spatial grid of vertical and horizontal lines against which the time-based movements of his subjects were plotted. This grid of evenly spaced and numbered vertical lines, intersecting with the horizontal rules across the bottom of the frame, was designed

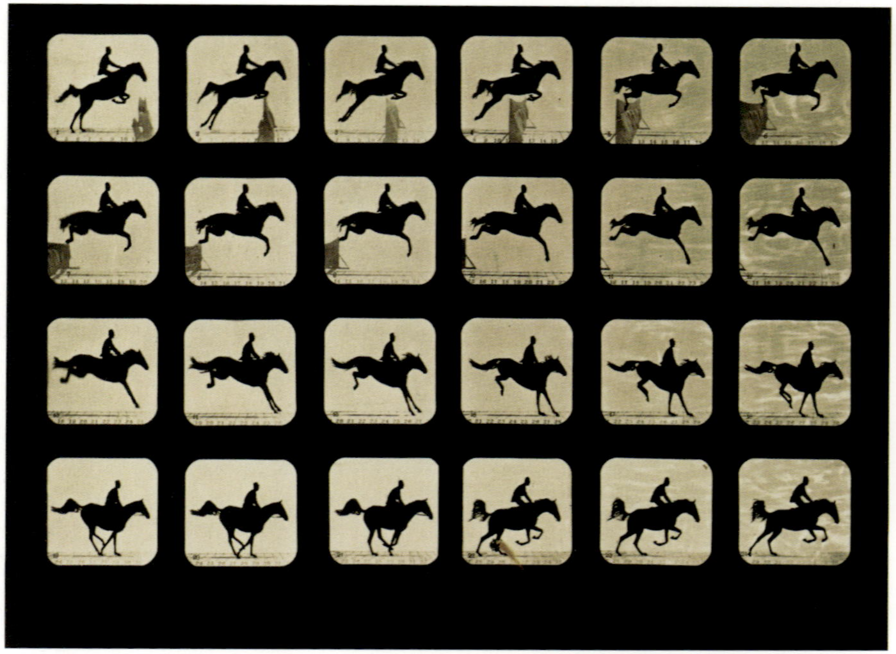

Fig. 74. *Horses. Leaping a 3 Ft. 6 Hurdle. Phryne L. Plate 49*, 1879, from *The Attitudes of Animals in Motion*, 1881

specifically to mark the space so as to enable a scientific and visual interpretation of how his subjects moved through a defined fragment of space and time. Muybridge's strategy of presenting photographic information against the backdrop of a numbered grid was helpful in a variety of ways. The photographer used it to order his myriad images, and to mirror the grid structure of individual pictures in his final presentation of an entire sequence. It also allowed others to accurately study the movements in great detail, which had been Stanford's motivation from the start. The grid was a visual device to quantify and calibrate information, but the conceptual nature of this work—its ability synthesize movement from a regular set of constituent parts—was also apparent to many who saw it (plate 126). "This interaction between the fixity of the moment and the mobility of time—the imputation of continuous movement across a chain of discrete moments—becomes clear in the precinematic motion studies of

Fig. 75. *Studies of Foreshortenings. Horses. Running. Mahomet. Plates 143–144*, 1879, from *The Attitudes of Animals in Motion*, 1881

Eadweard Muybridge and Étienne-Jules Marey in the 1870s and 1880s," concludes historian Leo Charney.[213]

Marey quickly envisioned Muybridge's technique as an improved tool for examining the locomotion of a bird in flight. Philadelphia-based painter Thomas Eakins, who taught at the Pennsylvania Academy of the Fine Arts, began using Muybridge's *Horse in Motion* images and "took them up for examination" to help him depict proper trajectories of equine movement in his own art.[214] "The horse played a major role in pictorial art, and his anatomy and muscular actions were subjects of scientific and artistic study second only to those of man," explains art historian Lloyd Goodrich. "In the 1870s and 1880s the traditional representation of the gallop, with all four legs extended like a hobbyhorse, was giving way to scientific investigation."[215] Philadelphia horse enthusiast and art patron Fairman Rogers constructed a zoetrope, and Eakins made intricate drawings from Muybridge's motion studies to animate the pictures using Rogers's machine. "To do this, Mr. Eakins plotted carefully, with due attention to all the conditions of the problem, the successive positions of the photographs and constructed, most ingeniously, the trajectories, or paths, of a number of points of the animal," said Rogers. "Although familiarity with the anatomical construction of the animal enables the artist thus to draw each of the twelve positions or attitudes, and when the figures thus made are put into the zoetrope, a perfect representation of the motion is obtained."[216] For Eakins, Muybridge's grid was inaccurate since the lines drawn on the background screen were some distance behind the horse, and he suggested a different system to Muybridge. Eakins wrote to the photographer in May 1879, asking, "Do you not find that in your old way the perspective [is] very troublesome. The lines being further off than the horse, a calculation was necessary to establish the size of your measure."[217] Eakins's painting *A May Morning in the Park (The Fairman Rogers Four-in-Hand)*, 1879–80, is one of the very first to use Muybridge's studies as a model for the correct visual representation of horses in motion (fig. 76).

Fig. 76. Thomas Eakins, *A May Morning in the Park (The Fairman Rogers Four-in-Hand)*, 1879–80

Motion Pictures

In 1879, Muybridge began to experiment with animating his own photographs, effectively combining the images to create the illusion of movement. He and Stanford had already used a zoetrope for this, and Muybridge eventually produced zoetrope strips of printed silhouettes from a variety of his pictures, which he made available for sale. The results, however, were not altogether adequate. In June, a writer for *The Field* in London described the outcome they had achieved with a praxinoscope: "It is obvious that if the figures in the ten stages in the stride of the racer could be as it were fused together, so as to be seen in rapid succession, the appearance of the horse in action should, if they are correct, be reproduced. ... We tried the effect of placing the photos in an improved Wheel of Life called the Praxinoscope, constructed with revolving mirrors. The effect was simply marvelous. On revolving the instrument the dead grotesque figures, that have been derided by so many as impossible absurdities, started into life, and such a perfect representation of racehorses at full speed as was never before witnessed was immediately visible."[218] Invented in 1877 by Charles-Émile Reynaud in France, the praxinoscope worked like a zoetrope without slots in its side for viewing. Mirrors were positioned around an inner cylinder, which reflected a strip of animated pictures as the machine was turned. Its image was brighter and more continuous than that produced by a zoetrope. Muybridge's first experiments at synthesizing motion used two stereo halves from his Palo Alto photographs, each placed in a separate zoetrope. "These two zoetropes were geared, and caused to revolve at the same rate of speed," he later wrote. "The respective halves of the stereographs were made simultaneously visible, by means of mirrors—arranged on the principle of Wheatstone's reflecting stereoscope—successively and intermittently, through the perforations in the cylinders of the instruments, with the result of a very satisfactory reproduction of an apparently solid miniature horse trotting, and another galloping."[219] Then, in another stroke of conceptual inventiveness, Muybridge modified a lantern slide projector to operate with a spinning glass disc displaying his images in succession in order to project them in motion onto a wall.

Muybridge's new machine, which he first called a "zoogyroscope," worked on the principle of the phenakistoscope, an optical toy developed independently in 1832 by Joseph Plateau in Belgium and Simon von Stampfer in Austria. Plateau, who had published his work on persistence of vision, created his device from a slotted disc. A number of individual drawings were sequenced around the perimeter of the disc to depict a simple action. When a viewer rotated the disc while looking at the drawings in a mirror through the slots, the reflected picture would appear to be in motion. Muybridge's projector, developed in 1879 and finally named "zoopraxiscope," which means animal action viewer, shined a quick succession of his instantaneous photographs onto a wall, one after the other, to animate them in a short narrative at most a few seconds long (fig. 77). While Muybridge's moving pictures have roots in experiments by others, they were likely the first animated projections to be derived from photographs.[220] Unlike all previous devices, the zoopraxiscope represented original action captured by a battery of cameras. Like the phenakistoscope, the zoopraxiscope worked on the theory of persistence of vision to create the illusion of movement. A series of photographs were used; they had to be hand-drawn by an artist directly onto the perimeter of a transparent glass disc (fig. 78). The disc was mounted behind the lens of an adapted lantern slide projector and rotated with an elaborate, geared cranking mechanism. The shutter—a black metal disc the same size as the transparent one with slots cut to correspond with each picture—rotated in the opposite direction. Images flashed onto the wall in sequence through the spinning slots, which acted like a strobe. The pictures had to be distorted to compensate for the deformation created by projecting them through a spinning shutter. At first they were drawn rather than copied photographically so they could be elongated. Later, Muybridge said he printed them onto flexible film that was "conically bent inwards, and inclined at the necessary angle from the lens of the copying camera to ensure the required horizontal elongation of the animal."[221] His early 16-inch discs, used from 1879 to 1892, show no evidence of photographic production techniques—they are outlined by hand.[222] His later 12-inch discs made in 1893 for display at the World's Columbian Exposition were created from glass photographic negatives (fig. 79) and often hand tinted in bright colors; these may have been produced from flexible film to produce the needed distortions.

Fig. 77. *Zoopraxiscope*, 1879–85

Muybridge first demonstrated his zoopraxiscope for a private gathering at Stanford's Palo Alto home in the fall of 1879.[223] Sharing his animated still images in this way helped him prove their veracity to those who still were not convinced. On May 4, 1880, he exhibited the projections for the press at the San Francisco Art Association. The *San Francisco Call* described the powerful effect of seeing on the screen "apparently the living, moving horse. Nothing was wanting but the clatter of the hoofs upon the turf and an occasional breath of steam from the nostrils to make the spectator believe that he had before him genuine flesh-and-blood steeds."[224] It is striking to consider today how lifelike these images must have appeared to viewers when they first were presented as part of Muybridge's lectures. And yet, despite its implied realism, the zoopraxiscope

Fig. 78. *Mule Bucking and Kicking, 12 phases*, 1893

image is fraught with many distortions. The horse, for example, is seen from numerous points of view—one for each of the twenty-four cameras lining the side of the track—like a long tracking shot that travels with the subject of the sequence. The photographs are manipulated when transferred to the disc to compensate for the anorthoscopic distortion inherent in the design of the projector, and therefore the image appears removed from its original photographic source when displayed. And the length of the sequence itself is limited to just a fragment of time: one shard of movement is stretched into an endless loop showing a horse running in place with no beginning or end as the disc spins. It portrays the *concept* of movement, restructured from still images.

In this body of work, Muybridge exploded the conviction that photographs could only record and document a moment of time—a technical achievement he had worked a decade to perfect—and a new notion emerged. For him, the illusion of time passing he had embedded within his still pictures—the landscapes and architecture as well as the motion studies—had been transformed into a representation of time itself. His zoopraxiscope discs, with their subtly changing icons marching around a glass rim, are analogous to the face of a clock. Like the numbers that delineate the geometry of a chronometer, each image on a zoopraxiscope disc symbolizes a moment in space and time. The space between pictures represents the distance traveled by the subject between shots in the sequence. The time it takes for the pictures to circumnavigate the perimeter as it spins represents the duration of the action as the subject moved along the racetrack. The images are like markers on the discs—when put in motion and projected, they are synthesized by the mind of the viewer to vivify the "stuff of the world" recorded by Muybridge's cameras.

The Attitudes of Animals in Motion

Throughout 1880, Muybridge continued to lecture about his work, dissecting the intricacies of animal and human motion that he had photographed in Palo Alto. He showed a mix of lantern slides and zoopraxiscope discs, often stopping the action to focus on details that were relevant to understanding both scientific and aesthetic aspects of the work (figs. 1–5). He also began printing for an album of photographs summarizing what he had accomplished in Palo Alto. He made several presentation copies of original prints that begin with images of horses hauling, walking, cantering, and pacing, then progressing through photographs of trotting, running, leaping, and backing, to several additional series depicting the other animal movements and human athletic feats (fig. 81). He also included documentary pictures of the stock farm, track, camera shed,

Fig. 79. *Mule Bucking and Kicking, 12 phases, No. 29*, 1893

and shutters, along with an index and technical overview of his process (plates 127, 128). Titled *The Attitudes of Animals in Motion: A Series of Photographs Illustrating the Consecutive Positions Assumed by Animals in Performing Various Movements*, Muybridge completed and submitted the album to Stanford and his wife on May 15, 1881, and the Library of Congress received it for copyright on May 17 (fig. 80). Stanford was also working on a book that would present a serious analysis of the motion experiments conducted in Palo Alto. For this he commissioned Stillman to summarize and evaluate the results in scientific language, using Muybridge's photographs to illustrate the text. Muybridge approved of Stillman's plan, assuming he would be fully credited, and on May 30, Stanford paid Muybridge a total of $2,000.00 for his work on the project and agreed to sell him for one dollar all the equipment, patents, and copyrights "that have been employed in, and about the representation of animals upon my premises at Palo Alto."[225]

Supported by Stanford, Muybridge traveled to Paris in September 1881, where he met Marey for the first time, as well as other members of the scientific and artistic communities, including the academic painter Jean-Louis-Ernest Meissonier. At the time, Meissonier had just completed a commissioned portrait of Stanford, which he had agreed to paint only when Stanford presented him with a set of Muybridge's horse photographs. After lecturing and showing his work at the homes of Marey and Meissonier, Muybridge was received in Europe like a visiting hero and was finally getting credit from his peers for his artistic ability and his innovations in instantaneous photography. He was introduced to the new and faster dry-plate technology then emerging in France, and spent much of his Parisian visit working to find financial support for a larger, more complex project he envisioned on animal locomotion. Marey was beginning work on his own new techniques, which would go beyond Muybridge's multicamera system by superimposing one-atop-the-next sequential images of motion, shot on a single negative through a rotating shutter. "The work of Étienne-Jules Marey, a scientist who switched from graphic to photographic notations of animal movement under Muybridge's tutelage, summarizes the point of disjunction between the still photograph and cinema," writes Hollis Frampton. "The photograph could no longer contain the contradictory pressures to affirm time and to deny it. It split sharply into an illusionistic cinema of incessant motion and a static photographic art that remained frozen solid for decades."[226]

In February 1882, Muybridge traveled from Paris to London, where he arranged to lecture to the Royal Society and the Royal Academy. Then in April he abruptly learned that Stillman's book, *The Horse in Motion as Shown by Instantaneous Photography, with a Study on Animal Mechanics, Founded on Anatomy and the Revelations of the Camera, in Which Is Demonstrated the Theory of Quadrupedal*

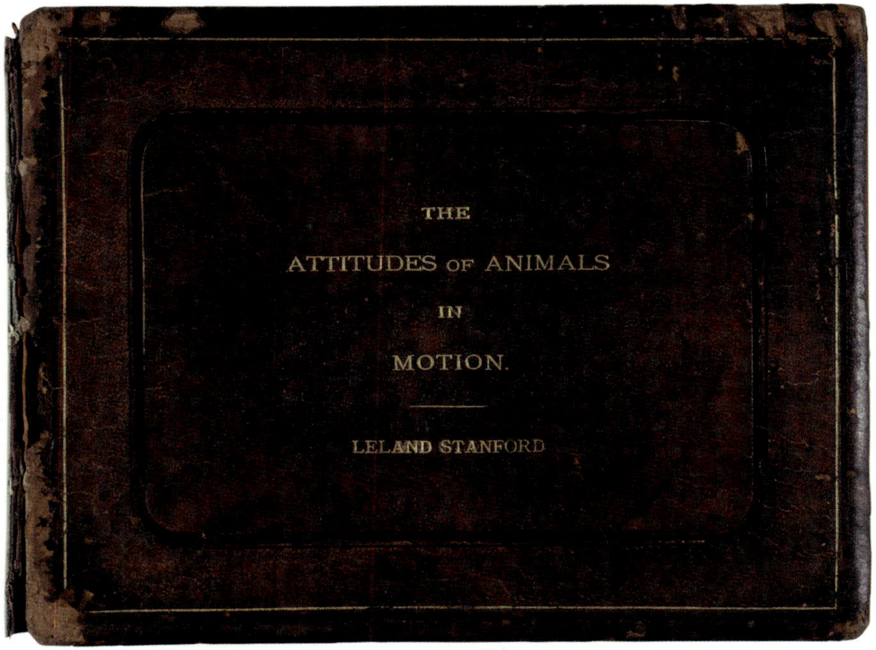

Fig. 80. *The Attitudes of Animals in Motion: A Series of Photographs Illustrating the Consecutive Positions Assumed by Animals in Performing Various Movements* (cover), 1881

Locomotion, which included five of Muybridge's photographs and ninety-one lithographs made from his Palo Alto pictures, had been published without giving him copyright credit or acknowledging his contributions to the project (fig. 82). It also omitted an introduction Muybridge had written, with Stillman's agreement. Mindful of his limited credentials as a scientist and possibly upset by the adulation Muybridge had garnered in Paris, Stanford had relegated Muybridge's creative contributions to the *Horse in Motion* project to those of a hired hand. Shunned by the Royal Society, Muybridge returned to the United States in June, embarrassed, deflated, and without funding to continue his work. He later filed a lawsuit against Stanford asking for $50,000.00 in damages, which Stanford handily defeated.[227]

Fig. 81. *Horses. Running. Phryne L. Plate 40*, 1879, from *The Attitudes of Animals in Motion*, 1881

Animal Locomotion, Philadelphia

In early 1883, Muybridge produced a prospectus for a new project, promising an "elaborate work upon the attitudes of man, the horse and other animals in motion."[228] He continued to travel in the United States, lecturing with renewed success. His stops included the Franklin Institute and the Pennsylvania Academy of the Fine Arts, where he met Eakins, with whom he had corresponded in 1878, at the start of his work on sequential motion. Eakins was a photographer as well as a painter, and his sophisticated use of the camera as an aid to his painting process was indicative of his modern, realist vision. Eakins's patron Fairman Rogers soon convinced William Pepper, the provost of the University of Pennsylvania, to sponsor a new project on animal locomotion. Pepper was interested in the research and scientific applications of such an endeavor, and he understood its broad connection to medicine, physiology, anthropology, fine arts, and sports.[229] In August 1883, he offered Muybridge both facilities and

Fig. 82. *Phryne L—Running. Stride. 19 ft. 9 in. Plate XVI* (in J. D. B. Stillman, *The Horse in Motion as Shown by Instantaneous Photography*), 1881

financing to begin work in Philadelphia. An advisory commission was established in March 1884, and Muybridge began photographing on the grounds of the veterinary department in June, in a new outdoor studio constructed by the university. "An extensive photographic plant, elaborate electrical and mechanical apparatus, access to typical animals of many kinds, much space for the erection of laboratories, and a large fund for current expenses were needed," wrote Pepper in his introduction to a university report on the project.[230] With the incentive to organize a much broader study based in both science and art, Muybridge set out to understand the myriad conceptual connections between the mechanics of his own process and the workings of animal locomotion. For a time he worked side-by-side with Eakins, who was photographing human motion with an adapted Marey-wheel camera, which made a series of superimposed images from a single point of view (fig. 83). Eakins was especially interested in the calculations needed to translate the motion and special perspective depicted in his photographs to the flat visual plane of a canvas. Muybridge, on the other hand, insisted that his use of multiple cameras looking at a subject from different perspectives gave a broader and more complex analysis of various types of locomotion. He, too, tested Marey-wheel cameras during the summer, but his first successful photographs, made in late August or September 1884, used multiple cameras and dispensed with the background grid of lines and numbers he had first implemented in Palo Alto. Muybridge began by photographing a variety of subjects from multiple angles, but depicting just one moment of action. *Plate 523* of *Animal Locomotion*, for example, shows an athlete in a series of different poses—striking a blow, throwing a disk, hurling a 75-pound stone, throwing a ball—and each pose is frozen in time (fig. 84). A bank of six cameras surrounds the subject, recording the same instant from equidistant points around an arc of about 160 degrees. The result, similar to viewing a sculpture in the round, is the opposite of the Marey-wheel approach, yielding multiple images of an instant in time rather than a single image that documents a span of time.

For his work in Philadelphia, Muybridge used the new gelatin dry-plate technology, which was simpler to implement and more sensitive to light, allowing for faster shutter speeds. But he was still having trouble with the complex array of multiple cameras and shutters. He spent the winter of 1884–85 making improvements to his process and started to photograph again in June 1885. In another significant departure from his Palo Alto technique, he now primarily employed a clockwork-driven electronic timer to trip his shutters (fig. 85). Rather than rely on wires, which were useful on only a limited array of actions and lacked the precision of a Marey-wheel shutter, Muybridge developed a regulating device that would connect electronically to his banks of shutters and trigger them in an accurately timed sequence. Through the summer of 1885, he worked to complete photography for the project that would become *Animal Locomotion: An Electro-Photographic Investigation of Consecutive Phases of Animal Movements*. He worked first at his studio near the university hospital, making numerous images of people and domesticated animals (fig. 86). In August, he moved his set-up to the Philadelphia zoological gardens to photograph wild animals and then to the racetrack at Gentlemen's Driving Park in Belmont for pictures

Fig. 83. Thomas Eakins, *Marey Wheel Photographs of Unidentified Model, with Eadweard Muybridge Notations*, 1884

of running and trotting horses. All told, Muybridge produced 781 collotype plates from about 20,000 individual photographs, and in 1887 he began to sell finished prints of his elaborate sequences by subscription. He had achieved his ambition of creating a "Standard Work of Reference for the Painter, the Sculptor, the Anatomist, the Physiologist … of interest to the general public, and of especial value to owners and trainers of horses."[231] *Animal Locomotion* appeared as a veritable atlas of imagery about movement and time, organized from an almost evolutionary perspective. The plates were ordered hierarchically from nude males to nude females, semi-nude and draped males, females, and children to abnormal movements of males and females, and then horses, domestic animals, wild animals, and birds. His animal subjects included horses, dogs, deer, birds, elephants, monkeys, and lions. In keeping with the univer-

Fig. 84. *A, striking a blow; B, throwing disk; C, heaving a 75-lb. stone; D, throwing a ball; E, throwing disk; F, heaving 75-lb. stone. Plate 523*, 1887

Fig. 85. *Electronic Timing Device Used by Eadweard Muybridge at University of Pennsylvania for Animal Locomotion*, 1885

sity's agenda, Muybridge also photographed many different types of people: athletes, laborers, women of leisure, working women, children, male and female nudes, self-portraits, and people with disabilities effecting "abnormal" movements (fig. 87). These pictures are at once humorous and solemn, artful and mundane, sensuous and dogmatic, sexualized and patrician, heartfelt and sad. *Animal Locomotion* is a typological archive that summarizes and expresses a great deal about how people and animals were viewed and represented in 1880s America.

Quoted extensively by artists and scientists alike, *Animal Locomotion* was at the time—and remains to this day—one of the most influential publications of its kind. Much has been written about the meaning of the work in terms of its social and narrative implications, which went well beyond the scientific and aesthetic umbrella initiated by the university. With *Animal Locomotion*, the "industrial gaze" of Muybridge's panoramas and the Palo Alto pictures became more objectified and focused on coding human types and cultural behavior, perhaps the result of the artist's typological rather than methodological agenda. "Like Marey, Muybridge used his camera to disclose those aspects of human activity that usually remain unseen: not because they were invisible to the naked eye, as in Marey's case, but because social conventions and morality dictated that they remain concealed except in the imaginative world of private fantasy," writes Marta Braun. "The photographs objectify erotic impulses and extend voyeuristic curiosity in language we now recognize as taken from the standard pornographic vocabulary."[232] Referencing *Animal Locomotion*, film historian Linda Williams cites "the construction of better machines of observation to measure and record bodies now conceived themselves as machines" and the "unanticipated pleasure attached to the visual spectacle of lifelike moving bodies" as key factors in the evolution of a cinematic form that reveals the mysteries of human motion while forging kinesthetic connections, through photography, that fetishize women.[233] As the project's producer, director, and photographer, Muybridge created scenarios for women to perform in roles that engender domestic or idyllic narrative fantasies, such as "ironing clothes" (plate 163), "spanking a child," "throwing self on a heap of hay," "crossing brook on step-stones with fishing-pole and can" (plate 172), and "getting into bed" (plate 165). In *Animal Locomotion*, men and women are never shown together except for one sequence of a couple waltzing. In contrast to the women, Muybridge portrays men, alone or in pairs, primarily as athletes, gymnasts, industrial laborers, and soldiers—subjects related to speed, virility, violence, and war. There are elabo-

Fig. 86. *Eadweard Muybridge's outdoor camera house, 36th and Pine Streets, Philadelphia*, ca. 1885

rate groups of images depicting men boxing, wrestling, or drilling with a rifle, sometimes described with a barely concealed homoerotic aura that spins the stories toward a more emotional drama. Men, too, are objectified in the pictures, often presented as powerful and desirable actors within a choreographed script. *Plate 333* shows the outcome of a staged fistfight between two almost nude men, one falling backward onto a mattress as the other looms over him menacingly (fig. 88). This work is as much about the human body and its existential position in the world as it is about its motion through space. Braun concludes, "Muybridge's concern then, is with narration, not with movement, and we are invited to share in this concern."[234]

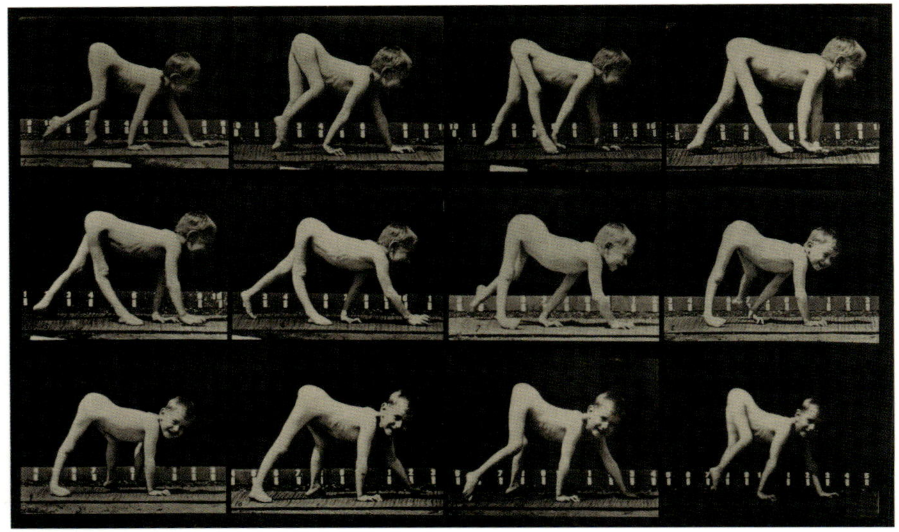

Fig. 87. *Infantile paralysis; child, walking on hands and feet. Plate 539*, 1887

The Lecture Circuit

Following the publication of *Animal Locomotion* in 1887, Muybridge continued to travel and lecture about his work, actively publicizing it and promoting its sale to wealthy patrons, museums, libraries, and scientific institutions around the world. For example, on May 16, 1887, he wrote to Major John Wesley Powell, then the director of the Smithsonian Institution's Bureau of Ethnology, with the following invitation: "By the courtesy of the Committee on Art a complete series of the plates in *Animal Locomotion* will be accessible to subscribers for a short time in the Corcoran Gallery. Will you kindly avail yourself of an early opportunity to call there, and make a memorandum of the serial numbers of the plates you require for the Library."[235] Muybridge lectured in New York and Chicago, as well as London, Berlin, Munich, Rome, and Turin, and numerous other cities throughout the United States and Europe, and he succeeded in selling subscriptions to a variety of institutions and individuals.[236]

In his lectures, Muybridge analyzed different sorts of movement and presented a quasi-scientific view of the physiology of locomotion, arguing that a visual

Fig. 88. *Boxing; 53 knocking 52 down (shoes). Plate 333*, 1887

understanding of the attitudes of motion could lead to improvements in art, racing, athletics, and even engineering. The subtext of his talks was complex and sometimes contradicted his own pictures. Muybridge often showed images from the history of art—everything from prehistoric cave paintings to details of the Parthenon frieze to Rosa Bonheur's *The Horse Fair*, 1853–55, and Meissonier's *Campagne de France, 1814* of 1864 (fig. 89)—and he critiqued the portrayal of animals in art before his own photography changed everything. He considered why some artists conveyed motion correctly and others indifferently. Referring to Meissonier's *Campagne de France* in a lecture about the quadrupedal walk of horses, Muybridge asserted, "Few designs of modern artists are so familiar to us as this ... the exhibition of the original of which provoked a storm of merciless criticism that would have overwhelmed an artist of lower rank than that which Meissonier held." Recalling his presentation of *The Attitudes of Animals in Motion* before "a large number of men eminent in Art, Literature, and Science" assembled at Meissonier's home in 1881, Muybridge noted the inspiration the painter took from his pictures: he had credited Muybridge's Palo Alto photographs for shedding light on true animal locomotion. "You see here the Sun himself has demonstrated to you the truth of Meissonier's interpretation of a walking horse," Muybridge boasted, reconnecting to his old mythic affinity with Helios—giver of light—as an arbiter of truth in art.[237] A diverse group of artists, from Frederic Remington to Edgar Degas, took note of his work as it continued to garner widespread attention, and they used it as an aid in modeling the correct positions of moving animals and humans in their work (fig. 90).

Muybridge continued to include zoopraxiscope projections in his lectures—as he had at Meissonier's home in 1881—using his moving images of horses and athletes to demonstrate the veracity of his photographic sequences. In 1888, he met Thomas Edison during a stop in Orange, New Jersey, and he was invited to introduce his zoopraxiscope and animal locomotion photographs at a special presentation in the Edison laboratory. According to newspaper accounts of the meeting, Muybridge envisioned an applied connection between the phonograph and the zoopraxiscope, "a scheme which, if carried to completion, will afford an almost endless field of instruction and amusement."[238] This was essentially a rudimentary postulation of talking pictures, though Edison soon realized the limitations of the zoopraxiscope; it could only project a very short segment of time, and its images were hand-drawn from photographs to compensate for the distortion introduced in the process. Soon thereafter, Edison acquired some of the *Animal Locomotion* photographs, though little came of their initial idea to connect moving pictures with sound. The following year, at the Paris Exposition Universelle, Edison was introduced to far more effective technologies, including Marey's chronophotography sequences made using one camera and George

Fig. 89. Jean-Louis-Ernest Meissonier, *Campagne de France, 1814*, 1864

Eastman's new paper film, which came in a flexible roll. This was a significant breakthrough that led to cinema as we understand it today, and Edison was instrumental in synthesizing many different contributions in his kinetoscope, which finally became commercially viable in 1894. Edison wrote to Muybridge on February 21 of that year, describing "a little instrument which I call a kinetograph with a nickel & slot but I am very doubtful if there is any commercial feature in it.... These Zootropic devices are of too sentimental a character to get the public to invest in."[239] By 1895, the Lumière brothers in France combined these technologies to record and project longer motion pictures using flexible, perforated celluloid film in their cinematograph.

The advent of cinema, inspired in part by Muybridge's pioneering work in photographing and projecting sequences of motion, grew from a complex series of technological innovations that began to pass him by as he prepared his zoopraxiscope for an extensive lecture tour in Japan and Australia, set to commence in October 1892. In advance of an extended stopover in San Francisco en route to the Orient, Muybridge began corresponding with officials at the new Leland Stanford Junior University, established in Palo Alto to memorialize Stanford's son, who had died tragically in 1884. Muybridge wrote to university president David Starr Jordan, inquiring about support for new research on photographing the flight of insects, and then about the possibility of lecturing at the school. From San Francisco he also drafted letters to Stanford, again spelling out in detail to the recently elected U.S. senator from California his decade-old embarrassment over the publication of Stillman's *The Horse in Motion* without getting credit for his work.[240] Muybridge was trying to redress some of the anger that resurfaced on his final visit to California, but he was rebuffed by Stanford and received no invitation to return to the site of his earlier photo-

Fig. 90. Frederic Remington, *Off the Range (Coming Through the Rye)*, 1902

graphic breakthroughs. He quickly changed course again when it was announced, "By invitation of the FINE ARTS COMMISSION of the World's Columbian Exposition, Mr. EADWEARD MUYBRIDGE will give at intervals from May to October, 1893, in the Zoopraxographical Hall of the Exposition, a series of Lectures on the Science of Animal Locomotion, especially in its relation to Design in Art.... It is a subject of the most profound importance to the Scientist and the Artist, and has been universally recognized as of the greatest interest to the general public."[241] With the artist Erwin F. Faber, Muybridge worked on a new set of hand-colored 12-inch zoopraxiscope discs from his *Animal Locomotion* negatives and signed a contract to construct a building on the Midway Plaisance in Chicago.

In the spring of 1893, Muybridge published a detailed analysis of his motion studies, based on his talks, with the quasi-scientific title *Descriptive Zoopraxography, or the Science of Animal Locomotion Made Popular* (fig. 91), and on April 12, the *Chicago Daily Tribune* announced that "Prof. Muybridge is engaged to deliver 300 lectures at the World's Fair under the auspices of the National Board of Education."[242] His presentations were called "The Science of Animal Locomotion in Its Relation to Design in Art," intended "merely to describe the common methods of limb action adopted by quadrupeds—especially by the horse—in their various acts of progressive motion, and to illustrate the most important phases of these movements by tracings from the original photogravures of the Author's work."[243] In part because the emerging technology of motion pictures had already advanced well beyond his zoopraxiscope, and in part because the public was hungry for more popular entertainment, his presentations failed to gain an audience, and the Zoopraxographical Hall soon closed its doors.

Concurrently, on July 12, historian Frederick Jackson Turner read his influential paper "The Significance of the Frontier in American History" to an audience of the American Historical Society, which was meeting that summer at the World's Columbian Exposition. Turner's lecture presented for the first time his "frontier thesis," proposing that during the course of American history, settlers distinguished themselves from their roots as a result of the natural world in which they found themselves; they developed new identities, customs, laws, and institutions from their desire to settle a frontier environment, which drove them progressively westward, generation by generation. "The first ideal of the pioneers was conquest," wrote Turner in his subsequent book *The Frontier in American History*.[244] Therefore, the tension between the wilderness and civilization became a major force in determining the character of the American people. Muybridge's technological decoding of animal and human locomotion—particularly his mechanical analysis of the running horse so coherently presented at the fair and in *Descriptive Zoopraxography*—symbolically links to Turner's theory that the taming and conquest of nature led to a need for locomotive speed and the subsequent development of the railroad that quickly pushed the frontier westward to its farthest limit. New financial and legal constructs that grew out of this systematic conquest of space and time can be connected on the one hand to Stanford's ambitions and on the other to Muybridge's realization of his goal to represent the horse (and human locomotion) from a mechanized, industrialized point of view. Muybridge's art, from his earliest landscapes to his final series of locomotion studies, defined the western frontier for public consumption and embodied the industrial points of view that made its occupation complete.

The Movements of Everyday Life, Kingston upon Thames

By mid-summer of 1894, Muybridge returned to live in Kingston upon Thames, not far from the three-story merchant's home by the river where he was born sixty-four years prior to his homecoming. He had effectively retired from photography and proposed to produce a new book of his recent work that would bring together his animal locomotion studies in one volume. With the zoopraxiscope, he lectured extensively throughout Great Britain during this time. In May 1895, he advertised a course of talks, stating, "although the training of the eye to a just appreciation of the movements of everyday life is their main object, they will be entirely free from technicalities."[245] Muybridge returned to the United States just once more, in 1896–97, to help William Pepper and the

University of Pennsylvania resolve the serious financial problems arising from the debt owed to the New York Photogravure Company for production of the *Animal Locomotion* plates. Pepper, who was no longer provost, agreed to personally pay the debt to absolve the university of responsibility. Muybridge retained control of his negatives, but all other plates and preparatory materials were given to Pepper to sell or donate, with the caveat that any future income be used to reimburse those who had guaranteed the project on behalf of the university. Back in Kingston, Muybridge finally produced two new books about his work, both published by Chapman and Hall in London: *Animals in Motion, an Electro-Photographic Investigation of Consecutive Phases of Animal Progressive Movements, etc.*, 1899, about the "quadrupedal gait," and *The Human Figure in Motion, an Electro-Photographic Investigation of Consecutive Phases of Muscular Actions*, 1901, about "bipedal movement."

Muybridge's inspired preface to *Animals in Motion*, written in the third person as though looking through someone else's eyes, is the closest he ever came to penning a memoir about the years he spent analyzing motion. He presents an objective, dispassionate view of this aspect of his career, documenting his own version of the highlights and sequence of events that led from his first vague attempts to photograph a running horse in 1872 to the publication of *Animal Locomotion*, his 1887 masterpiece containing "more than 20,000 figures of moving men, women, children, beasts, and birds."[246] He gives himself credit for his work on *The Horse in Motion* and mentions the "Proceedings of the Royal Institution of Great Britain," which was published following the release of Stillman's book. His text reveals a lot about how he wanted to be remembered: for his contributions to science, and specifically, for stopping time with photography and reanimating it to scrutinize the truth of his observations. Referring back to his discussions with Edison about the zoopraxiscope, Muybridge now connected his work directly to the birth of cinema: "The combination of such an instrument with the phonograph has not, at the time of writing, been satisfactorily accomplished; there can, however, be but little doubt that in the—perhaps not far distant—future, instruments will be constructed that will not only reproduce visible actions simultaneously with audible words, but an entire opera, with the gestures, facial

Fig. 91. *Descriptive Zoopraxography, or the Science of Animal Locomotion Made Popular* (frontispiece), 1893

expressions, and songs of the performers, with all the accompanying music.... And if the photographs should have been made stereoscopically, and if projections from each series be independently and synchronously projected on a screen, a perfectly realistic image of the original performance will be seen, in the apparent 'round,' by the use of properly constructed binocular glasses."[247]

Growth, progress, industrialization, and the mechanization of life are all intrinsically about change, and these became Muybridge's principal subjects. Despite his meteoric career as a "photographic view artist"—a creator of landscape views, surveys, documentary essays, architectural pictures, stereographic archives, war photographs, panoramas, narratives, albums, and advertisements—he wanted to go down in history for the technical innovations, inventive thinking, and image sequences that revolutionized how we see and conceptualize motion. His point of view and his craving for recognition of his expertise were intimately entwined with the time in which he lived and the people with which he worked. Industrialists like Stanford and educators like Pepper—the patrons who supported his work—understood that only by changing how we see and comprehend time and space are we able to embrace and buy into the ongoing modernization of life. An enhanced concept of speed—the swiftness of motion and communication across vast distances—would underpin a new economy and engender new ways to think about science and art. As one of the first artists to fully grasp the means of describing speed in relation to human and animal locomotion, Muybridge proposed a synthesis of people and machines in his work, reimagining the body and its movements through industrial systems of cameras, shutters, electrical wiring, batteries, and telegraphic apparatus. So Muybridge's photographs came to represent horses and people as if they were machines, and he pictured them with inventiveness to codify their speed. It was this correlation between his still images and their animations—between the static sequence and the projected, fluid embodiment of real time—that defined his modern style.

IV – Fade to Black

> The Photograph does not necessarily say *what is no longer*, but only and for certain *what has been.* This distinction is decisive. In front of a photograph, our consciousness does not necessarily take the nostalgic path of memory (how many photographs are outside of individual time), but for every photograph existing in the world, the path of certainty: the Photograph's essence is to ratify what it represents.[248]
>
> Roland Barthes

By the time Muybridge died of prostate cancer in Kingston on May 8, 1904 (Marey would die in France just a week later), the cinema was beginning to flourish as a new form of entertainment, a signifier of motion and modernity. In the United States, railroads crisscrossed the country, dominating the transportation industries; motorcars replaced horses on city streets and country roads; and the Wright Brothers had recently flown their first airplane hundreds of feet across a wind-swept dune in North Carolina, a development that would soon change the face of the world once again. The notion of speed as a commodity that fills the economic sails of business blew strongly across the wake of the industrial revolution. In this context—the dawning twentieth century—Muybridge's art represents an integral component of the "remaking of perception," as Jonathan Crary calls it.[249] Muybridge's work and biography have since been fraught with justifiable debate over his influence in the development of cinematic form and his representation of the human body as a carrier of Victorian colonial ethos. After all, his sophisticated vision became a bridge that connected an emerging, industrialized view of the global economy with a workforce that was rapidly becoming mechanized. Consequently, Muybridge's influence is pervasive in our arts and culture today; his true role in those times of change was that of a catalyst in a technological and aesthetic revolution that has continued, with ever-accelerating speed, to transform our representation of time and space.

[1] Solnit, Rebecca, *River of Shadows: Eadweard Muybridge and the Technological Wild West* (New York: Viking, 2003), 18.

[2] The invention of the daguerreotype, the first commercially successful photographic process, was announced in January 1839. Louis-Jacques-Mandé Daguerre was a talented painter who had an abiding interest in the use of perspective, optics, light, and chemistry to create grand, entertaining, and historically relevant dioramas that transported viewers into other worlds by means of elaborate visual illusions. From the mid-1820s through the mid-1830s, he worked with Joseph-Nicéphore Niépce to perfect the chemical and photographic process. Daguerre's process produced unique images on highly polished metal plates. The first photographs depicted scenes of Paris and works of art, such as still-life images of plasters and architectural elements. Following his publication of technical details, photography grew into a popular art, first in France and then in England and the United States. Just weeks after Daguerre's unveiling, William Henry Fox Talbot in Britain announced his calotype process of fixing a photographic image on paper.

[3] *The Pencil of Nature* was the first photographic sequence or book using actual photographs. Published by British photographer William Henry Fox Talbot in a series of subscribed chapters, it was produced between 1844 and 1846 and became the model for artistic exposition in photography.

[4] William Henry Fox Talbot to John Frederick William Herschel, January 29, 1839. Collection of the Royal Society, London.

[5] Cited in Larry J. Schaaf, *Out of the Shadows: Herschel, Talbot, and the Invention of Photography* (New Haven, CT: Yale University Press, 1992), 51. Schaaf notes that Talbot's full manuscript is in the Royal Society Library and a summary was published in the Society's *Proceedings* 4, no. 36 (January 31, 1839).

[6] Ibid. The Royal Society soon published Talbot's talk under the rubric "Some Account of the Art of Photogenic Drawing, or the Process by Which May Be Made to Delineate Themselves without the Aid of the Artist's Pencil."

[7] See Larry J. Schaaf, "Sir John Herschel's 1839 Royal Society Paper on Photography," *History of Photography* 3, no. 1 (January 1979): 47–60, which presents for the first time a transcript of Herschel's paper of March 14, 1839.

[8] To create a calotype negative, the paper was prepared by washing it in a solution of silver nitrate, drying it, and soaking it in potassium iodide. Then shortly before the paper was exposed, it was dipped in a solution of silver nitrate and gallic acid, placed in a camera, and exposed to light while still wet. The negative was then developed in a dark room with gallo-nitrate of silver, and when the image was sufficiently visible, it was fixed with a solution of potassium bromide or hypo, then washed and dried. The negative could be printed on sensitized paper. Talbot's original salted-paper process produced a better print. The printing paper was made by first soaking a sheet of writing paper in a solution of table salt. It was then dried and coated with silver nitrate on one side. Once dry, this paper was then placed under the paper negative and a sheet of glass, and then contact printed under strong natural light until the image developed out to a rich brown tone. The paper was then fixed with hypo, washed in water, and dried to produce the finished print.

[9] Maybanke Susannah Anderson, "My Sprig of Rosemary," unpublished memoir, 1915 (now in the collection of The State Library, New South Wales), quoted in Robert Bartlett Haas, *Muybridge: Man in Motion* (Berkeley: University of California Press, 1976), 1.

[10] John Tallis and J. G. Strutt, eds., *Tallis's History and Description of the Crystal Palace and the Exhibition of the World's Industry in 1851* (London: London Printing and Publishing Company, 1852), 18.

[11] Anderson, "My Sprig of Rosemary," 4. Anderson's father, Muybridge's uncle Henry Selfe, was an inventor and plumber whose high-pressure firehouse was displayed at the Great Exhibition of 1851 in London.

[12] Roger Taylor, *Impressed by Light: British Photographs from Paper Negatives, 1840–1860* (New Haven, CT: Yale University Press, 2007), 37.

[13] The wet-plate collodion process transformed photography following its introduction by Frederick Scott Archer in 1851. A collodion base (guncotton dissolved in ether and alcohol) was poured onto a sheet of glass, sensitized with silver nitrate, and exposed to light in a camera while still wet. When processed and fixed, this method produced a beautiful, continuous, and sharp negative that could be printed as a positive image onto sensitized paper. The collodion emulsion also allowed for much shorter exposure times, making portraits, outdoor photography, and instantaneous photography a possibility. The plate had to be created, exposed, and processed within about ten minutes, since collodion lost its sensitivity as it dried out. This became the dominant photographic process from the early 1850s until 1880, when the dry-plate silver gelatin process took precedence.

[14] James Glaisher, "Class X: Philosophical Instruments and Processes Depending on their Use," in *Exhibition of the Works of Industry of All Nations, 1851, Reports by the Juries on the Subjects in the Thirty Classes into Which the Exhibition Was Divided* (London: Printed for the Royal Commission by William Clowes and Sons, 1852), 244.

[15] Ibid., 275.

[16] Paul Young writes, "From its inception, the world's first international industrial display was presented as a truly global event, one which aimed at once to celebrate the material progress humankind had made and coordinate those advances in order that the world could work together." Paul Young, "Mission Impossible: Globalization and the Great Exhibition," in Jeffrey A. Auerbach and Peter H. Hoffenberg, eds., *Britain, the Empire, and the World at the Great Exhibition of 1851* (Farnham: Ashgate Publishing, 2008), 4.

[17] Glaisher, "Class X: Philosophical Instruments and Processes Depending on their Use," 244.

[18] On July 30, 1852, Talbot wrote to the Earl of Rosse regarding his willingness to give up rights to his patent on the calotype process: "Ever since the Great Exhibition I have felt that a new era has commenced for photography, as it has for many other useful arts and inventions. Thousands of persons have now become acquainted with the art, and, from having seen such beautiful specimens of it produced both in England and France, have naturally felt a wish to practice it themselves—A variety of new applications of it have been imagined, and doubtless many more remain to be discovered." Quoted in Taylor, *Impressed by Light*, 49.

[19] Julie K. Brown, *Contesting Images: Photography and the World's Columbian Exposition* (Tucson: University of Arizona Press, 1994), xvi.

[20] Ibid., 104.

[21] The exact date and circumstances of Muybridge's departure from Kingston are not known.

[22] Eadweard Muybridge, "Primitive Vision and Color Blindness," *The Standard*, March 28, 1902.

[23] "Silas Wright Selleck," in Peter E. Palmquist and Thomas R. Kailbourn, *Pioneer Photographers of the Far West: A Biographical Dictionary, 1840–1865* (Stanford, CA: Stanford University Press, 2000), 485.

[24] Silas Selleck testified in 1875 that he and Muybridge had again "met in California in the Autumn of 1855." *Sacramento Union*, February 5, 1875.

[25] *San Francisco Daily Evening Bulletin*, April 28, 1856, quoted in Gordon Hendricks, *Eadweard Muybridge: The Father of the Motion Picture* (New York: Grossman Publishers, 1975), 5.

[26] James Wilson, *Entomology* (Edinburgh: A. and C. Black; San Francisco: E. J. Muygridge: 1855), in the collection of the Bancroft Library, University of California, Berkeley.

[27] Peter Palmquist, *Carleton Watkins: Photographer of the American West* (Albuquerque: University of New Mexico Press, 1983), 5.

[28] Robert A. Carden, "Photography in California," *Photographic and Fine Art Journal*, no. 10 (April 1857): 112.

[29] Kevin Starr, *California: A History* (New York: Modern Library, 2007), 107.

[30] *San Francisco Daily Evening Bulletin*, August 13, 1856, quoted in Hendricks, *Eadweard Muybridge: The Father of the Motion Pictures*, 7.
[31] Vance, Shew, and a number of other San Francisco photographers began to offer the ambrotype process around this time. This relatively new technique used a wet-plate collodion negative placed against a black backing or varnish, which reversed the negative to create a positive image. Ambrotypes were then sealed behind another sheet of glass and mounted in a case like a daguerreotype. Ambrotypes were not mirrored like daguerreotypes and were far less expensive to produce. In 1854, James Ambrose Cutting of Boston was granted several patents for the ambrotype process, and Vance argued that he had the exclusive right to use Cutting's patented techniques.
[32] "Finden's Beauties of Moore," *Chambers' Edinburgh Journal*, no. 104 (December 27, 1845): 412. British engraver Edward Finden made the engravings from paintings by other artists. *Lalla Rookh* is a romantic 1817 narrative poem on an Oriental theme by Thomas Moore.
[33] Copies of this advertisement on the back cover of an edition of Spencer's *History of the United States* are in the collections of the California State Library, Sacramento, and the California Historical Society, San Francisco.
[34] "City Items," *Daily Alta California*, December 13, 1858. Beginning in 1849 in Sacramento, and then in San Francisco, the *Alta California* grew into a major daily paper and became one of the best sources of news in the region. Also pertaining to the business relationship between Watkins and Muybridge at this time, on July 20, 1859, Watkins wrote to Henry Laurencel designating Edward Muygridge as his agent to receive payments from Laurencel in his absence. Laurencel had been part owner of a quicksilver mine in Santa Clara County that Watkins had photographed for evidence used in a court trial. This letter is in the Special Collections, Stanford University Library, Stanford University, Stanford, California.
[35] Muybridge's brother George Muygridge (as listed in the 1858 San Francisco directory) had contracted tuberculosis and according to the family tree died in 1858, possibly in San Francisco. See Haas, *Muybridge: Man in Motion*, 8.
[36] *Daily Evening Post*, February 6, 1875.
[37] Haas, *Muybridge: Man in Motion*, 10.
[38] Brian Clegg, *The Man Who Stopped Time: The Illuminating Story of Eadweard Muybridge—Pioneer Photographer, Father of the Motion Picture, Murderer* (Washington, DC: Joseph Henry Press, 2007), 29.
[39] *International Exhibition 1862: Official Catalogue of the Industrial Department* (London: Truscott, Son, and Simmons, 1862), 82, catalogue number 5310A. See page 33 for catalogue number 2037 listing Muybridge's "washing machine, on the principal [*sic*] of a fulling mill."
[40] Haas, *Muybridge: Man in Motion*, 10–11.
[41] *The Times* (London), July 1, 1865. Stephen Herbert, whose ongoing research on Muybridge's life and work is one key resource for studying the artist, first uncovered details of Muybridge's role as a stockholder of the Austin Consolidated Silver Mines Company and the Bank of Turkey. See *The Compleat Eadweard Muybridge*, http://www.stephenherbert.co.uk/muybCOMPLEAT.htm.
[42] Austin, Nevada, and the surrounding Reese River Mining District were the site of a major mid-century silver rush: by 1862, more than ten thousand people had descended on the small town.
[43] *Glasgow Herald*, November 23, 1865, 8.
[44] Ibid.
[45] On April 24, 1866, Muybridge chaired a special meeting of the Bank of Turkey's shareholders, "for the purpose of determining whether the company should be wound up voluntarily." *Bankers' Magazine and Journal of the Money Market, and Commercial Digest* (London: Groombridge and Sons, 1866) 26 (May 1866): 596.
[46] *San Francisco Examiner*, February 6, 1881. Muybridge himself wrote this article, which was adapted from his unpublished introduction to J. D. B. Stillman's book *The Horse in Motion*.
[47] Robert Hunt, *Handbook to the Industrial Department of the International Exhibition 1862* (London: Edward Stanford, 1862), 2:249.
[48] Sarah Greenough writes of the Photographic Society: "Their exhibitions were routinely criticized for their lack of novelty.... At the same time they were excoriated for commercialization." Sarah Greenough, "'A New Starting Point': Roger Fenton's Life," in Gordon Baldwin et al., *All the Mighty World: The Photographs of Roger Fenton, 1852–1860* (New Haven, CT: Yale University Press, 2004), 30.
[49] Steve Edwards, *The Making of English Photography: Allegories* (University Park: Pennsylvania State University Press, 2006), 183.
[50] "Photography During the Past Year," *Photographic News: A Weekly Record of the Progress of Photography* 7, no. 226 (January 2, 1863): 1.
[51] Ibid., 2.
[52] Ibid.
[53] Hugh Welch Diamond, in *Reports by the Juries on the Subjects in the Thirty-six Classes into Which the Exhibition Was Divided* (London: Society for the Encouragement of Arts, Manufactures, and Commerce, 1863), quoted in Phillip Prodger, *Time Stands Still: Muybridge and the Instantaneous Photography Movement* (New York: New York: The Iris & B. Gerald Cantor Center for Visual Arts at Stanford University in association with Oxford University Press, 2003), 83.
[54] "Exhibition," *Journal of the Photographic Society*, May 21, 1858, 208, quoted in Malcolm Daniel, "'On Nature's Invitation Do I Come': Roger Fenton's Landscapes," in Baldwin et al., *All the Mighty World*, 33.
[55] Talbot patented his new calotype process for a time in Great Britain but not in France, so artists there were free to experiment with and improve on the technique without a costly license. Talbot argued that his patent covered all silver-based photographic processes, including wet-plate collodion.
[56] Hollis Frampton, "Eadweard Muybridge: Fragments of a Tesseract," in Hollis Frampton, *Circles of Confusion: Film, Photography, Video, Texts, 1968–1980* (Rochester, NY: Visual Studies Workshop Press, 1983), 74–75.
[57] Starr, *California: A History*, 117.
[58] Thomas Cole, "Essay on American Scenery," *American Monthly Magazine* 1 (January 1836): 1–12.
[59] See Franklin Kelly, "Niagara," in Eleanor Heartney et al., *A Capital Collection: Masterworks from the Corcoran Gallery of Art* (London: Third Millennium Publishing, 2002), 130.
[60] See Weston Naef, James N. Wood, and Therese Thau Heyman, *Era of Exploration: The Rise of Landscape Photography in the American West, 1860–1885* (Buffalo, NY: Albright Knox Art Gallery; Boston: New York Graphic Society, 1975), 21.
[61] *New York Times* review, 1867, quoted in John K. Howat, *Frederic Church* (New Haven, CT: Yale University Press, 2005), 72.
[62] Walt Whitman, "Rise, O Days, from Your Fathomless Deeps," in *Leaves of Grass* (Washington, DC: Sold by the author, 1871), 267–68.
[63] Weston Naef speculates that Watkins photographed Yosemite in 1858 or 1859, before Weed. Watkins's *Vernal Fall, 300 feet*, 1858–59, an albumen-on-glass stereograph in the collection of the California Historical Society, was used as the source image for an engraved illustration in *Hutchings' Illustrated California Magazine*, November 1859, 193. See Weston Naef, *Carleton Watkins in Yosemite* (Los Angeles: J. Paul Getty Museum, 2008), 4–5. Weed is commonly held to be the first photographer in Yosemite, for his images sponsored by Hutchings made in June 1859. Some of these were used as the basis for *Hutchings' Illustrated* engravings, beginning with the lead story in the October 1859 issue. Weed worked for Robert Vance at the time as his wet-plate photographer, and some of Weed's stereographs were sold to a New York gallery and distributed internationally. Photographs of Yosemite were scarce through the mid-1860s. See Mary V. Hood, "Charles L. Weed, Yosemite's

First Photographer," *Yosemite Nature Notes* 38, no. 6 (June 1959): 76–87, and Gary F. Kurutz, "Yosemite on Glass," in Amy Scott, ed., *Yosemite: Art of an American Icon* (Berkeley: University of California Press, 2006), 55–59.

[64] Josiah D. Whitney, *The Yosemite Book: A Description of the Yosemite Valley and the Adjacent Region of the Sierra Nevada, and of the Big Trees of California, Illustrated by Maps and Photographs* (New York: Julius Bien, 1868).

[65] Martha Sandweiss, *Print the Legend: Photography and the American West* (New Haven, CT: Yale University Press, 2002), 182.

[66] Bret Harte, *The Luck of Roaring Camp and Other Sketches* (Boston: Houghton Mifflin, 1899), 15–16.

[67] Following the Great Exhibition in 1851, the Crystal Palace was rebuilt in Sydenham, outside of London.

[68] Mary V. Jessup Hood and Robert Bartlett Haas, "Eadweard Muybridge's Yosemite Valley Photographs, 1867–1872," *California Historical Society Quarterly* 43 (March, 1963): 6.

[69] Frampton, "Eadweard Muybridge: Fragments of a Tesseract," 76.

[70] For a geopolitical discussion of Niagara Falls as symbol of death in American culture, see Patrick V. McGreevy, *Imagining Niagara: The Meaning and Making of Niagara Falls* (Amherst: University of Massachusetts Press, 1994).

[71] James M. Hutchings, *Scenes of Wonder and Curiosity in California, Illustrated with Over One Hundred Engravings: A Tourist's Guide to the Yo-Semite Valley* (New York: A. Roman, 1871), 130–31. The same engraving that appeared in *Hutchings' Illustrated California Magazine* in November 1859, likely made from an 1859 glass stereograph by Watkins, is used again on page 131 of *Scenes of Wonder and Curiosity in California* to illustrate Hutchings's text.

[72] Ibid., 142–43. Hutchings gives the date as 1868, which is the copyright date on the stereograph of *Fissure in Eagle Rock.* The photograph was taken in 1867.

[73] Ibid., 111.

[74] *Daily Alta California*, February 19, 1868, quoted in Hendricks, *Eadweard Muybridge: The Father of the Motion Picture*, 19.

[75] Eadweard Muybridge, "Yo-sem-i-te Valley," Cosmopolitan Gallery of Photographic Art, 415 Montgomery Street, San Francisco, February 1868. The Eadweard Muybridge Bequest, Kingston Museum, Kingston upon Thames, newspaper cuttings book (hereafter called Kingston Museum scrapbook), 15.

[76] "Sun Pictures of Yosemite," *San Francisco Call*, February 17, 1868.

[77] Muybridge, "Yo-Sem-i-te Valley." Kingston Museum scrapbook, 15. On May 14, 1868, Muybridge wrote on a copy of this brochure to the Directors of the Mercantile Library, offering to donate some of these photograhs.

[78] John S. Hittell, *Yosemite: Its Wonders and Its Beauties* (San Francisco: H. H. Bancroft, 1868). Hittell had previously been a writer and editor for the *Daily Alta California* newspaper.

[79] Edward L. Wilson, "Our Picture," *Philadelphia Photographer* 6, no. 71 (November 1869): 374.

[80] Muybridge, "Yo-Sem-i-te Valley." Kingston Museum scrapbook, 15.

[81] In March 1869, Muybridge sent his design for the sky-shade to the *Philadelphia Photographer* and a detailed description was published under the name Helios in May. See Helios, "A New Sky Shade," *Philadelphia Photographer* 6, no. 65 (May 1869): 142–44.

[82] Muybridge's early work was distributed at first through Silas Selleck's Cosmopolitan Gallery of Photographic Art on Montgomery Street and also through Wilson, Hood, and Company, a photographic wholesale and retail supply company in Philadelphia run by Edward Wilson and John Hood.

[83] Fardon's *San Francisco Album* is considered one of the very first books documenting an American city that was illustrated entirely with photographs. See George Robinson Fardon, *San Francisco Album: Photographs of the Most Beautiful Views and Public Buildings of San Francisco* (San Francisco: Herre & Bauer, 1856); Fardon's book was thoroughly documented in a modern facsimile edition, with contributions by Rodger C. Birt, Marvin Nathan, Peter Palmquist, and Joan Schwartz (San Francisco: Chronicle Books and Fraenkel Gallery, 1999). Fardon showed this work at the Mechanics Institute Industrial Exhibitions of 1857 and 1858, which Muybridge also attended as a book dealer. Muybridge knew Fardon's work and may have been inspired by the scale and complexity of his San Francisco panorama.

[84] David Harris, with Eric Sandweiss, *Eadweard Muybridge and the Photographic Panorama of San Francisco, 1850–1880* (Montreal: Canadian Centre for Architecture; Cambridge, MA: MIT Press, 1993), 84.

[85] Stereographic panoramas were made up of a group of stereo cards often viewed in sequence in a Beckers Viewer, designed by Alexander Beckers. This was a stereo viewer built into a tabletop cabinet that held about thirty-six stereo cards on a continuous belt. The cards were flipped one after another by turning a knob, giving the effect of panning across the panorama.

[86] Sharon Bohn Gmelch, *The Tlingit Encounter with Photography* (Philadelphia: University of Pennsylvania Museum of Archaeology and Anthropology, 2008), 22.

[87] Muybridge, "Alaska," in *Catalogue of Photographic Views Illustrating the Yosemite, Mammoth Trees, Geyser Springs, and Other Remarkable and Interesting Scenery of the Far West by Muybridge* (San Francisco: Bradley & Rulofson Gallery of Portrait and Landscape Photographic Art, 1873), 39.

[88] Gmelch, *The Tlingit Encounter*, 149.

[89] Henry W. Halleck to Eadweard Muybridge, October 13, 1868, in Henry Halleck letter book, page 450, Record Group 94, National Archives and Records Administration, Washington, DC.

[90] William Henry Seward Sr., quoted in "Extracts from Criticism on Muybridge's Photographs," a brochure announcing plans to produce new series of Yosemite photographs (San Francisco: Nahl Brothers Studio, May 1872). Kingston Museum scrapbook, 15.

[91] Plate 13, Haas, *Muybridge: Man in Motion*, 24.

[92] John Arturo Martini, *Fortress Alcatraz: Guardian of the Golden Gate* (Berkeley: Ten Speed Press, 2004), 61.

[93] "East and West: Completion of the Great Line Spanning the Continent," *New York Times*, May 10, 1869, 1.

[94] Andrew Joseph Russell was the official photographer for the Union Pacific Railroad, Alfred A. Hart was the official photographer for the Central Pacific Railroad, and Charles R. Savage owned a commercial photography studio in Salt Lake City, not far from Promontory Summit.

[95] "East and West: Completion of the Great Line Spanning the Continent," 1.

[96] Hart was trained as a painter, which influenced his sense of subject mater and composition. Other photographers, including Watkins, sometimes published Hart's stereographs without giving him credit. He returned to painting later in his life.

[97] Muybridge, *Catalogue of Photographic Views*, 13.

[98] J. D. B. Stillman, "The Last Tie," *Overland Monthly* 3 (1869): 77.

[99] See Rebecca Solnit's excellent summary of the railroad's impact on the perception of time and space during the nineteenth century in Solnit, *River of Shadows*, 57–62.

[100] A news clipping taped to the back of Muybridge's *carte-de-visite* photograph of Chang Woo Gow (fig. 38) in the Special Collections, University of Nevada, Reno, reads: "The Chinese Giant.—The agent of Chang Woo Gow, the celebrated Chinese Giant, arrived in town yesterday. The Giant, his wife and the baby giant, will arrive here next Wednesday, and will remain here on exhibition four days. The 'Local' of the Marysville 'Standard' gobbled the Giant upon his appearance in that city, and interviewed him pretty thoroughly upon all points. Chang Woo Gow gave the following answers to questions put to him by the 'Standard' man: 'I am 24 summers old. I was born in Fychow, China. My ancestors came from Japan, and are all noted for their size. I am eight feet high, and weigh 324

pounds. My father was nine feet high, and died in his 65th year. I have a sister more tall than I am. My baby was born on the Fourth of last July—I call it Fire Works. I am a Catholic, and go to church whenever I can—nearly every Sunday. My father was a Catholic. I like California very much, and am trying to buy a farm at San Jose….' This would seem to cover pretty nearly the whole ground, but we may add that the mammoth Celestial is said to be well educated, and able to converse fluently in English, German and French." There is no indication of the source or when it was printed. Muybridge's photographs of Chang Woo Gow and his wife were taken in August 1870.
[101] *San Francisco Daily Evening Bulletin*, August 10, 1870.
[102] Eadweard Muybridge to Colonel R. S. Williamson, Lighthouse Board, January 13, 1871, Lighthouse Board letter book, National Archives and Records Administration, Washington, DC.
[103] Verso of *Light House at Punta de los Reyes, Coast of California (8074)*, 1871, Record Group 26, National Archives and Records Administration, Washington, DC.
[104] American painter Albert Bierstadt also visited the Farallon Islands in early 1872 and made sketches of some of the same landscapes that Muybridge photographed. Later that year, Bierstadt developed these in a series of paintings infused by the narrative drama of crashing waves and hunting seals that take elements from photographs by both Muybridge and Watkins.
[105] Hutchings, *Scenes of Wonder and Curiosity in California*, 228.
[106] Muybridge, *Catalogue of Photographic Views*, 37.
[107] The *San Francisco Daily Evening Bulletin* reported that the two met at an Art Association reception where Muybridge was exhibiting "some very effective views of the Russian River Valley," where the Buena Vista Vineyard was located. Solnit, *River of Shadows*, 263.
[108] Richard Steven Street, *Photographing Farmworkers in California* (Stanford, CA: Stanford University Press, 2004), 4.
[109] H. H. (Helen Hunt Jackson), *Bits of Travel at Home* (Boston: Roberts Brothers, 1878), 86. First published in the *Independent* (New York), August 29, 1872. Revealing a similar sensitivity to California farm imagery as she did in discussing Muybridge's photographs of the Buena Vista Vineyard, Jackson went on to write the popular and heartrending 1884 novel *Ramona*, about the love between a Native American field hand and a young woman of mixed Scottish and Native American heritage.
[110] Eadweard Muybridge, *Animals in Motion* (New York: Dover Publications, 1957), 13. Muybridge wrote, "Having constructed some special exposing apparatus and bestowed more than usual care in the preparation of the materials he was accustomed to use for ordinary quick work, the author commenced his investigation on the racetrack at Sacramento, California, in May, 1872, where he in a few days made several negatives of a celebrated horse named Occident, while trotting, laterally, in front of his camera, at rates of speed varying from two minutes and twenty-five seconds to two minutes and eighteen seconds per mile." This date is contested, since the landmark event was not reported in the press until the following year and no pictures survived from the sessions.
[111] Muybridge, in "12 Montgomery Street, San Francisco, May, 1872," his printed brochure announcing plans to produce a new series of Yosemite photographs. Kingston Museum scrapbook, 15.
[112] H. H. (Helen Hunt Jackson), "Bits of Travel," *The Independent* (New York), August 29, 1872. Kingston Museum scrapbook, 9.
[113] Ibid.
[114] "Photograph Studies: Eight Hundred Views of Yosemite Valley and the Big Trees," *Daily Alta California*, April 7, 1873.
[115] H. H. (Helen Hunt Jackson), *Bits of Travel at Home*, 86.
[116] John Ruskin, "The Storm-Cloud of the Nineteenth Century," lecture to the London Institution, February 4, 1884, in John Ruskin, *The Storm-Cloud of the Nineteenth Century: Two Lectures Delivered at the London Institution, February 4 and 11, 1884* (New York: John Wiley and Sons, 1884), 41.
[117] Scott, *Yosemite: Art of an American Icon*, 113.
[118] Albert Bierstadt's painting *Indians in Council, California*, 1872, is now in the collection of the Smithsonian American Art Museum, Washington, DC.
[119] Muybridge, *Catalogue of Photographic Views*, 30.
[120] "Industrial Condition of the Slope," *Daily Alta California*, November 4, 1872.
[121] Muybridge and Towne together signed the hotel register of the Casa Nevada in Little Yosemite Canyon as "photographic artists" on July 20, 1872; Solnit, *River of Shadows*, 97. Also, Mary Hood and Robert Haas suggest that Towne may be responsible for some of Muybridge's 1872 Yosemite stereographs. See Hood and Haas, "Eadweard Muybridge's Yosemite Valley Photographs," 19.
[122] Weed had also shot a significant series of mammoth-plate landscapes of Yosemite and the sequoias in 1864.
[123] Peter E. Palmquist, "Charles Leander Weed, California's Peripatetic Photographer," *California History* 58, no. 3 (Fall 1979): 216. Palmquist also notes, "A series of coincidences point to the possibility that Weed may have collaborated … on Muybridge's 1872 photographic expedition to take mammoth-plate views in Yosemite. In 1872, Weed and Muybridge both had affiliations with Thomas Houseworth and Company, which initially may have supported Muybridge's expedition. However, after returning from Yosemite, Muybridge forged a new alliance with Bradley and Rulofson. At the same time, Weed left Houseworth for Bradley and Rulofson. Although Weed's presence on the 1872 Yosemite trip is conjectural, his experience with large format photography in the Valley would have made him an ideal assistant for Muybridge, who apparently took his first mammoth views on that expedition. Weed was not adverse to working anonymously on other photographers' projects." Palmquist and Kailbourn, *Pioneer Photographers of the Far West*, 587–88. Also see Palmquist's comments in Solnit, *River of Shadows*, 97, 281.
[124] Palmquist and Kailbourn, *Pioneer Photographers of the Far West*, 230. Fiske worked as a printer for Carleton Watkins in the late 1860s and lived and worked in Yosemite for much of his life after 1879.
[125] There are no confirmed portraits of Muybridge by Watkins. Watkins scholar Weston Naef has attributed to Watkins the Mariposa Grove portrait of Muybridge sitting at the base of the General Grant, based on its composition and the past relationship between the two photographers. Some of Watkins's earlier photographs of the base of the Grizzly Giant are similarly composed. In addition, one Watkins stereograph titled *The Prince of the Forest, Calaveras Grove* (895), in the collection of the New York Public Library, depicts two small figures—a bearded man seen from behind and a woman looking at the camera—at the base of a giant tree. On close examination, it is possible the two are Muybridge and his wife, Flora.
[126] In 1826, George William Featherstonhaugh commissioned Thomas Cole to create a number of paintings of his developing estate in northern New York state. Featherstonhaugh improved farming techniques, helped create the New York Board of Agriculture, and promoted the construction of a steam railroad system to connect disparate parts of the growing United States. He was later appointed government geologist and was assigned to survey the resources between the Missouri and Red rivers.
[127] Muybridge, *Catalogue of Photographic Views*, 2.
[128] Peter E. Palmquist, *Lawrence and Houseworth/Thomas Houseworth and Co.: A Unique View of the West, 1860–1886* (Columbus, OH: National Stereoscopic Association, 1980), 36.
[129] *San Francisco Daily Evening Bulletin*, November 26 and November 29, 1873. The exchange appears in Kingston Museum scrapbook, 9.
[130] Thomas Houseworth and Company, *Sun Pictures of the Yo-Semite Valley, Cal.* (Chicago: Coyne and Relyea, 1874).
[131] Muybridge's Bradley and Rulofson photograph of *The Domes. Valley of the Yosemite. No. 20* was shot from the same standpoint but probably earlier in the season, as the water in the Merced

River is higher, covering several rocks and more of the sandbar than in the later Houseworth print.

[132] Hermann Wilhelm Vogel, "German Correspondence. Photography—The Vienna Exhibition—North America—Brazil—England," *Philadelphia Photographer* 10, no. 117 (September 1873). Vogel is known for developing the panchromatic properties of film emulsions, making them sensitive to a wider spectrum of color. He was also Alfred Stieglitz's teacher in Berlin from 1882 to 1886.

[133] Hermann Wilhelm Vogel, "Medals Awarded at the Vienna Exhibition," *Philadelphia Photographer* 10, no. 118 (October 1873): 487–88.

[134] "Photograph Studies: Quick Work," *Daily Alta California*, April 7, 1873.

[135] "General Notes," *New York Times*, May 2, 1873, 5. See also *Atlanta Constitution*, May 8, 1873; *Le Moniteur de la Photographie: International Journal of the Progress of the New Art* (Paris), June 1, 1873; and *Bruce Herald* (Milton, New Zealand), September 19, 1873, 7.

[136] Peter Cozzens, "Introduction," in Jefferson C. Davis Riddle, *The Indian History of the Modoc War* (1914; Mechanicsburg, PA: Stackpole Books, 2004), v.

[137] "Photographs of the Lava Beds," *San Francisco Chronicle*, April 29, 1873.

[138] Loa-kum Ar-nuk appears in several other group portraits of the Warm Springs scouts, hired to aid the U.S. Army in its fight against the Modocs. A variant image of the *Modoc Brave* picture was published in an eye-witness history of the Modoc War, captioned *Warm Springs U.S. Indian Scout on the Lookout for the Modocs during the Modoc War*. See Riddle, *The Indian History of the Modoc War*, 123.

[139] Eadweard Muybridge, "Leland Stanford's Gift to Art and Science: Mr. Muybridge's Inventions of Instant Photography and the Marvelous Zoogyroscope," *San Francisco Examiner*, February 6, 1881. Muybridge initially wrote this text about the history of his work with Stanford as an introduction to *The Horse in Motion* by J. D. B. Stillman. It was rejected by Stillman as being "ungrammatical, redundant, full of hyperbole." See Anita Ventura Mozley, Robert Bartlett Haas, and Françoise Forster-Hahn, *Eadweard Muybridge: The Stanford Years, 1872–1882* (Stanford, CA: Stanford University Department of Art, 1972), 119–23.

[140] Rosalind Krauss, "Photography's Discursive Spaces: Landscape/View," *Art Journal* 42, no. 4 (Winter 1982): 312.

[141] *Notes on Muybridge Given to Harry C. Peterson by Mr. Nash, June 10, 1917, at the Odd Fellow's Home in Los Gatos*. Harry C. Peterson Papers, Wilson Centre for Photography, London. Peterson was the first director of the Stanford University Museum and he spent years researching Muybridge's life and work for a screenplay about the artist. There are many inaccuracies in the various correspondence and interviews conducted by Peterson. Some of these result from the imprecision of people's memories and some from differing points of view about the information. In citing the Peterson materials, facts were checked against other sources.

[142] Jeanne Schinto, "Deviled Ham Untouched by Hands: Food-Related Vintage Stereoviews," *Gastronomica: The Journal of Food and Culture* 2, no. 4 (Fall 2002): 53–54. According to Schinto, Peter Palmquist says about this picture: "The whole idea is that California can grow anything both faster and bigger than anyplace else."

[143] Robert Taft, *Photography and the American Scene: A Social History, 1839–1889* (New York: MacMillan, 1938), 332.

[144] Testimony of Susan Smith, *San Francisco Chronicle*, February 6, 1875.

[145] George Smith, "The Fatal Amour: An Interview with Muybridge, the Slayer of Major Larkyns," *San Francisco Chronicle*, December 21, 1874.

[146] *San Francisco Chronicle*, January 21, 1875. Cited in Haas, *Muybridge: Man in Motion*, 73.

[147] Peter Galassi, *Before Photography: Painting and the Invention of Photography* (New York: Museum of Modern Art, 1981), 17.

[148] Lyman L. Palmer, *History of Napa and Lake Counties, California* (San Francisco: Slocum, Bowen, 1881), 152–53.

[149] *San Francisco Daily Evening Post*, February 2, 1875. Cited in Hendricks, *Eadweard Muybridge: The Father of the Motion Picture*, 76.

[150] *Daily Register* (Napa, California), February 6, 1875. Cited in Haas, *Muybridge: Man in Motion*, 77.

[151] *San Francisco Daily Examiner*, May 1, 1875.

[152] According to a letter from the California State Library to Gordon Hendricks, Muybridge sent Floredo to the Protestant Orphan Asylum on September 16, 1876; cited in Solnit, *River of Shadows*, 266. A handwritten receipt from Mrs. J. Durbrow dated December 30, 1877, states, "Received from Mr John H Dall, Sixty ($60) on account of Mr E J Muybridge for six months board, of his son, in the P O Asylum." Collection of the Wilson Centre for Photography, London.

[153] George Montague Wheeler, *Geographical and Geological Explorations and Surveys West of the 100th Meridian, Seasons of 1871, 1872, and 1873* (Washington, DC: War Department, Corps of Engineers, U.S. Army, 1874). Collection of the Corcoran Gallery of Art, Washington.

[154] Robin Kelsey, *Archive Style, Photographs and Illustrations for U.S. Surveys, 1850–1890* (Berkeley: University of California Press, 2007), 191.

[155] In December 1874, the Central Pacific Railroad, headed in part by Stanford, Huntington, and Hopkins, and the Union Pacific, then controlled in part by robber baron Jay Gould, invested in a competing steamship company to break the monopoly of the Pacific Mail Steamship Company on their routes between New York, San Francisco, and the Orient. The directors of the new Occidental and Oriental Steamship Company of California chartered several steamers from England with plans to begin operations in February 1875. A *New York Times* report implied that the new steamship company had been formed because the Pacific Mail Steamship Company had broken its contract with the Central Pacific and Union Pacific railroads to use only their lines in hauling goods coming east from China. See "Pacific Mail Rival," *New York Times*, December 17, 1874. In fact, the impending competition from the new company actually lowered stock prices for the Pacific Mail Steamship Company and by January 1875, with the help of Stanford and Huntington, Gould acquired a majority stake to take control of the company, helping to ensure that shipping rates by sea would not undercut the rates charged by the railroads. By spring, when Muybridge was already working in Panama, stock prices had jumped for both Pacific Mail and the Union Pacific Railroad. See Maury Klein, *The Life and Legend of Jay Gould* (Baltimore: Johns Hopkins University Press, 1997), 143–44.

[156] Hendricks, *Eadweard Muybridge: The Father of the Motion Picture*, 81.

[157] E. Bradford Burns, *Eadweard Muybridge in Guatemala, 1875: The Photographer as Social Recorder* (Berkeley: University of California Press, 1986), 14.

[158] *Panama Star*, November 1, 1875. Kingston Museum scrapbook, 14.

[159] Katherine Manthorne, "Plantation Pictures in the Americas, circa 1880: Land, Power, and Resistance," in *Nepantla: Views from the South* 2, no. 2 (2001): 331.

[160] Jonathan Crary, *Suspensions of Perception: Attention, Spectacle, and Modern Culture* (Cambridge, MA: MIT Press, 1999), 145.

[161] In 1877, *The Graphic* in London published an article about the Las Nubes plantation, declaring, "Although Guatemala is a republic, the Government is in effect despotic; and the Indians, although nominally free, are really slaves attached to the plantation. This anomaly is caused by the universal system of the Indians drawing pay in advance for work to be performed. They always insist on being in the planter's debt, and consequently are prohibited by law from working upon any other plantation until the debt is cancelled. Each plantation of any extent employs a detective to hunt up runaways, who, unless they elect to return to their plantation labours, are set to work in the chain-gang, making and repairing roads." "Coffee Cultivation in Guatemala," *The Graphic*, October 20, 1877.

[162] Eadweard Muybridge, "Announcement," Guatemala, October 1, 1875. Kingston Museum scrapbook, 15.

[163] Katherine Manthorne calls this "a reminder of the constructed nature of all images—and of [Muybridge's] own role in their manipulation." See Manthorne, "Plantation Pictures in the Americas," 336.
[164] "Report of the Judges," Eleventh Industrial Exhibition, San Francisco, 1876. Kingston Museum scrapbook, 18, cited in Mozley et al., *Eadweard Muybridge: The Stanford Years*, 59.
[165] Eadweard Muybridge to Mrs. W. W. Pendegast, Napa, California, May 25, 1876. California State Library, Sacramento.
[166] *San Francisco Chronicle*, February 7, 1875.
[167] One album documenting the Stanford family residence is now in the Special Collections of the Green Library at Stanford University. Another, once owned by Stanford's wife, Jane, is in the collection of the San Francisco Public Library. That album contains fifty-six photographs of the interior rooms, two exterior views, and five views of San Francisco taken from the house. The album was given by Jane Stanford to her brother-in-law Josiah Stanford and contains descriptive captions in her hand of room styles and colors.
[168] Muybridge may have received the commission because of the Johnsons's acquaintance with Stanford. They had previously owned land attached to the Buena Vista Vineyard in Sonoma. In a letter to Étienne-Jules Marey, written in July 1882, Muybridge mentioned Robert Johnson as a potential funder of a major new animal locomotion project in Europe. See Braun, *Picturing Time: The Work of Étienne-Jules Marey*, 231. The *Daily Alta California* reported on May 21, 1890, that the Johnsons sold water rights along San Francisquito Creek in Menlo Park to Stanford to support the new Leland Stanford Junior University. The *Johnson Photo Album* is now in the collection of the Clements Library, University of Michigan, Ann Arbor. A second, larger album of Muybridge's photographs of the Johnson homes, including portraits and several images of the Buena Vista Vineyard, is in the collection of the Achenbach Foundation for Graphic Arts, Fine Arts Museums of San Francisco.
[169] Frampton, "Eadweard Muybridge: Fragments of a Tesseract," 77.
[170] *San Francisco Daily Evening Bulletin*, August 22, 1877, cited in Solnit, *River of Shadows*, 266.
[171] Though noted in his catalogues and reviews, Vance's panoramas of San Francisco have disappeared and are thought to be lost.
[172] David Harris, "Eadweard Muybridge and the Photographic Panorama of San Francisco, 1850–1880," in Harris, *Eadweard Muybridge and the Photographic Panorama of San Francisco*, 38.
[173] *San Francisco Chronicle*, May 19, 1875, cited in Mozley et al., *Eadweard Muybridge: The Stanford Years*, 64.
[174] Plate 7 of his 1878 panorama was damaged and remade by Muybridge at a later time. The angle of the light in prints of plate 7 reveals that it was shot at a different time of day from the others in the sequence.
[175] Vanessa R. Schwartz, "Cinematic Spectatorship before the Apparatus: The Public Taste for Reality in Fin-de-Siècle Paris," in Leo Charney and Vanessa R. Schwartz, eds., *Cinema and the Invention of Modern Life* (Berkeley: University of California Press, 1995), 311.
[176] Referring to 1872, when Muybridge first attempted to photograph a running horse, Jacob Stillman discussed instantaneous photography in the appendix to *The Horse in Motion* as follows: "Instantaneous pictures were defined to be, at that time, in ordinary photographic parlance, when the exposure time has been very brief, or under half a second. In the British Journal Photographic Almanac for 1868 it is stated that good street views had been taken in a twelfth part of a second. The conditions, as there given, for extreme rapidity of exposure are a good and quick acting shutter, a lens with a large angular aperture, and chemicals in perfect condition." "Appendix," in J. D. B. Stillman, *The Horse in Motion as Shown by Instantaneous Photography, with a Study on Animal Mechanics Founded on Anatomy and the Revelation of the Camera in Which Is Demonstrated the Theory of Quadrupedal Locomotion* (Boston: James R. Osgood, 1881–1882), 123.
[177] Brian Coe, *Muybridge and the Chronophotographers* (London: British Film Institute, Museum of the Moving Image, 1992), 6.
[178] Oscar Gustave Rejlander, "On Photographing Horses," *British Journal Photographic Almanac* (1872–1873): 115. In his essay, "Eadweard Muybridge, 1830–1804," Robert Haas proposes that Muybridge's "method adopted for the serial photographs at the Palo Alto farm was a practical elaboration of Rejlander's scheme of 1872/73, which envisioned the use of a battery of cameras to illustrate the varied positions of the animal's feet in a sequential series of photographs." See Mozley et al., *Eadweard Muybridge: The Stanford Years*, 21.
[179] Quoted in Marta Braun, *Picturing Time: The Work of Étienne-Jules Marey (1830–1904)* (Chicago: University of Chicago Press, 1992), 39.
[180] Prodger, *Time Stands Still*, 143.
[181] Haas, *Muybridge: Man in Motion*, 49.
[182] A written receipt is in the Collection of the Wilson Centre for Photography, London.
[183] Eadweard Muybridge to Alfred Poett, Esq., Menlo Park, Cal., May 21, 1877. Collection of the Wilson Centre for Photography, London.
[184] Sherman Blake to Walter R. Miles, May 6, 1929, Stanford University Library Special Collections, cited in Solnit, *River of Shadows*, 184.
[185] Tom Gunning, "Never Seen This Picture Before: Muybridge in Multiplicity," in Prodger, *Time Stands Still*, 244.
[186] *San Francisco Bulletin*, August 3, 1877. Kingston Museum scrapbook, 19.
[187] "'Occident' Photographed at Full Speed," *Daily Alta California*, August 3, 1877. Kingston Museum scrapbook, 18.
[188] Eadweard Muybridge to Fred McCrellish, publisher, San Francisco, August 2, 1877, published in *Daily Alta California*, August 3, 1877. Kingston Museum scrapbook, 19. Also cited in Haas, *Muybridge: Man in Motion*, 94.
[189] Library of Congress, Washington, DC, catalogue number PH-Muybridge (E.), no. 3.
[190] "Report of Judges," Twelfth Industrial Exhibition, San Francisco, 1877, 197. The report states, "The negative for this photograph was executed for the Hon. Leland Stanford, the owner, while the horse was trotting at the rate of 2:40 or about thirty-six feet in a second. As an illustration of the marvelous resources of photography, this is a wonderful production, the duration of the exposure having been less than one-thousandth part of a second. Kingston Museum scrapbook, 18.
[191] John Isaacs, July 18, 1883, Edward J. Muybridge v. Leland Stanford, deposition, Collis P. Huntington Papers, Special Collections, George Arents Research Library, Syracuse University.
[192] Eadweard Muybridge to Alfred Poett, Esq., Palo Alto, California, September, 1877. Collection of the Wilson Centre for Photography, London.
[193] Eadweard Muybridge to Alfred Poett (Friend Poett), October 11, 1877. Collection of the Wilson Centre for Photography, London.
[194] Receipt from the San Francisco Telegraph Supply Company to Mr. Muybridge, November 3, 1877. Collection of the Wilson Centre for Photography, London.
[195] Muybridge confirms, "We began our experiments in May…," in a letter published in *La Nature*, February 1879, cited in Solnit, *River of Shadows*, 189.
[196] "The Stride of a Trotting Horse: Photographing a Horse while Traveling Forty Feet a Second," *California Rural Press*, June 22, 1878.
[197] Stillman, *The Horse in Motion*, 125.
[198] Anticipating potential problems, Muybridge had written to Poett on October 29, 1877: "I think it will probably be as well to have the benches made in two pieces, each 15 feet long, with the tops movable so that they can be taken off with the cameras, and shutters attached and put into the dark room at the rear of the background, in the event of rain." Eadweard Muybridge to Alfred Poett

(Friend Poett), October 29, 1877. Collection of the Wilson Centre for Photography, London.
[199] Stillman, *The Horse in Motion*, 124.
[200] "A Wonder of the Century," *San Francisco Morning Call*, June 16, 1878. Kingston Museum scrapbook, 21.
[201] Stillman described the clockwork mechanism as follows: "For the purpose of making the exposures at the proper intervals of time, a machine was constructed on the principle of a Swiss music box, having a cylinder with a row of twelve pins arranged spirally. This was put in motion by a spring, and, as it revolved, each pin in succession established a magnetic circuit, with the magnet connected with each of the twelve cameras in succession, and the whole series of exposures was made in the time occupied by a single complete stride of the horse." *The Horse in Motion*, 124.
[202] A receipt from Jero. L. Boone to Edward J. Muybridge dated June 14, 1878, shows that Muybridge paid Boone $67.50, "in full except last Govt fee on his app for patent on Instantaneous Photography and for tracing of drawing and copy of specifications." Collection of the Wilson Centre for Photography, London. U.S. patent number 212865 was granted on March 4, 1879. Muybridge's patent model for the shutter mechanism is housed in the National Museum of American History, Smithsonian Institution, Washington, DC.
[203] Receipt from the San Francisco Telegraph Supply Company to E. J. Muybridge, November 3, 1877. Collection of the Wilson Centre for Photography, London.
[204] "Automatic-Electro-Photography: The Horse in Motion," *Resources of California*, August 1878. Kingston Museum scrapbook, 28.
[205] John Ott, "Iron Horses: Leland Stanford, Eadweard Muybridge, and the Industrialised Eye," *Oxford Art Journal* 28, no. 3 (2005): 407. Ott's concise description in this article of the "industrial gaze" is very important in clarifying the relationship between Stanford and Muybridge and their production of the Palo Alto motion studies.
[206] William H. Rulofson, "The California Horse—Electric Feat," *Philadelphia Photographer* 15, no. 176 (August 1878). Kingston Museum scrapbook, 32. In January 1879, editors of the *Philadelphia Photographer* responded to Rulofson's claim, stating, "Surely Mr. Rulofson could not have given much thought to the subject when he called these pictures 'silhouettes' only, and the claims made for them as 'bosh.'" Kingston Museum scrapbook, 33.
[207] "A Horse's Motion Scientifically Determined," *Scientific American*, October 19, 1878, 241. Kingston Museum scrapbook, 33–35.
[208] Étienne-Jules Marey to Gaston Tissandier, December 18, 1878 (published in *La Nature* on December 28, 1878). Quoted in Braun, *Picturing Time: The Work of Étienne-Jules Marey*, 47.
[209] Eadweard Muybridge to Gaston Tissandier, February 17, 1879 (published in *La Nature* on March 22, 1879). Quoted in Braun, *Picturing Time: The Work of Étienne-Jules Marey*, 393–94.
[210] Eadweard Muybridge to Alfred Poett, San Francisco, January 13, 1879. Collection of the Wilson Centre for Photography, London.
[211] The *San Francisco Call* reported on January 20, 1880, that "Doubling the number of cameras gives every portion of the stride with absolute fidelity, and nothing is left for conjecture." Kingston Museum scrapbook, 57.
[212] "Photographing Men in Motion: Instantaneous Pictures of Athletic Exercises," *San Francisco Chronicle*, August 9, 1879, also quoted in *Photographic Times*, October 1879, 227. Kingston Museum scrapbook, 57.
[213] Leo Charney, "In a Moment: Film and the Philosophy of Modernity," in Charney and Schwartz, *Cinema and the Invention of Modern Life*, 289.
[214] Fairman Rogers, "The Zoötrope: Action of Animals in Motion—The Muybridge Photographs of Horses—The Instrument as a Factor in Art Studies," *Art Interchange* 3, no. 1 (July 9, 1879). Kingston Museum scrapbook, 55.
[215] Lloyd Goodrich, *Thomas Eakins* (Cambridge, MA: Harvard University Press, 1982), 260.
[216] Rogers, "The Zoötrope." Kingston Museum scrapbook, 55.
[217] Goodrich, *Thomas Eakins*, 263.
[218] W. B. Tegetmeier, "The Paces of the Horse," *The Field: The Country Gentleman's Newspaper*, June 28, 1879. Kingston Museum scrapbook, 62.
[219] Eadweard Muybridge, *Animals in Motion* (New York: Dover Publications, 1957), 14.
[220] Other examples of nineteenth-century image animating devices include John Ayrton Paris's thaumatrope, demonstrated in 1825; Franz Freiherr von Uchatius's projecting phenakistoscope in 1845; another version of the projecting phenakistoscope by Louis Jules Duboscq from the 1850s; the choreutoscope, invented in 1866 by L. S. Beale; A. B. Brown's projecting phenakistoscope from 1869; the phasmotrope designed by Henry Heyl in 1870; and Reynaud's praxinoscope, patented in 1877.
[221] Muybridge, *Animals in Motion*, 15.
[222] Stephen Herbert, "Projecting the Living Image," in Stephen Herbert, ed., *Eadweard Muybridge: The Kingston Museum Bequest* (Hastings, UK: Projection Box, 2004), 113. For an excellent, detailed discussion of the technical complexity of the zoopraxiscope, the production of the discs, and Muybridge's subsequent lectures, see pages 107–53. Some glass negatives for the 12-inch zoopraxiscope discs are now in the collection of the National Museum of American History, Smithsonian Institution, Washington, DC.
[223] George T. Clark, *Leland Stanford: War Governor of California, Railroad Builder, and Founder of Stanford University* (Stanford, CA: Stanford University Press, 1931), 367–68.
[224] "The Zoogyroscope: Photographs Illustrating Animals in Motion," *San Francisco Call*, May 8, 1880. Kingston Museum scrapbook, 60.
[225] Leland Stanford to E. J. Muybridge, Bill of Sale and Assignment, May 30, 1881. Bancroft Library, University of California, Berkeley.
[226] Frampton, "Eadweard Muybridge: Fragments of a Tesseract," 75.
[227] Depositions from the Muybridge v. Stanford lawsuit have provided a great deal of information about the technological development of the Palo Alto project. These are included in the Collis P. Huntington Papers, Special Collections, George Arents Research Library, Syracuse University.
[228] Eadweard Muybridge, *Prospectus of a New and Elaborate Work upon the Attitudes of Man, the Horse and Other Animals in Motion* (New York: Scovill Manufacturing Company, 1883).
[229] William Pepper, "Note," in W. D. Marks, H. Allen, and F. X Dercum, *Animal Locomotion: The Muybridge Work at the University of Pennsylvania—The Method and the Result* (Philadelphia: J. B. Lippincott, 1888), 5–6.
[230] Ibid., 5.
[231] Muybridge, *Prospectus of a New and Elaborate Work upon the Attitudes of Man, the Horse and Other Animals in Motion*, cited in Anita Ventura Mozley, "Introduction to the Dover Edition," in Eadweard Muybridge, *Muybridge's Complete Human and Animal Locomotion: All 781 Plates from the 1887 Animal Locomotion* (Mineola, NY: Dover Publications, 1979), 1: xxiv–xxv.
[232] Braun, *Picturing Time: The Work of Étienne-Jules Marey*, 249.
[233] Linda Williams, *Hard Core: Power, Pleasure, and the "Frenzy of the Visible"* (Berkeley: University of California Press, 1999), 38–42.
[234] Braun, *Picturing Time: The Work of Étienne-Jules Marey*, 249.
[235] Eadweard Muybridge to Major John Wesley Powell, May 16, 1887. Collection of the National Anthropological Archives, Smithsonian Institution, Washington, DC. The Corcoran Gallery of Art acquired a complete set of *Animal Locomotion* plates from Muybridge in 1887, and they were used by the artist as examination copies for other potential subscribers from the region. These are the plates on view in the exhibition *Helios: Eadweard Muybridge in a Time of Change.*

[236] A series of letters from Muybridge to Reverend Jesse Burke, secretary of the University of Pennsylvania, written between 1887 and 1895, describe some of Muybridge's extensive travels and his complex involvement in the business of producing and selling *Animal Locomotion.* Collection of the University Archives and Records Center, University of Pennsylvania, Philadelphia. Despite some strong sales, *Animal Locomotion* was not a financial success for the University or for Muybridge, and in 1895 a lawsuit was filed against Muybridge by the New York Photogravure Company to recoup unpaid bills.

[237] Eadweard Muybridge, "Lecture by Eadweard Muybridge: Quadrupedal Walk," typescript in the collection of the National Museum of American History, Washington, DC.

[238] *New York World,* June 3, 1888, quoted in Charles Musser, *The Emergence of Cinema: The American Screen to 1907* (Berkeley: University of California Press, 1990), 62.

[239] Thomas Edison to Eadweard Muybridge, February 21, 1894. Edison Laboratory National Monument, West Orange, New Jersey.

[240] Muybridge's handwritten drafts and subsequent letters to Stanford, quoted in Haas, *Muybridge: Man in Motion,* 170–172, are in the collection of the Bancroft Library, University of California, Berkeley.

[241] Newton H. Carpenter, "World's Columbian Exposition: Department of Zoopraxography, Announcement," January 1893. Kingston Museum scrapbook, 239.

[242] Eadweard Muybridge, *Descriptive Zoopraxography; or, The Science of Animal Locomotion Made Popular...* (Philadelphia: University of Pennsylvania, 1893); *Chicago Daily Tribune,* April 12, 1893.

[243] Muybridge, *Descriptive Zoopraxography,* 2.

[244] Frederick Jackson Turner, *The Frontier in American History* (Tucson: University of Arizona Press, 1997). Cited in Richard Kluger, *Seizing Destiny: How American Grew from Sea to Shining Sea* (New York: Alfred A. Knopf, 2007), 44.

[245] Eadweard Muybridge, "Announcement," May 1895. Kingston Museum scrapbook, 239.

[246] Muybridge, *Animals in Motion,* 17.

[247] Muybridge, *Animals in Motion,* 16.

[248] Roland Barthes, *Camera Lucida* (New York: Hill and Wang, 1983), 85.

[249] Crary, *Suspensions of Perception,* 138.

6. *Father of the Forest (Est). 450 feet high, 112 feet circumference (130)*, 1867

The Infinite Landscape

7. *Yosemite Tourists*, 1867

8. *Yosemite Hotel (59)*, 1867

9. *Pioneers' Cabin, 318 ft. high, 73 ft. circumference (125)*, 1867

10. *Fissure in Eagle Rock. 1100 ft. deep*, 1867

11. *California—Yosemite Valley, Iris from the Summit of Pi-wy-ack* (297), 1867

12. *Summit of Third Fall of the Yosemite (4035)*, 1867

13. *Kahchoomah, Wild Cat Fall, 30 feet high, on the main fork of the Merced (4061)*, 1867

14. *An Hour Before Sunset, Bay of San Francisco*, 1868

15. *First ME Church, Powell Street, San Francisco*, 1871

16. *Panorama, from Rincon Hill (428)*, 1868

17. *Panorama, from Rincon Hill (434)*, 1868

18. *Military Post at Goat Island (4108)*, 1870

19. *The Steamer "Golden City" on the California Dry Dock (4123)*, 1868

20. *Fort Wrangle*, 1868

21. *Fort Tongass*, 1868

22. *Group of Indians, Alaska (479)*, 1868

23. *Alaska Ter.—Sitka. Russo Greek Priests, Alaska (472)*, 1868

24. *Cut at Alta, looking East (787)*, 1869

25. *Long Ravine Trestle and Bridge—113 feet high, 878 feet long, looking East (766)*, 1869

26. *View from Cape Horn, looking South, Mineral Bar Bridge 2,500 feet below (779)*, 1869
27. *Coming storm—Wasatch Mountains, from the Weber River, near Ogden (756)*, 1869

28. *William Herman Rulofson and Family Group at His San Francisco Home*, 1871

29. *Progress of Construction, U.S. Branch Mint*, 1870

30. *Progress of Construction, U.S. Branch Mint*, 1871

31. *Progress of Construction, U.S. Branch Mint,* 1871

32. *Geysers of Cal.*, 1870, from *Photo Album No. 9*, page 27, ca. 1873

33. *The "Heathen Chinee" Prospecting (848)*, 1868

34. *Photo Album No. 9*, page 36, ca. 1873

35. *Photo Album No. 9*, page 43, ca. 1873

36. *Animals at Woodward's Gardens (589)*, 1870

37. *Views in Woodward's Gardens*, 1870

38. *Point Shubrick, Tower Head, Lighthouse, Parrot Rock, and Gull Peak from the Hole in the Wall, South Farallon Island*, 1871

39. *The Ramparts, Funnel Rock, Hole in the Wall, Pyramid, Sugar Loaf, Oil House, and Landing Cove on Fisherman's Bay, South Farallon Island,* 1871

40. *Cape Disappointment Lighthouse, Mouth of Columbia River, First Order Fixed Light, 232 feet above Sea level (4164)*, 1871

41. *Cape Blanco, Lighthouse and Dwelling (4161)*, 1871

42. *Point Loma Lighthouse*, 1871

43. *Light House at Punta de los Reyes, Coast of California (8074)*, 1871

44. *Pigeon Point Lighthouse*, 1873

45. *Pigeon Point Lighthouse*, 1873

46. *City Hall Under Construction*, 1873

47. *Mills Seminary, Seminary Park, Alameda Co., Cal.*, 1873

48. *View of South Park, San Francisco*, ca. 1873

49. *Saucilito from the N.P.C.R.R. Wharf, looking South (1664)*, 1873

50. *Buena Vista Vineyard, Sonoma. Fertilizing the Soil (1091)*, 1871–72
51. *Buena Vista Vineyard, Sonoma. Plowing (1092)*, 1871–72

52. *Buena Vista Vineyard, Sonoma (4167)*, 1871–72, from *Lone Mountain College Collection, Album 3*, 1873

53. *An Octogenarian (1580)*, 1872

54. *Valley of the Yosemite, from Mariposa Trail. No. 3*, 1872

55. *Tutokanula, "The Great Chief El Capitan" from Rocky Ford (4178)*, 1872

56. *Valley of the Yosemite, from Rocky Ford. No. 4*, 1872

57. Tutokanula. Valley of the Yosemite. The Great Chief "El Capitan." 3,300 feet high. No. 9, 1872

58. *Tutokanula. Valley of the Yosemite. No. 11*, 1872, from *Yosemite Views*, 1873

59. *Valley of the Yosemite. Confluence of the Merced, and Yosemite Creek. No. 21*, 1872

60. *Cottonwood Bend. Valley of the Yo-semite. No. 15*, 1872

61. *Mirror Lake*, 1872, from *Photo Album No. 9*, page 10, ca. 1873

62. *Mirror Lake, (Reflection of Mount Waiya). "Pine Mountain." No. 25*, 1872

63. *Falls of the Yosemite. From Glacier Rock. (Great Grizzly Bear) 2600 feet fall. No. 36*, 1872

64. *Valley of the Yosemite. From Mosquito Camp. No. 22*, 1872

65. *Vernal Fall*, 1872, from *Photo Album No. 9*, page 11, ca. 1873

66. *Pi-Wi-Ack. Valley of the Yosemite. (Shower of Stars) "Vernal Fall." 400 Feet Fall. No. 29*, 1872

67. *Piwayack (Shower of Stars) Early Morning (1272)*, 1872

68. *Wild Cat Fall. Valley of the Yosemite. No. 30*, 1872

69. *Valley of the Yosemite from Union Point. No. 33*, 1872

70. *Tenaya Canyon. Valley of the Yosemite. From Union Point. No. 35*, 1872

71. *Ancient Glacier Channel. Lake Tenaya, No. 47*, 1872, from *Yosemite Views*, page 38, 1873

72. *Glacier Channels. Valley of the Yosemite. From Panorama Rock. No. 41*, 1872

73. *Wm. H. Seward. 85 feet in Circumference, 268 feet high. Mariposa Grove of Mammoth Trees. No. 51*, 1872

74. Charles Leander Weed (attributed), *U.S. Grant, Mariposa Grove, portrait of Eadweard Muybridge seated at base of tree*, 1872

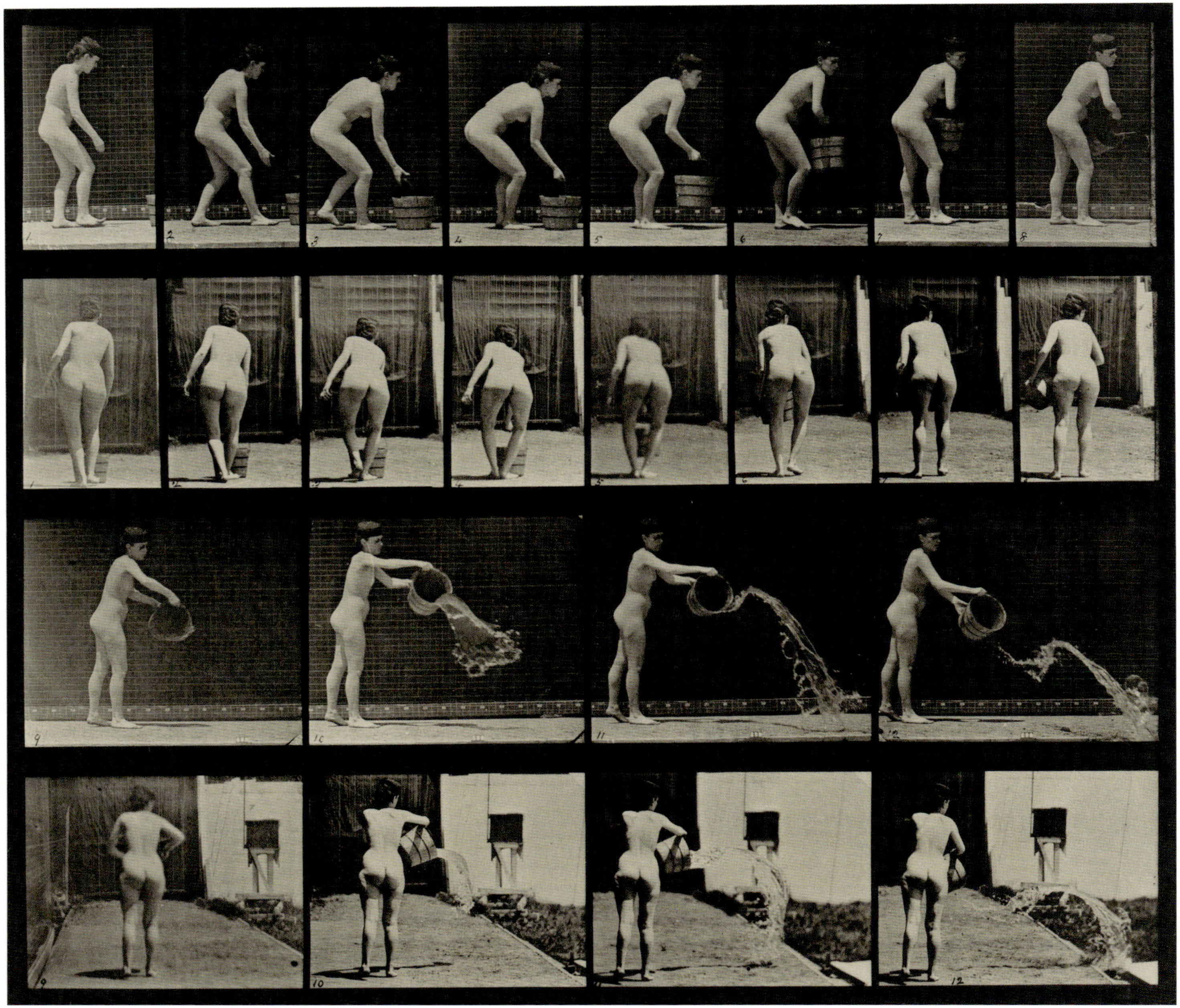

Fig. 92. *Emptying bucket of water. Plate 401*, 1887

Tangles, Time, Solitude, Transformation: Continuities in Eadweard Muybridge's River of Images

Rebecca Solnit

In the first sixteen frames of Eadweard Muybridge's motion study *Emptying bucket of water. Plate 401*, a nude woman with her dark hair tied back and a funny little flat cap atop it stoops and picks up the bucket and then straightens again (fig. 92). In the eight wider frames in two more rows below, consisting of four images each from two vantage points, she tosses the water high into the air. By the second of these frames, the water is a sort of flying wedge with a frill of droplets along the bottom edges. In the third, a fragment of a second later, the water arches from the bucket held at chest level to the ground, a long figure with a lot of fringes almost like legs or loops on the underside, striations along its upper side, and a few stray droplets all about. By the fourth, the water has leapt free of the bucket like a wild animal, and its tail has begun to catch up with its head. The far end of this liquid zigzag is at the very corner of the photographic frame, as if it were escaping. In another version of this final frame, taken at the same moment from a different angle, from behind the nude water thrower, you see an even wilder figure like a translucent fish, all fins and curves. This one looks almost as though it is jumping at the woman, not away from her.

The water is even more dynamically alive than she is. The term "bodies of water" is usually used for larger, more stable ponds, lakes, and seas, but in these hydrological motion studies, this water is itself a body in motion. Her body moves only slightly, while the water leaps like some liquid dragon, a Proteus that appears, transforms itself, and turns from a volume of water to a puddle on the floor too fast for the human eye to apprehend, a wraith that had remained essentially invisible. This is why the photographer is so intent upon capturing it for what must have been the first time in all consciousness and history. As Phillip Prodger remarks, "It was Muybridge, more than any other figure, who introduced what Walter Benjamin, decades later, termed 'the optical unconscious,' revealing that much of everyday life takes place beneath the threshold of our conscious awareness."[1] Horses' legs arranged themselves in surprising ways even at a walk, let alone at a trot or a gallop; men leaping and somersaulting were seen in hitherto unsuspected positions; the flapping wings of doves and vultures were made visible in all their phases; and then there were the pours and splashes.

Water appears again and again in Muybridge's great 781-plate opus that might have been titled "Human, Animal, and Liquid Locomotion." *Pouring basin of water over head. Plate 409* shows a woman doing so while standing in a small tub, the kind the painter Edgar Degas' dancers washed themselves in. In the first frame, the basin is level with and obscures her head. By the fifth it is raised high, and in the two rows of images in which she is seen from the front and the back, her bent elbows form a sort of diamond: it looks as though the big blank round of the tilted basin is now the head of a supernatural figure, a strange awkward deity that has appeared, a hydrological kachina. And like a divinity, it could not and cannot be seen by the ordinary human eye, for in this frame the deity wears a wide cloak or wings of splashing water that obscure the face and part of the body and seem like some solid substance. You can imagine a contemporary artist casting this water in glass so that what was frozen forever in the flat realm of a photograph might live again in three dimensions. Documented simultaneously from three angles, these splashes from fleeting moments of the 1880s are startlingly reconstructable.

Each watery motion study is an astonishing, eerie image of something no one had ever seen before. In *Pouring water in basin and washing face. Plate 413*, a shining water snake stretches from a pitcher to a basin. Another one stretches from a wet piece of laundry to the ground in *Wringing clothes. Plate 433*; this most banal of acts reveals its own poetry of motion, as do they all. The kachinas, snakes, and dragons are the hitherto invisible presences of the most quotidian acts with one of the most common substances on earth. By *Stooping, and rinsing tumbler, Plate 440*, the water breaks into big droplets like a string of pearls flying apart

Fig. 93. *Emptying bucket of water. Plate 399*, 1887

or a figure delineated by scattered dots. *Two models, 1 pouring bucket of water over 8. Plate 408* shows two nudes (identified by their model numbers: 1 and 8). The one over whom the water is being poured in one frame disappears under the glassy veil but for her startled, defensive hands. In *Emptying bucket of water. Plate 399*, a balding man in his middle years tosses water even higher than the woman in *Plate 401* does: in the penultimate frame it occupies more space than he does, a dangerous-looking creature that could be leaping either way (fig. 93). Other figures haul buckets, dip water from containers, pour it, splash it, wash in it. In *A, sitting down; B, sitting down; C, sprinkling water; D, stooping for cup and drinking. Plate 490*, a nude Muybridge himself, white-bearded but strong in his mid-fifties, bends to fill a cup and straightens to drink from it (fig. 115).

Eadweard Muybridge's work has long been treated as though it were made by a succession of different artists. Some of these artists have eclipsed the rest, particularly the last one who left San Francisco for Philadelphia in the 1880s, the motion-study artist who made those astonishing 781 plates of people and animals (and water) in motion. This body of work expanded and deepened the 1877–79 Bay Area motion studies that revolutionized photography and perception and laid the groundwork for cinema. The ethnographer, survey, war, and architectural photographer that Muybridge had been has almost disappeared from view, hidden behind his San Francisco panoramas and Yosemite mammoth-plate landscapes and other great landmarks in the history of photography. This fragmentation may be a consequence of how descriptions and lists are generated, for a conventional catalogue of his work reveals a dizzying array of genres and subjects but not their common threads, and the artist himself never spoke of these continuities. "HELIOS is prepared to photograph private residences, animals, or views in the city or any part of the coast," said an advertisement he took out in April of 1868; the back of one of his Central America photographs taken in 1875 declared, "Muybridge, landscape, marine, architectural and engineering photographer, official photographer of the U.S. Government." Curators and scholars of cinema and of photography have tended to divide up the work: a major scholarly show singled out the panoramas of San Francisco, for example, while others have dealt only with the Yosemite work or the motion studies or "the Stanford years." The caliber of Muybridge's work too moves from merely adequate for the market to dazzlingly inventive and eccentric, though strange and disturbing beauties show up everywhere.

Fig. 94. *Helios Flying Studio*, 1870

But a consistent set of themes or preoccupations runs through all the work. He was concerned with water, with clouds, with the complex textures of trees and tangles of foliage or stone. He sought out unusual angles and uncomfortable vantage points, preferring the interesting to the harmonious. Many of his photographs were disconcerting, destabilizing, showing the world in unfamiliar ways long before he dissected and reassembled motions made visible as never before. Also evident throughout is an interest in motion and the representation of time, a willingness to break rules and make images that defy conventional aesthetics, a restlessness, a taste for complexity, and an interest in pushing the possibilities of the medium. He also photographed from the very beginning in series and sequences, building complex portraits of places and processes out of many images, and these too prefigure the motion studies. It may privilege the motion studies too much to call the rest of the work a prefiguration; you might say equally that the motion studies follow in many ways the mindset and methodology established long before. And the motion studies developed in three stages, the first two simultaneously with other important projects. In 1872 the railroad magnate Leland Stanford first contacted him about photographing a horse trotting so that the exact motion of its legs could

Fig. 95. *Yowiye, Nevada Fall, 600 feet fall (4201)*, 1872

be frozen and seen for the first time. Muybridge worked on this challenge while making his second major suite of Yosemite photographs and photographing the Modoc War. In 1877 and 1878 he returned with new technical prowess to the challenge and succeeded as he had not before. During those two years, he completed his two 360-degree panoramas of San Francisco, the second one a superlative example of the genre made like the 1872 Yosemite work with a mammoth-plate camera. It is all one body of work, the fruit of one exceptional and eccentric mind.

The preoccupation with water was there in his very first known image series, the Yosemite photographs of 1867. Though Yosemite was celebrated for its vertical stone faces and waterfalls, Muybridge often chose to celebrate the quotidian landscape instead—and waterscape, for the Merced River dominates many of his 1872 mammoth-plate photographs as well. But it is not an ordinary river. In it, Muybridge often showed the surrounding monumental edifices of stone so that their solidity dissolved into upside-down and fleeting surface reflection, a fine way to destabilize them. He had grown up on the banks of the Thames, near London, and perhaps his fascination with water was begotten then and there. On February 19, 1868, the *Daily Alta California*, then one of San Francisco's leading newspapers, reported, "He presents the gorgeous scenes of that locality from points of view entirely different from any heretofore taken.... The view of the Yuwi-ha or Nevada Fall is a fine piece of Instantaneous photographing. It seems as though the artist had arrested the descending sheet of water until its mottled and foamy surface had paid tribute to his genius. A special rainbow spans the gorge where sunlight catches the cataract's spray."

In most of the Yosemite photographs of 1867 and 1872, water is not captured anywhere near instantaneously; the slow exposures capture something as alien to the human eye as those frozen droplets. This water is neither arrested nor frozen but moves so much during the exposure that it becomes a filmy, smoky substance, a little like the ectoplasm so popular in faked photographs from spiritualist séances at the time. In his mammoth plate of Wildcat Falls, the cataract is an hourglass-shaped volume of white vapor (plate 68); in another mammoth plate taken in the remote high country, of Helmet Dome and Little Grizzly Falls, the diagonal chute of water is diaphanous whiteness that becomes a white tongue lapping at the dark pool of still water below. Which is to say that water seen slower than the human eye is phantasmagorical, eerie, and enchanting, especially to this artist who let it come to life thus in his photographs. So too is water captured at a speed the eye cannot match: in both cases water becomes something unsettling, the mystery that surrounds us but eludes our limited senses all the time. As the filmmaker and photographer Hollis Frampton put it in his paean to the strangeness of the man at the birth of his medium, "Yet Muybridge, in some of his earliest landscape work, seems positively to seek, of all things, waterfalls; long exposures of which produce images of a strange, ghostly substance that is in fact the tesseract of water: what is to be seen is not water itself, but the virtual volume it occupies during the whole time-interval of the exposure ... he seems to have been the first to accept the 'error,' and then systematically, to cherish it."[2]

One more consistent element throughout his work is loneliness, the great distance always evident between artist and subject, when the subject was human.

He photographed other human beings at a certain distance both literally and emotionally. Nearly all his human figures are genre figures. In the wide-ranging early work, the subjects include prisoners, Native American warriors, white soldiers, artisans, coffee pickers, paraders, laundresses, and so forth; they are never close enough in any sense to be individual portraits, though there are some splendid moody figures out in the landscape that seem to represent sensibility rather than category. The motion studies, which represent drawers of water and hewers of wood, shoers of horses, jumpers, walkers, and dancers, reduce the genres to their essences. And like the rest, these do not look back at the photographer with an air of recognition or engagement; they stand apart in their own task or place, seemingly unaware. This aloofness is present all through the Central America photographs as well, where the figures in groups seem sealed off in their own world, the lone ones in a solitude that the photographer nearby does not violate. The most direct gaze back at the artist comes from a group of Shoshone warriors in the Central Pacific Railroad survey (carried out entirely in stereograph): they stare unwaveringly and the central figure bends a bow as if to shoot the photographer (fig. 36).

Muybridge's photograph of the mad Joshua Norton of 1860s and 1870s San Francisco comes closer than any to being an exception, and there is an echo of himself in this man who made his psychological disability into a successful identity as self-proclaimed Emperor of the United States and Protector of Mexico. But even the bicycle-mounted, epauletted Emperor Norton looks down, not back, from a middle distance that gives equal weight to the unpainted picket fence behind him. Muybridge might have been perceived as an important ethnographic photographer had he not eclipsed his images of Alaskan, Californian, and Central American indigenous people with his far more extensive work on other subjects. These people too have the stern aloofness found in all his work. Perhaps the most personal of all are the stereoscope images of a subject both foreign and peculiar: Chang Woo Gow, the Chinese giant, in Woodward's Gardens, an amusement park in San Francisco, though that may be due to the mild bearing of the subject himself and the soft presence of his average-height wife. Muybridge was more than content to include figures in his landscape; he was not, as

Fig. 96. *San Rafael, Ross' Landing (1723)*, 1873

Carleton Watkins was, in any way a precursor of the "wilderness" aesthetic of the unpeopled paradisiacal imagery Ansel Adams would raise to an apogee; but his figures, particularly in the 1872 Yosemite mammoth-plate photographs, are so tiny, so lost in vastness, that they often disappear in reproduction.

The physiologist Edward T. Reichert testified that Muybridge in the 1880s "was very much of a recluse, so that few got to know the man, and likely no one ever learned hidden secrets that must have radically influenced his life."[3] And at his 1875 trial for murdering his wife's lover, his friends repeatedly described him as an isolated, eccentric man. That brief marriage that ended so badly is about all that seems to have existed of a personal life for him. In one of the motion studies, a woman walks with a basin of water and then turns and splashes it: in it, you can almost hear Muybridge's voice saying, "Turn now, splash now." Still, the figures in his photographs are usually solitary and withdrawn, sometimes as romantic in their isolation in the pictures' foreground as those in certain paintings of Caspar David Friedrich (figs. 96, 97). Take the figure on Goat Island from early in Muybridge's oeuvre, staring across the San Francisco Bay, his back to us: the picture is all dense chaparral foreground and vast watery expanse, and loneliness. Even the pairs and trios are often still and

Fig. 97. Caspar David Friedrich, *Moonrise over the Sea*, 1822

silent, turned away from the camera to contemplate the mysterious world; a great many such groupings appear in his early genre pictures of the San Francisco Bay. There is one disorienting smile I have spotted in the motion studies, on the face of a nude woman getting out of bed, perhaps grinning with embarrassment. The rest are blank-faced, models rather than actors, "the human figure" rather than personalities. The special ability of the camera to capture the distinctness of each person is not relevant to Muybridge.

It is important to remember when considering the motion studies and all of Muybridge's work that art and science weren't so separate then. Artists were still for the most part attempting to be accurate in some objective sense, to provide reliable information about the visible world, and scientists had license to make their language and imagery beautiful; the two then collaborated. Muybridge was as much an artist for scientists as he was a scientist for artists. The motion studies were meant at least in part to provide an encyclopedia of poses for artists to draw from (and many still do so; Dover Books reprinted many of the studies and filed them under "Anatomy for Artists"), while in other bodies of work he intended to aid science. One notable example is in the 1872 images that show and in their titles mention glaciers and glaciation in Yosemite Valley. As the *Daily Alta California* reported on April 7, 1873, "At the suggestion of Mr. Clarence King, the artist has made several pictures of the ancient glacier channel at Lake Tenaya, and other glacier indications." King was caught up in the theological and geological controversies of the time: How old was the world? Where in the fossil and geological record and Darwin's new theory of evolution was there room for God? Were the Sierra Nevada in general and the Yosemite region in particular shaped by the huge glaciers that were a particular theological challenge, positing as they did harsh ice ages and extinctions in contrast to the old picture of a steady-state world? Yosemite, as John Muir recognized (and over-emphasized) and other more distinguished geologists of the time denied, was indeed carved out in part by glaciers, whose traces lie in the grooves and planes of polished granite Muybridge photographed. The pictures are evidence that he understood the geology well enough to recognize glacial traces and acknowledge them in the titles of at least three stereo pictures and two mammoth-plate images (*Ancient Glacier Channel, from Panorama Rock* and *Ancient Glacier Channel, at Lake Tenaya*; many more of his photographs in the series also show the results of glaciation).

The point is, on the Yosemite expedition of 1872, Muybridge made photographs both for the artist Albert Bierstadt—who evidently requested he photograph the Native Americans still quietly living in the touristed valley—and for the scientist Clarence King. Later his work would pose a major challenge for artists. No one painted water the way Muybridge photographed it, but he revolutionized the position of horses' legs and other figures in motion and pushed artists to decide whether to paint the world as it appeared to the human eye or as superhuman visual technologies revealed it to be. The art historian Marc Gotlieb puts Muybridge at the beginning of the split between art and science, saying of the prominent academic painter Jean-Louis-Ernest Meissonier after his exposure to Muybridge's work, "His commitment to the veracity of his representations forced him to ignore what he could see in favor of what he could not, but Meissonier remained faithful to a standard of truth unavailable to human perception, a standard that nullified his project. From this perspective, Meissonier's choice could be said to have brought the tradition of French Salon painting to its terminus."[4] It was killed off by the new gap between the human eye and the camera's revelations, a coup de grace delivered by Muybridge.

Early in his career, Muybridge produced stereoscopic studies of trees and clouds, each a series of several cards. It is unlikely that they would have appealed to the average person looking for something to view through their parlor stereo viewer. Instead, they seem to have been intended for artists and to have been made in much the same way as artists themselves made sketches and studies to try to understand a phenomenon and describe its variations. That is, they were resources for the art of others as much as or more than art themselves, as were the motion studies. Seldom discussed is what kind of science Muybridge thought these latter works were, but all the neoclassical studies of people carrying demijohns of water and genre images of walking and waltzing, as well as the grids behind them—a method used since at least Albrecht Dürer to block out two-dimensional renditions of figures—make it clear that painters were an intended audience, if not the only one.

However valuable for painters, Muybridge's photographic work was not at all painterly. In Yosemite, he photographed from vertiginous precipices, from under waterfalls, and often with flood flotsam, dead trees, and other debris in the foreground and jagged, destabilizing horizons in the background. He seemed to be actively avoiding the notions of good and harmonious composition many photographers had learned from paintings and prints. My own sense of these 1872 images is that they show Muybridge to be deeply immersed in the scientific debates of the time, photographing the chaotic, erratic state of nature perceived by those who believed science had eliminated God and his harmony from the creation. They are rough, turbulent, romantic works as far as imaginable from the comparatively serene classicism of Carleton Watkins, who often photographed the same general views. (An earlier American artist, Thomas Cole, painted blasted, shattered trees and apocalyptic landscapes, but always within the conventions of landscape composition which Muybridge confidently violated and Watkins did not.) One partial explanation for some of Muybridge's eccentric originality is that Watkins, the other prodigy of nineteenth-century California photography, had photographed Yosemite and made panoramas of San Francisco earlier. Muybridge may have been in part seeking to diverge from the other artist not only by taking photographs of Yosemite from new vantage points but by making them with a new sensibility and by making a superlative panorama that has no precedents. But disposition more than competition explains his vision and his tendency to go to extremes.

All through his work he showed affection for rough surfaces, irregular lines, and complexities verging on chaos. His tree studies from the late 1860s revel in the intricate zigzag branchings of groves of California oaks, and in 1875 he took many pictures in Central America where tangles of vines, trees, leaves, and more created an astonishingly intricate tapestry. In 1872, in the Lava Beds of northeasternmost California, where the Modoc War unfolded, he clearly liked the stark volcanic landscape with its heaps and crevices, a place entirely at odds with Victorian ideas of the beautiful, but much like the rocky landscapes of the Farallon Islands and the steamy Sonoma geysers he had earlier photographed. In Yosemite he photographed not only dead and blasted trees but slopes dense with foliage, rivers choked with snags and fallen trees, exfoliating walls of granite, and the sheer oddity of a glacier-carved landscape of pale stone and dark pine. There he seemed to take pleasure in disconcerting perspectives, in the sudden dropping away of the foreground when that foreground was a cliff, in jagged horizons that did not supply the stability pictorial horizons usually do, and in steepness so extreme that it defies pictorial logic. In his urban photographs, particularly his late panoramas, he seems to have reveled in the camera's ability to catch an astonishing amount of detail, so that the rigging of tiny ships in the harbor and the laundry on rooftops in the distance are visible, along with the patchwork of thousands of buildings crowded together. This density of information did not resolve readily into coherent composition.

He often took multiple photographs of a subject, trying to describe it comprehensively, whether it was a lighthouse or a trail or a parade. He produced many photo-essays of sorts. *A Vintage in California* investigated a winery in depth; another series documented the many aspects of coffee production on a Guatemalan plantation (with some startlingly strange and beautiful images of half-nude indigenous workers in the groves); San Quentin Prison, the Missions of California, the Farallon Islands off the coast of San Francisco, and the Bay Area military installations are among other subjects portrayed in detail. These works were more in the mode of the great documentary photographers of the mid-twentieth century, from Dorothea Lange to W. Eugene Smith, than that of his peers. He also worked for the government on a survey of Pacific Coast lighthouses, mostly in stereo and full-plate prints (but several startlingly beautiful mammoth plates of lighthouses exist; see plates 44, 45).

In many cases the photographs are not just serial, but explicitly sequential. Some are panoramic: a sequence of images traveling across space. Others are chronological: a sequence traveling through time. James L. Sheldon in his essay for the 1991 exhibition *Motion and Document, Sequence and Time: Eadweard Muybridge and Contemporary American Photography* writes, "It was not until the 1878 Palo Alto experiments that Muybridge saw the possibilities for these single 'instantaneous' photographs to become documents of a series of images captured over time. The concept of serial photography would follow logically."[5] In fact, Muybridge had been exploring serial photography since the beginning of his career. By 1868 he had made a seven-image stereo panorama from Rincon Hill. The genre of stereo panoramas is an odd one, and even in San Francisco it goes back to before 1860: Charles Leander Weed, the third significant landscape photographer working in California in that era, made just such a panorama at the end of the 1850s. The cards would be seen sequentially, rather than all at once (and by 1860 a cabinet stereo viewer had been designed that could hold up to thirty-six stereo cards at once and flip them in a set sequence—a precursor if not of cinema exactly then at least of Muybridge's precursors thereof). That means that the viewer would travel visually through both time and space, turning around the horizon in a series of views. In contemporary cinema we speak of the panning shot, whereby the movie camera travels through space (and time, of course): these stereo sequences were rough equivalents, allowing a viewer to explore an expanse of space that would unfold over time.

Muybridge made several such panoramas before his epic 1877 and 1878 panoramas of his city seen from atop Nob Hill (then known as California Street Hill). In 1873 he made two panoramas while at the Lava Beds under contract with the United States government to photograph the Modoc War there. One is a fairly conventional panorama within the stereo genre, five cards showing the expanse from the military camp to the Lava Beds—describing, in other words, the distance between two sides in a very peculiar and unequal war. (A small group of Modocs had left their forced exile on a reservation among a more numerous and hostile tribe in Oregon to return to their beloved homeland, where they quickly ran afoul of the settlers and then the government that backed them. A war was launched in November 1872 and dragged on during the first half of 1873. Though greatly outnumbered, the Modocs situated themselves in the labyrinth of the Lava Beds, and this stronghold and their tactical superiority allowed them to hold out for many months.) A far more peculiar twelve-card panorama of the Lava Beds region itself was also made, and could only have been made by Muybridge, who apparently relished the dense complexity of the volcanic rock in its battlements, crests, crevices, and piles of loose stone and bare ground, unbeautiful by any Victorian standard (plates 79–90).

This panorama has hitherto never been detected, but I assembled eleven of its twelve cards in 2002, after one Bay Area archive kindly delivered their cards to another for me. It is quite evidently one by the way the images are listed as subsets of the same number in Muybridge's numbering system and the fact that the cards fit together (though not as numbered) into a continuous expanse. He made what appears to be his first nonstereographic panorama, a rough scan of the horizon around the view from Cerro del Carmen, Guatemala, in 1875. Another eleven-part stereo panorama of that nation was made on the same journey. The artist Byron Wolfe, who has been rephotographing this Guatemalan work and retracing Muybridge's travels, says the photographs fit together like "pieces of a puzzle." Through rephotography the hidden spatial and temporal relationships between distinct images can be detected (as Wolfe, Mark Klett, and I explored when we rephotographed Muybridge's Yosemite work during 2001–3). Wolfe expands on this comment: "Sometimes I think there was a clearly thought-out method to his approach. For example, in the Antigua [Guatemala] city plaza he made views looking to and from each cardinal direction. You simply wouldn't know that by looking at the pictures, but when considered as a group, they form an interesting three-dimensional model of the plaza and the surrounding landscape."[6]

Muybridge's first successful venture into spatial rather than temporal panorama came in 1877. The usual 180-degree photographic panoramas describe a sweeping view that is little more than what a viewer might take in with a slight turn of the head or a clarification of all that exists at the far ends of peripheral vision. Muybridge took this perspective far beyond the realm of human vision when he chose to make an almost unprecedented 360-degree panorama of San Francisco. It is typical that he pushed the possibilities past the familiar to the disconcerting and disorienting. He did so twice, with the large-format panorama in eleven plates published in July 1877 and in the astonishing 1878 panorama made up of thirteen vertical mammoth plates that together compose an almost seamless view more than seventeen feet long, a view that can be apprehended all at once only from a distance (plates 117, 118). To see it in detail—and the detail is astonishingly sharp, including elements that can hardly be seen without a magnifying glass—means to travel through it somewhat as viewers travel through long Chinese landscape scrolls (which explicitly encourage the sense of journeying and getting lost in the long unfolding passage of detail; unlike most Western works of art, they are not meant to be taken in in one view or from one perspective). A legend exists that he made the monumental, unprecedented 1878 panorama because his studio burned down and therefore his negatives of the 1877 panorama were lost. The *San Francisco Chronicle* and other sources from this time appear to mention no fire, nor does any of Muybridge's admittedly scanty writing. Other negatives from the 1870s apparently existed much later.[7] Even to say that he made the superlative work to replace the good is to miss that it was not a replacement but a leap forward for which the earlier works may have been preparatory studies.

Like the dissolved and dissected water of Yosemite and the motion studies, the 360-degree panorama was an exercise in impossible seeing, transcending the bounds of ordinary human vision. And like much of his work, it is unsettling. Unless the panorama is set up as a circle inside which the viewer stands, its spatial arrangements are very odd. The viewer is simultaneously looking north, east, south, and west, and seeing California Street as it stretches in opposite directions from the vantage point atop Nob Hill. The ring of the horizon has been flattened into a linear expanse. Analysis makes it clear that these streets stretching in various directions are one street, but this is not evident to the eye. The space is almost incomprehensible and certainly counterintuitive. Muybridge made these last landscape photographs as he began, in 1877, the first round of his investigations into instantaneous sequential photography of motion, under Leland Stanford's patronage. They were made from the tower of the Nob Hill mansion of Stanford's neighbor and business partner, Mark Hopkins. And they resemble the motion studies intimately. They consist of several images taken over time assembled to be seen simultaneously, though the motion studies would show a few seconds of motion whereas the panoramas folded together hours of time. And both bodies of work defy and transcend human vision.

Marta Braun has noted that the motion-study sequences are sometimes not as continuous as the grid format makes them seem: frames are missing or arranged nonsequentially. This also happened in the big panoramas. In the 1877 version, Muybridge made and used two different versions of panel 5, the view down California Street, easily detectable because the clock of St. Mary's church is at quarter to two in one and at nearly half past five in the afternoon in the other and the shadows have shifted. The photographer Mark Klett, who rephotographed the 1878 mammoth-plate panorama in 1990, was struck by this version's discontinuity of shadows in the seventh panel, the one just north of St. Mary's church. Its shadows clash with those on either side of it. He used to speculate that it had perhaps been made to replace a damaged or unsatisfactory panel taken earlier in the day, but the discontinuous panels I noted in the earlier panorama suggest that Muybridge may have had motives other than necessity for slipping out of chronological sequence. He was always a trickster,

Fig. 98. *Lake Tenaya. Sierra Nevada Mountains. No. 46*, 1872

making sly and strange things happen in his work. He himself appears quietly in a number of his photographs. In a mammoth-plate photograph of Lake Tenaya in Yosemite made in 1872, there are several rocks in the foreground; most of them can be seen today, but one never existed and was added by the photographer, perhaps to cover up negative damage—or perhaps not (fig. 98). Klett detected it when he, Wolfe, and I went to the site in 2001. The added rock with the odd shadows certainly created a photograph that cannot be repeated exactly. In another Yosemite photograph, he removed a mountain from the background to make a starker composition.

Muybridge early on took pleasure in adding clouds to his photographs. The wet-plate film emulsions of his era were very sensitive to blue and tended to turn the sky into a blank white expanse. Watkins worked with this and made it almost elegant; Muybridge changed it. He even published an account of his "sky shade," which allowed him to expose the ground more than the sky and so preserve some detail in the former, but he also sometimes used two negatives to put together a landscape with satisfactory celestial atmospherics. By the standards of mid-twentieth-century "straight" photography, this would be seen as unacceptable or at least manipulative, but photography in Muybridge's era was such a messy, inexact, improvisational, novel, and demanding pursuit that such touches were more than acceptable. The criteria for truth were also different, or at least the rules of photographic truth had not yet been compiled (though falsehoods such as spirit photography were denounced, and Thomas Hardy has a novel whose plot turns on a manipulated photograph that makes the hero look inebriated). So, Muybridge embraced clouds, as cloud studies and as clouds added to landscapes. Individual photographs in his Yosemite mammoth-plate images exist in both cloudy and cloudless versions, and the former may have been the deluxe edition.

Earlier on, he had photographed the San Francisco Bay Area "by moonlight" many times, with added clouds and an invented moon, a method a little like cinema's "day for night," in which the former simulates the latter with underexposure, filters, or other techniques. Muybridge's sky photographs are also part of his ongoing body of sequential work—there are three successive stereographs of the sun (or an addition made to look like the sun) going down over Mount Tamalpais in the North Bay from early in his photographic career (plate 14), as well as a series of cloud studies. A mark of how fond he was of these clouds and how blurry the lines were between art and scientific documentation can be found in his photographs of his motion-studies set-up at Stanford's racetrack in 1879, nearly the last of his landscape photographs (plate 127). He added a different kind of cloud to each of several images of the racetrack, the stable, and the equipment. It is a truly eccentric decision. The clouds add emotion to the landscape, inflecting it with romance or foreboding. In one sense they make a more "real" photograph, one whose sky looks like those visible to the eye; in another they make the view more subjective, more expressive of the artist's own mood. But they are also documents of a scientific laboratory, not the most likely subject for this romantic enhancement.

Perhaps the most direct precedents of the motion studies come in the very different field of architectural photography. In 1870 and 1871, he photographed the construction of the bulky stone U.S. Mint in downtown San Francisco (plates 29–31). The site is depicted from various angles and at various stages, but at least three of the images are taken from one vantage point, while two others show development from another spot, and a final image returns to a place near the earlier vantage point to show the almost finished building. Thus, as with the motion studies, you see the process of becoming. Just as water leaping from a bucket in a split second is broken into three photographic images, so the slow rise of a building is made visible and comprehensible by a series of images over time. If the creation of a building can be imagined as motion—a slow, orchestrated arranging of materials—then these are indeed motion studies. So are the four or more images of the erection of the United States Appraisers' Building in a sequence made between 1874 and 1877 (figs. 99–106). These too feature multiple images from the same vantage point. Like the motion studies, the building sequences demonstrate that the world exists as processes and changes, as something far less stable and certain than the solid objects seen in a single instant that are the common subject of photography.

Klett remarks that Watkins's popularity in recent decades may be because his work so embodies the values of modernism. Certainly, the world in his vision is made up of solid objects clearly seen and framed and appraised. "Composed" could refer to both their aesthetic and emotional content, for Watkins's work often has a classical calm. Like the subjects of modernist photography, they stand apart from time in an eternal moment decisively frozen. Muybridge, even when photographing almost exactly the same subjects, could not be more different. In his sensibility, the world is all but *dis*composed, constantly in flux. Even something as solid as a government building reveals itself to be a creature of change; the water we think we know becomes eerily unfamiliar whether seen too slowly, as ghostly films, or too quickly, as leaping beasts; his late portraits are not portraits of human beings but of their actions, of movement itself; some of his landscapes, via their clouds, are really two different moments spliced together; and his panoramas often promise a single sweeping glance while actually being made up of several nonconsecutive moments. Even his taste in surfaces and textures runs to the intricate, elaborate, dense, and tangled: they are not smooth, stable, or easily deciphered. Finally, his tricksterish moments of subversion of the supposed truth or continuity of a given work of art render more uncertain and less stable the subject at hand. Muybridge's vision of a world in constant change binds all the work into a radically original and deeply coherent achievement.

[1] Phillip Prodger, *Time Stands Still: Muybridge and the Instantaneous Photography Movement* (New York: Iris & B. Gerald Cantor Center for Visual Arts at Stanford University in association with Oxford University Press, 2003), 224.

[2] Hollis Frampton, "Eadweard Muybridge: Fragments of a Tesseract," in *Circles of Confusion: Film, Photography, Video, Texts, 1968–1980* (Rochester, NY: Visual Studies Workshop Press, 1983), 77.

[3] Reichert quoted in Anita Ventura Mozley, "Introduction to the Dover Edition," in *Muybridge's Complete Human and Animal Locomotion: All 781 Plates from the 1887 Animal Locomotion* (New York: Dover Publications, 1979), 1:xxx.

[4] Marc Gotlieb, *The Plight of Emulation: Ernest Meissonier and French Salon Painting* (Princeton: Princeton University Press, 1996), 182–83.

[5] James L. Sheldon, "A Man Beyond His Time," in James L. Sheldon and Jock Reynolds, *Motion and Document, Sequence and Time: Eadweard Muybridge and Contemporary American Photography* (Andover, MA: Addison Gallery of American Art, Phillips Academy, 1991), 13.

[6] Byron Wolfe, e-mail to the author, December 3, 2008.

[7] Regarding the fire in Muybridge's studio, I reviewed the case against this at length in a footnote in my book *River of Shadows: Eadweard Muybridge and the Technological Wild West* (New York: Viking, 2003), 273.

Fig. 99. *United States Appraisers' Building*, 1874; Fig. 100. *United States Appraisers' Building*, 1875; Fig. 101. *United States Appraisers' Building*, 1876; Fig. 102. *United States Appraisers' Building*, 1876
Fig. 103. *United States Appraisers' Building*, 1876; Fig. 104. *United States Appraisers' Building*, 1876; Fig. 105. *United States Appraisers' Building*, 1877; Fig. 106. *United States Appraisers' Building*, 1877

Stopping Time

75. *The Brandenburg Album of Bradley & Rulofson "Celebrities" and Muybridge Photographs*, page 45, 1874

76. *The Brandenburg Album of Bradley & Rulofson "Celebrities" and Muybridge Photographs*, page 104, 1874

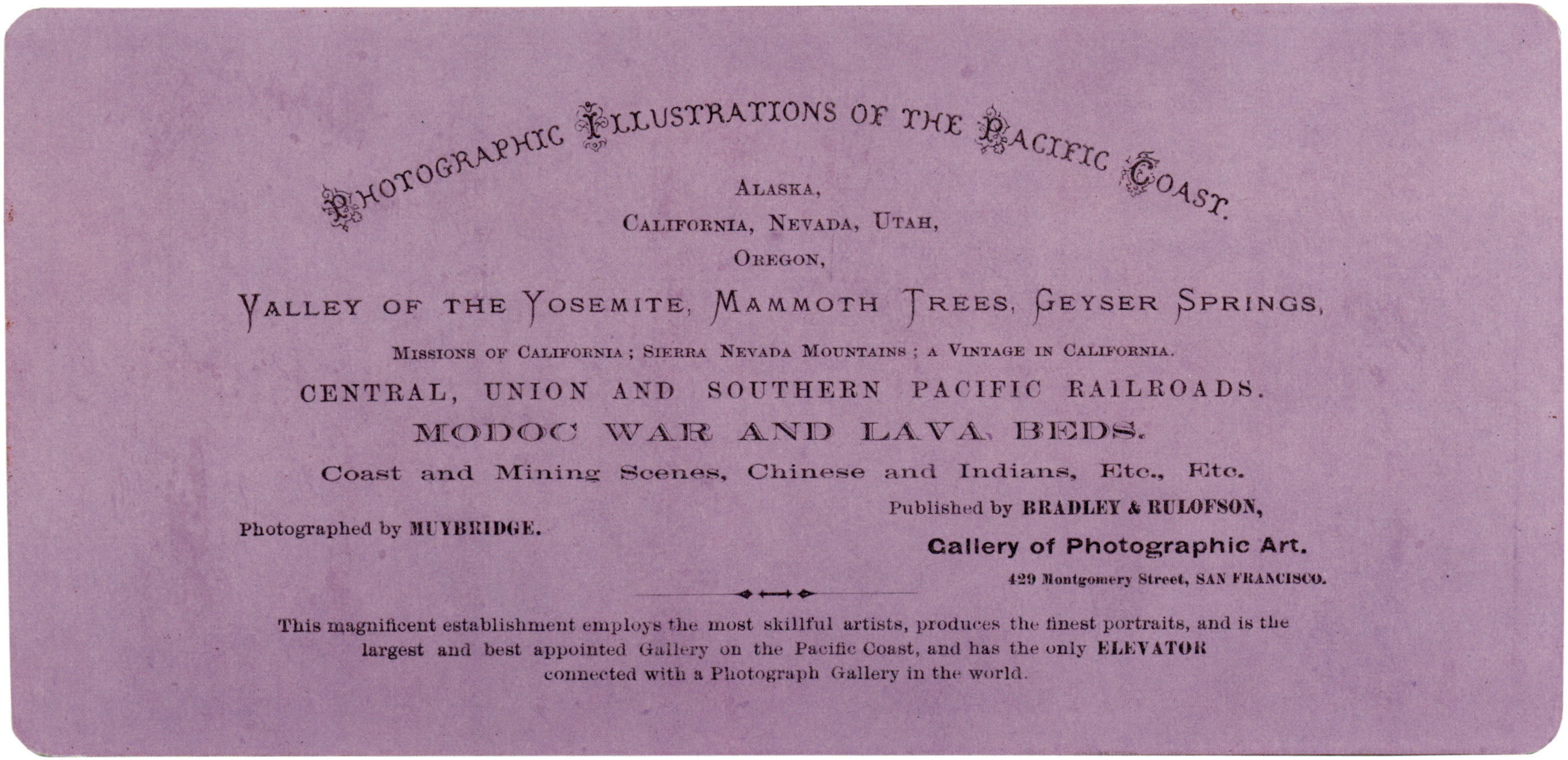

77. The Lava Beds, No. 8 (1619) (verso), 1873

78. *A Modoc Brave on the War Path (1627)*, 1873

79. *The Lava Beds, No. 1 (1619)*, 1873; 80. *The Lava Beds, No. 2 (1619)*, 1873
81. *The Lava Beds, No. 3 (1619)*, 1873; 82. *The Lava Beds, No. 4 (1619)*, 1873

83. *The Lava Beds, No. 5 (1619)*, 1873; 84. *The Lava Beds, No. 6 (1619)*, 1873
85. *The Lava Beds, No. 7 (1619)*, 1873; 86. *The Lava Beds, No. 8 (1619)*, 1873

87. *The Lava Beds, No. 9 (1619)*, 1873; 88. *The Lava Beds, No. 10 (1619)*, 1873
89. *The Lava Beds, No. 11 (1619)*, 1873; 90. *The Lava Beds, No. 12 (1619)*, 1873

91. *Donald Mc Ky, the celebrated Warm Springs Indian Scout and his chief men (1623)*, 1873

92. *Generals Jeff C. Davis, Hardie, Gillem, and Officers of the Modoc Campaign (1621)*, 1873

93. *The Peace Commission Tent, and Stone on which Gen. Canby was sitting when Shot by Capt. Jack (1601)*, 1873

94. *Panorama of Lava Beds from Signal Station at Tule Lake Camp, South (1605)*, 1873
95. *Panorama of Lava Beds from Signal Station at Tule Lake Camp, South (1606)*, 1873

96. *Panorama of Lava Beds from Signal Station at Tule Lake Camp, South (1607)*, 1873

97. *Warm Spring Indian Scouts on Picket Duty (1630)*, 1873

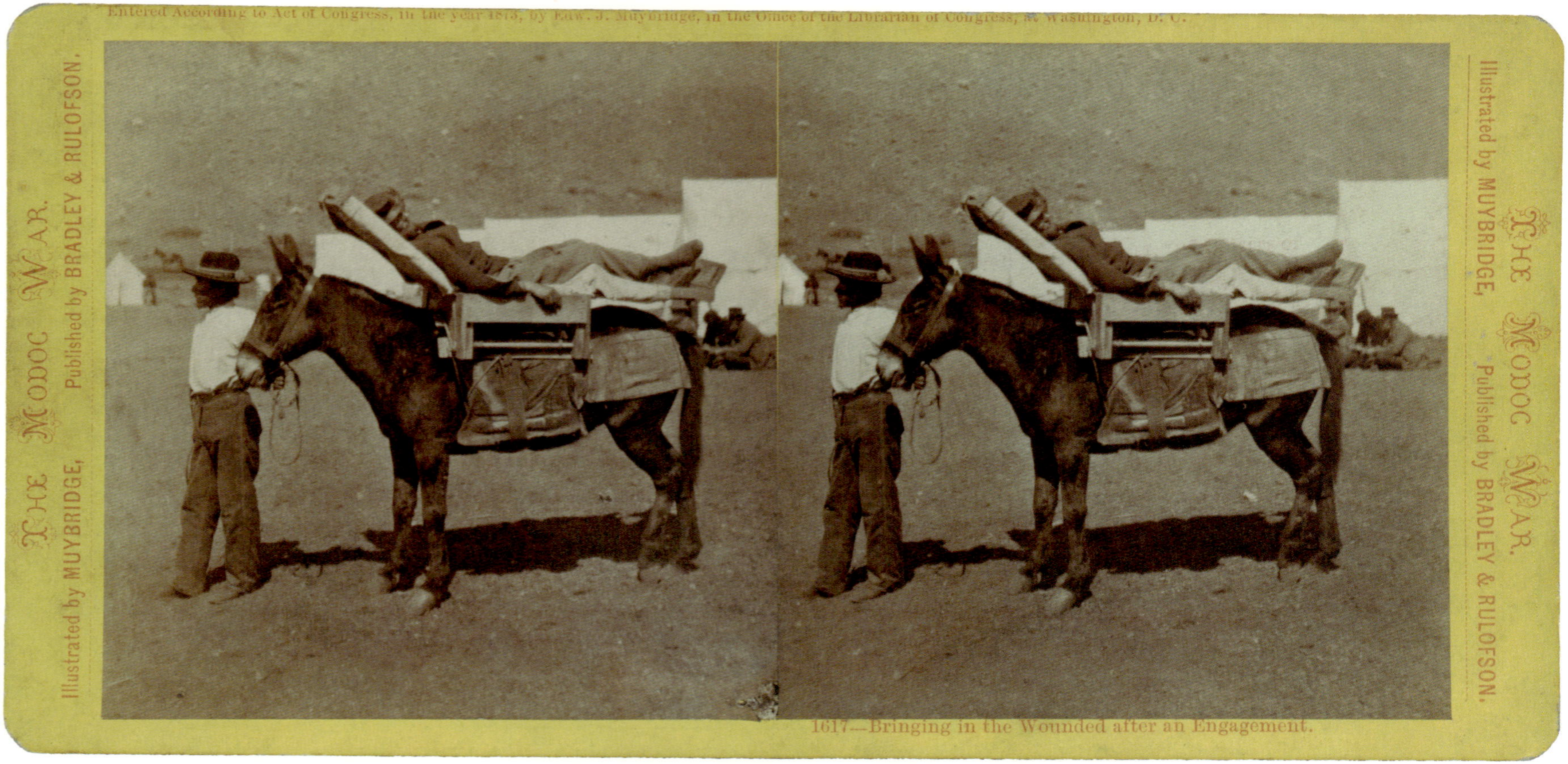

98. *Bringing in the Wounded after an Engagement (1617)*, 1873

99. *Trade card of Muybridge for Bradley & Rulofson*, 1874

100. *Reception of the Artist at Panama (4262)*, 1875, from *The Pacific Coast of Central America and Mexico, The Isthmus of Panama, Guatemala, and the Cultivation and Shipment of Coffee*, 1876

101. *Church of San Filipe (4243)*, 1875

102. *Burial of a Native—Panama*, 1875

103. *Ruins of a Church, Antigua, Guatemala*, 1875

104. *Ruins of the Church of San Domingo, Panama,* 1875

105. *Laundry of Calvario, Guatemala (4347)*, 1875

106. *Bridge on the Puerto Bello, Panama (4267)*, 1875

107. *Solola, Guatemala*, 1875

108. *Quetzaltenango, from the Crater of a Volcano*, 1875

109. *Machinery for Pulping Coffee, Las Nubes*, 1875

110. *Reception of the Artist, Quetzaltenango, Guatemala,* 1875

Fig. 107. J. Walter Wilson, *Mr. Muybridge Shows His Instantaneous Photographs of Animal Motion at the Royal Society*, *Illustrated London News*, May 25, 1889

Magnificent Entertainment: The Spectacular Eadweard Muybridge

Corey Keller

Few stories in the history of photography can match that of Eadweard Muybridge. Born Edward James Muggeridge in Kingston upon Thames to a middle-class mercantile family, he set off sometime around 1852 to make his fortune in the United States.[1] Like so many ambitious young men in America, Muybridge ended up in San Francisco, an "instant" city on the country's farthest frontier, a mythic land of gold and limitless possibility for self-invention. By the time he returned to England as Eadweard J. Muybridge some fifty years later, he had undergone a total self-transformation. Having left a young man of unremarkable origins and talents, he came back an acclaimed photographer and international celebrity.

Historians have been hard-pressed to construct a cohesive narrative out of his diverse photographic career and gripping biography. Typically, they have focused on a singular aspect of his work, most often the well-known motion studies with which his name is now popularly synonymous. Many of these same examinations are also undisguised attempts to rehabilitate Muybridge's place in history, not necessarily within the history of photography, but more often in that of the moving picture.[2] The heterogeneity of his photographic practice, seen through the lens of his dramatic life story and inflected by the agendas of his biographers, has served, one might argue, to obscure rather than elucidate his place in history. The result has been to portray him as a figure so exceptional that it is hard to see him as anything but an outlier.

It is certain that Muybridge's contributions were essential to the development of cinematic technology. And it is indisputable that he was one of the most inventive and colorful figures in the history of the photographic medium. But there is an arena in which we might consider his kaleidoscopic career as typical of his historical moment, rather than exceptional, and that is in its thorough engagement with the spectacular. Muybridge's entire oeuvre can be seen as part of a larger transformation of visual culture of which cinema is but a facet. In her study of nineteenth-century Parisian entertainments, *Spectacular Realities*, Vanessa Schwartz argues that the birth and rise of cinema must be understood as "part of a broader cultural climate that demanded 'the real' as a spectacle."[3] For her, cinema evolved within a wide network of viewing habits and visual entertainments that emerged over the nineteenth century and flourished in its last quarter.[4] Muybridge's stereographs, panoramas, and illustrated lectures were viewed, purchased, and attended by people who also took part of the host of nineteenth-century public attractions—phantasmagoria, panoramic paintings, wax museums, expositions, department store displays, and eventually, the cinema—that produced their effects by transforming the real world into spectacular images. By framing Muybridge's work within this context, it becomes less urgent to secure his historical place within a teleological progression that culminates in the invention of cinema and becomes instead a question of seeing his work, much like the emergence of cinema itself, within a modernity characterized by a new kind of spectatorship.

Muybridge catered to the demands and desires of this new audience on every level of his photographic practice. From his choice of subjects to his use of the stereographic, mammoth-plate, and panoramic formats, he combined the authority of the photographic image with sensitivity to what we might call "effect," producing images whose affective power derived both from their immediacy and from their direct and deliberate appeal to the spectator. Furthermore, Muybridge recognized that his viewers were also purchasers of photographs. He was a showman in a century that prized showmanship, and a self-made celebrity in the era that invented the concept. He turned not only the photograph's subject but also its making—and perhaps more important, its maker—into a spectacle, and thereby created a market for his work. By considering four different bodies of work across his career—his stereographs of Woodward's Gardens, the mammoth-plate views of Yosemite, the panoramas of San

Francisco, and the motion studies—we can begin to understand Muybridge's particular genius within a broader cultural context.

Crafting a Public Persona

From the earliest years of his adulthood, Muybridge had a keen sense of the importance of building what we would today call a brand. No sooner had he left England than he began to fashion his public persona. Shedding first one iteration of his last name, then another, by 1867 Muggeridge had become Muybridge. Some fifteen years later, he changed his first name as well, borrowing the idiosyncratic spelling from two tenth-century Saxon kings honored on a monument in his hometown, and from that point on, the conventional spelling was used at one's own peril. As his projectionist Ernest Webster later recalled, "He was always announced as Professor Eadweard Muybridge and woe betide any one who spelt his Christian name 'Edward.'"[5] For his brief stint in Central America in 1875, he even adopted a Latinized version: the work from that trip is signed Eduardo Santiago Muybridge.

Aside from changing his name, for the first years of his career Muybridge also adopted a flamboyant professional alter ego: Helios. Photographs were often referred to as "sun pictures" in the nineteenth century, alluding to the chemical—as well as to the quasi-alchemical—nature of the photographic process. Not only did the conceptualization of the medium as nature "drawing her own picture" seem to ensure the photograph's truthfulness by removing any human agency in the production of the image, it also problematized the idea of photographic authorship, effectively writing out the role of the photographer.[6] Muybridge, however, was not content to relinquish authorship to the sun (or to the company publishing his photographic views).[7] Instead, he scratched his pseudonym into the negatives, adorning rocks and walls with the sun's name like so much graffiti. Though he jettisoned Helios as a professional moniker in 1872, he was apparently reluctant to part with it altogether: two years later he gave the unwieldy—and immodest—name to his newborn son, Floredo Helios Muybridge.

Fig. 108. *Savings and Loan Society, Clay street (340)*, 1869

Even Muybridge's mobile darkroom was whimsically named: The Flying Studio. At a time when photography was a tremendously cumbersome process requiring the transportation of heavy equipment, glass plates, and a laboratory's worth of chemicals, the name was decidedly tongue-in-cheek. Certainly the photograph he took in 1867 of his equipment strewn about a Yosemite campsite has little about it to suggest lightness or ease (plate 2). Emblazoned across the back of the small carriage he procured and outfitted for photographic equipment in 1870 and accompanied by a logo of a winged camera, the name was no doubt intended to convey a sense of agility and speed (fig. 108). It also, however, evokes the aerial gymnastics of the trapeze. The last quarter of the century marked the heyday of American circuses, increasingly extravagant spectacles that began traveling across the country with the completion of the transcontinental railroad. At the time Muybridge named his wagon, one of the most popular songs in both Europe and the United States was "The Daring Young Man on the Flying Trapeze" (1867), written to commemorate Jules Léotard's death-defying performances on the apparatus he had pioneered in 1859.[8] The connection to the circus, whether deliberate or accidental, seems appropriate for a man with a keen awareness of the importance of putting on a good show.

Muybridge cultivated his professional and public reputation: he applied for several patents (including a new method for book printing and a "sky shade" for producing more evenly exposed negatives); wrote effusive prospectuses for his photographic series; and authored pamphlets and letters to the editor of the local newspaper. (The publicity he would receive in 1875 as a result of his trial for the murder of his wife's lover was no doubt unintentional, but nonetheless it had the fortuitous effect of raising his profile in San Francisco, where after his acquittal he continued to find many of his clients.) Over the course of his career, he paid attention to his associations, whether the various galleries that represented his Yosemite views, the wealthy clients who commissioned work from him, or the institutions that sponsored his motion studies and public lectures. He also carefully tended to his legacy: he compiled a large scrapbook of positive clippings about his work and his public appearances, an exceptionally useful sourcebook of primary materials.[9] The copious historical record Muybridge left behind can be contrasted with the utter absence of such material about his photographic rival in Yosemite, Carleton E. Watkins. When Watkins died, he left behind no written clues for posterity, only the wordless testimony of his pictures. As a result, Watkins's photographs—though universally recognized for their technical and aesthetic excellence—have been an interpretive challenge for historians.[10] Muybridge appears to have been determined to ensure that his own career posed no such difficulties.

Nature as Spectacle: Domesticated and Wild

When Muybridge—or Helios—launched his photographic enterprise in San Francisco in 1867, the American public was cultivating a voracious appetite for views of the American West. The most affordable and popular photographic format was the stereograph, a pair of photographs produced by a camera whose twin lenses were positioned eye-distance apart. The photographs were mounted to a card which, when seen through a special viewer—a device that was by this time a regular fixture of most middle- and upper-class parlors—caused the two pictures to resolve into a single, startlingly three-dimensional image. There was an enormous market for such views in the second half of the nineteenth century, especially of far-off and exotic locales. Muybridge made hundreds of stereographs, the vast majority of which focus on San Francisco and the landscape of northern California (including Yosemite), and which were intended for both San Francisco and East Coast urban audiences.[11]

Fig. 109. *Chinese Women Outside, Woodward's Gardens*, 1870

Among Muybridge's most photographed San Franciscan subjects was the famous Woodward's Gardens, a sprawling, four-acre park bounded by Mission, Valencia, Thirteenth, and Fifteenth Streets. It was the eclectic creation of one R. B. Woodward, who had made his fortune during the gold rush with a very successful temperance hotel named What Cheer House. In 1866 Woodward opened his private home and gardens to the public, charging twenty-five cents admission. In addition to the main residence, which housed Woodward's holdings of European and American paintings and sculpture and a natural history collection, the property was much admired for its magnificent grounds, exotic plant specimens and animal displays, fountains, and meandering paths. At the height of its popularity in the 1870s, Woodward's Gardens also featured a large zoo, a steam-heated conservatory, a large library of rare books, a skating rink, and, beginning in 1873, the first public aquarium in the American West.[12]

Muybridge had a retail operation at the entrance to the gardens (while also maintaining his space at 121 Montgomery Street), where he sold souvenir views

Fig. 110. *A, walking; B, ascending step; C, throwing disk; D, using shovel; E, F, using pick. Plate 521,* 1887

to the gardens' visitors.[13] The backs of the stereograph cards he made during this period bear both addresses. That Muybridge should choose to open a second outpost here is testament to the enormous popularity of Woodward's Gardens, as well as to the volume of sales he anticipated to make there. It also places his photographic production squarely within a popular tradition of spectacular entertainment. The work he made there is filled with imagery of the grounds' pavilions and rides, visitors interacting with the startling animal displays, and portraits of the side-show entertainers who regularly performed there animal displays (plate 36).

Pleasure gardens became an increasingly important site for urban leisure over the course of the nineteenth century. As Barbara Berglund has argued, in creating his gardens Woodward aspired to emulate Frederick Law Olmsted's famous Central Park in New York City (the construction of which had just been completed), offering a tamed and uplifting experience of nature as a balm to the jangled nerves of modern city dwellers, as well as a wholesome alternative to the seamier amusements of San Francisco's Barbary Coast.[14] In Woodward's Gardens, nature was landscaped into a bucolic "rural" experience and carefully ordered for visual consumption. The Gardens combined the pastoral respite of the pleasure garden with a vigorous educational program centered around visual displays ranging from coins and weapons to local and exotic wildlife—live and taxidermied—placed judiciously throughout the grounds. The park also offered—in nearly equal parts—a variety of circuslike entertainments, including tightrope artists, mind readers, trained bears, Japanese acrobats, and a Chinese giant named Chang Woo Gow, earning Woodward the sobriquet of "the Barnum of the West."[15] Muybridge's three-dimensional stereographs of Woodward's Gardens represent an intersection between a culture of visual entertainment and a technology of visual spectacle that is distinctively nineteenth-century.

In his views of Woodward's Gardens, Muybridge gives us a little foretaste of the able showman he would become over the course of his career. Though he photographed under the name Helios, he seemed loath to lose sight of himself entirely and repeatedly insinuated his own figure into his pictures. One view of the attraction's art galleries shows a visitor caught nodding off on a bench in the exhibition (fig. 39). The boorish patron is none other than the photographer himself. Muybridge shows up as a visitor to the gardens in several other photographs as well. While his appearance might easily be explained as a matter of convenience—where else could a photographer find a subject more sensitive to the long exposure such a picture required?—a work such as *View in Woodward's*

Fig. 111. Carleton Watkins, *Primitive Mining; The Rocker, Calaveras Co., Cal.* (self-portrait), ca. 1876

Gardens can only be read as a mischievous self-portrait (plate 37). The photograph shows a view into a grove of trees with a polished decorative globe cleverly placed in the foreground, creating a disorienting spatial recession. In the mirrored surface we can see not only the distorted reflection of the buildings behind the photographer but also a dark, three-footed object that turns out to be Muybridge himself, hunched over his camera and draped with a dark cloth. The pictures from Woodward's Gardens are not exceptional in this regard. Muybridge appeared in his own photographs with surprising regularity throughout his career: he is the figure sitting on a San Francisco wharf,[16] and the nude, pick-wielding model in *Plate 521* from *Animal Locomotion* (fig. 110).

Muybridge's stereographs of the tamed and disciplined nature of Woodward's Gardens are a neat foil to the pictures he made in Yosemite Valley, which had emerged as a potent symbol of the wild beauty and boundless resources of the American West. In 1872 Muybridge made his second trip to Yosemite. In addition to making more small views and stereographs, as he had done on his first visit in 1867, this time he planned to produce a series of mammoth-plate views.

He was not the first to recognize either the suitability of this large format to the majestic scenery of Yosemite or the market potential for the clear and dramatic views such large negatives could provide. Indeed, his chief rival, Watkins, had already earned great acclaim for his large-scale views of Yosemite Valley in 1861, and by 1872 his career was in full bloom. Muybridge's deliberate positioning of himself as a challenger to Watkins is barely concealed between the lines of the prospectus for a "series of large-size photographic negatives" that he issued before departing for Yosemite that summer:

> I am encouraged in this undertaking from the generally expressed opinion, especially of our best Art Critics, that although many carefully executed large-size photographs of our scenery have already been published, yet the wonderful improvement in the scene of photographic manipulation, and a judicious selection of points of view, with an aim at the highest artistic treatment the subject affords, will result in a more complete realization than has hitherto been accomplished of the vast grandeur and pictorial beauty for which our State and Coast have so world-wide a reputation.[17]

Acknowledging his predecessor's achievements, Muybridge—backed by a chorus of unnamed experts—shows himself to be engaged in a contest of photographic one-upmanship. His hyperbolic marketing found its equivalent in the breathless response of the press, which similarly compared his work favorably to that of an unnamed but easily identifiable rival.

According to the newspapers, Muybridge's magnificent mammoth-plate views were matched only by the feats of derring-do required to obtain them. A review in the *Daily Alta California* of an 1873 exhibition at the galleries of Bradley and Rulofson reported on the extreme lengths to which Muybridge went to get his shot, from cutting down trees that interfered with his view to scaling precipices while laden with heavy and cumbersome equipment. On some occasions,

Fig. 112. Carleton Watkins, *Cathedral Rock—Yosemite. No. 20*, 1861

according to the reviewer, he insisted on venturing out to vantage points so per ilous that his assistants were too frightened to accompany him.[18] Muybridge documented these displays of bravado, enlisting his assistants to make stereographs of him roughhousing in the snow or perched at the edge of a vertiginous cliff (fig. 6). (It is notable that this latter picture was later introduced as evidence of insanity at his murder trial.[19]) Muybridge's daring often resulted in depictions of the valley that provoke a dizzying sense of instability for the viewer while also calling attention to the position of the photographer. The press hailed his achievements in glowing terms, calling his works "the most perfect photographs ever offered for public inspection—some of them gems of art."[20] The eminent photochemist Dr. Hermann Wilhelm Vogel, upon seeing Muybridge's images exhibited in Germany, drew the connection between the man and his photographs, praising them both in a compliment that conflates the two: "Landscapes of this size are the exception here, and the thought that Muybridge, with his mammoth camera for plates of twenty-two inches, climbed mountains, fills many a one with respect and admiration."[21] Muybridge the photographer and Muybridge the performer developed in lockstep.

Besides climbing mountains and cutting down obstructing foliage to make his pictures, Muybridge also manipulated the negatives themselves. He punched up the drama by adding rocks and clouds (using multiple plates), and sometimes retouched the negatives to remove errant tree branches or other compositional distractions (fig. 98). The photographic interventions he made on his mammoth plates were, however, far more subtle than those he had performed on many of his earlier stereographs. A number of these include a fake full or crescent "moon" and are printed unusually dark, creating a day-for-night feeling he called a "moonlight effect" (fig. 28). Muybridge's photographic constructions were calculated to produce maximum theatrical impact. It is certainly a different view of nature than that put forth by Watkins (fig. 111). A side-by-side comparison of the two men's works, even when taken from similar vantage points, reveals key differences in their styles: whereas Watkins in his balanced compositions evokes the beauty of pristine and majestic nature, Muybridge conjures a dramatic and turbulent sublime, all craggy rock, crashing waterfalls, and gnarled tree limbs. In Watkins's 1861 view of Cathedral Rocks, for example, he centered the rock formation within a domed frame and filled the picture with its monumental and imposing form (fig. 112). Just more then ten years later, when Muybridge photographed the same rugged peaks, he pulled back to a more distant vantage point (fig. 113). Without the dome that gracefully frames Watkins's scene and the geometry of the foliage at the rock's base, Muybridge's Cathedral Rocks evokes nature's irregularity, even unruliness. Muybridge's peaks are barely contained by the photograph's frame, and the tangled branches on the river's closest shore seem to form a deliberate disturbance to the placid expanse of river in the foreground.[22]

By the time Muybridge began photographing in Yosemite, it had already been granted to California as an inalienable public trust (thanks in no small part to

Fig. 113. *Cathedral Rocks, Valley of the Yosemite, 2,600 Feet High. No. 20*, 1872

Watkins, whose photographs convinced Congress of the site's extraordinary natural beauty): an oasis of inviolate nature as a buffer against the encroaching forces of modernization. As a result, however, it was also growing more accessible. Even as Muybridge was portraying Yosemite as an untrammeled wilderness—quite distinct from the classical proportions and stately grandeur of Watkins's views—new roads were under construction to allow greater numbers of tourists to visit the site.[23] Muybridge was ever mindful of the audience for his pictures. His depiction of Yosemite might have anticipated the inevitable transformation that increased human presence would bring; or perhaps he was producing pictures that conformed to the mythological image of Yosemite that had already taken shape in the American imagination.

Panoramic Visions

Muybridge's experiments in Yosemite coalesced in his series of panoramas of San Francisco, in particular his spectacular 1878 view, composed of thirteen mammoth plates (plate 118). (They are also evidence of his robust and ongoing competition with Watkins, who made several panoramas from the same point, California Street Hill.) Muybridge's panorama was taken from the turret of the newly constructed home of one of the city's wealthiest men, Mark Hopkins. The hill (now known as Nob Hill), where Hopkins and the city's elite lived, was the most popular vantage point, due to its unique combination of physical and socioeconomic geography.[24] It is both a corporeal and an ideological point of view, providing a commanding vista over the burgeoning metropolis, the distant hills and suburbs, and the busy port bristling with the masts of sailing ships headed for faraway ports. The hallucinatory clarity of detail afforded by the large negatives conveys a sense of mastery of the city, bringing all of it to sight at once.

The mammoth-plate panorama of 1878 was Muybridge's second documented effort to record this particular view; he had made a smaller panorama the previous year, of which there were several different printings. In one, the set of eleven 7 1/2-by-9-inch full-size prints were accordion bound (plate 117); in another, the prints were cropped and mounted onto boudoir cards. The mammoth-plate panorama (also accordion bound) was far more impressive than either of the earlier formats. At twenty-five dollars a copy (more than twice the price of the smaller version), it was intended for a different clientele. Only nine extant examples of the mammoth-plate version are known, all of which were either given as gifts to wealthy patrons (including Mrs. Mark Hopkins, Mrs. Leland Stanford, and Stanford's private secretary) or purchased by institutions. No record exists of the number of prints made of the smaller version, but numerous copies survive today, which indicates that it was printed in a much larger quantity.[25] The panoptic view of the city afforded by the panorama surely had different implications for its various audiences. For a consumer of more modest means, it offered a voyeuristic experience of life at the crest of California Street Hill; for the wealthy recipients of the deluxe edition, a reaffirmation of their dominion over all they surveyed. To accompany the 1877 panorama, Muybridge had also published an illustrated legend that identified many of the homes and businesses visible in the pictures. It was a clever marketing strategy: serving at once as a map of the city's most prominent residences for the curious onlooker and as an appeal to the vanity of those mentioned on the key—all potential clients for the photographer.

The panorama depends upon and reinforces the position and psychology of the individual spectator. It also reflects a culture of mass spectatorship. Painted panoramas—large-scale, moving pictures shown in the round, accompanied by light and sound effects—had been a form of popular entertainment since the beginning of the nineteenth century. Fashionable throughout Europe and America, panoramic entertainments had a particular character in the United States and peaked in the 1840s and 1850s. Most focused on the American West—its dramatic landscape and exotic native inhabitants; following the gold rush of 1849, the development of San Francisco was also a popular theme.[26] The economic and public success of these painted panoramas inspired photographers to try to adapt their own medium to the production of such views. Though expensive and difficult to produce, as early as the 1850s photographic—or, more frequently, photographically based—panoramas began to rival those of the painters. Some, like John Wesley Jones's *Pantoscope of California, Nebraska, Utah, and the Mormons*, were highly elaborate affairs that bore little apparent relationship to the original photographs on which they were based; others were more traditional, drawing on their photographic origins as proof of authenticity. Martha Sandweiss argues that the popularity of photographic panoramas, or painted panoramas based on photographs, derived from "their simultaneous ability to tap into bigger cultural myths and their capacity to convey a certain specificity of information."[27]

San Francisco's unique topography as well as the mythological place it occupied in the American imagination made it a particularly desirable subject for panoramas. Muybridge's mammoth-plate version is arguably the most remarkable of the many such photographs made from the top of California Street Hill, primarily because of its size (fully extended it measures some seventeen feet), but also because of its 360-degree view. By photographing in the round, Muybridge created an experience for the viewer that is distinctly at odds with that of physical vision, mapping a circular view onto a flat plane. The resulting distortion causes the parallel lines of the streets to converge as they radiate outward from Muybridge's camera, a visual effect that is as disorienting as it is pleasurable. The sense of spatial confusion is all the more pronounced because the accordion album binding dictates that the panorama be seen in a linear fashion—perhaps on a long table—rather than in the round.[28] The individual panels, rich with detail, invite the viewer to examine each closely, but it is impossible to reconcile such a lingering, close-up examination with the dizzying experience of the panorama's full sweep. In Muybridge's panorama we have the entangling of two positions: an oscillation between the commanding view and the pleasure of the gaze, between document and spectacle.

Magnificent Entertainment

The dramatic approach that Muybridge employed in his landscape photographs seems irreconcilable with the "scientific" work he undertook first under the patronage of Leland Stanford and then under the auspices of the University of Pennsylvania. Muybridge went to great lengths in his motion studies to counter any impression that the subjectivity of either the photographer or the viewer was in play. In Philadelphia, he used a black backdrop marked with white lines at regular intervals, carefully numbered the photographs in sequence, and arranged them in a grid. He also invented and employed highly complex and sophisticated equipment. But as Marta Braun has painstakingly and convincingly demonstrated, Muybridge's motion studies are "inconsistent with what we understand to be a scientific analysis of locomotion," and they were regularly reordered and otherwise manipulated, resulting in merely the *appearance* of scientific objectivity.[29]

Whatever his motivations, Muybridge's interests clearly extended beyond the realm of strictly scientific inquiry. As in his landscape work, he maintained a balance, however tenuous, between the informative value of his photographs and their affective power. Critics have paid careful attention to the disparity between Muybridge's treatment of male and female subjects, pointing out that the women's activities are frequently inflected with narrative overtones and erotic fantasy, leading them to compare these pictures to pornography.[30] Without discounting this reading, there is a larger argument to be made about visual pleasure in his motion studies, even in the least titillating of his subjects. In rearranging his pictures to form a visually—if not objectively or scientifically—compelling narrative, Muybridge was once again privileging his audience's experience over any rigid fidelity to the real. He took (mostly) everyday actions and broke them into fragments of time too brief to be appreciated by the human eye, allowing the viewer the opportunity for a lingering look at what was otherwise invisible. As in his panorama of San Francisco, the desire for knowledge and the pleasure of looking bump up against each other.

It is worth considering the audience Muybridge anticipated for this body of work and the formats in which the photographs were displayed and distributed. The original motion studies of horses, done for Leland Stanford in Palo Alto, were largely unseen outside Stanford's circle prior to 1878, according to Muybridge himself.[31] It was not until they were reproduced in *Scientific American* in October 1878 and then in *La Nature* in December of that same year that any of the pictures were seen publicly. He wanted his second and far more ambitious undertaking, *Animal Locomotion,* to have a broader distribution. Though only a handful of the complete sets of 781 prints were sold, a great many 100-print sets were bought, primarily by wealthy patrons, institutions, and above all, artists. But at $100 for a 100-plate subscription (the smallest set one could purchase), it was clearly not intended for the general public.

Muybridge took great pride in referring to the wealthy and influential people who had subscribed to his work. In *Descriptive Zoopraxography* (1893), a slim volume published to supplement his lectures at his Zoopraxographical Hall in Chicago, Muybridge prefaced the text with eight pages of facsimile autographs of the subscribers to *Animal Locomotion.*[32] Among the many still familiar names are former president Rutherford B. Hayes, Gaston Tissandier (editor of the journal *La Nature*), scientist Rudolf Virchow, architects Henry Hobson Richardson and Daniel Burnham, and the art critic John Ruskin. The list functions as an intellectual equivalent to the key that accompanied Muybridge's 1877 panorama of San Francisco: if by identifying the homes of the wealthy he appealed simultaneously to the egos of prospective patrons and to the voyeuristic impulses of the public, by printing the names of his influential subscribers he demonstrated not only their imprimatur but also his own elevated status.[33]

At the same time that he was cultivating this exclusive clientele, he also was courting the public at large. The majority of people who experienced Muybridge's motion studies did so through his illustrated lectures. Beginning in the late 1870s, Muybridge embarked on a series of presentations about his discoveries illustrated with images projected by magic lantern. In 1879 he introduced into his lectures the zoogyroscope (later renamed the zoopraxiscope), an innovative modification of the optical toys the zoetrope and the phenakistoscope, with the added capacity to project images. The result was the reanimation of his fragmented still moments into a remarkable simulation of continuous motion. His presentations came to be lively and elaborate affairs, involving alternating projections of still images (lantern slides of both his own photographs and examples from the history of art) with moving pictures. He delighted his audiences by demonstrating the inaccuracies in traditional representations of moving horses by contrasting them with his own pictures, and astonished them with moving images of horses, athletes, and women dancing. "Nothing was wanting," said a reporter for the *San Francisco Call,* "but the clatter of hoofs upon the turf and an occasional breath of steam from the nostrils, to make the spectator believe that he had before him genuine flesh-and-blood steeds."[34] Muybridge gave his talks to artist salons and wealthy patrons, scientific and artistic institutions. He also spoke at lecture halls and high schools, and his titles—such as "The Romance and Realities of Animal Locomotion"—were designed to appeal to a wide audience.[35]

Muybridge's performances fit easily within a thriving tradition of popular science lectures in the nineteenth century. Much as Woodward's Gardens aimed to inform visitors about scientific principles through enticing visual displays, popular scientific lecturers captured their audiences' imagination with animated talks and vibrant special effects, often courtesy of the magic lantern.[36] Leading scientists of the day promoted the education of the public; one advocate was Michael Faraday, whose many scientific achievements include the discovery of electromagnetism and whose regular Friday night lectures (as well as his 1859–60 lectures for children) at the Royal Institution in London were exceptionally well attended. Though technically innovative and even—to judge by contemporary accounts—astonishing, Muybridge's moving pictures still conform to a well-established model of popular scientific entertainment.

No small part of the attraction of Muybridge's lectures was the man himself. An engaging speaker, he earned fees for his appearances, but also relied on his charisma to sell pictures; a number of the institutions he visited bought subscriptions to *Animal Locomotion* as a result.[37] Phillip Prodger has argued that Muybridge served as a kind of expert witness, testifying to the accuracy and authenticity of the awkwardly frozen movements his photographs depicted.[38] He either feared the pictures could not speak for themselves or was unwilling to let them do so. Just as in his photographs of Woodward's Gardens and Yosemite, Muybridge was not content to disappear behind his motion studies. He promoted his person as thoroughly as he did his lectures, right down to his carefully cultivated artistic appearance (fig. 107), as described by one reporter at the time: "With gray hair carelessly tied back from an intellectual forehead, bright flashing eyes and a pleasant mouth, Mr. Muybridge must himself make an interesting subject for a photograph, whether in motion or in repose."[39] Though his motion studies have often been considered as distinct from his earlier landscape photography, Muybridge's illustrated lectures are clearly the culmination of the concerns that shaped his entire three-decade-long career. His performances—part evidence, part entertainment, and part sheer visual extravagance—were, as *Le Figaro* reported, a "most extraordinary spectacle."[40]

End of an Era

It is perhaps fitting that the twilight of Muybridge's career coincided with the 1893 World's Columbian Exposition in Chicago, the greatest spectacle of the nineteenth century. Ostensibly organized to celebrate the 400th anniversary of Columbus's discovery of the New World, the exposition was a defiant display of American ingenuity, a nationalistic rejoinder to the success of the Paris Internationale of 1889.[41] It was also a barometer of national politics: the pitched battle that erupted between New York and Chicago for the right to host the fair brought to the surface the tensions that roiled between east and west, urban and provincial populations. Chicago, the winner, positioned itself as representing the civilization of the American West. At that same fair, Frederick Jackson Turner delivered his famous frontier thesis, arguing that the distinctive character of the American people was a direct result of their constant struggle toward westward expansion.[42] Yet this frontier, which according to Turner was the "meeting point between savagery and civilization," no longer existed. The West, as they say, had been won: its wilderness tamed, its landscape crisscrossed with railroad tracks, its natural resources commodified, and its native populations subdued. The transformation of the West to which Muybridge's photographs had borne such vivid witness, was now complete.

Muybridge bought a concession at the fair, where he intended to present his illustrated lectures. Although the official name of his building was the Zoopraxographical Hall,[43] he must have worried that it was too pedantic and off-putting for the entertainment-seeking visitors to the Midway Plaisance: photographs reveal a sign over the doors that reads "Zoopraxiscopical Circus."[44] Despite the impressive edifice he built to house his performances and his reputation as an engaging showman, the lectures were poorly attended. Perhaps the fair-goers were distracted by the myriad entertainments along the Midway vying for their attention, from the belly dancers in the Street in Cairo to the Electric Scenic Theater, featuring illuminated scenes from the Alps, to the peepshow-style motion pictures of Ottomar Anschütz's electrical tachyscope. Or maybe the public, in its insatiable quest for the spectacular, had simply moved on to something new.

[1] According to Muybridge biographer Gordon Hendricks, no record of Muybridge's sea passage to the United States has been found, and the date 1852 is supported only by anecdotal evidence. See Gordon Hendricks, *Eadweard Muybridge: The Father of the Motion Picture* (New York: Grossman Publishers, 1975), 5. Official records show him to have been in San Francisco no later than 1856.
[2] In addition to Hendricks's seminal study above, see also Kevin MacDonnell, *Eadweard Muybridge: The Man Who Invented the Moving Picture* (Boston: Little, Brown, 1972), for examples. Tom Gunning makes an excellent argument against such assignment of "paternity" and contends that to do so is to limit both the readings of Muybridge's motion studies and our understanding of early cinema. See Gunning, "Never Seen This Picture Before: Muybridge in Multiplicity," in Phillip Prodger, *Time Stands Still: Muybridge and the Instantaneous Photography Movement* (New York: Iris & B. Gerald Cantor Center for Visual Arts at Stanford University in association with Oxford University Press, 2003), 222–72.
[3] Vanessa R. Schwartz, *Spectacular Realities: Early Mass Culture in Fin-de-Siècle Paris* (Berkeley: University of California Press, 1998), 178.
[4] Ibid., 179. In the introduction to *Cinema and the Invention of Modern Life* (Berkeley: University of California Press, 1995), the editors, Leo Charney and Vanessa R. Schwartz, identify the evolution of this charged relationship between the image and the real as a defining feature of the modern experience: "the blurring of representation and reality gave rise to one crucial aspect of modernity—the increasing tendency to understand the 'real' only as its re-presentations" (p. 7).
[5] In a conversation with Janet Leigh (the daughter of his defense attorney in his murder case and a vigorous defender of Muybridge's reputation). Cited in Hendricks, *Eadweard Muybridge: The Father of the Motion Picture*, 213.
[6] For a thorough discussion of the complexity of this metaphor, see Geoffrey Batchen, *Burning with Desire: The Conception of Photography* (Cambridge, MA: MIT Press, 1997), 57–69.
[7] Weston Naef suggests that Muybridge's inscription of the name Helios into the negative was an effort to prevent Thomas Houseworth from exhibiting or publishing his pictures without attributing authorship. See Weston J. Naef, James N. Wood, and Therese Thau Heyman, *Era of Exploration: The Rise of Landscape Photography in the American West, 1860–1885* (Buffalo, NY: Albright-Knox Art Gallery; Boston: New York Graphic Society, 1975), 170.
[8] The circus had been a staple of San Francisco entertainment since the gold rush; in 1850 the city boasted three circuses. See Barbara Berglund, *Making San Francisco American: Cultural Frontiers in the Urban West, 1846–1906* (Lawrence: University Press of Kansas, 2007), 58.
[9] Muybridge was very deliberate in what he included in the scrapbook: only positive reviews of his work and performances. There are no negative assessments, references to his murder trial, or any personal correspondence whatsoever. I thank Philip Brookman for this astute observation. The original album is in the collection of the Kingston Museum, but is available on microfilm at the Bancroft Library at the University of California, Berkeley.
[10] For the historiographical difficulties presented by Watkins's career, see Douglas R. Nickel, *Carleton Watkins: The Art of Perception* (San Francisco: San Francisco Museum of Modern Art; New York: Harry N. Abrams, 1999).
[11] Bancroft Library at the University of California, Berkeley, owns four bound volumes of Muybridge's half-stereo views. The first three albums serve as a kind of catalogue of the stereos Muybridge made in the American West. The fourth album contains only views of Woodward's Gardens and was apparently assembled (and annotated) by a member of the Woodward family. I thank James Eason for allowing me access to the collection.
[12] For an excellent overview and analysis of Woodward's Gardens, see Berglund, *Making San Francisco American*, 70–80. See also *Illustrated Guide and Catalogue of Woodward's Gardens*, San Francisco, 1875.
[13] Robert Bartlett Haas, *Muybridge: Man in Motion* (Berkeley: University of California Press, 1976), 23.
[14] Berglund, *Making San Francisco American*, 70.
[15] Phineas T. Barnum, founder of the American Museum in New York, pioneered the concept of the sideshow. A household name by the 1860s, he launched his "Greatest Show on Earth" in 1871; his name is still synonymous with the American circus.
[16] Though there is some debate about whether the man on the wharf is Muybridge himself, both Anita Ventura Mozley and Gordon Hendricks have identified him as such. Muybridge's penchant for including himself in his own pictures makes this a logical, if difficult to prove conclusively, conjecture.
[17] May 1872 prospectus. Cited in Haas, *Muybridge: Man in Motion*, 41.
[18] *Daily Alta California*, April 7, 1873. Cited in Haas, *Muybridge: Man in Motion*, 44–45.
[19] Testifying at Muybridge's trial, his publisher William Rulofson commented on this picture (at the behest of Muybridge's attorney, who was pursuing an insanity defense). His testimony was reported by the *San Francisco Chronicle* on February 6, 1875: "Among the strange freaks which Muybridge had committed was to have his picture taken on a rock at Yosemite valley, where a biscuit, if slightly tilted, would have fallen down 2,000 feet." Cited in Arthur P. Shimamura, "Muybridge in Motion: Travels in Art, Psychology, and Neurology," *History of Photography* 26, no. 4 (Winter 2002): 348.
[20] *Daily Alta California,* April 7, 1873. Cited in Haas, *Muybridge: Man in Motion*, 45.
[21] "German Correspondence," *Philadelphia Photographer*, February 1874. Cited in Haas, *Muybridge: Man in Motion*, 45.
[22] I thank Keith Davis, who makes a similar comparison in his recent book *The Origins of American Photography, 1839–1885: From Daguerreotype to Dry-Plate* (Kansas City, MO: Hall Family Foundation in association with the Nelson-Atkins Museum, 2007), plates 380 and 381. Davis beautifully pairs Muybridge's image of Cathedral Rocks with another Watkins took in 1861. Although the composition of these two images is quite similar, they nonetheless brilliantly exhibit the different approaches of the two photogaphers.
[23] The completion of the transcontinental railroad dramatically increased tourism to Yosemite, though it still required an arduous horseback journey into the valley itself. The construction of several stagecoach roads in the mid-1870s made tourist access far easier and led to a vastly greater number of visitors to the park.
[24] For a thorough and nuanced analysis of this panorama, as well as discussion of the various formats of the 1877 panorama, see David Harris with Eric Sandweiss, *Eadweard Muybridge and the Photographic Panorama of San Francisco, 1850–1880* (Montreal: Canadian Centre for Architecture; Cambridge, MA: MIT Press, 1993).
[25] As David Harris states: "Judging by the number of surviving copies in public institutions and private collections, this panorama must have been produced in fairly large numbers." Ibid., 118.
[26] See Martha Sandweiss, *Print the Legend: Photography and the American West* (New Haven, CT: Yale University Press, 2002), 57.
[27] Ibid., 61.
[28] According to Harris, there is no evidence that Muybridge ever displayed this panorama in the round, the format in which painted panoramas were shown. The New York Public Library's copy does have small hanging loops along the top edge, which suggests that it may have been exhibited on a wall at some point. Most of the surviving copies are in their original accordion-bound albums, a format that more or less dictates viewing the prints lying flat. See Harris, *Photographic Panorama of San Francisco*, 50.
[29] Marta Braun, *Picturing Time: The Work of Étienne-Jules Marey (1830–1904)* (Chicago: University of Chicago Press, 1992), 247. See also her essay in the present volume.

[30] One such example is *Hard Core: Power, Pleasure, and the "Frenzy of the Visible"* (Berkeley: University of California Press, 1999), 34–58, in which Linda Williams connects this aspect of Muybridge's motion studies to the prehistory of the pornographic film.
[31] Preface to Eadweard Muybridge, *Animals in Motion* [1898] (New York: Dover Publications, 1957), 14.
[32] Eadweard Muybridge, *Descriptive Zoopraxography; or, The Science of Animal Locomotion Made Popular, by Eadweard Muybridge. Published as a Memento of a Series of Lectures Given by the Author under the Auspices of the United States Government, Bureau of Education, at the World's Columbian Exposition, in Zoopraxographical Hall, 1893* (Philadelphia: University of Pennsylvania; Chicago: Lakeside Press, 1893).
[33] Muybridge was known to have exaggerated his success by adding famous names to this list, even if they were not subscribers, a marketing technique he had used earlier in his career by offering photographs of Idaho and Montana among other places he had certainly never been.
[34] *San Francisco Call*, May 5, 1880. Cited in Haas, *Muybridge: Man in Motion*, 120.
[35] For the most complete history of Muybridge's illustrated lectures, see Stephen Herbert, "Projecting the Living Image," in *Eadweard Muybridge: The Kingston Museum Bequest*, ed. Stephen Herbert (Hastings, UK: Projection Box, 2004), 107–45.
[36] See Gunning, "Never Seen This Picture Before," 252–53, for further discussion of the connection between Muybridge's lectures and popular scientific presentations. See also Laurent Mannoni, *The Great Art of Light and Shadow: Archaeology of Cinema*, trans. Richard Crangle (Exeter: University of Exeter Press, 2000), for the history of magic lantern shows.
[37] Haas, *Muybridge: Man in Motion*, 168.
[38] Phillip Prodger, "Comparing Muybridge to Muybridge: Proving the Invisible in Motion Photography," 45 (unpublished proceedings from the symposium "Eadweard Muybridge: Pioneer, But of What?" Stanford University, May 2–3, 2003). I thank Bernard Barryte for making the transcript of the symposium available to me.
[39] *Philadelphia Times*, March 26, 1882. Bancroft Library, University of California, Berkeley, Eadweard Muybridge Papers, 1876–1955, BANC MSS C-B 715 (hereafter cited as Muybridge Papers), Kingston Scrapbook, 78.
[40] *Le Figaro*, cited in Muybridge's prospectus for a lecture in October 1895. Muybridge Papers, Kingston Scrapbook, 235.
[41] See R. Reid Badger, *The Great American Fair: The World's Columbian Exposition and American Culture* (Chicago: Nelson-Hall, 1979), 46–47.
[42] See Alan Trachtenberg, *The Incorporation of America: Culture and Society in the Gilded Age* (1982; reprint, New York: Hill and Wang, 2007), 11–37.
[43] George E. Nitzsche, who worked at the University of Pennsylvania and devoted the better part of his retirement to the restoration of Muybridge's place in the history of cinema, wrote in a letter to Janet P. Leigh (daughter of William Wirt Pendegast, Muybridge's defense attorney) on December 30, 1949: "In the World's Fair in Chicago in 1893, as a young man, I had the opportunity of seeing the first moving picture house in the world (Muybridge's Zoophraxical [*sic*] Hall), but I am afraid that the scientific name was rather forbidding." Muybridge Papers, Part I, Folder 1949–51.
[44] Herbert, "Projecting the Living Image," 135.

Animals in Motion

111. *Edgington—Fast (Phase 1)*, 1878

112. *Thomas Day, Importer of Gas Fixtures, French Clocks, Fine Bronzes, Etc. 122 and 124 Sutter Street, San Francisco*, 1880

113. *Hall Looking towards the Back Mosaic Floor*, from *Leland Stanford's Residence Photograph Album*, 1878

114. *"Fountain of Love" in Pompeian Room*, from *Leland Stanford's Residence Photograph Album*, 1878

115. *View from Windows Looking South*, 1877

116. *View from Windows Looking South-west*, 1877

117. *Panorama of San Francisco from California Street Hill*, 1877

118. *Panorama of San Francisco from California Street Hill, San Francisco, California,* 1878

California St

Mechanics Exposition building
City Hall

119. *Robert C. Johnson in his study*, from *Kate and Robert Johnson Photograph Album*, 1880

120. *Robert C. Johnson in conservatory*, from *Kate and Robert Johnson Photograph Album*, 1880

121. *Kate Johnson with "spirit" of husband, Robert, adjacent*, from *Kate and Robert Johnson Photograph Album*, 1880

122. *Rosalind Johnson on her pony*, from *Kate and Robert Johnson Photograph Album*, 1880

123. *Rosalind Johnson and governess with Chinese parasols, Heartsease*, from *Kate and Robert Johnson Photograph Album*, 1880

124. *Heartsease*, from *Kate and Robert Johnson Photograph Album*, 1880

125. *Untitled (landscape near Watsonville, California)*, ca. 1879

126. *Leland Stanford, Jr. on his Pony "Gypsy"—Phases of a Stride by a Pony While Cantering,* 1879

127. *General view of experiment track, background and cameras. Plate F*, 1879, from *The Attitudes of Animals in Motion*, 1881

128. *Camera and shutter (Side View). Plate B,* 1879–80, from *The Attitudes of Animals in Motion*, 1881

129. *The Horse in Motion. "Abe Edgington," owned by Leland Stanford; driven by C. Marvin, trotting at a 2:24 gait over the Palo Alto track, 15th June 1878*, 1878

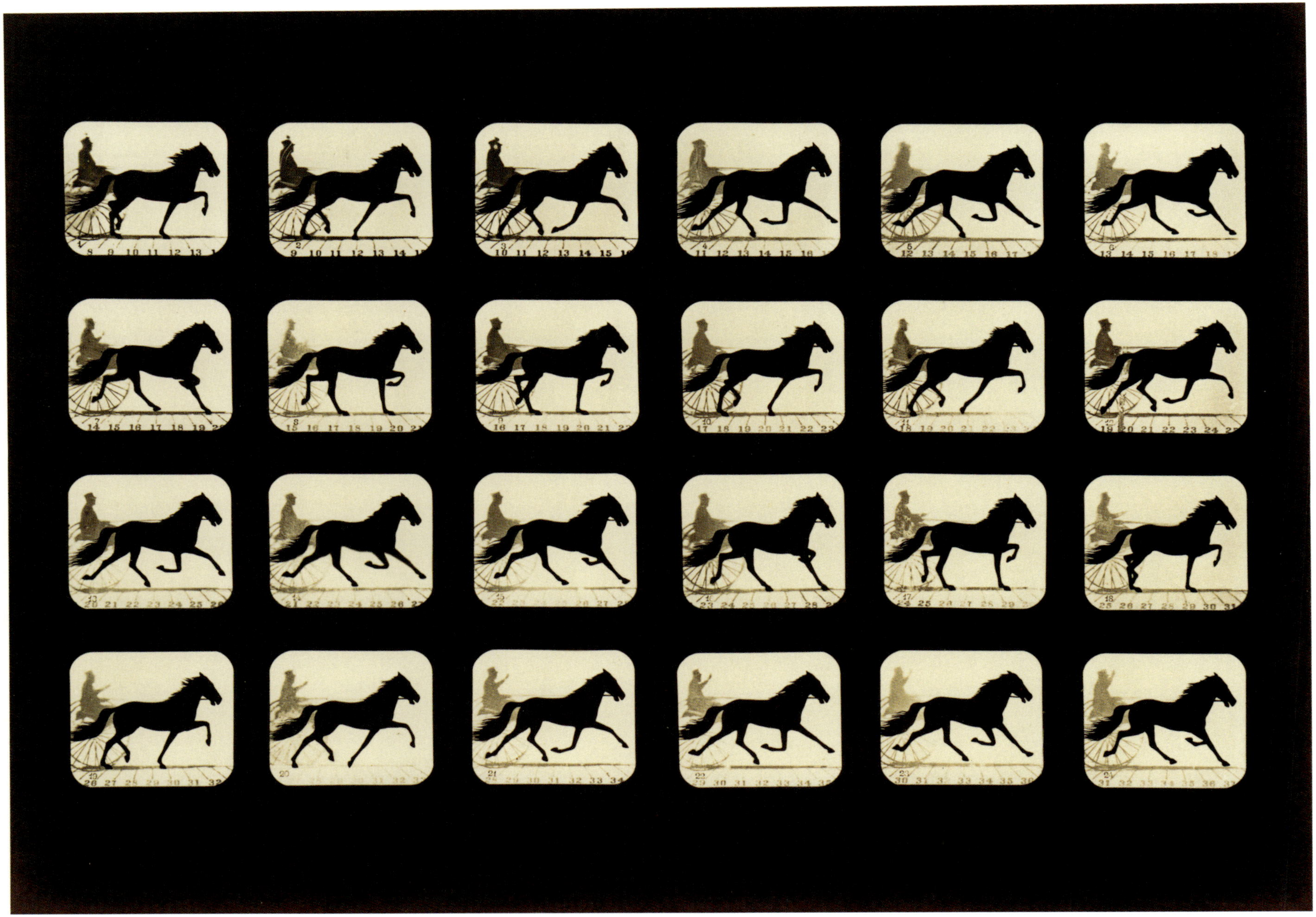

130. *Horses. Trotting. Occident. Plate 32*, 1879, from *The Attitudes of Animals in Motion*, 1881

131. *Horses. Cantering. Mahomet. Plate 18*, 1879, from *The Attitudes of Animals in Motion*, 1881

132. *Various Animals, Boar, Trotting. Plate 91*, 1879, from *The Attitudes of Animals in Motion*, 1881

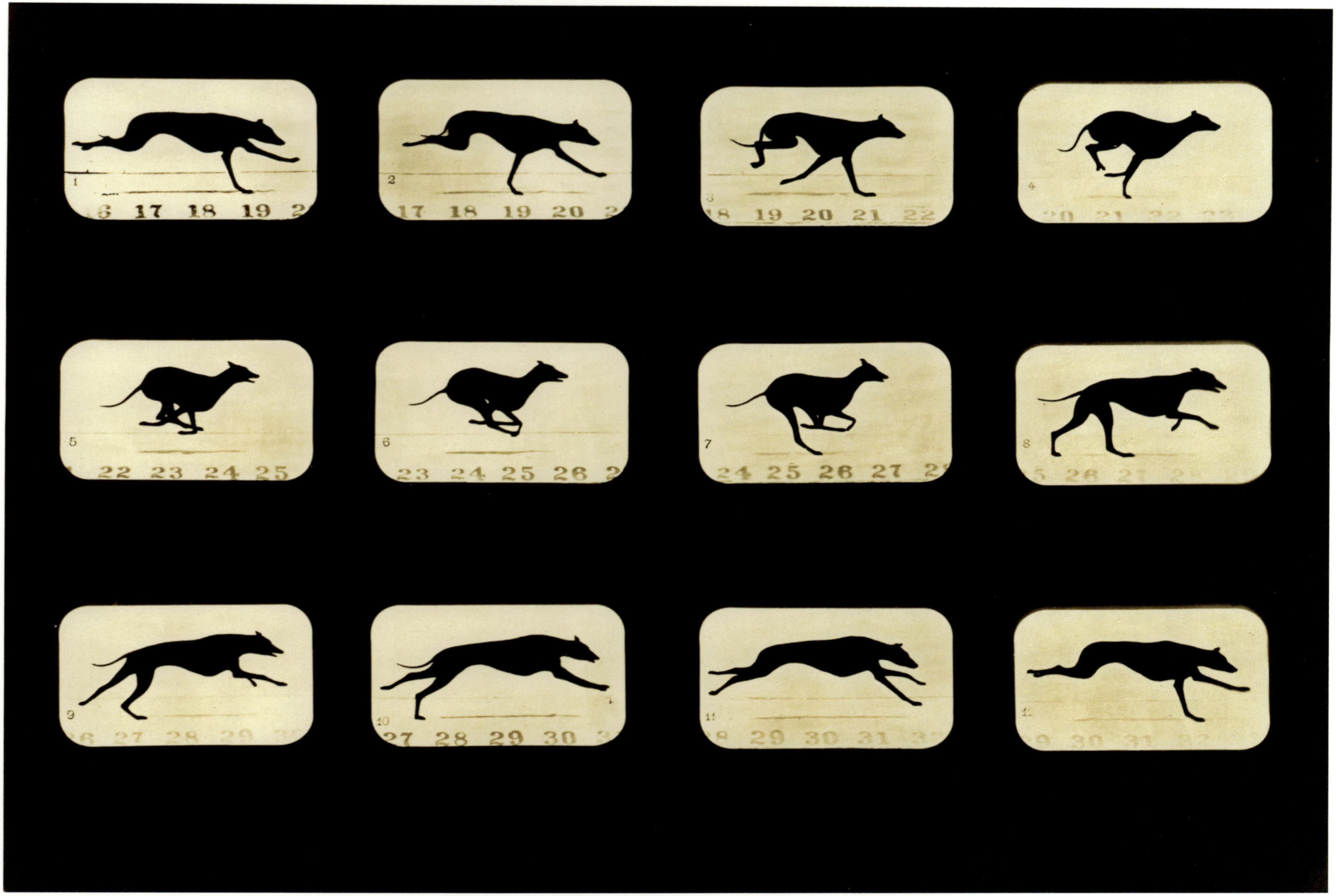

133. *Various Animals, Greyhound, Running. Plate 76*, 1879, from *The Attitudes of Animals in Motion*, 1881

134. *Athletes. Flip Flap. Plate 109*, 1879, from *The Attitudes of Animals in Motion*, 1881

135. *Athletes. Posturing. Plate 115*, 1879, from *The Attitudes of Animals in Motion*, 1881

136. *Athletes. Swinging Pick. Plate 110*, 1879, from *The Attitudes of Animals in Motion*, 1881

137. *Studies of Foreshortenings. Horses. Running. Mahomet. Plates 131–132*, 1879, from *The Attitudes of Animals in Motion*, 1881

138. *Nimrod Pacing, maquette for a zoopraxiscope disc*, ca. 1879

139. *Mule Bucking and Kicking, 12 phases, No. 29*, 1893

140. *Athlete, Running Long Jump, 14 phases,* 1893

141. *Four Horses with Riders, Dogs, and Spectators, 20/21 phases, No. 35,* ca. 1879–80

142. *Woman Dancing*, 1893

143. *Woman Dancing*, 1893

Fig. 114. *Eadweard Muybridge self-portrait*, 1883

Animal Locomotion

Marta Braun

Advertised by Eadweard Muybridge as "A work for the Art Connoisseur, the Scientist, the Artist and the Student of Art or of Nature," *Animal Locomotion* is all that and more.[1] First published in 1887, its 781 collotype prints are grouped into 11 volumes, in an order conforming to what a logical and progressive atlas of human and animal movement should look like at the end of the nineteenth century. The first category, the nude, constitutes the largest group of images and the first four volumes of the eleven. Within this category, walking precedes running, jumping, and then stooping; activities without props precede activities with props; and men precede women. Semi-nude follows the nude: men sport a "pelvis cloth" (Vol. 5) and women are draped in gauzy material (Vol. 6). The sequence of movements in the semi-nude category progresses from simple to more complex, as it did with the nudes. The order is completed by the movements of children, also part of the sixth volume and naked except for one, and then men and women fully dressed (Vol. 7), pathological locomotion (Vol. 8), and the movements of horses and other domestic and wild animals (Vols. 9–11). The scientific "logic," in other words, follows a covert sociological hierarchy, from the movements of the highest subject—the nude male—through women, children, and the crippled to the lowest, namely, animals. It also moves from the more visually intelligible to the more obscure: the more difficult pictures, in which the movement is hardest to determine, are placed toward the end of each volume, where they are less noticeable. Finally, each plate bears a number corresponding to a description of the action photographed published by Muybridge in the accompanying *Prospectus and Catalogue of Plates*. Some titles, for example, *Base-ball; batting. Plate 279*, seem almost superfluous. Others such as *Crossing brook on step-stones, with fishing-pole and basket. Plate 176* and *Arising from the ground with pamphlet in left hand. Plate 270* direct the viewer to imagine a narrative context for the actions.

But the published order conceals another, quite different one, created as Muybridge photographed the almost twenty thousand individual images that make up the plates. By reconstructing that order we can, I believe, discern Muybridge's interests in the project as opposed to those of the university committee that oversaw it. This separation, in turn, enables us to reveal the competing claims of art and science in the work as well as aspects of Muybridge's own personality: his theatrical bent, his desire for status and approbation, and his gifts as an entertainer.

Six years earlier, his photographs of the gaits of Leland Stanford's racehorses at Palo Alto had made Muybridge famous and launched him on a triumphal tour of Europe. In Paris, in the fall of 1881, he was fêted at the home of the eminent physiologist Étienne-Jules Marey, astonishing his host and the other scientists, artists, and Parisian luminaries in attendance with the moving pictures he projected on the wall with his zoopraxiscope. A second party, held in November at the home of the painter Jean-Louis-Ernest Meissonier, won Muybridge more accolades and underlined the importance of his photographs for artists. In February 1882, he left for London, where initially he enjoyed a similar success. Lecturing at the Royal Institution and the Royal Academy of Arts, Muybridge enthralled and delighted his audiences of notables, among them the Prince of Wales, with his pictures of galloping horses, trotting deer, and trundling pigs. "This pleasing display was the essence of life and reality," wrote an observer for the *Photographic News*. "[A] new world of sights and wonders was, indeed, opened by photography, which was not less astounding because it was truth itself."[2]

Muybridge's success went to his head. To be a mere—albeit famous—photographer was no longer enough for him. In the 1880s, photography was an uninspired medium: its practitioners devoted themselves to developing new processes for depicting their subjects, which were still primarily formulaic studio portraits or scenic wonders. The struggle to raise the status of photography to that of art, a struggle that began virtually at the birth of the medium in 1839, was temporarily abandoned after the commercialization of the dry plate, a fast, easy-to-use process that spawned the industrialization of the medium and favored new applications and techniques rather than aesthetics. By the 1880s, photography seemed an intractably mechanical process, and those who used it were generally regarded as technicians—mere machine operators. Scientists, on the other hand, had both prestige and authority. Unlike the gentlemen amateurs at the beginning of the century who carried out their experiments in private laboratories supported by family money, scientists in Muybridge's time had become professionalized and specialized. They worked in state-funded laboratories, and their results were disseminated by scientific academies and in popular journals (such as *La Nature*, in which Muybridge's photographs of Stanford's horses had first appeared in December 1878). Scientists were revered, and the influence of their work was felt at every level of educated society. It was this association with science that Muybridge sought. He imagined himself remade as a scientific investigator rather than a photographic artist. This is how he presented himself before Marey's guests, members of the Academy of Sciences in Paris, and the Royal Institution in London. And this is how he described himself in his letters to Frank Shay, Stanford's secretary, soon after arriving in Paris.[3] Muybridge wanted his new project to be a scientific study of movement. But for this he needed financial support. As a writer from the *Photographic Times* noted, Muybridge was lacking only the proper "encouragement":

> Mr. Muybridge modestly calls his series of animals in motion—they include horse, dog, deer, bull, pig, &c.—simply preliminary results. They contain little or no half tone, and are only proof of what may be done. What he desires now to secure, if he only received sufficient encouragement, is a series of photographic "pictures," and these, with the experience he has now acquired, and with the gelatine process to help him, should be well within his reach. We only trust this encouragement will be forthcoming, and that Mr. Muybridge will be tempted to carry on the difficult work he has commenced with such genuine success.[4]

And then there was a setback. In April, just as Muybridge's paper "on the attitudes of animals and motion" was read to the Royal Society—the most important scientific society in Britain—a book appeared in London called *The Horse in Motion*.[5] Its author was J. D. B. Stillman; it was executed and published under the auspices of Leland Stanford and was illustrated with lithographs made from Muybridge's photographs as well as five original photographs, but it appeared without Muybridge's name on the cover or title page. His role could not have been clearer; he was simply a technician. Muybridge, who had maintained contact with Marey, wrote a letter suggesting a project in collaboration with him and Meissonier, but received no answer. He set sail for America in June and by September had launched a lawsuit against both Stanford and the publisher of Stillman's book, J. R. Osgood and Company.[6]

However, in the United States, the publication of *The Horse in Motion* did little to diminish Muybridge's reputation. He was invited to demonstrate his zoopraxiscope in Newport, Boston, and New York. His lecture—as in Europe, it was on representations of the horse's gaits in art and photography—promoted photography as a scientific instrument with which to correct the errors of artists. In February 1883, he projected his images in Philadelphia at the Franklin Institute and the Pennsylvania Academy of the Fine Arts. In March he produced a prospectus for a "new and elaborate work upon the attitudes of man, the horse and other animals in motion," in which we can see the kernel of what would become *Animal Locomotion* with its fusion of popular entertainment, aesthetics, and science. The new work would include "photographs of actors performing their respective parts, works of art from the Museums, Picture Galleries and Libraries of Europe," photographs of "trajectory curves" that would be made with Marey's "photographic revolver," and an essay by Marey himself. To make his appeal as wide as possible, Muybridge added that the subscriber could "send a horse or other animal or subject to my studio for a special photographic analysis of its movements which will be illustrated without extra charge."[7]

By late spring, Muybridge had found his new patron: the University of Pennsylvania. Fairman Rogers, a sportsman and scientist who was the head of the Academy of the Fine Arts, and Thomas Eakins, who taught art both there and at the university, persuaded William Pepper, the provost of the university, of the value of a new Muybridge project.[8] By the beginning of August, Pepper had sent a formal invitation to Muybridge to begin the work, committed five thousand dollars to it, and put the grounds outside the veterinary hospital at the photographer's disposal.[9]

In March of 1884, Pepper put together a committee to "supervise the entire affair and thus insure its thorough scientific character."[10] It included Eakins, neurologist Francis X. Dercum, anatomist Joseph Leidy, physiologist Harrison Allen, and engineer William Marks. The members of the committee had different interests and different subjects they wanted Muybridge to photograph, but they all believed in the scientific capability of the camera—its proven capacity both to make permanent what the eye could see and to record what the eye could not grasp. Considering that they met only once and for the most part disappeared from the scene when Muybridge started working, they must have had faith in his ability as a technically proficient photographer. For his part, Muybridge appreciated the status the committee provided. It gave his work, he claimed, "additional weight and value," as he was "neither a physiologist nor an anatomist."[11] But such acknowledgments would not in any way stifle his creative impulses. He would serve two masters with his project: his scientific committee and his art. Associated now with a famous university and a group of reputable scientists, Muybridge finally won the credibility he had long desired. At the same time, he created an aesthetic repertoire for the camera that no one could ever have predicted.

Together with Eakins, Muybridge began to photograph in June 1884. Eakins had earlier criticized Muybridge's battery-of-cameras system as unscientific and urged him to use a variation of a Marey-wheel camera—a single camera with a single lens that remained open and a rotating slotted-disc shutter that alternately exposed and masked the plate behind it. Muybridge seems to have made some pictures with such an apparatus, but according to a letter written in August by Eakins's student Thomas Anshutz, the university, which had hired Muybridge to make serial photography, was not pleased with his progress:

> [Muybridge] is making some very nice photographs of men and women doing things. Such as throwing, jumping stepping down and up etc. These are all made as I crudely explained on one plate. So that it shows the figure at intervals of a few inches as it goes through the movement. These images, of course frequently overlap but do not seem to confuse on that account. This work is, of course, done with one lense [*sic*]. He has not yet done any work with his series of lenses and I hear they do not work. The shutters are too clumsy and slow. The university people are dissatisfied with the affair as he cannot give them the result they expected. Which was to photograph the walk of diseased people paralytics etc. so that by means of the zoopraxiscope (help!!!) they could show their peculiarities to the medical student. This it seems however cannot be done even with the best known contrivances. So they would like to fire the whole concern but they have gone too far to back out.[12]

Eakins, too, became fed up with Muybridge and left. With the rest of the committee on vacation, Muybridge was on his own.[13]

The earliest surviving photographs he made at the university are of a Mrs. Cooper, walking with and carrying a child and pretending to spank him, and three unidentified male models leaping, sparring, and throwing a rock (plates 145–147). Mrs. Cooper was an artist's model, a widow "aged 35, somewhat slender and above the medium height."[14] She and most of the female models we see in *Animal Locomotion* were introduced to Muybridge by J. Liberty Tadd, director of the Industrial School of Art. We do not know how the two men met, but Muybridge photographed Tadd's hands tracing a circle, beating time, and lifting a ball (plates 180, 183) and also took pictures of his wife and daughter (plate 176).[15] Mrs. Cooper and the athletes were photographed not with a battery of cameras, but with five or six cameras placed in a semicircle around them, the

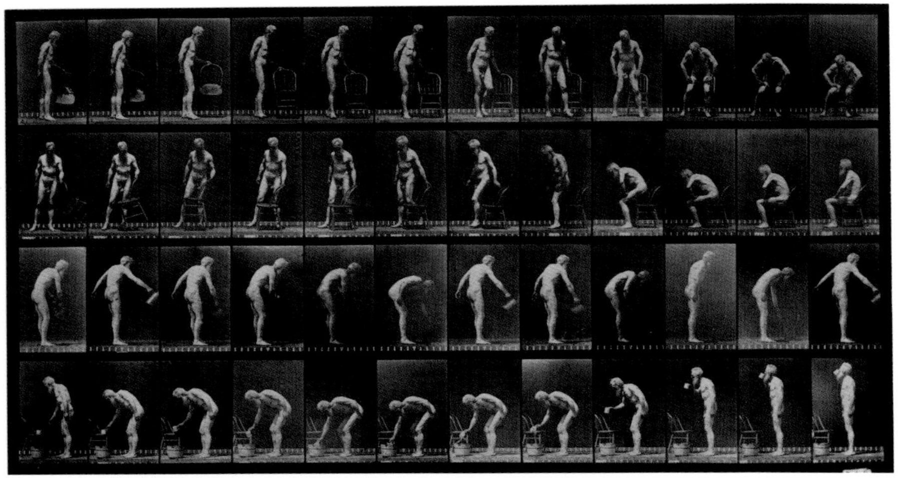

Fig. 115. *A, sitting down; B, sitting down; C, sprinkling water; D, stooping for cup and drinking. Plate 490,* 1887

shutters triggered simultaneously. Using the cameras this way produced a single frozen gesture seen from six different points of view, but not a sequence of motion. The effect is that of walking around the model, each picture adding to the view of the figure in three dimensions.

Muybridge first used this setup in 1879 in Palo Alto. There, according to the *San Francisco Examiner,* he had "arranged five cameras in a semi-circle and concentrating upon one point, galloped a horse over the point where the electric current was completed and produced a perfect picture of a horse at fullest speed, as seen from five different points of view all at the same instant of time and while, of course, the horse was in one and the same position" (plate 137).[16] Why did he revert to this method after five years? Perhaps it was taking longer than he expected to build the shutters for his battery of cameras. Meanwhile, he did make almost one hundred of these tracking shots. They focus on the spectacular nature of that single frozen moment only the camera can capture, that instant in which the laws of gravity no longer seem to prevail. In these first photographs made in Philadelphia, we can find all the aesthetic motifs that would reappear obsessively throughout *Animal Locomotion.* Not just the body suspended in midair (plate 148) but also the frozen arc of water thrown from a bucket or bowl, the fluttering of drapery, the handling of vases, jugs, and baskets (fig. 119, plate 166). Unlikely to be studied by scientists or doctors, these are the images and gestures Muybridge would return to whenever he was beyond the scrutiny of his committee.

When the professors returned from vacation that August, Muybridge was already working with a battery of twelve cameras, photographing university athletes, whose training was of special interest to Pepper (fig. 88), and disabled children, the first of Dercum's patients. (Here one can find the earliest method of measuring the space the models traverse: narrow strips of wood marked off vertically and numbered, placed on either side of the path.) The most frequently photographed subject, however, is Muybridge himself.[17] We don't know who tripped the shutters, but Muybridge, who at fifty-four describes himself as an "ex-athlete, aged about sixty," poses for the camera in the acts of walking, climbing a ladder, throwing a disk, swinging a pickax, digging, sawing and hammering, using a hatchet, sitting down, sprinkling water, and drinking (figs. 110, 115). He is naked. His body is lean, with sinewy muscles, and his long white beard is particularly distinct.[18] Muybridge obviously loved to be photographed. He appeared in some of his early landscapes, and in 1879 he photographed himself shaking hands with one of the athletes from the San Francisco Olympic Club. (plate 135). Though he was fully clothed in that picture, in another made at the same time he is naked, swinging a pickax, as he does for the Philadelphia cameras (plate 136). The insertion of his own naked body into the project tells us a lot about Muybridge's narcissism, and the pickax helps us imagine how he saw himself: strong and rugged, part of the Wild West where such implements served to build railroads and mines and to tame the wilderness for farming.

In late August, Muybridge took his battery of twelve cameras to the Philadelphia Zoological Garden, where he managed to make sequences of storks and swans, a sloth on a pole, an eland, a camel trotting and galloping—but photographed only from the rear—a crane flapping its wings, and deer leaping over one another (plate 189). His system was not yet perfect enough to photograph birds in flight, and most of the images he made, he confessed, were not publishable. The cameras were not consistently synchronized, and the shutter speeds, as Anshutz had complained, were probably too slow.

Muybridge spent the winter and spring lecturing in Philadelphia and other cities, including Baltimore. When he began photographing again on June 2, 1885, his subject was the "mulatto pugilist" Ben Bailey (plate 156). It is in the Bailey photographs that the familiar grid found in the background of all subsequent *Animal Locomotion* pictures first appears. The idea of using a grid made up of white threads (in many of the images, one or two of Muybridge's four assistants are visible behind it) was borrowed from the English ethnologist J. H. Lamprey, who had devised it in 1869 as an aid in making ethnographic photographs of Malay natives.[19] Though artists had long used grids to ensure the accuracy of bodily proportions in their drawings, for Lamprey, the grid aided comparative measurements of general body morphology, "which no verbal description can convey and but few artists could delineate."[20] In other words, by increasing the scientific readability of the image, the grid served to authenticate the subject of the photograph as an object of anthropological scrutiny. At this time, anthropological description was aimed at collecting comparative data in order to produce pictorial evidence of racial typologies. These descriptions, in turn, were used to support an evolutionary hierarchy of races of which whiteness was the apex. The presumed superiority of the white man to the nonwhite was based on assumptions that circulated with popularized Darwinism. The most fundamental of these was the particular definition of "race" itself, which joined the physical, biological nature of man to his cultural, moral, and intellectual nature. This conflation of physical and cultural traits allowed nineteenth-century scientists to claim culture as biologically determined and consequently to see the advanced technological culture of the West as proof of the biological superiority of the white man. In this construction, "civilization" was an explicitly racial concept. It denoted a precise stage in human racial evolution—following the more primitive stages of "savagery" and "barbarism."[21] The emergence of scientific anthropology at this time was supported by positivist beliefs about the validity of observation, recording, and classification, and photography helped transform these activities into a visible, permanent, and measurable form.

What was new in the Bailey photographs was the use of sequence photography rather than a single image to determine racial difference.[22] The presence of the grid in the images of Ben Bailey can be attributed to Pepper and Dercum and fellow committee members Harrison Allen and Joseph Leidy, the latter a member of the Ethnological Society of London. All these men were to be the founding members, in 1889, of the American Anthropometric Society, devoted to an anthropological subdiscipline that used the measurement of physical difference such as skull size to define racial types and hierarchies. Of the ninety-five models who appear in *Animal Locomotion*, Bailey is the only black man. And whereas Muybridge photographed the other pairs of boxers, fencers, and wrestlers in combat, he photographed Bailey alone, punching the air. Furthermore, unlike the others who only fight and fence (plate 168), Bailey also walks and runs for the cameras. The grid behind him cues the viewer to read his body as an example of "primitive" masculinity. It transforms him from a subject into an object, part of a taxonomy of race constructed by *Animal Locomotion*.

In the days following Bailey's appearance, Muybridge photographed students from the university (as well as anatomy professor A. J. Parker). They are less muscled than Bailey, and their bodies contrast sharply with the boxer's overdeveloped physiognomy. The athletes were the product of Pepper's new system of physical education at the university, meant to develop an overall fitness and vigor, key, according to Pepper, to the "cultivation of manly and courageous qualities" that defined white masculinity.[23] Beginning in the 1870s, elite American universities had attempted to reshape American sports on the English model, both as a way of establishing the social distinctiveness of the new American middle class and as a way of separating it from the growing working class (who also appeared in front of Muybridge's cameras), particularly immigrants. In the minds of pioneers like Pepper, the physical prowess of the university athlete would be a manifestation of his moral superiority, representative of the values of the American way of life—the emphasis being not on professional sportsmen but on amateur athletes, whose physical education would allow them to withstand neurasthenia, or "American nervousness." This "disease of civilization," seen by the medical profession as resulting from the inhu-

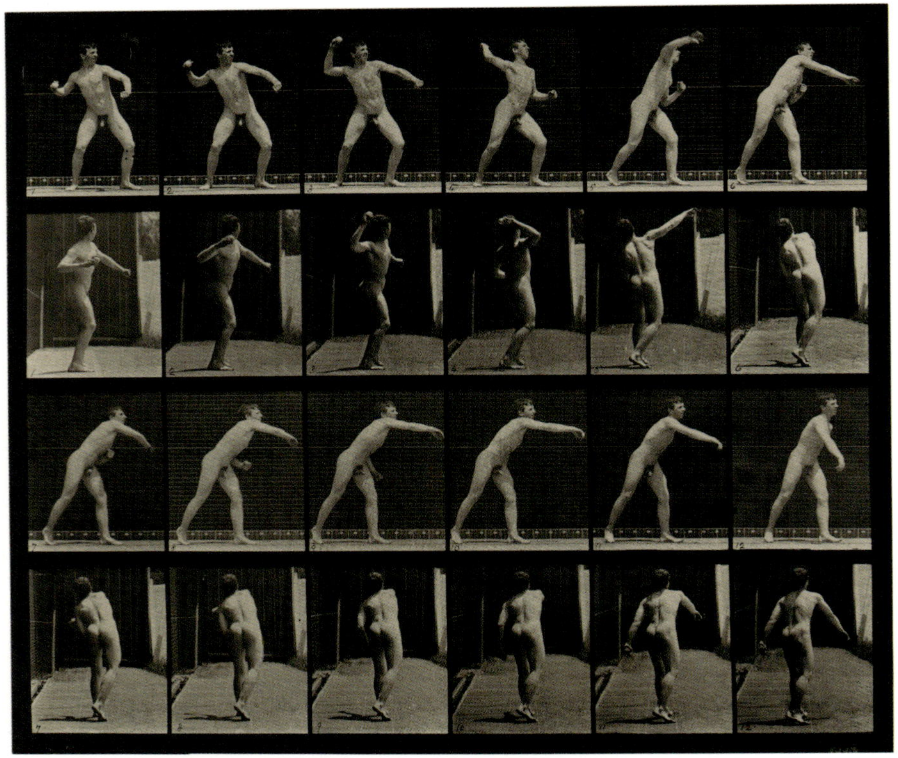

Fig. 116. *Base-ball; throwing. Plate 286*, 1887

man pace of modern life, was perceived as a threat to America's "brain workers," the college-educated professionals.[24] University of Pennsylvania students Peter Daley and Morris Hacker (fig. 116), noted in the university athletic club for their exceptional abilities, were two beneficiaries of Pepper's system as well as paradigms of the specimens he hoped to develop. Percy C. Maderia was the 1884 amateur one-mile champion; Thomas Grier, who leapfrogs over his brother Edward's back (plate 182), won the mile walk in the 1886 university track meet; and Frank Gummey, photographed wrestling (plate 347), was also the school's champion rower.

That first week in June, Muybridge also expanded his setup. He placed two additional cameras, each with a bank of twelve lenses, shutters, and plate holders, at sixty and ninety degrees to his subject. From this point on, he would use his battery of twelve single cameras parallel to the model to make what he called "lateral" views, at the same time as the other two cameras made the "foreshortened" views. He devised an electric circuit breaker, his "exposor motor," that triggered all thirty-six shutters (or as he called them, "electro-exposors") synchronously (fig. 85). At least that was the intention. But as we know from the *Prospectus and Catalogue of Plates*, Muybridge had difficulties with his apparatus. The cameras would not go off in the correct sequence or at all, negatives fogged or were broken, and some exposure times and intervals were not recorded. A close look at the plates shows unexpected inconsistencies. There are gaps between the individual images and missing phases of movements, some accounted for by Muybridge, some not. There is no movement at all in some images, while in others two contrasting sets of consecutive numbers can be seen. There are horizontal and vertical images, purportedly taken simultaneously, in which the positions of the figures do not match. Supposedly chronological sequences of images are marked by two different sets of consecutive numbers, one in the negative and one in the print, and what were clearly separate sequences taken at different times have been pieced together in a single plate. *Plate 504*, for example, shows Blanche Epler walking up and down stairs with a bowl of water in her hands. In the fourth image, a water jug has taken the place of the bowl (figs. 117–119). Two sets of numbers can be seen in some of the individual images in this sequence. The assemblages of sequences that constitute the published plates show how Muybridge compensated for such problems. The inconsistencies are invisible because the images are in sequences, and a sequence dictates a logical relation among its parts, in Muybridge's case, cuing us to perceive movement even when there is none.

If the assemblages of the laterals and foreshortenings in the final prints hide the problems Muybridge had with his apparatus, the *Animal Locomotion* cyanotypes make those problems visible. Made with iron rather than silver salts and similar to blueprints, they were found in storage in the Smithsonian Institution in 1999 (fig. 117, plates 149, 151, 177, 179). They are the contact proofs Muybridge made from his Pennsylvania negatives; they came to the museum

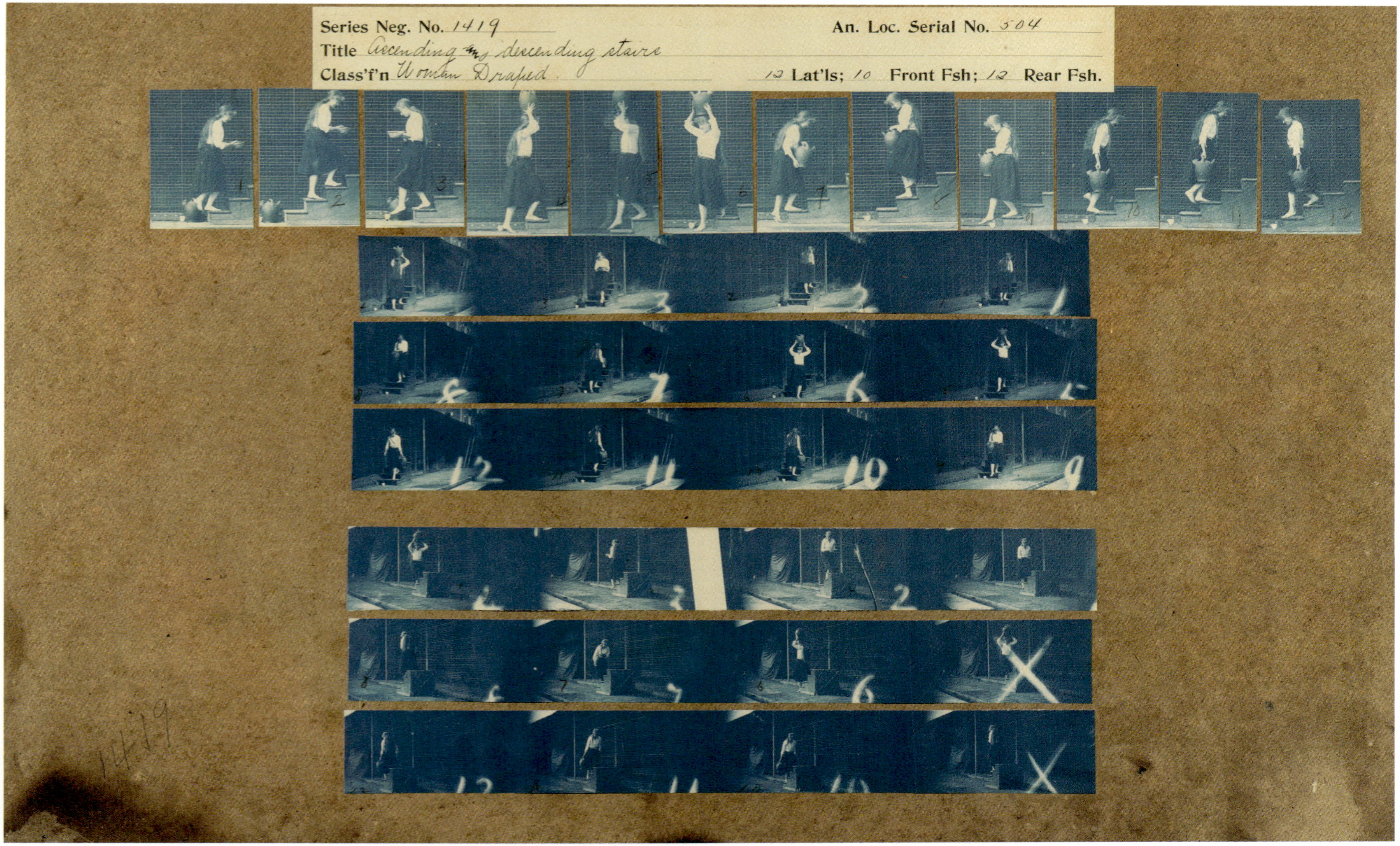

Fig. 117. *Cyanotype proof for Animal Locomotion. Ascending and Descending Stairs (Blanche Epler). Plate 504*, 1885

Fig. 118. *Glass positive for Animal Locomotion. Ascending and Descending Stairs. Plate 504,* 1887

Fig. 119. *Ascending and descending stairs. (Movements. Female). Plate 504*, 1887

from Colt's, the New York lantern slide manufacturer, and were intended to be used as references for making and selling slides of individual images from *Animal Locomotion* after its publication in 1887. The cyanotypes are mounted roughly on board: each board is labeled with a preprinted sheet inscribed with the *Animal Locomotion* plate number, a negative series number, the costume category (nude, pelvis cloth, draped), and the number of lateral and foreshortened views taken.

The cyanotypes represent the first and formulating stages of *Animal Locomotion.* Using them as a guide, Muybridge would create a composite glass positive from his negatives and from the glass positive, an internegative that was used to make a collotype plate from which the final print was struck (figs. 117–119). Muybridge's negatives are lost, so the cyanotypes constitute the only evidence we have of what he originally photographed. And what is most striking upon first viewing these ur-pictures is how much they differ from the published plates. The cyanotypes illustrate the painstaking labor Muybridge undertook to realize each plate of *Animal Locomotion*. He has worked over every single one of the twenty thousand glass negatives. He has enlarged some, cropped others—often radically—moved them around, and pieced them together into deceptively cohesive patterns of laterals and foreshortenings. In fact, he had thirty-six different ways of arranging the rows of images that constitute the finished plates. For example, *Plate 703* is assembled as two rows of front foreshortenings over two rows of rear foreshortenings; *Plate 709* shows two rows of lateral views over one row of rear foreshortenings; and *Plate 45* has twenty-four laterals in four rows (plates 174, 193). The cyanotypes, in short, show us that *Animal Locomotion* is a project whose every element has been subject to manipulation of one kind or another.

Muybridge's cropping and enlarging are the first differences we can see between the facts presented to the camera and the picture of them he offers the viewer. The cyanotypes make visible what has been eliminated in the collotypes, revealing the project's sites and working methods, camera placements and positioning, the details of the outdoor studio, the black-and-white "grid" (we can clearly see now the white threads), the assistants behind it, and the white reflectors. The whole technology of construction is exposed in the cyanotypes, and all of it is concealed in the final prints.

From June 2, 1885, the day Ben Bailey and the grid first appeared before his cameras, Muybridge began to keep notebook entries for each sequence of images he took. He recorded a negative series number, the model's name, a brief description of his or her movement, the number of images made, the time spent with the model, and the mechanical details of the session. Before the cyanotypes came to light, it was not possible—except in a few cases where the negative series number is visible in the published plate or where the subject is unmistakable ("Miss Cox, *aet.* 30, 340 lbs.")—to apply the notebook descriptions to the corresponding plate, because the published plates are numbered and ordered in a sequence that is unrelated to the negative series number, which, as explained above, is important in itself. But the cyanotypes are a kind of Rosetta stone, labeled with both the negative series number and its corresponding plate number, so the notebook entries can now be deciphered and the descriptions applied to the plates. The correlation has produced some fascinating results. For one thing, the notebook titles can differ from the plate titles. *Two models; one standing, the other sitting, crossing legs. Plate 239* is "Miss N[ellie] placing chair & Miss M[amie] sitting on it" in the notebooks, while *Turning around in surprise and running away. Plate 73* (plate 155) is "Ashamed." And we can now specify the order of the works and trace their progress. The chronology provided by the cyanotypes shows how the project evolved: how, for example, within the tracking shots, Muybridge quickly abandoned the spectacular attractions of suspended motion and began to construct embryonic narratives created through a vertical arrangement of poses, and how, in August 1884, the six single cameras gave way to the battery of twelve. The correlation of negative series, plate numbers, and notebooks yields another reward: the names of the individual models and their role in the project.

Much has been written about the social construction of gender in these images: how the women undertake activities not usually associated with science; how, proscribed by modesty and social norms, they use props that overdetermine the narrative function of the images; how the pictures of males engaged in sports such as baseball or professions such as bricklaying are paralleled by images of women engaged in dusting and washing or passive and eroticized movements, including crawling naked, getting in and out of beds and hammocks, smoking, and dancing half-naked.[25] With the identification of the models, we can make even finer distinctions. We can show, for example, how their representation indicates their social and material inequality. Muybridge identifies the artist's models only by their first names in the notebooks—Mamie, Nellie, and Lily. They are represented only in the nude. The wife of J. Liberty Tadd is always "Mrs. Tadd" and is only pictured dressed. Muybridge records the full name of Peter Daley, as he does for all the students, and he tends to represent them as idealized Greek nudes, as befits their youth and status. But the fencer Bonifou and the blacksmiths Redinger and Breen (plate 150) are known only by their last names and are never without the "pelvis cloth," out of concern, their own or the photographer's, for their working-class modesty.[26] When it comes to self-consciousness, however, the pleasure the artist's models take in performing for the camera mirrors that of the students, while the more proper wives and mothers go about their activities with the same solemn demeanor we find in the laborers and professionals.

Muybridge's photographs—of the university athletes; of the laborers, carpenters, bricklayers, and blacksmiths who were all "experts in their particular trades";[27] of the boxers, fencers, and other professionals who demonstrated their prowess for the cameras; and of the hospital patients with locomotive disorders—all make visible his committee's concerns with health, class, and race at the end of the nineteenth century. Such concerns are much less discernable in his photographs of women—mostly artist's models, whom he paid by the hour. Wanting, so he said, to photograph them in everyday activities, Muybridge complained that they lacked poise: "I have experienced a great deal of difficulty in securing proper models. In the first place, artists' models, as a rule are ignorant and not well-bred. As a consequence, their movements are not graceful and it is essential for the thorough execution of my work to have my models of a graceful bearing."[28] While the women's "everyday activities" often parallel those of the men—they walk, run, jump, lift objects, and throw buckets of water—the men are usually involved in feats of strength: they throw big rocks and heave logs so the movement of their muscles can be seen. The women, on the other hand, carry out domestic chores: they sweep, dust, scrub floors, and spend a remarkable amount of their time with cups, bowls, jugs, and vases. And even though both men and women engage in occupations deemed natural to their gender by contemporary social mores, the women's movements are sometimes difficult to interpret. They often wear clothes that hide their limbs and muscles, interfering with the study of their movement (plate 174), or they engage in awkward or ungainly actions and what at the time were forbidden activities such as smoking. What the women do most, however (and the men not at all), is participate in narrative scenarios that Muybridge invents, as evident in the notebook titles for *Plates 427–429*: "Inspecting a Slave, White" (fig. 120); *Plate 175*: "Crossing brook on step-stones with fishing-pole and can" (plate 172); *Plates 452* and *453*: "Woman kneels and drinks from the water jar of another woman and both walk off"; "Walking in a gale" (unpublished); and even *Plates 430* and *431*: "Relinquishing drapery for nature's garb" (his notebook title for almost every sequence of a woman undressing). Very few of these images depict activities appropriate to a woman in the industrialized, modern America of 1885. Instead, Muybridge directs his most "graceful" models—Blanche Epler, Kate Larrigan (whom Muybridge describes as a "danseuse from New York" in one pamphlet), and Catherine Aimer—to reenact figures from art (peasants or Greeks) or narratives (playing with hay, fishing rods, goblets, and "ancient" water jugs) from a distant pastoral age.

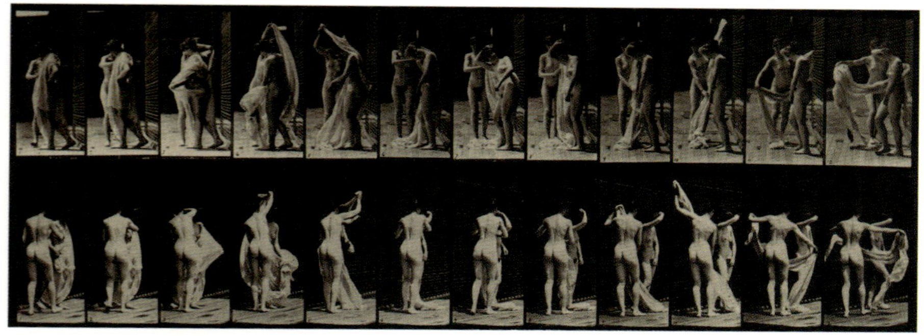

Fig. 120. *Toilet; two models, 1 disrobing 8. Plate 428*, 1887

Since 1887, *Animal Locomotion*, in one form or another, has never been out of print. Muybridge made zoopraxiscope discs from what he considered to be its best images, particularly of wild animals and birds on the wing (which he finally succeeded in photographing at the zoo in August and September 1885). For his Zoopraxographical Hall at the 1893 Chicago World's Fair, he published fifty images on paper discs that purchasers could color and animate at home.[29] After his return to England in 1894, he made selections that Chapman and Hall published as *Animals in Motion* in 1899, which was reprinted four times, and *The Human Figure in Motion* in 1901, which was reprinted six times. Other selections and editions have followed until the present. *Animal Locomotion*'s hold on our imagination remains secure. It has inspired artists from Francis Bacon to Cy Twombly and numerous contemporary filmmakers, playwrights, poets, and novelists. From the earnest athletes and the pirouetting women to the domestic and wild animals, the donkey taking off the professor's clothes, and the ponies ringing bells, Muybridge's photographs are a compendium of art, science, and entertainment in which we immediately recognize what we have never seen before.

[1] Eadweard Muybridge, *Animal Locomotion: An Electro-Photographic Investigation of Consecutive Phases of Animal Movements*, 11 vols. (Philadelphia: University of Pennsylvania, 1887). Plates printed by the Photo-Gravure Company, New York, published under the auspices of the University of Pennsylvania. The advertisement is in Muybridge's subscription offer of 1885, Archives, University of Pennsylvania, Philadelphia.

[2] "Mr. Muybridge at the Royal Institution," *Photographic News* 26, no. 1228 (March 17, 1882): 1.

[3] Muybridge letters to Frank Shay, November 28 and December 23, 1881; Collis P. Huntington Collection, George Arents Research Library, Syracuse University.

[4] "Mr. Muybridge at the Royal Institution," 1.

[5] J. D. B. Stillman, *The Horse in Motion as Shown by Instantaneous Photography, with a Study on Animal Mechanics Founded on Anatomy and the Revelations of the Camera, in Which Is Demonstrated the Theory of Quadrupedal Locomotion* (Boston: James R. Osgood, 1882).

[6] Muybridge lost both lawsuits, but never forgot the insult. The last paragraph of *Animal Locomotion: An Electro-Photographic Investigation of Consecutive Phases of Animal Movements—Prospectus and Catalogue of Plates* (Philadelphia: J. B. Lippincott, 1887), entitled "Retrospective," reminds the reader that "a number of his early experimental photographs of animal movements, and his original Title ... were copied, and published a few years ago."

[7] E. J. Muybridge, *Prospectus of a New and Elaborate Work upon the Attitudes of Man, the Horse and Other Animals in Motion* (New York: Scovill Manufacturing, 1883).

[8] A defender of photography as an aid to the accurate vision of the painter, Eakins made his own photographic studies for his pictures. He had bought Muybridge's "Horse in Motion" cards showing Stanford's racehorses running in front of a white screen and had even written to Muybridge about their use. Rogers, in turn, had asked Eakins to paint his carriage and horses in movement, basing the animals' gait on the photographs. See Anne McCauley, "'The Most Beautiful of Nature's Works': Thomas Eakins's Photographic Nudes in Their French and American Context," in Susan Danly and Cheryl Leibold, eds., *Eakins and the Photograph: Works by Thomas Eakins and His Circle in the Collection of the Pennsylvania Academy of the Fine Arts* (Washington, DC: Published for the Pennsylvania Academy of the Fine Arts by the Smithsonian Institution Press, 1994), 23–63.

[9] The project would end up costing approximately $30,000.00, the equivalent of more than $700,000.00 today.

[10] William Pepper, "Note," in W. D. Marks, H. Allen, and F. X. Dercum, *Animal Locomotion: The Muybridge Work at the University of Pennsylvania—The Method and the Result* (Philadelphia: J. B. Lippincott, 1888), 6.

[11] "Mr. Muybridge's Photographs: Interesting Pictures to Be Taken of Wild Birds and Beasts in Motion," *Philadelphia Ledger*, August 12, 1885.

[12] Thomas Anshutz to J. Laurie Wallace, August 1884. Archives of the Pennsylvania Academy of the Fine Arts, Philadelphia.

[13] Except for, it seems, Dr. Dercum: "I should add that when Muybridge was about to start his work, which was of course in the summer, the members of the commission which had been appointed to assist him, left on their various vacations and poor Muybridge was left along [*sic*]. Because I had become so much interested in his work, I voluntarily gave up my summer vacation for two successive years." Francis X. Dercum to Dr. Goodspeed, April 23, 1929. Gordon Hendricks Papers, Archives of American Art, Smithsonian Institution, Washington, DC (hereafter cited as Gordon Hendricks Papers).

[14] Muybridge, *Prospectus and Catalogue of Plates*, 12.

[15] According to Tadd's wife, "Mr. Tadd was very much interested in Mr. Muybridge's work from the first and worked at the University with him for several months and also helped him in his selection of models and in posing them." Margaret Tadd to unknown recipient, July 17, 1917. Gordon Hendricks Papers.

[16] "Leland Stanford's Gift to Art and to Science: Mr. Muybridge's Inventions of Instant Photography and the Marvelous Zoogyroscope," *San Francisco Examiner*, February 6, 1881.
[17] For a description of this work, see "Motion Photographed: Muybridge's Novel Work at the University of Pennsylvania," *Bangor Daily Whig & Courier*, August 1, 1884, col. F. The article seems to have been taken from an earlier item in the *Philadelphia Record*. And also: "Dear Sir, I was engaged by Mr. Muybridge in the Spring of 1885 as Photographic operator to do the darkroom work, the first work was to make positives for the zoopraxiscope. The subject was 12 pictures of Muybridge himself in nude Form walking up a stepladder.... I developed in my work more than 1800 dozen dry plates and from them 1800 dozen positives." H. L. Bell Sr. to G. E. Nitzsche, December 7, 1923. Gordon Hendricks Papers.
[18] "He was so busy that he did not have much time to go to a barber and therefore his hair and beard were at times unduly long." Francis X. Dercum to G. E. Nitzsche, May 20, 1929. Gordon Hendricks Papers
[19] Lamprey's grid, in turn, was based on earlier systems used by artists faced with the problem of depicting bodily proportions. See Frank Spencer, "Some Notes on the Attempt to Apply Photography to Anthropometry During the Second Half of the Nineteenth Century," in *Anthropology and Photography 1860–1920,* ed. Elizabeth Edwards (New Haven, CT: Yale University Press, 1992).
[20] J. H. Lamprey, "On a Method of Measuring the Human Form for the Use of Students of Ethnology," *Journal of the Ethnological Society of London* 1 (1869): 84–85. Cited in Spencer, "Some Notes on the Attempt to Apply Photography to Anthropometry," 103.
[21] See George W. Stocking Jr., *Victorian Anthropology* (New York: Free Press, 1987).
[22] The information on athletics and athletes at the university come from Elspeth Brown, "Racialising the Virile Body: Eadweard Muybridge's Locomotion Studies 1883–1887," *Gender & History* 17, no. 3 (November 2005): 627–56.
[23] Cited in Brown, "Racialising the Virile Body," 47. Pepper's remarks were made on the occasion of a University of Pennsylvania student setting the world record for the high jump; the original newspaper clipping (dated October 17, 1887) is in the William Pepper, MD, Collection, Van Pelt Library Special Collections, University of Pennsylvania.
[24] See Jayne Morgan, "Eadweard Muybridge and W. S. Playfair: An Aesthetics of Neurasthenia," *History of Photography* 23, no. 3 (Autumn 1999): 225–31; and Gail Bederman, *Manliness and Civilization: A Cultural History of Gender and Race in the United States, 1880–1917* (Chicago: University of Chicago Press, 1995).
[25] See, for example, Linda Williams, "Film Body: An Implantation of Perversions," *Cine-tracts* 3, no. 4 (Winter 1981): 19–36; Braun, "Muybridge's Scientific Fictions," *Studies in Visual Communication* 10, no. 3 (Summer 1984): 2–22; and Sarah Gordon, "Sanctioning the Nude: Production and Reception of Eadweard Muybridge's *Animal Locomotion*, 1887" (PhD diss., Northwestern University, 2006).
[26] "I have the greatest difficulty, however, in inducing mechanics, at any price, to go through the motions of their trade in a nude condition to the waist only." "Animal Motion," *Philadelphia Times*, August 2, 1885.
[27] Muybridge, *Prospectus and Catalogue of Plates,* 12.
[28] "Animal Motion," *Philadelphia Times*, August 2, 1885.
[29] "I sold to Colt—as I think you know—all of the sets of transparencies (mounted on large sheets of glass) from which the large gelatin negatives for printing were made also several thousand—20 or 30,000—plain, unperforated zoopraxiscopic cards; and some few hundreds of the colored perforated disks ready for use as a scientific toy (if I recollect right—you have a complete set of each—50 of the former in a portfolio, and a dozen of the latter)." Muybridge, from 161 King's Rd., Kingston upon Thames, to Irwin Faber, March 11, 1901. Archives, University of Pennsylvania, Philadelphia. UPT 50 M993; Box 1 Folder 37: Correspondence, 1887–1901.

Fig. 121. Frances Benjamin Johnston, *Portrait of Eadweard Muybridge*, 1890

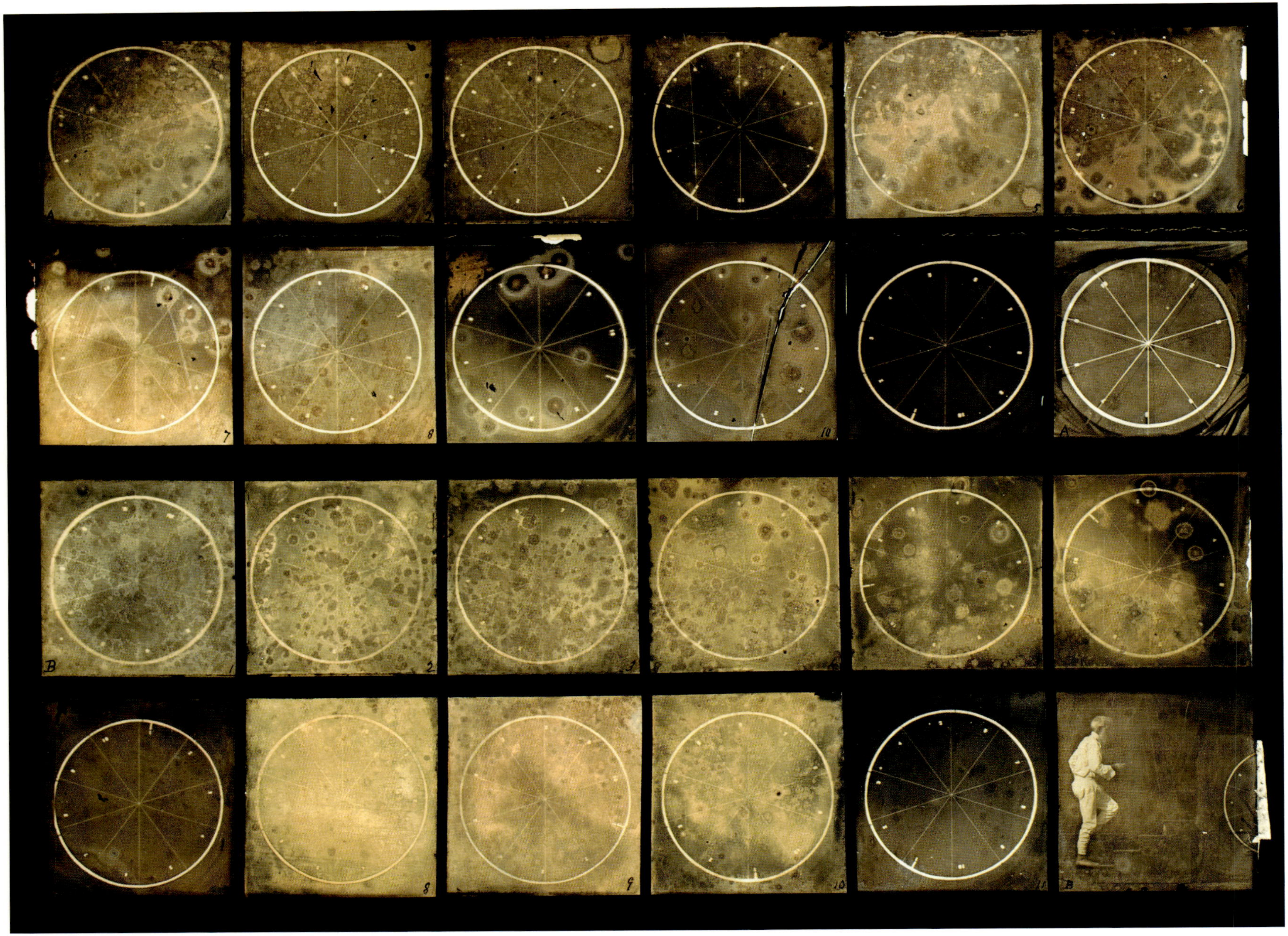

144. *Sequenced image of a rotating sulky wheel with Muybridge climbing steps*, ca. 1887

Animal Locomotion

145. *A, throwing water from bucket; B, descending step; C, ascending step; D, lawn tennis. Plate 524*, 1887

146. *Heaving a 20-lb. rock. Plate 315*, 1887

147. *A, 98 and 100, wrestling; B, 98 and 100, wrestling; C, 99 and 101, wrestling; D, 99 and 101, sparring without gloves. Plate 520,* 1887

148. *A, 97, jumping; B, 98, hand-spring; C, 98, somersault; D, 99, somersault; E, 99, spring over man's back. Plate 522,* 1887

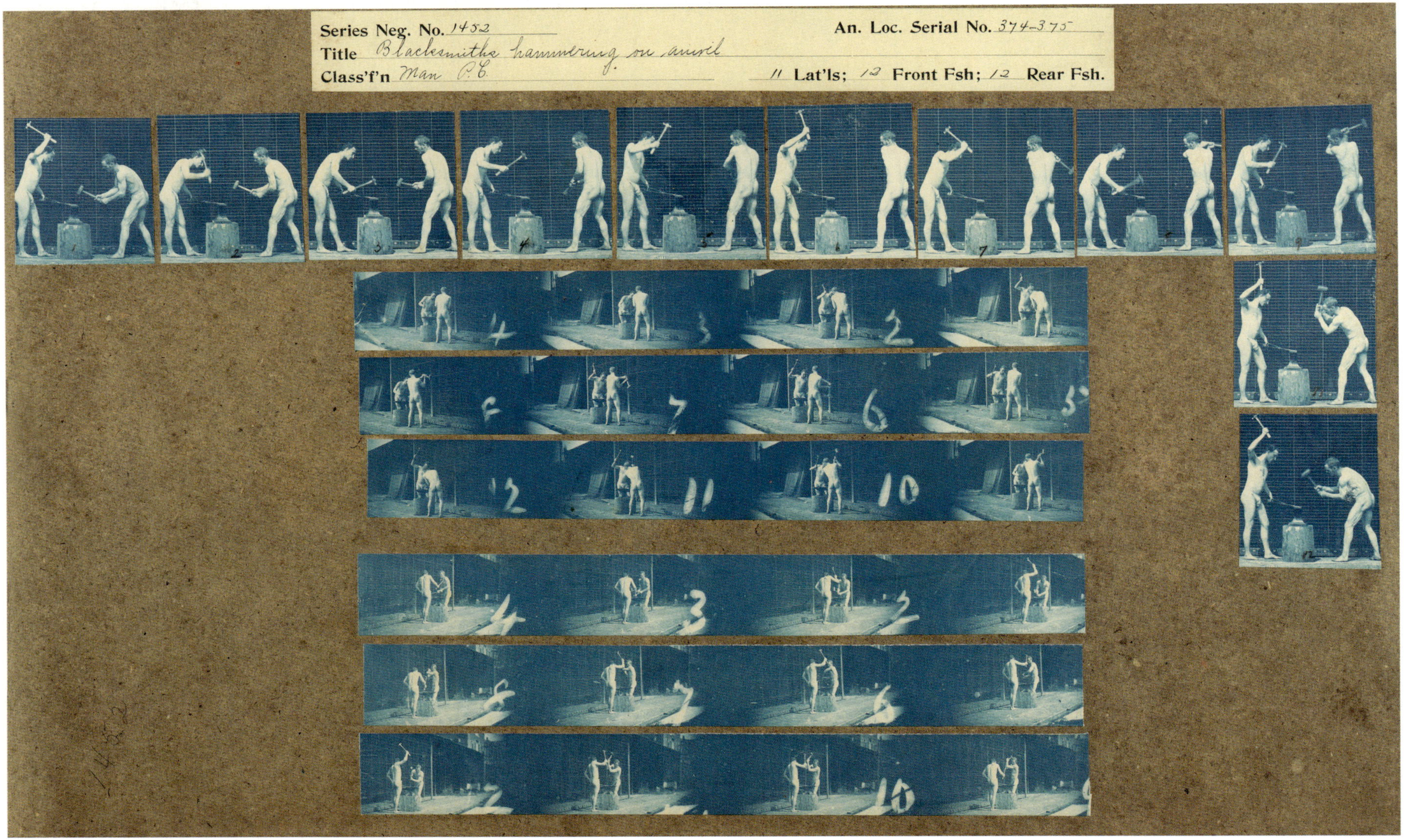

149. *Cyanotype proof for Animal Locomotion. Blacksmiths hammering an anvil, males, pelvis cloth (Messrs. Redinger and Breen). Plate 374,* 1885

150. *Blacksmiths, two models, hammering on anvil. (Movements. Male). Plate 374,* 1887

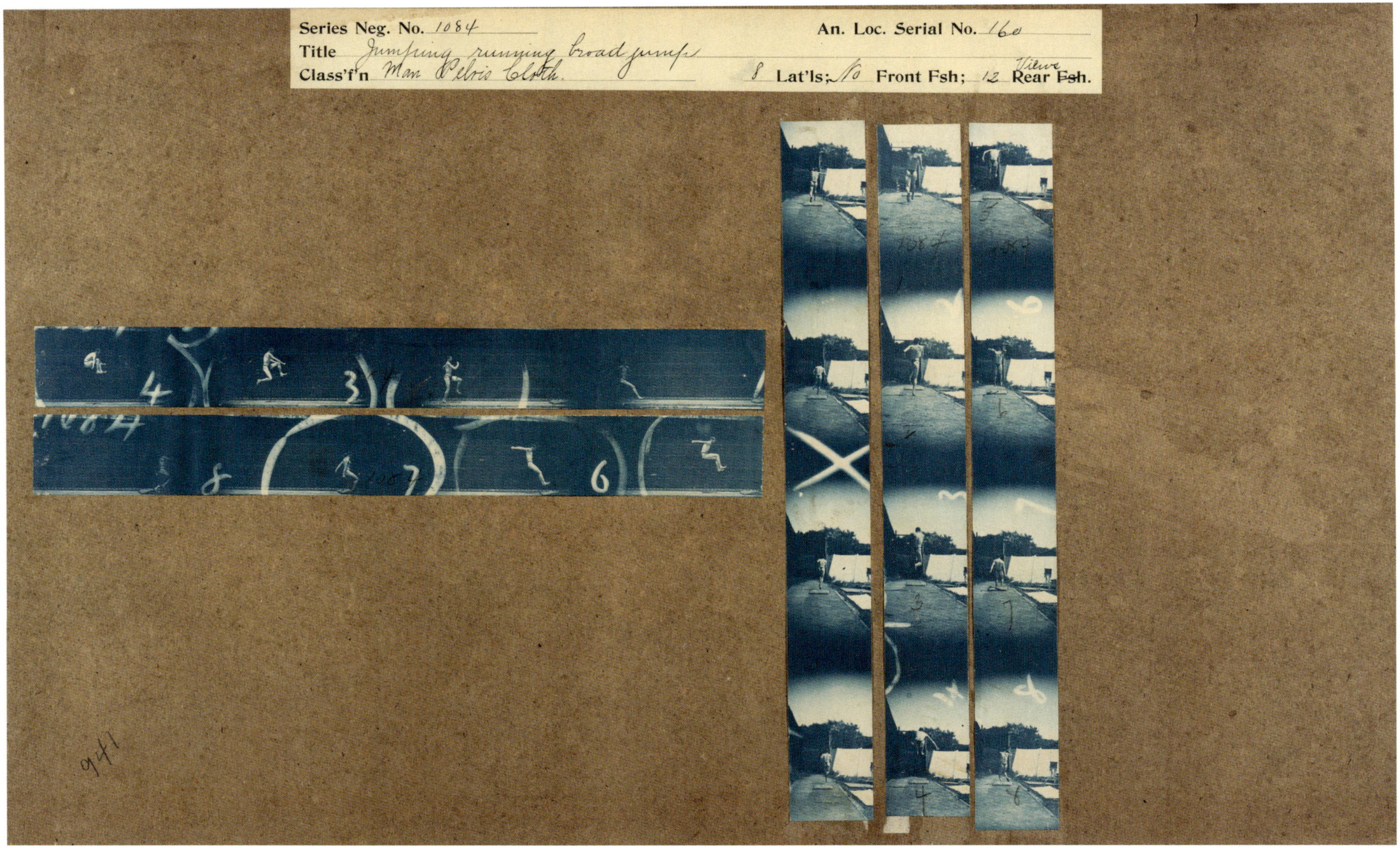

151. *Cyanotype proof for Animal Locomotion. Jumping; running broad jump. Plate 160*, 1885

152. *Jumping; running broad jump. (Movements. Male). Plate 160*, 1887

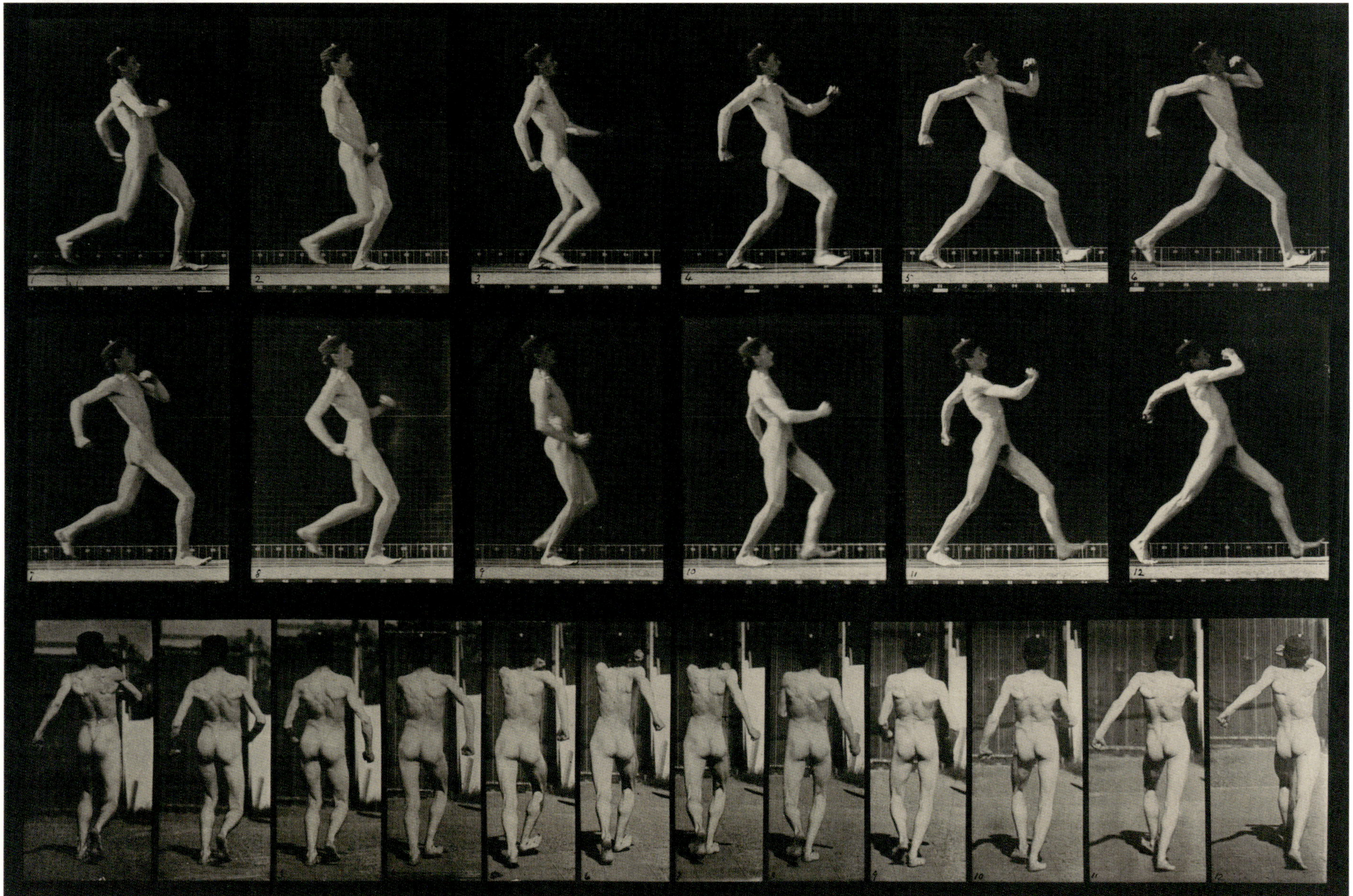

153. *Walking. (Movements. Male). Plate 7*, 1887

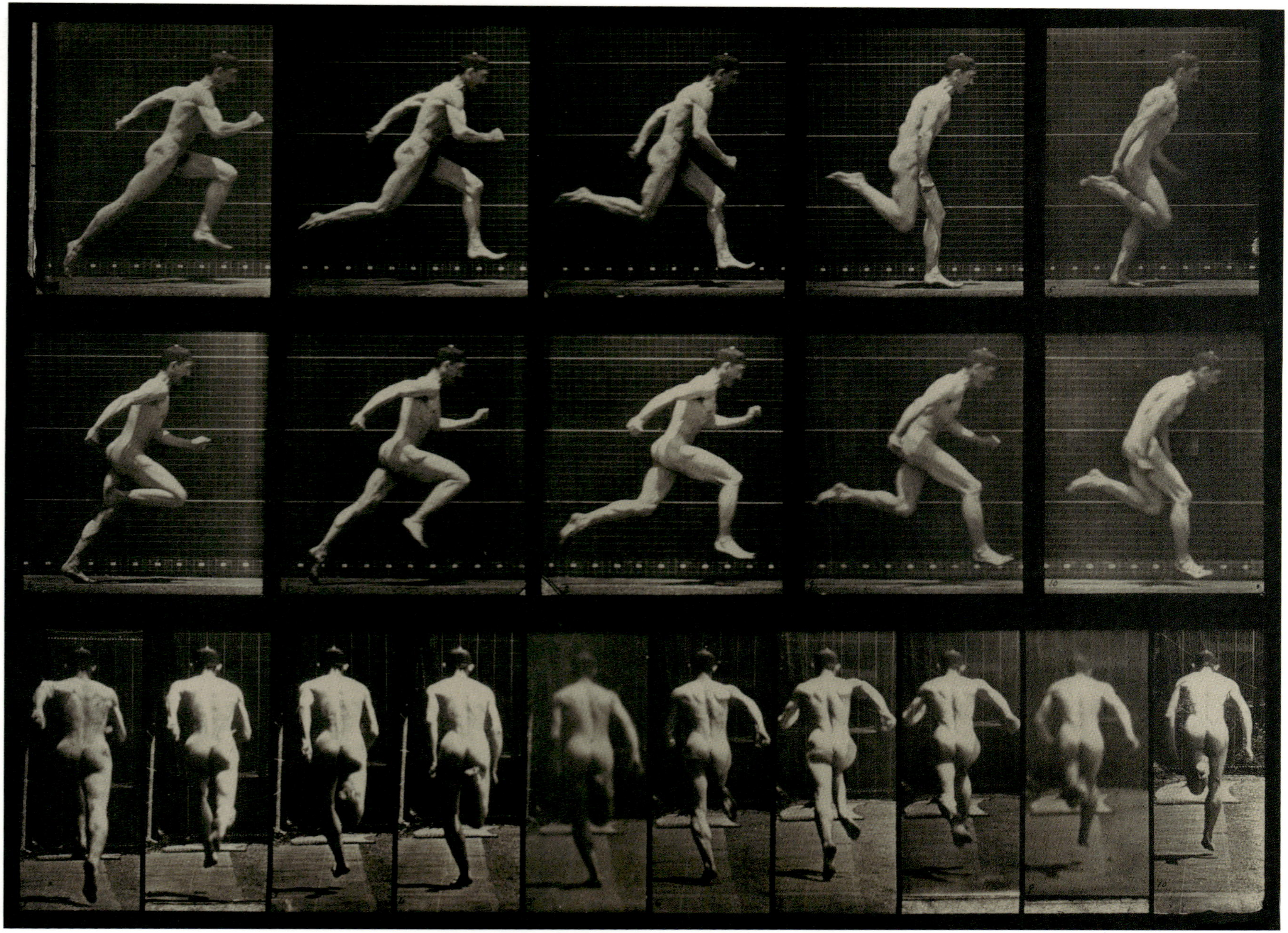

154. *Running full speed. (Movements. Male). Plate 63*, 1887

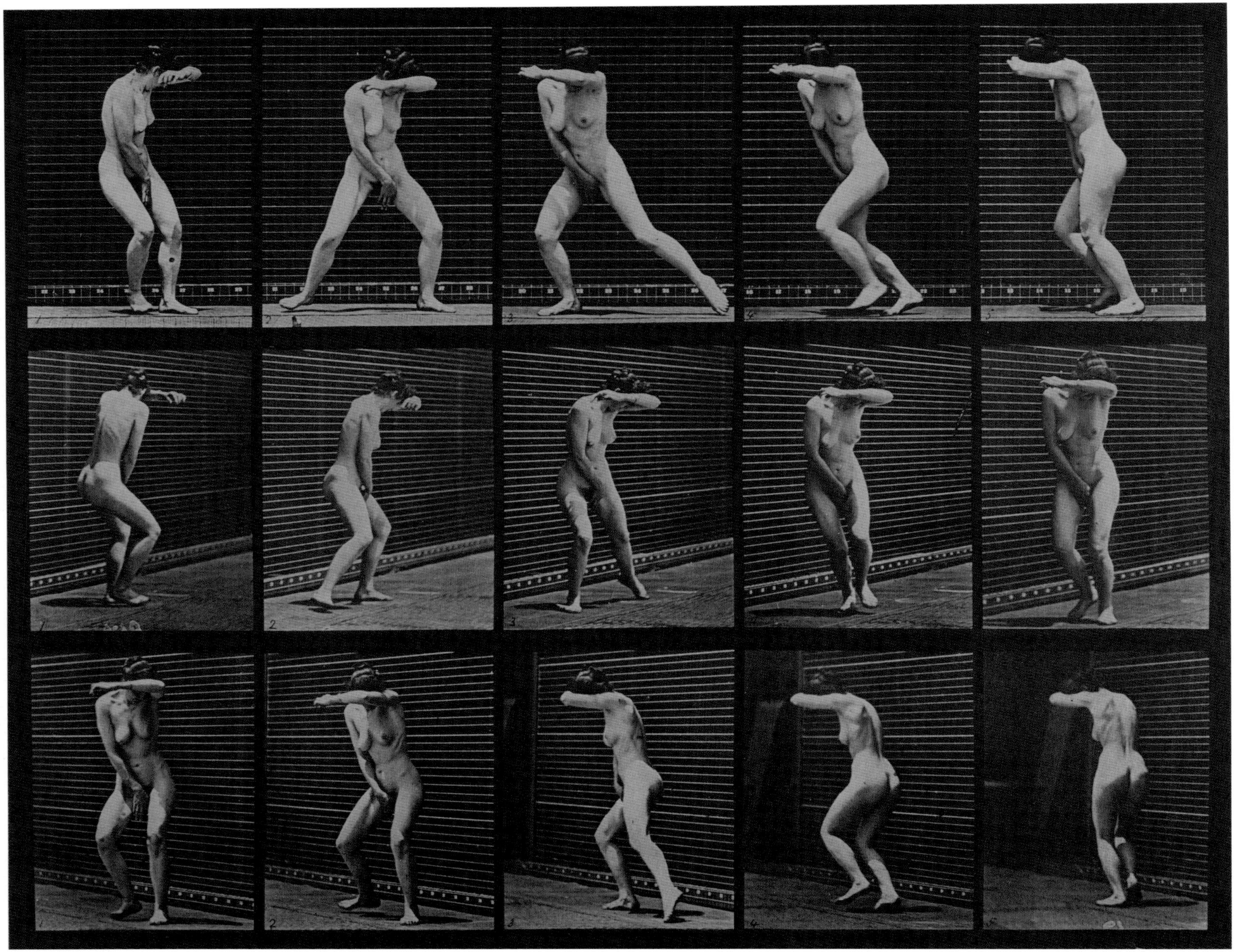

296

155. *Turning around in surprise and running away. (Movements. Female). Plate 73*, 1887

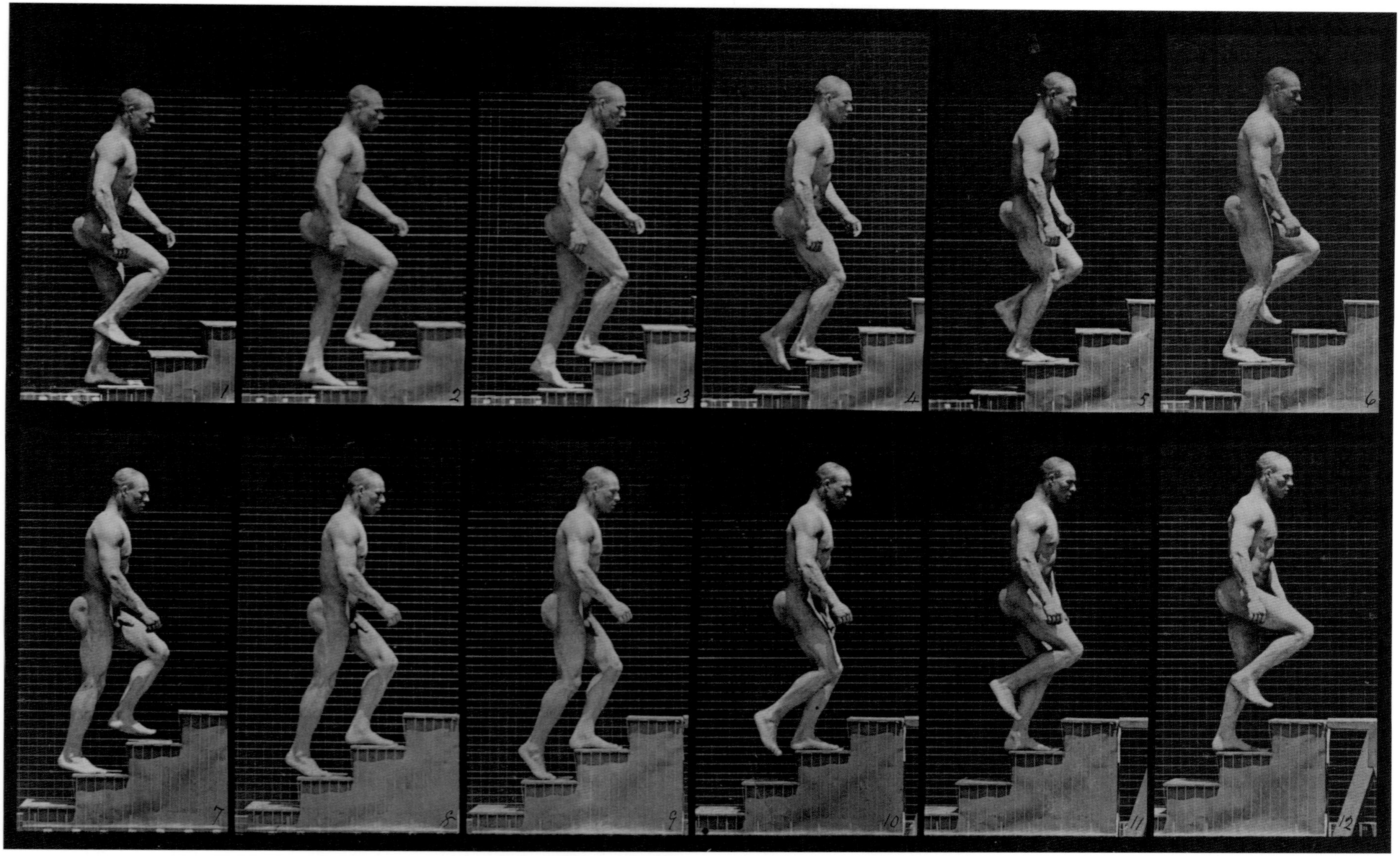

156. *Ascending stairs. (Movements. Male). Plate 91*, 1887

157. *Glass Positive for Animal Locomotion. Descending Stairs, a cup and saucer in right hand. Plate 143*, 1887

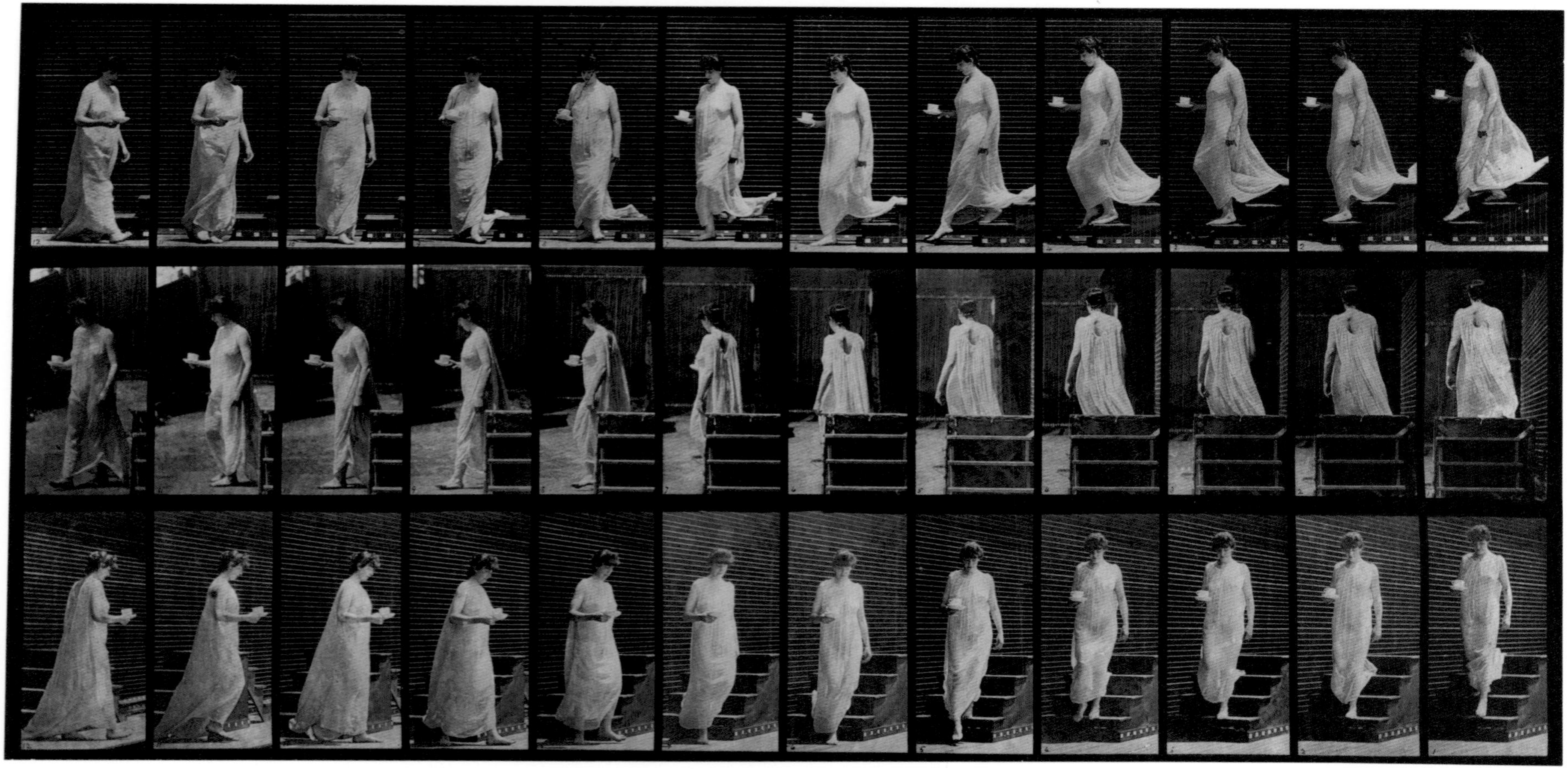

158. *Descending stairs, turning, cup and saucer in right hand. Plate 143,* 1887

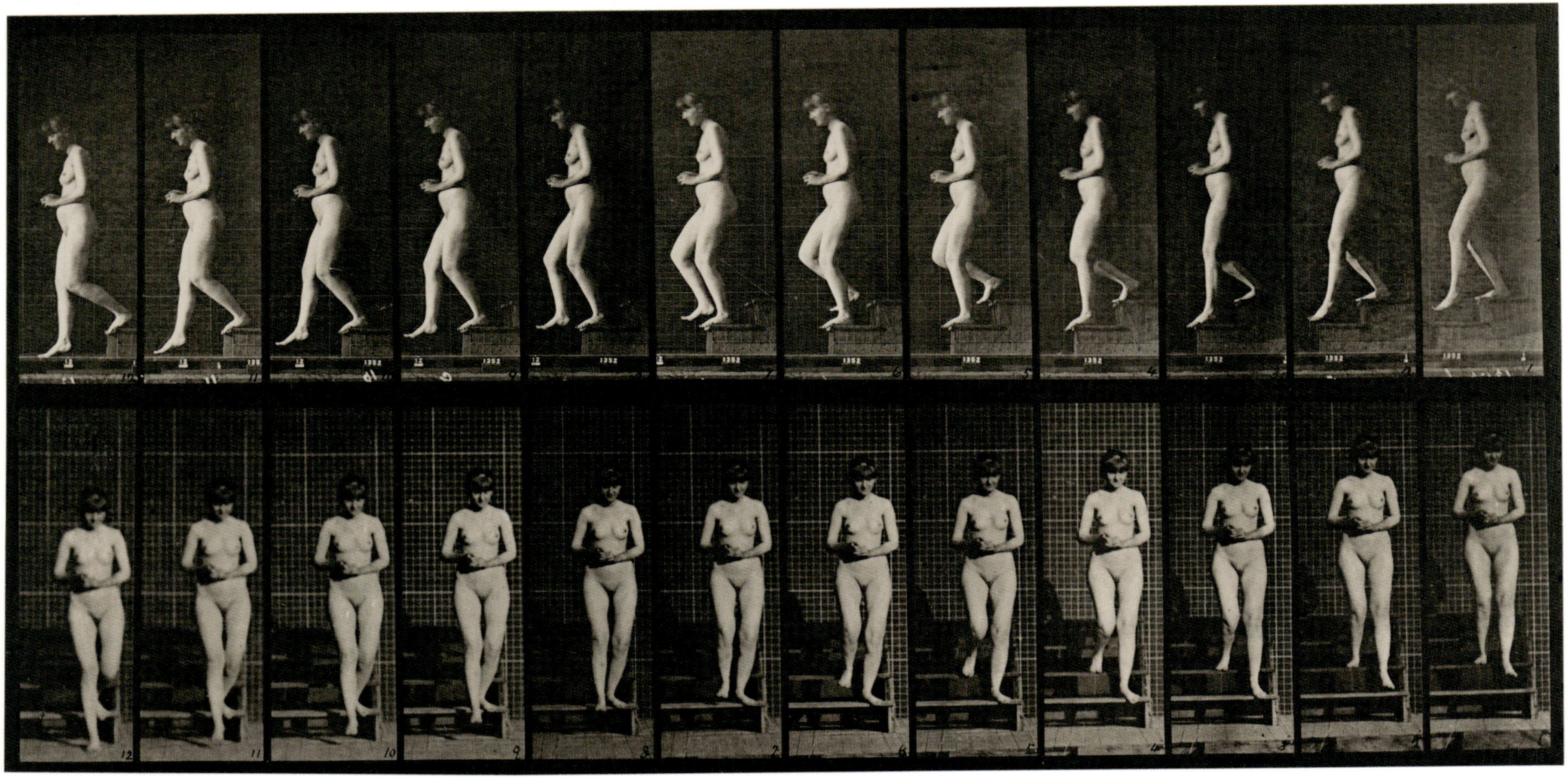

159. *Descending stairs with hands clasped. Plate 128,* 1887

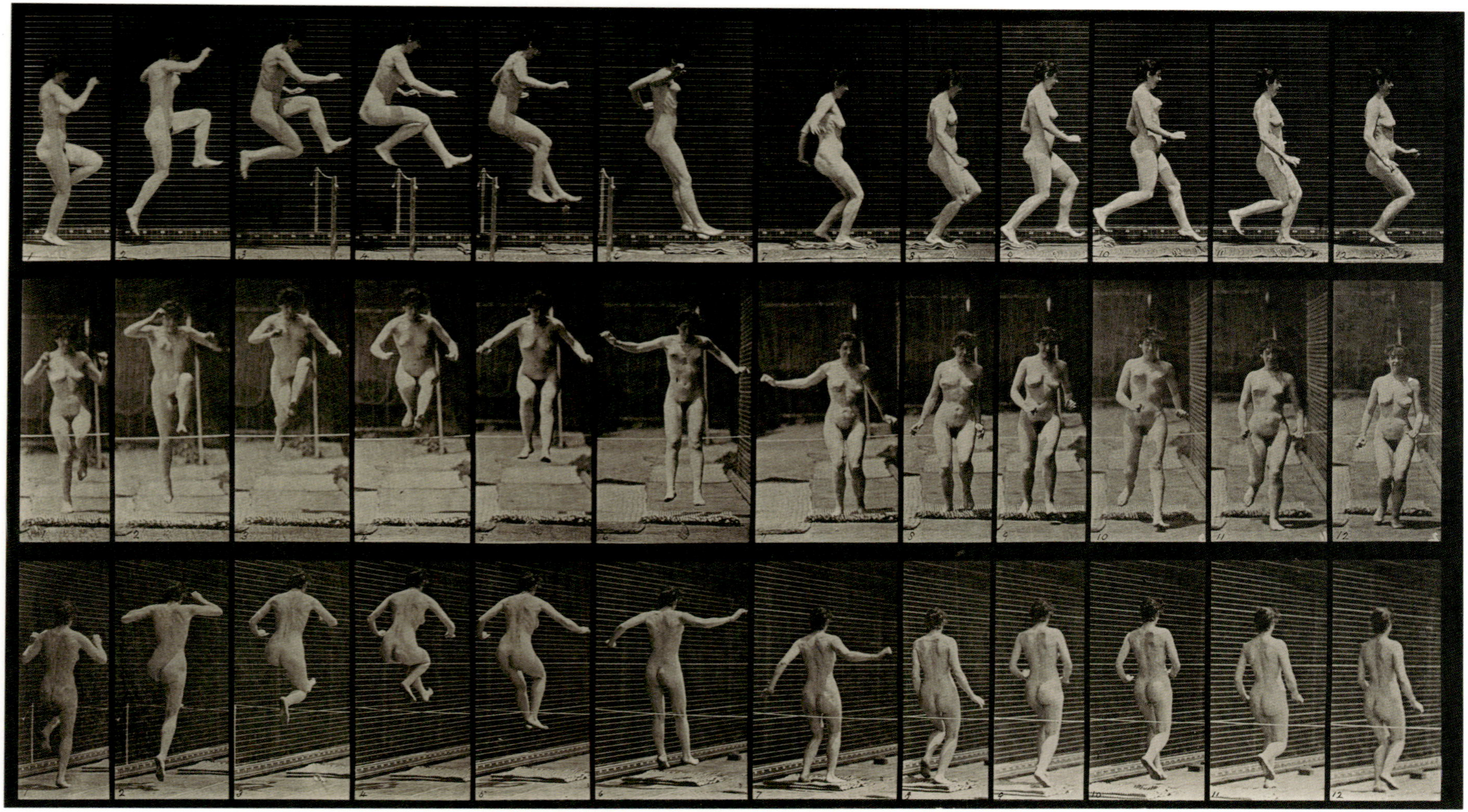

160. *Jumping; running straight high jump. Plate 155,* 1887

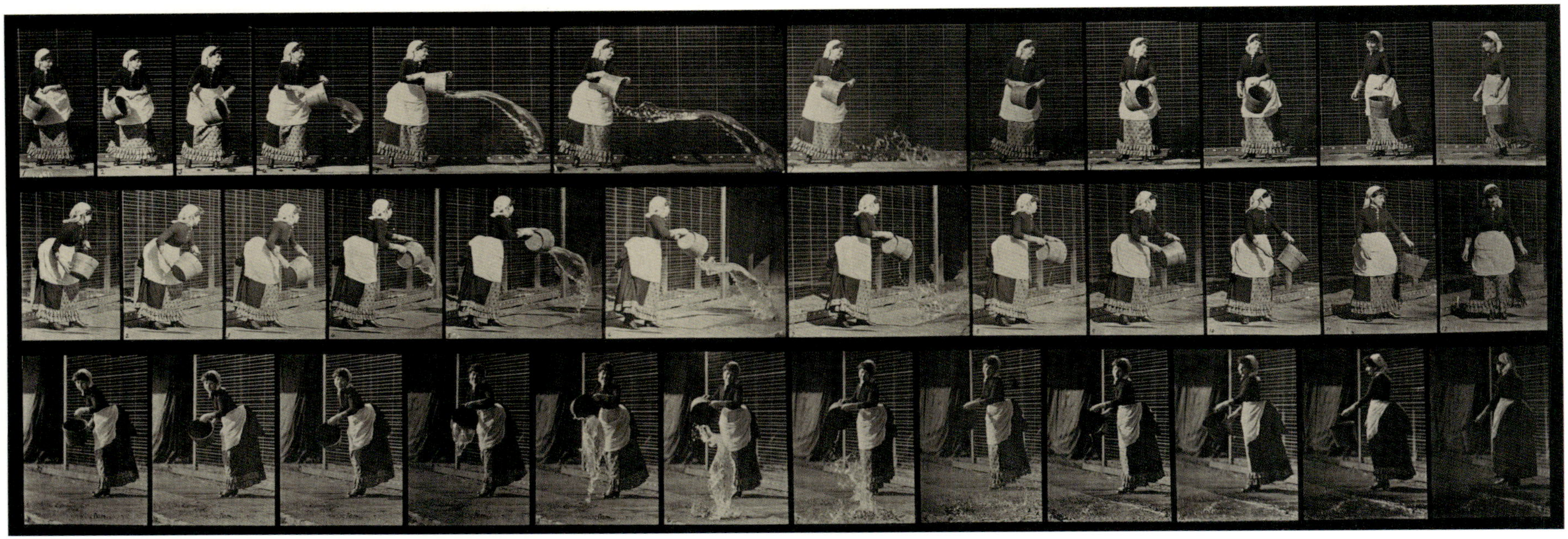

161. *Emptying bucket of water. Plate 404*, 1887

162. *Emptying bucket of water. Plate 403*, 1887

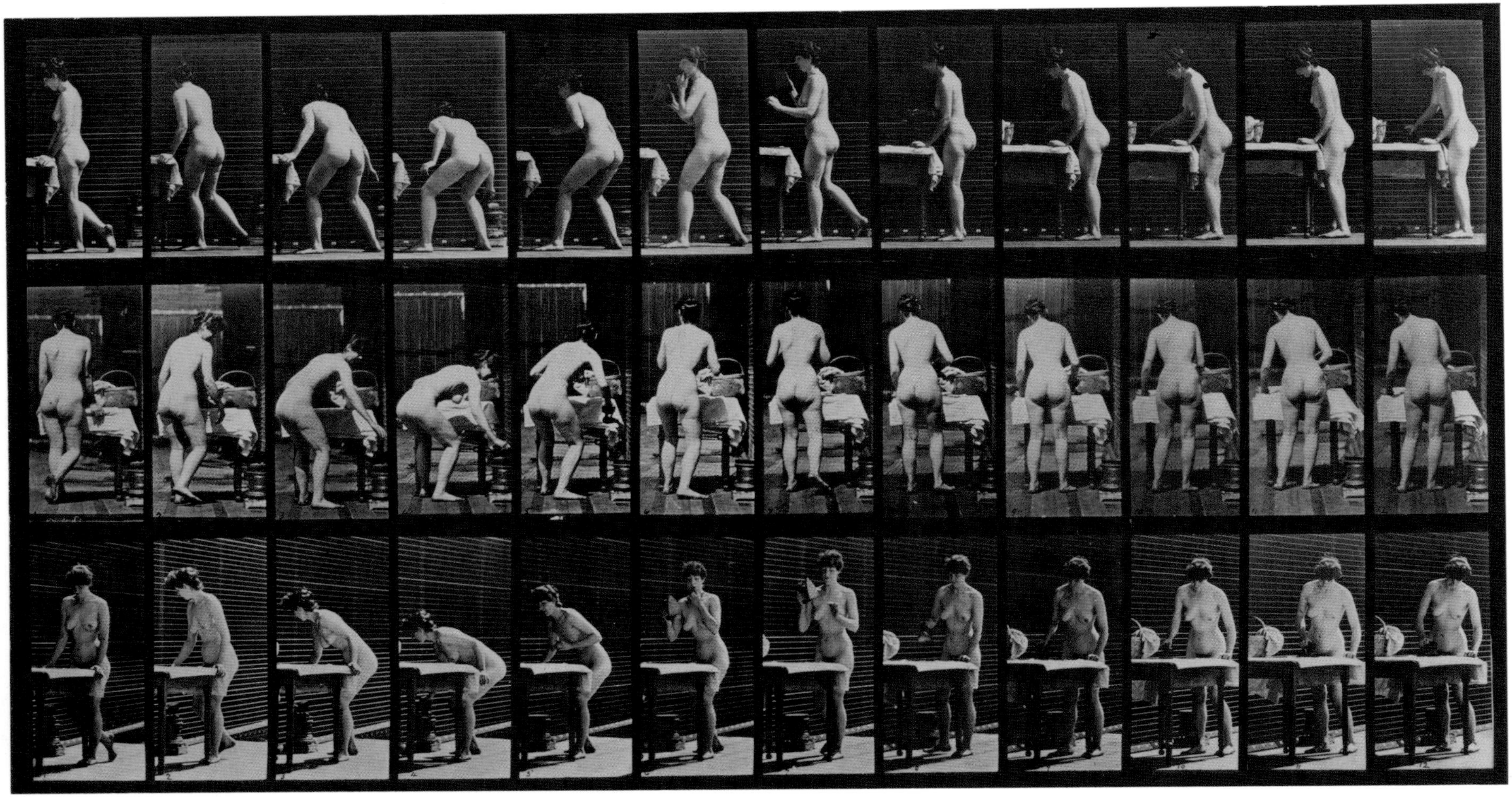

163. *Ironing clothes. Plate 435*, 1887

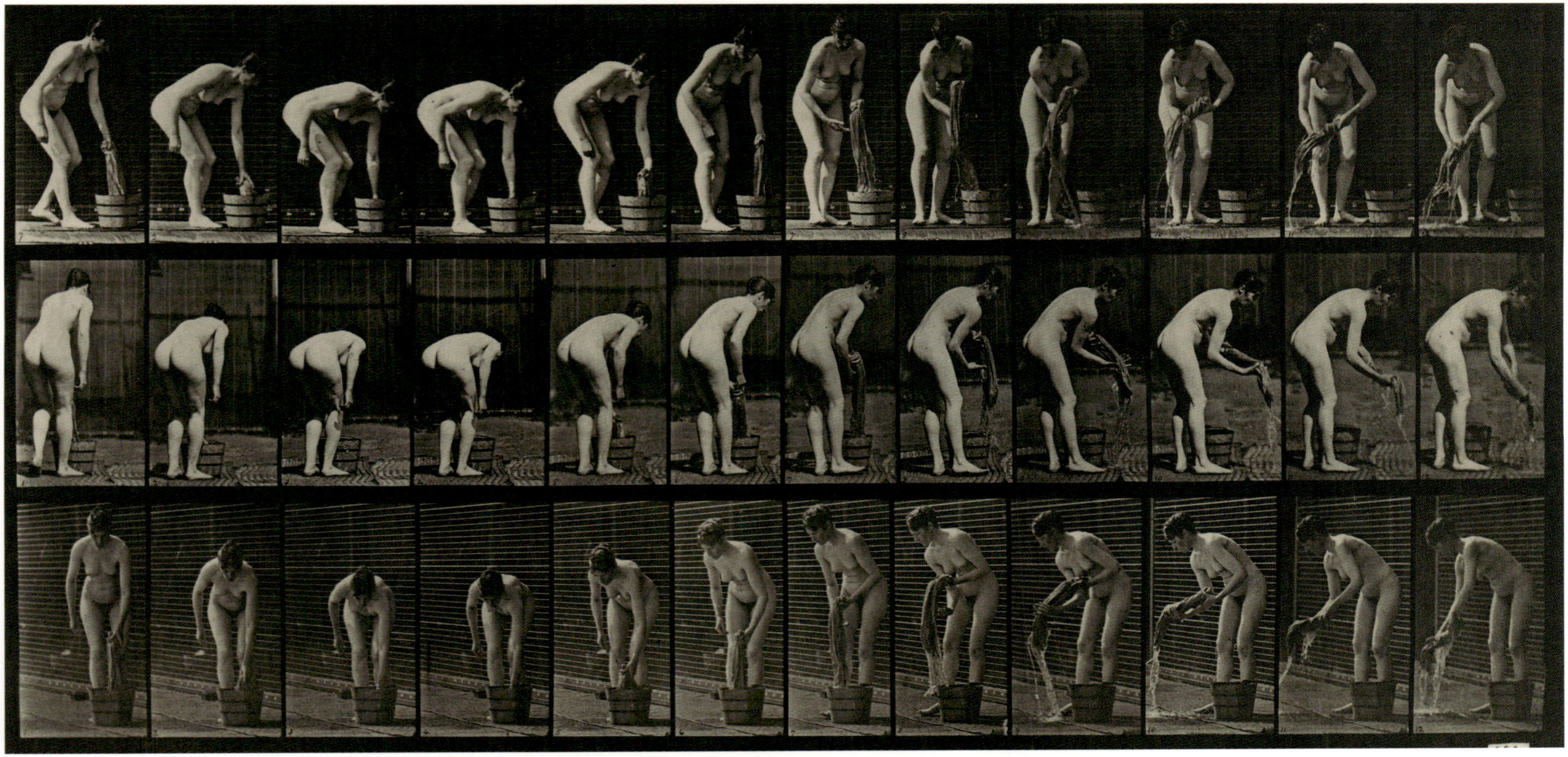

164. *Wringing clothes. Plate 433*, 1887

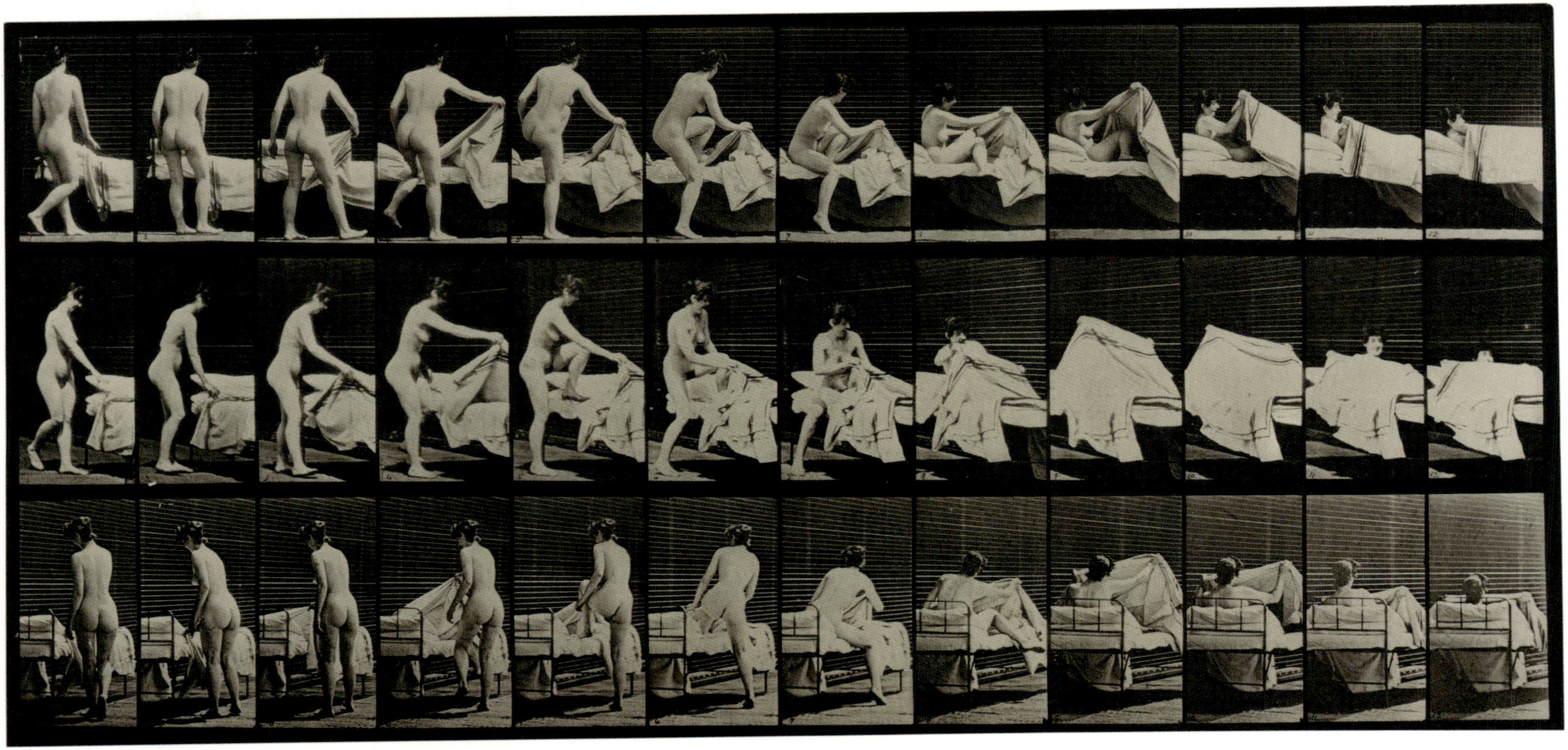

165. *Getting into bed. Plate 263*, 1887

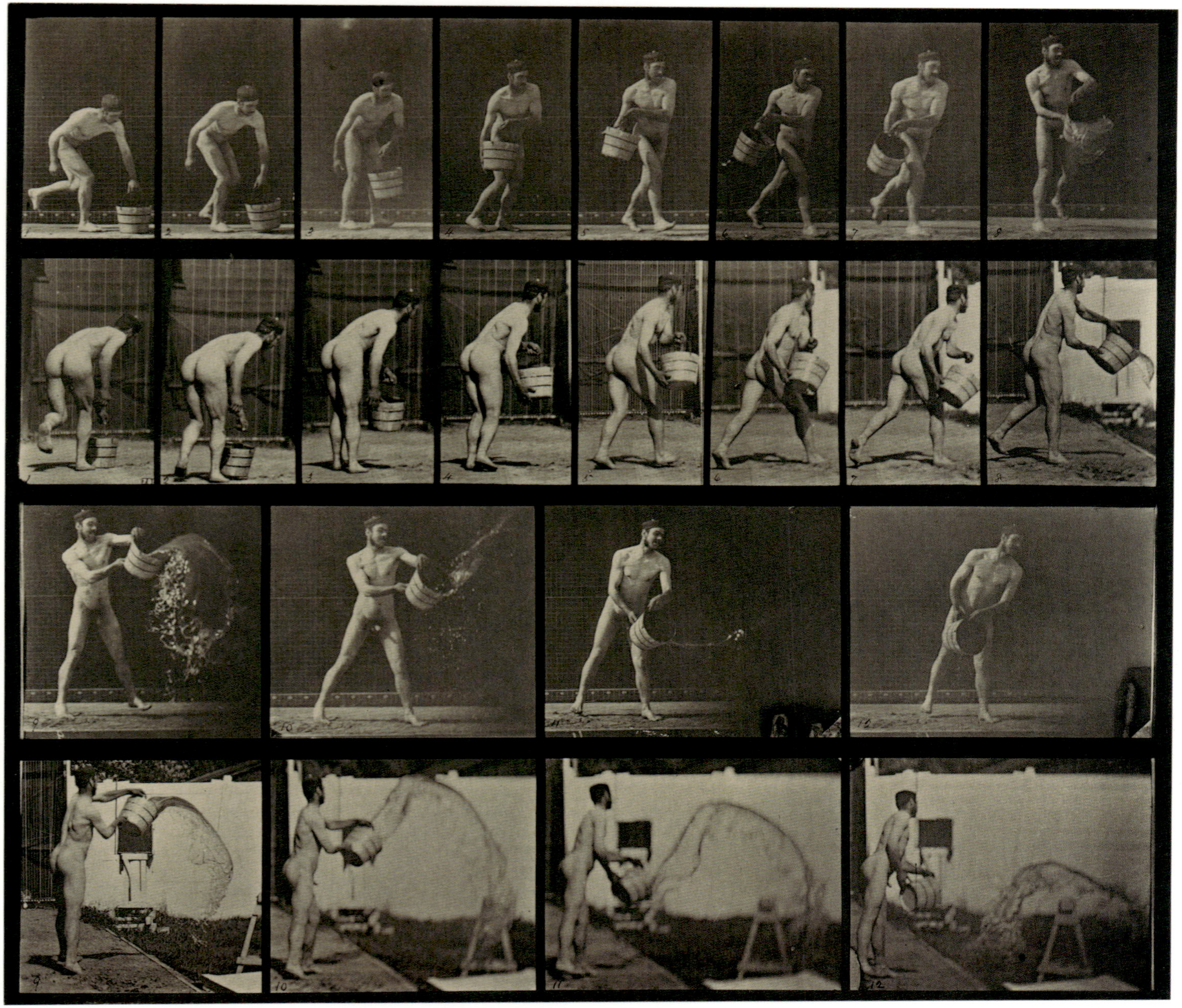

166. *Emptying bucket of water. Plate 400*, 1887

167. *Contortions on the ground. Plate 510*, 1887

168. *Fencing. (Movements. Male). Plate 349*, 1887

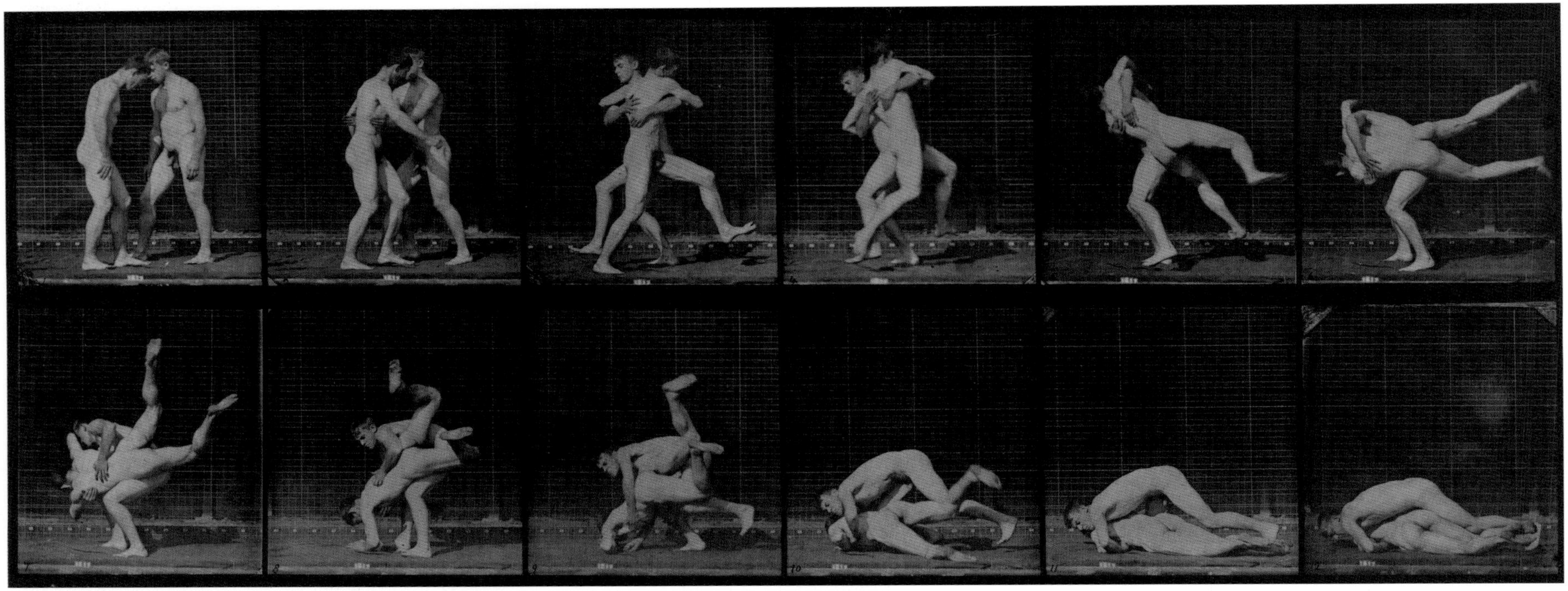

169. *Wrestling; Graeco-Roman. Plate 347*, 1887

170. *Boxing; open hand. Plate 340*, 1887

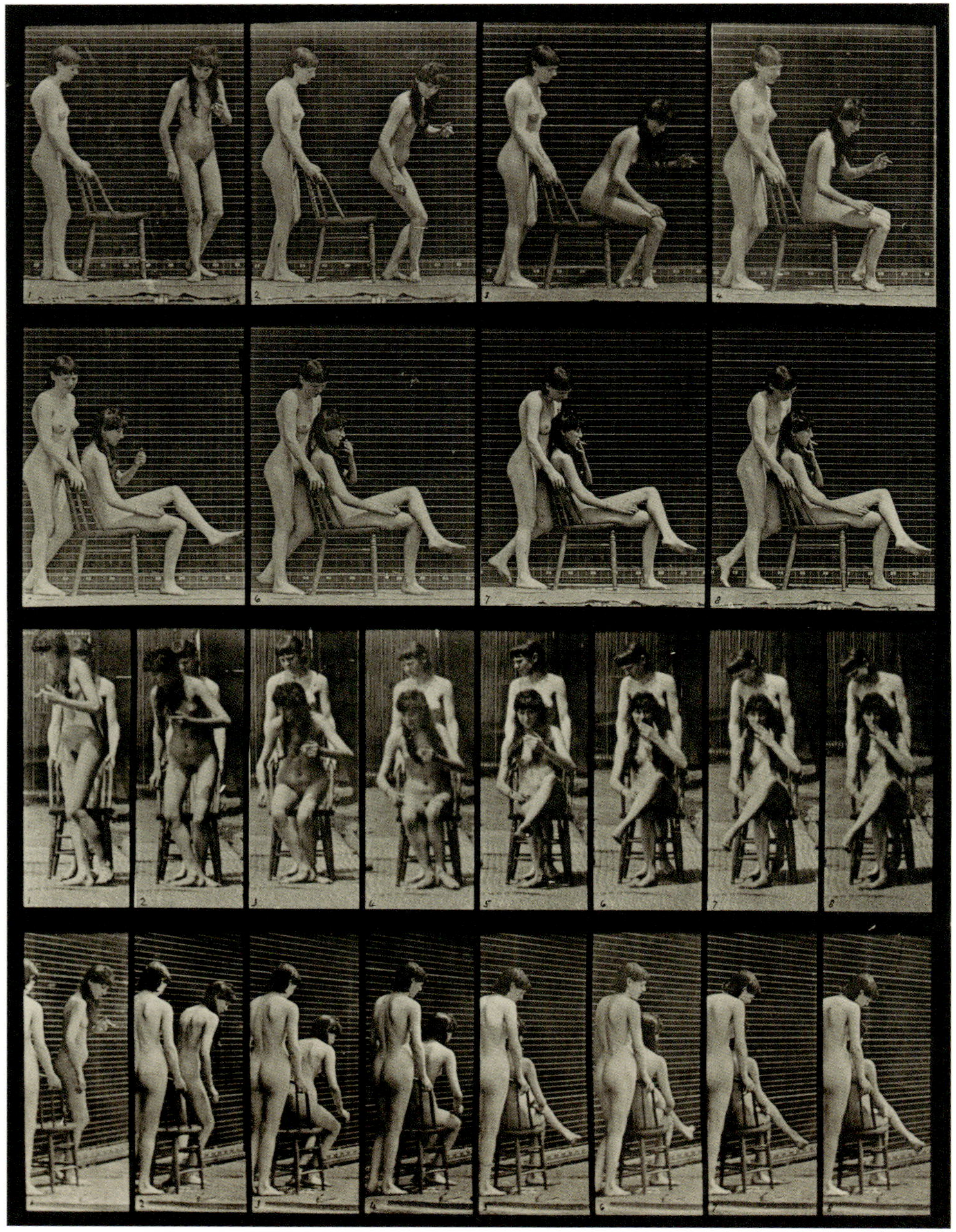

171. *Two models; one standing, the other sitting, crossing legs. Plate 239*, 1887

172. *Crossing brook on step-stones with fishing-pole and can. Plate 175,* 1887

173. *Jumping; running straight high jump. (Movements. Female). Plate 156*, 1887

174. *Walking, two models meeting, and partly turning. Plate 45*, 1887

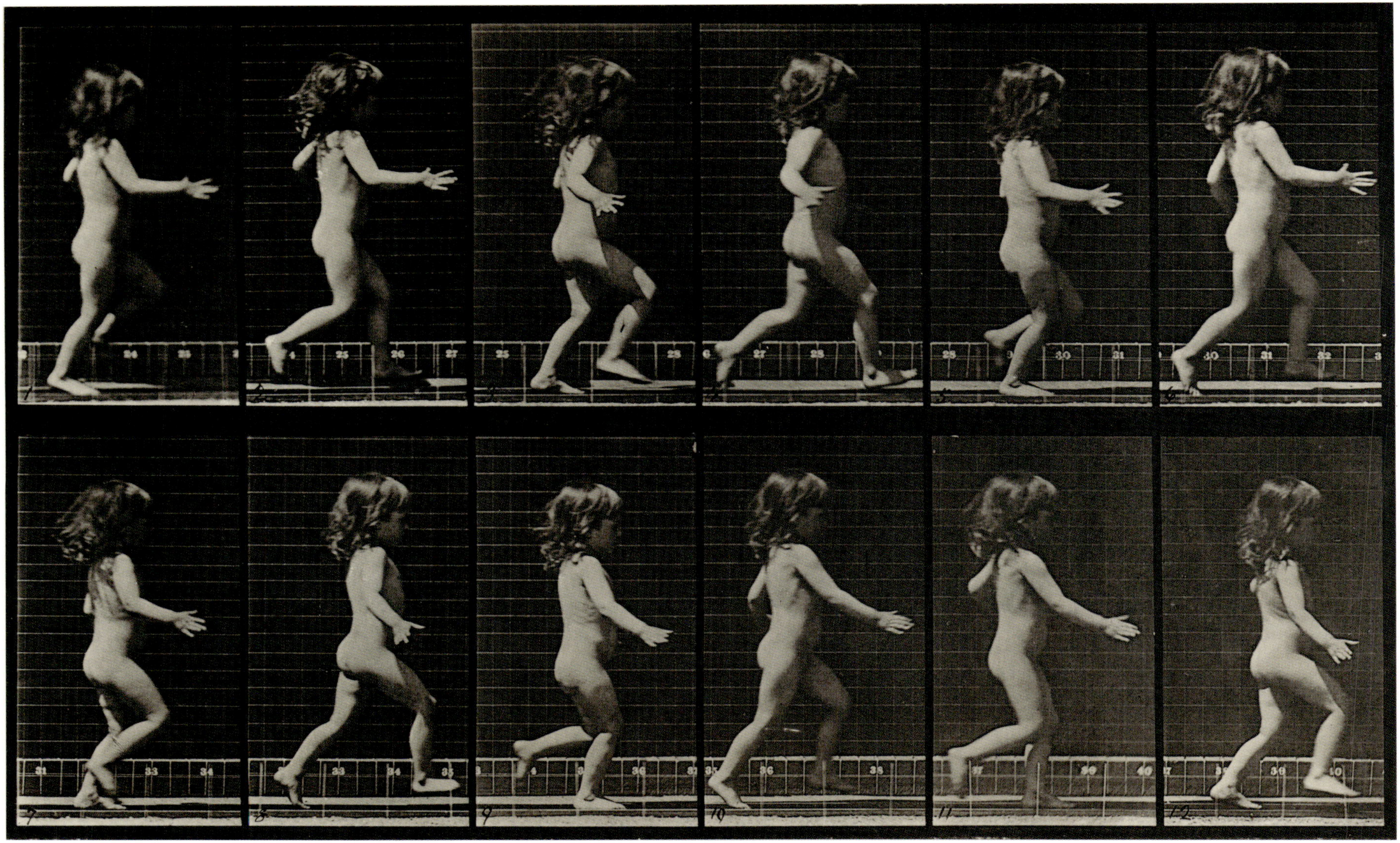

175. *Child, running. Plate 469*, 1887

176. *Two models, child, 70 N, bringing bouquet to 12. Plate 465*, 1887

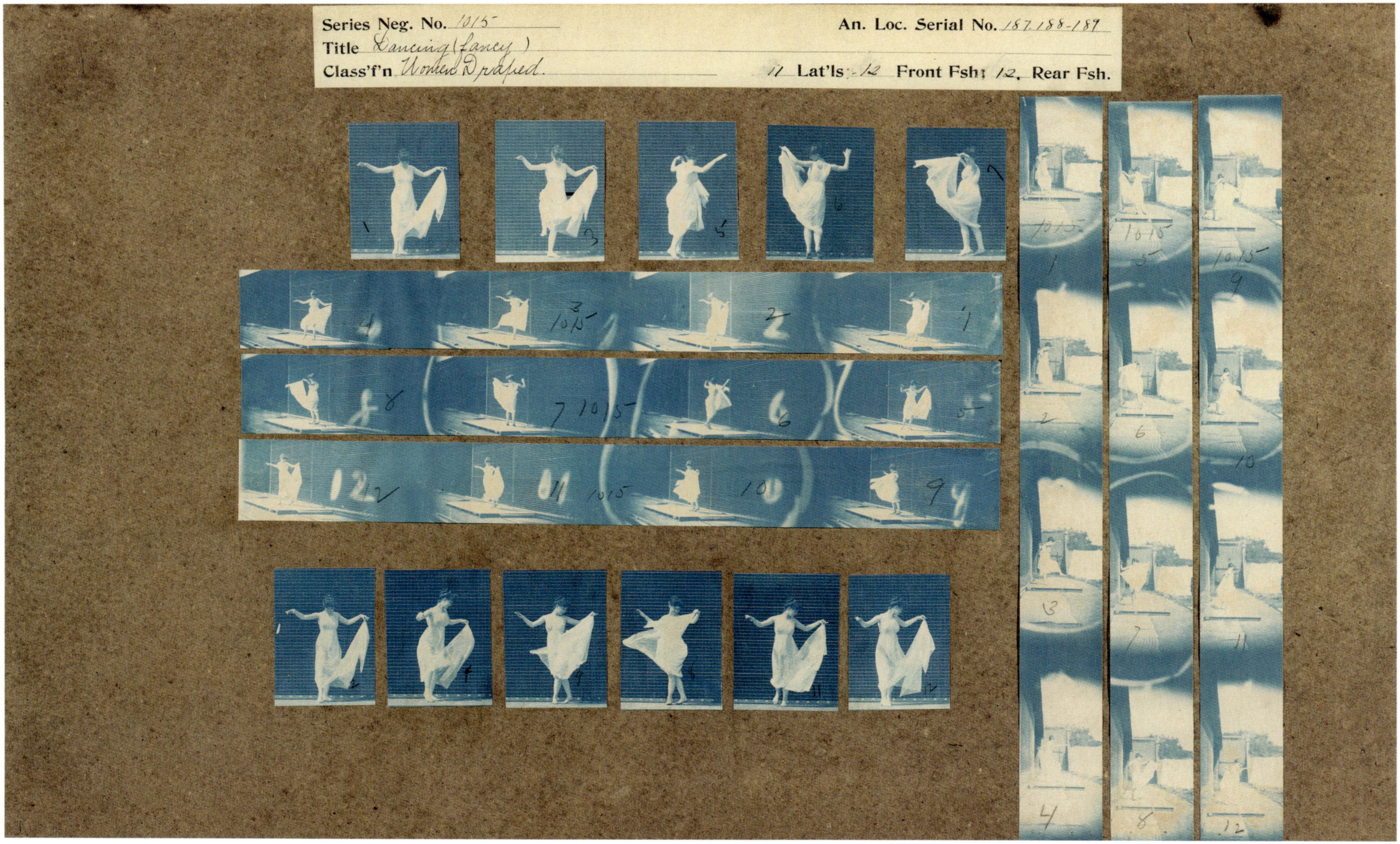

177. *Cyanotype proof for Animal Locomotion. Dancing (Fancy), Woman Draped (Miss Larrigan). Plate 188*, 1885

178. *Dancing (fancy). (Movements. Female). Plate 188,* 1887

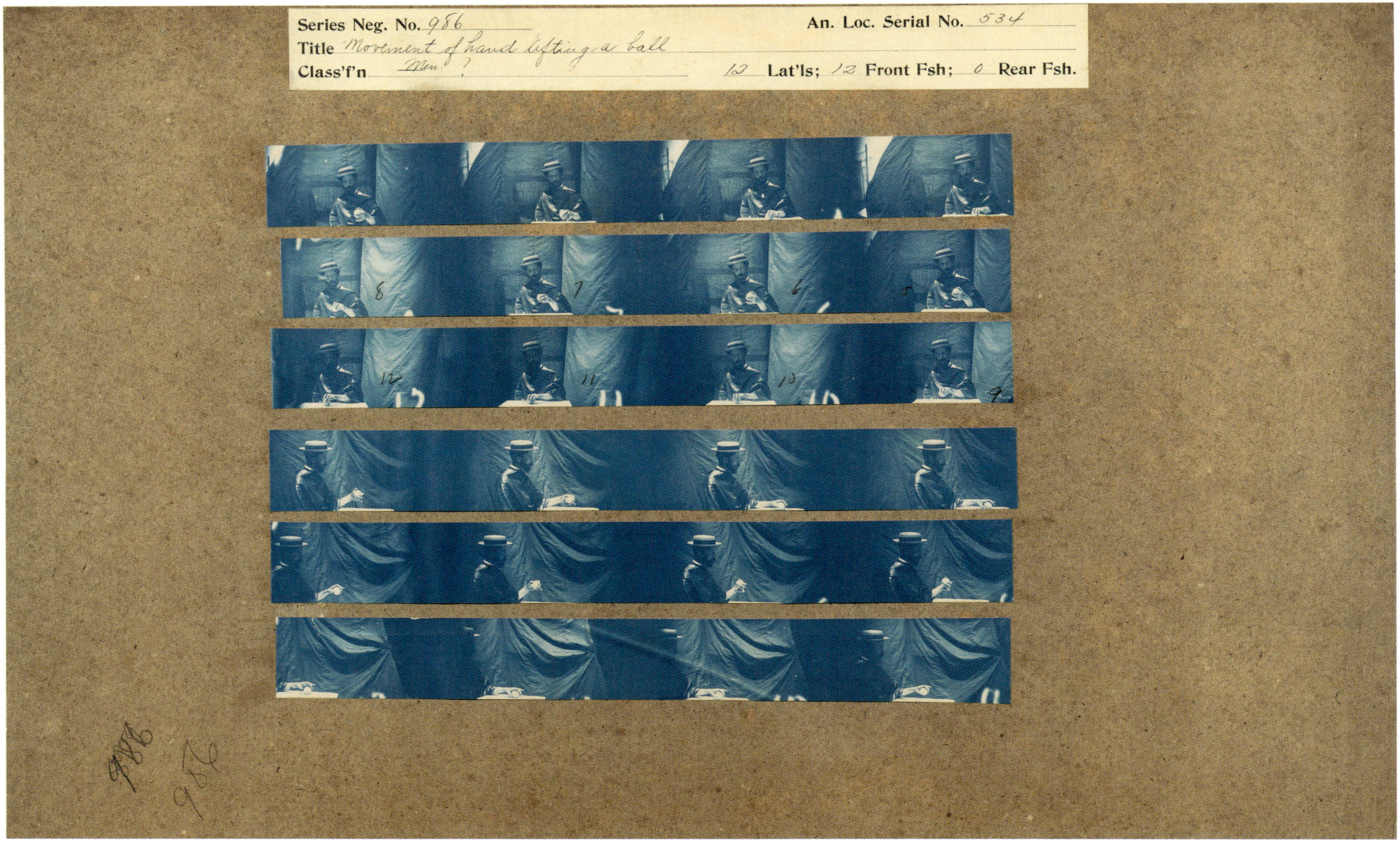

179. *Cyanotype proof for Animal Locomotion. Movement of the Hand, Lifting a Ball (Mr. Tadd). Plate 534*, 1885

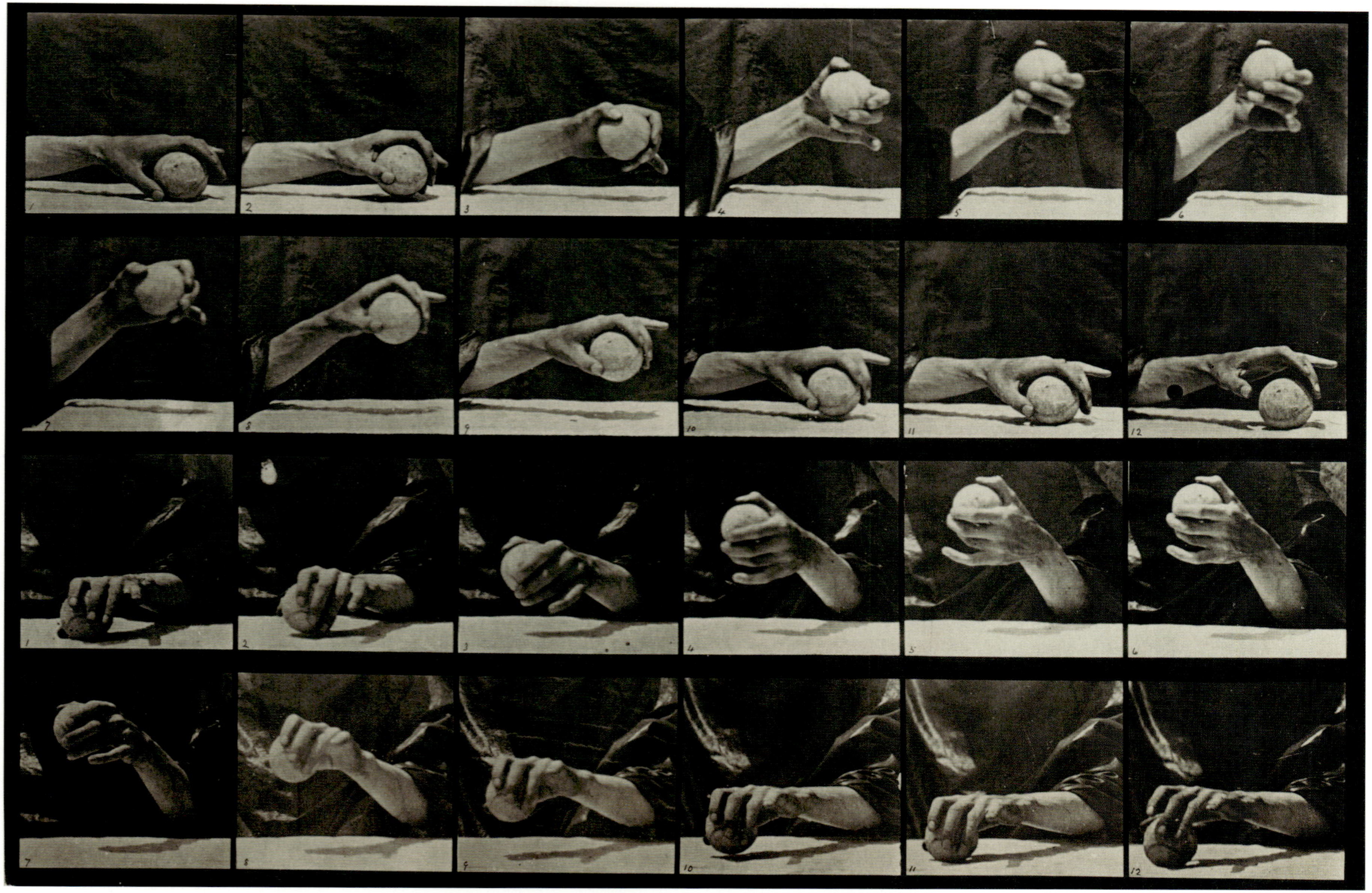

180. *Movement of the hand; lifting a ball. Plate 534*, 1887

181. *Glass Positive for Animal Locomotion. Jumping over boy's back (leap-frog). Plate 169*, 1887

182. *Jumping over boy's back (leap-frog). Plate 169*, 1887

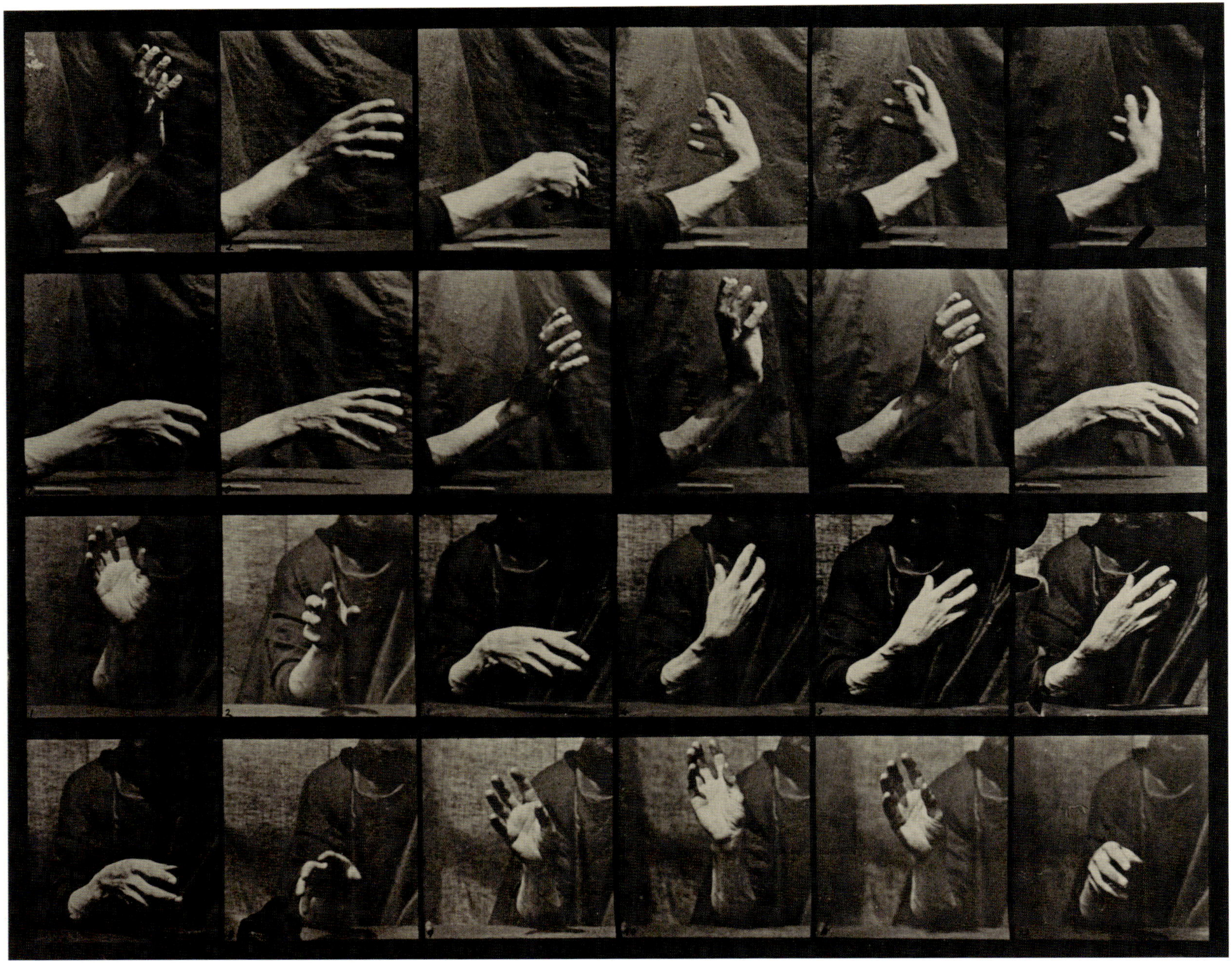

183. *Movement of the hand; beating time. Plate 535*, 1887

184. *Head-spring, a flying pigeon interfering. Plate 365*, 1887

185. *Trotting; bareback; dark-gray Belgian horse. Dusel. Plate 599*, 1887

186. *Trotting; sulky; bay mare Lizzie M. Plate 609*, 1887

187. *Gallop; saddle; thoroughbred bay horse Bouquet. Plate 632*, 1887

188. *Jumping a hurdle; saddle; bay horse Daisy. Plate 640*, 1887

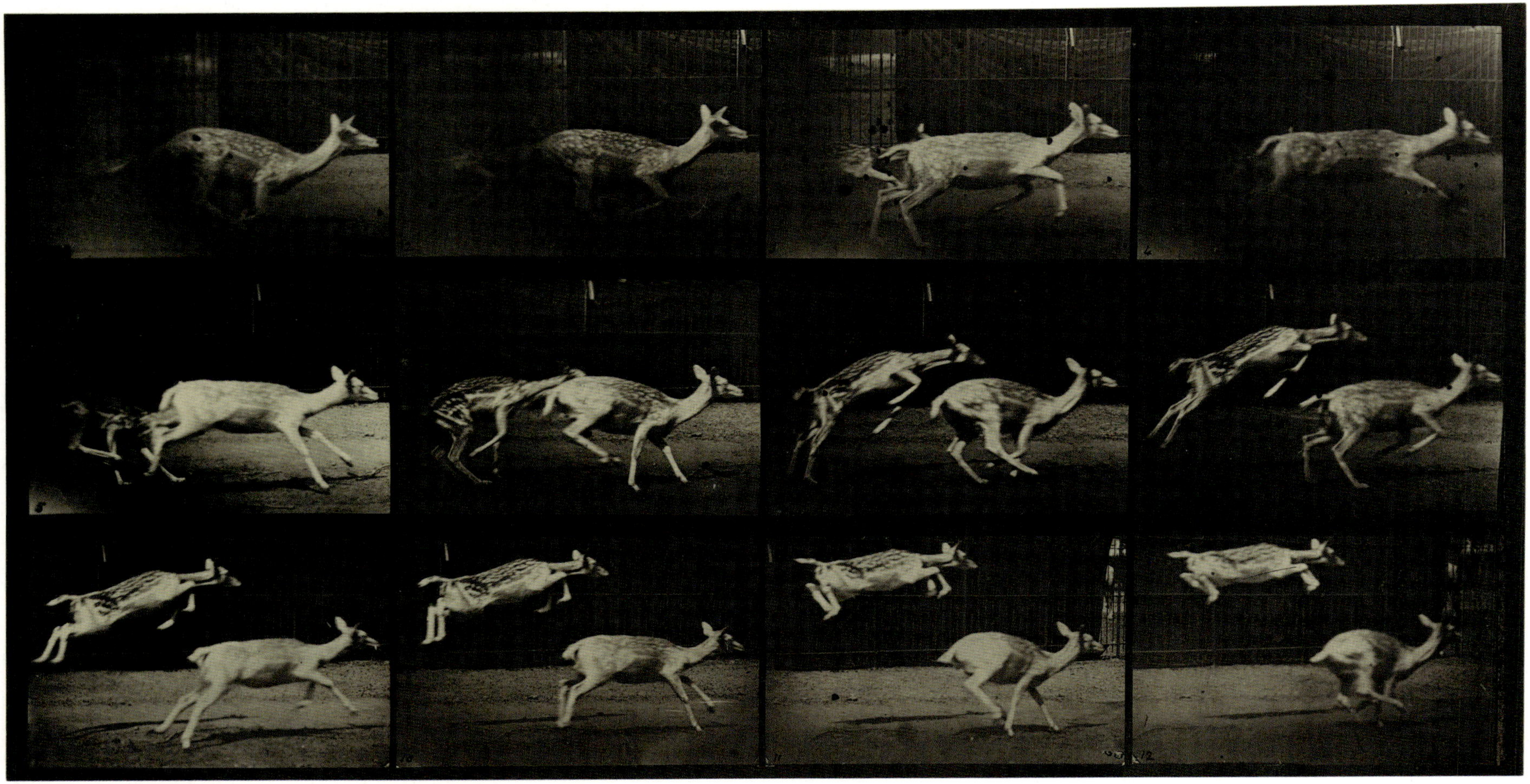

189. *Fallow deer; doe; galloping; and kid; jumping. Plate 691,* 1887

190. *Canter; bareback; rider, 106, nude; white horse Clinton. Plate 617*, 1887

191. *Baboon; walking on all-fours. Plate 748*, 1887

192. *Elephant; walking. Plate 733*, 1887

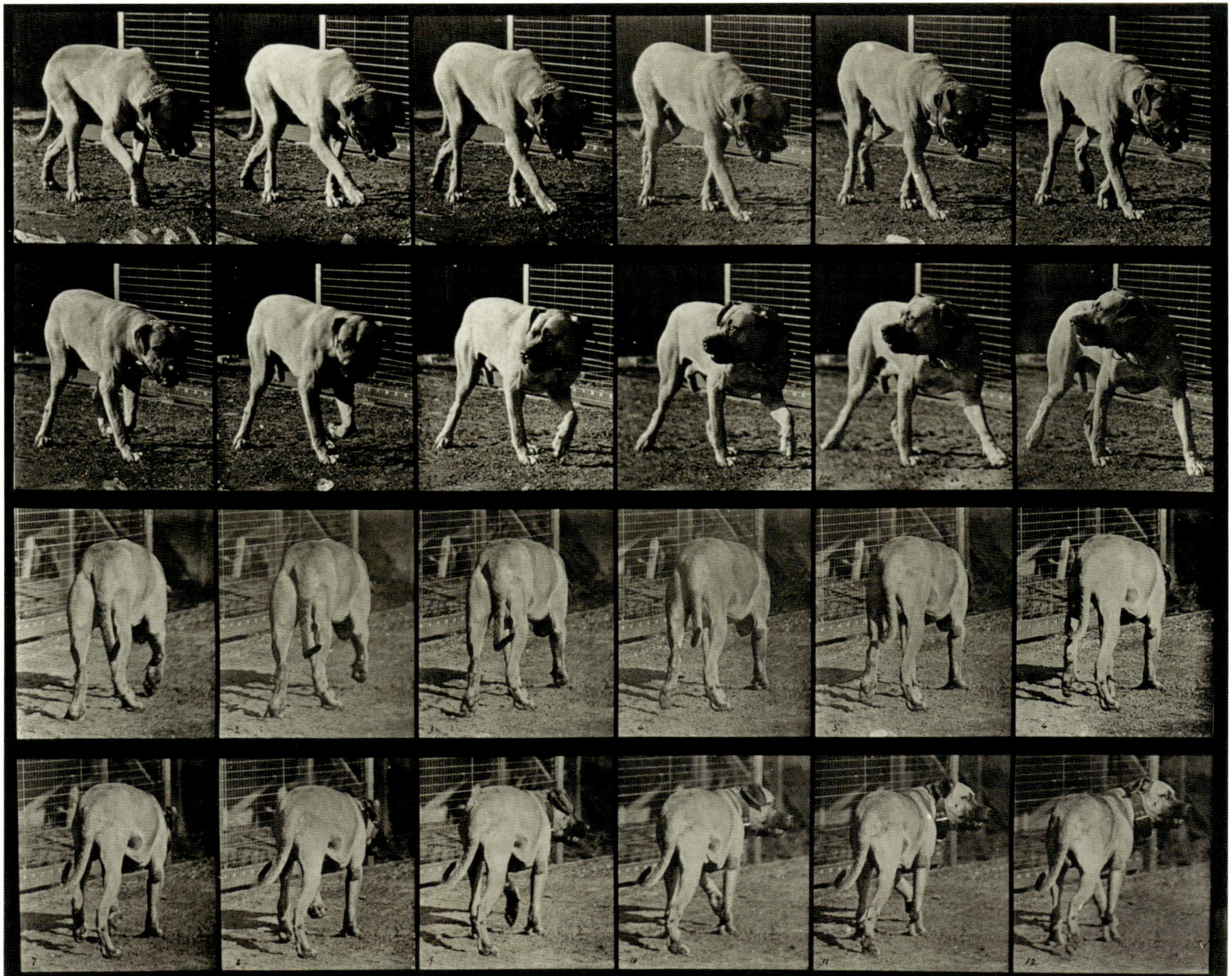

193. *Dog; Walking; Interrupted; Mastiff-Dread. Plate 703*, 1887

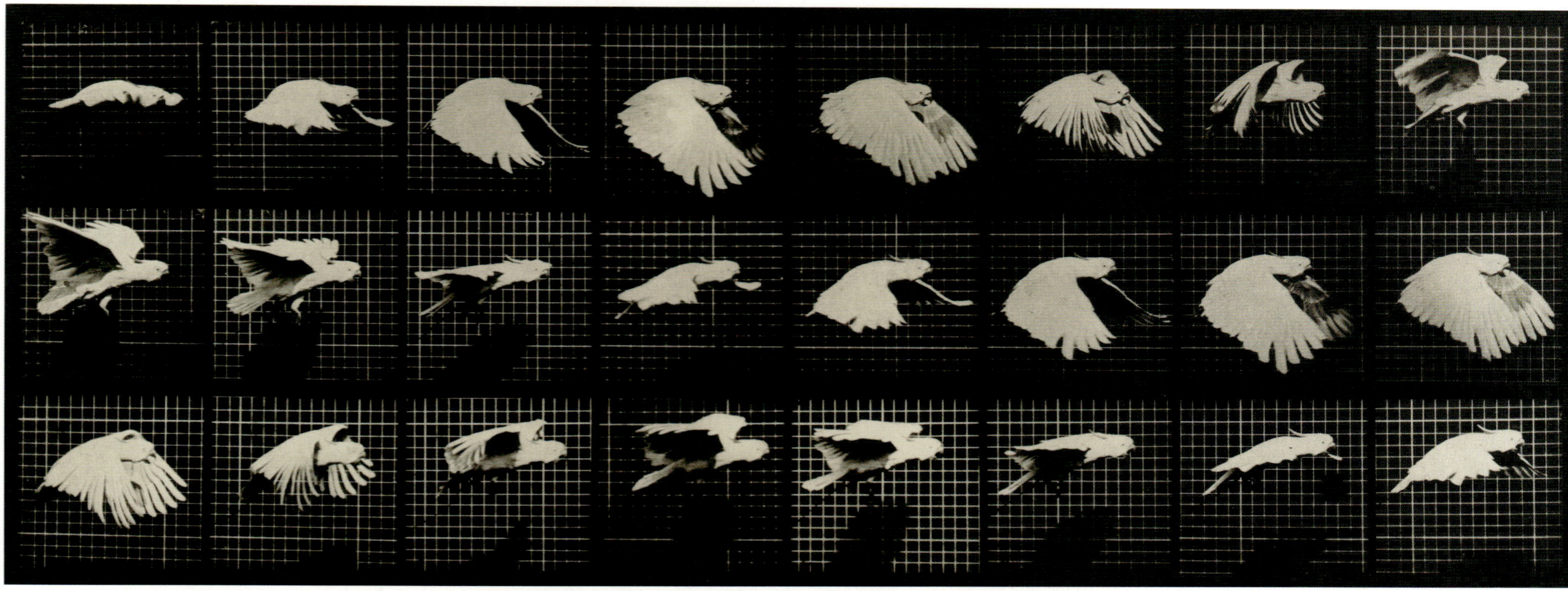

194. *Cockatoo; flying. Plate 759*, 1887

Selected Bibliography

Selected Books, Albums, Articles, Pamphlets, Other Publications, and Lectures by Eadweard Muybridge (published during his lifetime)

Animal Locomotion: An Electro-Photographic Investigation of Consecutive Phases of Animal Movements. 1872–1885. 11 vols. 781 plates. Philadelphia: University of Pennsylvania, 1887.

Animal Locomotion: An Electro-Photographic Investigation of Consecutive Phases of Animal Movements by Eadweard Muybridge Published under the Auspices of the University of Pennsylvania. Prospectus and Catalogue of Plates. Philadelphia: University of Pennsylvania, 1887.

Animals in Motion: An Electro-Photographic Investigation of Consecutive Phases of Animal Progressive Movements. London: Chapman and Hall, 1899, 1901, 1902, 1907, 1918, 1925.

The Attitudes of Animals in Motion. A Series of Photographs Illustrating the Consecutive Positions Assumed by Animals Performing Various Movements: Executed At Palo Alto, California, in 1878 and 1879. 1881. Album of original photographs.

"The Attitudes of Animals in Motion." *Journal of the Franklin Institute* 115, no. 4 (April 1883): 260–74. Transcript of lecture, Franklin Institute, February 13, 1883.

Catalogue of Photographic Views Illustrating the Yosemite, Mammoth Trees, Geyser Springs, and Other Remarkable and Interesting Scenery of the Far West by Muybridge. San Francisco: Bradley & Rulofson Gallery of Portrait and Landscape Photographic Art, 1873.

Descriptive Zoopraxography; or, The Science of Animal Locomotion Made Popular, by Eadweard Muybridge. Published as a Memento of a Series of Lectures Given by the Author under the Auspices of the United States Government, Bureau of Education, at the World's Columbian Exposition, in Zoopraxographical Hall, 1893. Philadelphia: University of Pennsylvania; Chicago: Lakeside Press, 1893.

The Human Figure in Motion: An Electro-Photographic Investigation of Consecutive Phases of Muscular Actions. London: Chapman & Hall, 1901, 1904, 1907, 1913, 1919, 1931.

"Leland Stanford's Gift to Art and to Science: Mr. Muybridge's Inventions of Instant Photography and the Marvelous Zoogyroscope." *San Francisco Examiner*, February 6, 1881, 3.

Letter. *Daily Alta California.* August 3, 1877.

Letter. *Journal of the Camera Club.* November 9, 1897.

"A New Sky Shade." *Philadelphia Photographer* 6, no. 65 (May 1869): 142–44.

"'Occident' Photographed at Full Speed." *Daily Alta California*, August 3, 1877.

The Pacific Coast of Central America and Mexico: The Isthmus of Panama, Guatemala, and the Cultivation and Shipment of Coffee. 1877. Album of original photographs.

Panorama of San Francisco from California Street Hill. By Muybridge. San Francisco: Morse's Gallery, 1877. Quarto album of original photographs.

Panorama of San Francisco from California Street Hill. San Francisco: Morse's Gallery, 1878. Elephant folio of original photographs.

Panorama of San Francisco from California Street Hill. San Francisco: C. P. Heininger, 1886.

"Photographing Animals in Action." *Philadelphia Photographer* 16, no. 181 (January 1879): 71.

"Pneumatic Clocks." Letter. *Scientific American* 40, no. 13 (March 29, 1879): 196.

"Primitive Vision and Color Blindness." *The Standard*, March 28, 1902.

Prospectus of a New and Elaborate Work upon the Attitudes of Man, the Horse and Other Animals in Motion. New York: Scovill Manufacturing, 1883.

The Science of Animal Locomotion (Zoopraxography): An Electro-Photographic Investigation of Consecutive Phases of Animal Movements by Eadweard Muybridge Executed and Published under the Auspices of the University of Pennsylvania. Philadelphia: University of Pennsylvania, 1891–92.

Syllabus of a Course of Two Lectures on the Science of Animal Locomotion in Its Relation to Design in Art. London: Royal Institution, April 1882.

Selected Books by Eadweard Muybridge (published posthumously)

Animal Locomotion. Vol. 1. "Males (Nude)." New York: Da Capo Press, 1969.

Animals in Motion. Edited with essay by Lewis S. Brown. New York: Dover Publications, 1957.

Eadweard Muybridge: The Complete Locomotion Photographs. Edited with essay by Hans-Christian Adam. Cologne: Taschen, 2010.

Eadweard Muybridge: Electra Editrice Portfolio. Edited with essay by Daniela Palazzoli. Milan: Gruppo Editoriale Electra/Rizzoli, 1982.

Eadweard Muybridge: Yosemite Photographs 1872. Chicago: Chicago Albumen Works, 1977. Portfolio of new albumen prints.

Horses and Other Animals in Motion: 45 Classic Photographic Sequences. New York: Dover Publications, 1985.

The Human Figure in Motion. Introduction by Robert A. Taft. New York: Dover Publications, 1955.

The Human Figure in Motion: A Sourcebook of Sequential Action Images by a Master Photographer. Edited by B. L. Masterson. New York: Random House, 1989.

La locomozione animale. Bologna: Grafis, 1979.

The Male and Female Figure in Motion: 60 Classic Photographic Sequences. New York: Dover Publications. 1984.

Muybridge's Animals in Motion. Electronic Clip Art Series. New York: Dover Publications, 2007.

Muybridge's Complete Human and Animal Locomotion: All 781 Plates from the 1887 Animal Locomotion. 3 vols. Introduction by Anita Ventura Mozley. Mineola, NY: Dover Publications, 1979.

Panoramic San Francisco, from California Street Hill, 1877. San Francisco: Thomas C. Russell, 1911.

Selected Books about Eadweard Muybridge

Burns, E. Bradford. *Eadweard Muybridge in Guatemala, 1875: The Photographer as Social Recorder.* Berkeley: University of California Press, 1986.

Clegg, Brian. *The Man Who Stopped Time: The Illuminating Story of Eadweard Muybridge—Pioneer Photographer, Father of the Motion Picture, Murderer.* Washington, DC: Joseph Henry Press, 2007.

Coe, Brian. *Muybridge and the Chronophotographers.* London: British Film Institute, Museum of the Moving Image, 1992.

Delimata, Joyce, ed. *Marey/Muybridge: pionniers du cinéma, rencontre Beaune/Stanford.* Beaune: Conseil régional de Bourgogne, 1996.

Gilardi, Ando. *Muybridge il magnifico voyeur.* Milan: Gabriele Mazzotta Editore, 1980.

Haas, Robert Bartlett. *Muybridge: Man in Motion.* Berkeley: University of California Press, 1976.

Harris, David, with Eric Sandweiss. *Eadweard Muybridge and the Photographic Panorama of San Francisco, 1850–1880.* Montreal: Canadian Centre for Architecture; Cambridge, MA: MIT Press, 1993.

Hendricks, Gordon. *Eadweard Muybridge: The Father of the Motion Picture.* New York: Grossman Publishers, 1975.

Herbert, Stephen, ed. *Eadweard Muybridge: The Kingston Museum Bequest.* Hastings, UK: Projection Box, 2004.

Hill, Paul. *Eadweard Muybridge.* London: Phaidon, 2001.

Hinton, Marion. *Eadweard Muybridge of Kingston upon Thames.* Kingston, UK: Kingston Museum, 1984.

Kerr, Kevin. *Studies in Motion.* Vancouver: Talonbooks, 2008.

Klett, Mark. *One City/Two Visions: San Francisco Panoramas 1878 and 1990.* San Francisco: Bedford Arts. 1990.

Klett, Mark, Rebecca Solnit, and Byron Wolfe. *Yosemite in Time: Ice Ages, Tree Clocks, Ghost Rivers.* San Antonio: Trinity University Press, 2005.

Leuba, Marion. *E. J. Muybridge (1830–1904), E. J. Marey (1830–1904). Histoires parallèles.* Beaune: Musée de la Ville de Beaune, 2000.

MacDonnell, Kevin. *Eadweard Muybridge: The Man Who Invented the Moving Picture.* Boston: Little, Brown, 1972.

Manes, Stephen. *Pictures of Motion and Pictures that Move: Eadweard Muybridge and the Photography of Motion.* New York: Coward, McCann and Geoghegan, 1982.

Marks, William Dennis, Harrison Allen, and Francis X. Dercum. *Animal Locomotion: The Muybridge Work at the University of Pennsylvania—The Method and the Result.* Philadelphia: J. B. Lippincott, 1888.

Martini, John A. *Eadweard Muybridge & James D. Givens: Photographers of the Golden Gate.* Fairfax, CA: J. Martini, 2000.

______. *Inventory of Eadweard Muybridge's Photographs of Lands within the Future Golden Gate National Recreation Area, 1868–1873: Images from the Bancroft Library and Other Archival Collections.* Fairfax, CA: J. Martini, 2000.

McBain, Robert S. *San Francisco 1878: Portrait of the City by Eadweard Muybridge.* San Francisco: Wells Fargo Bank History Department, 1979.

Metres, Philip. *Instants.* Brooklyn: Ugly Duckling Presse, 2006.

Mozley, Anita Ventura. *Eadweard Muybridge.* Stuttgart: Württembergischer Kunstverein, 1976.

Mozley, Anita Ventura, Robert Bartlett Haas, and Françoise Forster-Hahn. *Eadweard Muybridge: The Stanford Years, 1872–1882.* Stanford, CA: Stanford University Department of Art, 1972.

Muñoz, Luis Luján. *Fotografías de Eduardo Santiago Muybridge en Guatemala—1875.* Guatemala City: CENALTEX, Ministerio de Educación, 1984.

______. *Guatemala—1875: Fotografías de Eduardo Santiago Muybridge*. Guatemala City: CENALTEX, Ministerio de Educación, 1975.

Nygren, Edward J., and Frances Fralin. *Eadweard Muybridge: Extraordinary Motion*. Washington, DC: Corcoran Gallery of Art, 1986.

Osterwold, Tilman. *Eadweard Muybridge*. Stuttgart: Wurttembergischer Kunstverein, 1976.

Phelan, Robert J., and Thomas Weston Fels. *Animal Locomotion: Images from the Philadelphia Years: A Selection of Collotypes from the Series "Animal Locomotion: An Electro-Photographic Investigation of Consecutive Phases of Animal Movements, 1872–1885," from the Collection of the Williams College Museum of Art*. Albany: University Art Gallery, State University of New York at Albany, 1985.

Prodger, Phillip. *Time Stands Still: Muybridge and the Instantaneous Photography Movement*. New York: Iris & B. Gerald Cantor Center for Visual Arts at Stanford University in association with Oxford University Press, 2003.

Robertson, David. *The Art and Science of Eadweard Muybridge*. Carlisle, PA: Dickinson College, 1985.

Sheldon, James L., and Jock Reynolds. *Motion and Document, Sequence and Time: Eadweard Muybridge and Contemporary American Photography*. Andover, MA: Addison Gallery of American Art, Phillips Academy, 1991.

Solnit, Rebecca. *River of Shadows: Eadweard Muybridge and the Technological Wild West*. New York: Viking, 2003.

Stillman, Jacob Davis Babcock. *The Horse in Motion as Shown by Instantaneous Photography, with a Study on Animal Mechanics Founded on Anatomy and the Revelations of the Camera, in Which Is Demonstrated the Theory of Quadrupedal Locomotion*. Executed under the auspices of Leland Stanford. Boston: James R. Osgood, 1882.

Winger, Rob. *Muybridge's Horse*. Gibsons, BC: Nightwood Editions, 2007.

Selected Articles about Eadweard Muybridge

"The Actual Movements of Horses." *New York Times*, December 24, 1878, 3.

Adams, W. I. Lincoln. "Some Modern Achievements of the Camera." *Outing, an Illustrated Monthly Magazine of Recreation* 18, no. 6 (September 1891): 481–82.

Alfieri, Bernard. "Eadweard Muybridge: The Real Father of Cinematography." *Courier* 36, no.1 (January 1961): 28–31.

_____. "The Work of Eadweard Muybridge." *American Annual of Photography 1950* 64 (1949): 70–73.

Andersen, Thom. "Eadweard Muybridge." *Film Culture* (Summer 1966): 22–24.

"Animal Locomotion." *Art Amateur* 18, no. 4 (March 1888): 85.

"Animal Locomotion." *Nation* 46 (January 19, 1888): 55.

"Animal Motion," *Philadelphia Times*, August 2, 1885.

"Animal Progression in Court." *New York Times*, February 2, 1883, 5.

Anthony's Photographic Bulletin 2 (December 1871): 402.

Baetens, Jean. "Muybridge with(out) Marey. Rereading Aaron Scharf's *Art and Photography*." *Image [&] Narrative—Online Magazine of the Visual Narrative* (July 2000).

Beard, W. H. "Motion as Art Shows It." *New York Times*, January 7, 1894, 23.

Braun, Marta. "Muybridge's *Animal Locomotion*: The Director's Cut." *History of Photography* 24, no. 1 (Spring 2000): 52–54.

______. "Muybridge's Scientific Fictions." *Studies in Visual Communication* 10, no. 3 (Summer 1984): 2–22.

Braun, Marta, and Elizabeth Whitcombe. "Marey, Muybridge, and Londe: The Photography of Pathological Locomotion." *History of Photography* 23, no. 3 (Autumn 1999): 218–24.

Brown, Elspeth H. "Racialising the Virile Body: Eadweard Muybridge's Locomotion Studies 1883–1887." *Gender & History* 17, no. 3 (November 2005): 627–56.

"Caught in the Act." *New York Times*, January 10, 1883, 5.

Conkling, F. A. "Transactions of the Society for the Year 1873." *Journal of the American Geographical Society of New York* 5 (1874): 35–45.

Croffut, William A. "Inventors." In *An American Procession, 1855–1914: A Personal Chronicle of Famous Men*, 294–303. 1931. Reprint, Freeport, NY: Books for Libraries Press, 1968. De la Sizeranne, Robert. "The Photographer and the Artist." *The Chautauquan* 17, no. 2 (May 1893): 193–97.

Dercum, Francis X. "The Walk and Some of Its Phases in Disease, Together with Other Studies Based on the Muybridge Investigation." *Transactions of the College of Physicians of Philadelphia* 10 (1888): 308–38.

Detheux, Jean. "Notes from the Underground, Part Five: Escaping Muybridge's Curse (Can We?)." *Animation World* 24 (January 2003).

"A Dog's Movements Photographed." *Anthony's Photographic Bulletin* 11 (January 1880): 13.

Falconer, Paul A. "Muybridge's Window to the Past: A Wet-Plate View of San Francisco." *California History* 57, no. 2 (Summer 1978): 130–57.

Finny, W. E. St. Lawrence. "Eadweard James Muybridge: A Famous Kingstonian, Scientist, Inventor, Benefactor." *The Concentric* (Summer 1931).

"The First Motion Pictures." *Moving Picture World* 15 (January 11, 1913): 143.

Frampton, Hollis. "Eadweard Muybridge: Fragments of a Tesseract." In *Circles of Confusion: Film, Photography, Video, Texts, 1968–1980*, 69–80. Rochester, NY: Visual Studies Workshop Press, 1983.

Fuller P. "Eadweard James Muybridge, 1830–1904." *Journal of Audiovisual Media in Medicine* 21, no. 4 (1998): 140–42.

______. "Life and Work of Eadweard Muybridge." *Proceedings—SPIE: The International Society for Optical Engineering*, no. 3516, Part 1 (1998): 4–14.

Gibson, H. L. "The Muybridge Pictures of Motion." *Medical Radiography and Photography* 26, no. 1 (1950): 18–26.

Glazier, Jeremy. "Freud, on the Orient Express: Images from Muybridge." *Western Humanities Review* 61, no. 1 (2007): 96–97.

Gopnik, Blake. "Muybridge's Canter through Central America." *Washington Post*, March 1, 2007.

Gordon, Sarah. "Prestige, Professionalism, and the Paradox of Eadweard Muybridge's *Animal Locomotion* Nudes." *Pennsylvania Magazine of History and Biography* 130, no. 1 (January 2006): 79–104.

______. "Sanctioning the Nude: Production and Reception of Eadweard Muybridge's *Animal Locomotion*, 1887." PhD diss., Northwestern University, 2006.

Gurtner, H. "Étienne-Jules Marey and Edward Muybridge, Pioneers in Motion Analysis." *Ciba Symposia* (August–September 1942): 1356–59.

Hagen, Charles. "Eadweard Muybridge Revisited, with Others." *New York Times*, March 6, 1992.

Hamilton, Harlan. "'Les Allures du Cheval': Eadweard James Muybridge's Contribution to the Motion Picture." *Film Comment* 5, no. 3 (Fall 1969): 16–35.

Hastings, Charles Edward. "Cinematic Beginnings." *Moving Picture World*, March 26, 1927, 289–300.

Hecht, Hermann. "Muybridge and the Stereoscopic Moving Picture." *Stereoscopic Society Bulletin*, no. 54 (June 1976).

Heininger, Charles Peter. *Historical Souvenir of San Francisco, Cal. With Views of Prominent Buildings, the Bay, Islands, Etc.* San Francisco: C. P. Heininger, 1887.

Homer, William I. "Concerning Muybridge, Marey, and Seurat." *Burlington Magazine* 104 (September 1962): 391–92.

Homer, William I., with John Talbot. "Eakins, Muybridge, and the Motion Picture Process." *Art Quarterly* 26, no. 2 (Summer 1963): 194–216.

Hood, Mary V. Jessup, and Robert Bartlett Haas. "Eadweard Muybridge's Yosemite Valley Photographs, 1867–1872." *California Historical Society Quarterly* 43 (March 1963): 5–26.

"The Horse in Motion." *Philadelphia Photographer* 16, no. 183 (March 1879): 22–23.

"A Horse's Notion Scientifically Determined." *Scientific American* 39, no. 16 (October 19, 1878): 241.

"Instant Photography." *New York Times*, February 19, 1881, 2.

Jaffe, Irma B., with Gernando Colombardo. "The Flying Gallop: East and West." *Art Bulletin* 65, no. 2 (June 1983): 183–200.

Kantor, J. R. K. "Muybridge Views of Berkeley in 1874." *California History* 57, no. 4 (Winter 1978–79).

"The Last of the Modocs." *Harper's Weekly*, June 21, 1873, 532–34.

Lawrence, Amy. "Counterfeit Motion: The Animated Films of Eadweard Muybridge." *Film Quarterly* 57, no. 2 (Winter 2003–04): 15–25.

The Library Company of Philadelphia. "Muybridge and Zoopraxography." *Annual Report … 1961* (1961): 19–24.

"The Man Who First Made Pictures Move and Who Revolutionized Photography." *Current Opinion*, no. 1 (July 1917): 31.

Marable, Darwin. "Photography and Human Behavior in the Nineteenth Century." *History of Photography* 9, no. 2 (April–June 1985): 141–47.

Marey, E. H. J. "The Mechanism of the Flight of Birds." *Frank Leslie's Popular Monthly* 25, no. 5 (May 1888): 632–34.

Mason, O. G. "Report of the Photographic Section of the American Institute." *Anthony's Photographic Bulletin* 12 (April 1881): 121–22.

Mileaf, Janine A. "Poses for the Camera: Eadweard Muybridge's Studies of the Human Figure." *American Art* 16, no. 3 (Fall 2002): 31–53.

Miles, Walter R. "Leland Stanford and Motion Pictures." *Stanford Illustrated Review* (June 1929).

_____. "The Stanford-Muybridge Motion Pictures of 1878–79." *Bulletin of the Minnesota Federation of Architectural and Engineering Societies* (September 1929): 1–5.

Morgan, Jayne. "Eadweard Muybridge and W. S. Playfair: An Aesthetics of Neurasthenia." *History of Photography* 23, no. 3 (Autumn 1999): 225–31.

"Motion Photographed: Muybridge's Novel Work at the University of Pennsylvania." *Bangor Daily Whig & Courier*, August 1, 1884, col. F.

Mozley, Anita Ventura. "Introduction." In *Muybridge's Complete Human and Animal Locomotion*, vol. 1, vii–xxxviii. New York: Dover, 1979.

"Mr. Muybridge at the Royal Institution." *Photographic News* 26, no. 1228 (March 17, 1882): 1.

"Mr. Muybridge Showing His Instantaneous Photographs of Animal Motion at the Royal Academy." *Illustrated London News* 94, no. 2614 (May 25, 1889): cover, 648.

"Mr. Muybridge's Photographs." *New York Times*, August 16, 1885, 5.

"Mr. Muybridge's Photographs: Interesting Pictures to Be Taken of Wild Birds and Beasts in Motion." *Philadelphia Ledger*, August 12, 1885.

"Mr. Muybridge's Photographs of Animals in Motion." *Anthony's Photographic Bulletin* 13, no. 1 (January 1882): 6–7.

"The Muybridge Case: Great Interest Manifested in the Trial." *San Francisco Examiner*, February 4, 1875.

"The Muybridge Trial: The Jury Locked Out All Night." *San Francisco Chronicle*, February 6, 1875.

"Muybridge's Zoogyroscope." *Scientific American* 42, no. 23 (June 5, 1880): 353.

Newhall, Beaumont. "The George E. Nitzsche Collection of Muybridge Relics." *Medical Radiography and Photography* 26, no. 1 (1950): 24–26.

_____. "Muybridge and the First Motion Picture." *U.S. Camera Annual* (1957): 235–42.

_____. "Muybridge and the First Motion Picture: The Horse in the History of the Movies." *Image* 5, no. 1 (January 1956): 4–11.

"New Publications." *New York Times*, June 20, 1882, 5.

Nitzsche, George E. "The Evolution of Moving Pictures." *Old Penn*, April 19, 1913.

_____. "The Muybridge Moving Picture Experiments at the University of Pennsylvania." *General Magazine and Historical Chronicle* 11, no. 3 (April 1929): 323–32.

_____. "Philadelphia: Birthplace of Moving Pictures." *Germantown Crier* (March 1950).

"Notes In and Out of the Studio: Testing Iodides in Collodion—American Stereographs—Durable Sensitive Paper." *Philadelphia Photographer* 9, no. 100 (April 1872): 119–20.

Ott, John. "Iron Horses: Leland Stanford, Eadweard Muybridge, and the Industrialised Eye." *Oxford Art Journal* 28, no. 3 (2005): 407–28.

"Our Picture." *Philadelphia Photographer* 6, no. 71 (November 1869): 373–75.

Pajunen, Timo Tauno. "Eadweard Muybridge." *Camera* (January–February 1973).

Palmquist, Peter. "Imagemakers of the Modoc War: Louis Heller and Eadweard Muybridge." *Journal of California Anthropology* 4, no. 2 (Winter 1977): 206–41.

Peterson, H. C. "The Birthplace of the Motion Picture." *Sunset Magazine* (November 1915).

"Photograph Studies." *Daily Alta California*, April 7, 1873.

"Photographic Notes." *Scientific American* 53, no. 20 (November 14, 1885): 305.

"Photographing a Horse in Action by the Instantaneous Process." *Wallace's Monthly* 4, no. 7 (August 1878): 551–52.

"Photographing a Race-Horse." *New York Times*, June 23, 1878, 5.

"The Photographing of Motion." *Scientific American* 47, no. 11 (September 9, 1882): 167.

"Photographs of the Lava Beds." *San Francisco Chronicle*, April 29, 1873.

Pizzanelli, David. "Multiplex Holograms from Eadweard Muybridge's Studies of the Human Figure in Motion." *Proceedings—SPIE: The International Society for Optical Engineering*, no. 2333 (1994): 156.

Premeaux, Sonya F. "The Flying Horse: Eadweard Muybridge's Contribution to Motion Study." *Journal of Applied Management and Entrepreneurship* (October 2003).

Proctor, Richard A. "Photographs of a Galloping Horse." *Gentleman's Magazine* 27 (December 1881): 666–89.

Prodger, Phillip. "How Did Muybridge Do It?" *Aperture*, no. 158 (2000): 12–21.

"Rapid Photographing." *Scientific American* 41, no. 11 (September 13, 1879): 160.

"Realities of Animal Motion." *Anthony's Photographic Bulletin* 13, no. 24 (November 1882): 375–76.

"Realities of Animal Motion." *New York Times*, November 18, 1882, 2.

Rogers, Fairman. "The Zoötrope: Action of Animals in Motion—The Muybridge Photographs of Horses—The Instrument as a Factor in Art Studies." *Art Interchange* 3, no. 1 (July 9, 1879): 2–3.

"Romance and Reality of Animal Motion." *Scientific American* 47, no. 23 (December 2, 1882): 352.

Rondinella, L. F. "More about Muybridge's Work." *General Magazine and Historical Chronicle*, July 1929.

_____. "Muybridge's Motion Pictures." *Journal of the Franklin Institute* 208 (September 1929): 417–20.

Rulofson, William H. "The California Horse—Electric Feat." *Philadelphia Photographer* 15, no. 176 (August 1878): 247.

San Francisco Morning Call, April 9, 1882.

"Set at Rest: The True Action of a Horse in Trotting Determined." *San Francisco Sunday Chronicle*, June 16, 1878, 5.

Shimamura, Arthur P. "Muybridge in Motion: Travels in Art, Psychology, and Neurology." *History of Photography* 26, no. 4 (Winter 2002): 341–50.

Smith, George. "The Fatal Amour: An Interview with Muybridge, the Slayer of Major Larkyns." *San Francisco Chronicle*, December 21, 1874.

Snell, George. "On the Galloping Horse in Art." *The Century* 26, no. 2 (June 1883): 315–17.

"Some Applications of Photography." *Scientific American* 64, no. 11 (March 14, 1891): 168.

"Some of the July Magazines: *The Century*." *New York Times*, June 26, 1882, 3.

"Sporting News." *National Police Gazette* 39, no. 223 (December 31, 1881): 14.

"The Study of Animal Motions." *Scientific American* 44, no. 18 (April 30, 1881): 272.

Taylor, R. "A Talk about Photography." *Frank Leslie's Popular Monthly* 29, no. 2 (February 1890).

Tegetmeier, W. B. "Mr. Muybridge's Photographs of Animals in Motion." *Anthony's Photographic Bulletin* 13, no. 11 (June 1882): 174.

Thomas, D. B. "The Lantern Slides of Eadweard Muybridge." *British Journal of Photography Annual* (1967).

Towler, J. "Practical Landscape Photography." *Philadelphia Photographer* 8 (1871): 352–54.

Van Guysling, G. G. "The Horse in Motion." *International Photographer* (January 1933).

"Views and Maps That Came Too Late." *San Francisco Chronicle*, May 10, 1873.

Vogel, Hermann Wilhelm. "German Correspondence: Photography—The Vienna Exhibition—North America—Brazil—England." *Philadelphia Photographer* 10, no. 117 (September 1873).

_____. "Medals Awarded at the Vienna Exhibition." *Philadelphia Photographer* 10, no. 118 (October 1873): 487–88.

Waring, George E. Jr. "The Horse in Motion." *The Century* 24, no. 3 (July 1882): 381–88.

Williams, Talcott. "Animal Locomotion in the Muybridge Photographs." *The Century* 34, no. 3 (July 1887): 356–68.

"The Zoogyroscope; Interesting Exhibition of Photographs of Animals in Motion." *San Francisco Call*, May 5, 1880.

Selected Books and Articles Related to Eadweard Muybridge

Albera, François, Marta Braun, and André Gaudreault, eds. *Stop Motion, Fragmentation of Time: Exploring the Roots of Modern Visual Culture*. Lausanne: Éditions Payot Lausanne, 2002.

Allen, Harrison. "On the Teaching of Anatomy to Advanced Medical Students." *Science* 19, no. 471 (February 12, 1892): 85–87.

AlSayyad, Nezar. *Consuming Tradition, Manufacturing Heritage: Global Norms and Urban Forms in the Age of Tourism*. London: Routledge, 2001.

Avery, Kevin J. "Movies for Manifest Destiny: The Moving Panorama Phenomenon in America." In *The Grand Moving Panorama of "Pilgrim's Progress"*. Montclair, NJ: Montclair Art Museum, 1999.

Badger, R. Reid. *The Great American Fair: The World's Columbian Exposition and American Culture*. Chicago: Nelson-Hall, 1979.

Bailey, Richard W. *Rogue Scholar: The Sinister Life and Celebrated Death of Edward H. Rulloff*. Ann Arbor, MI: University of Michigan Press, 2003.

Baird, Joseph Armstrong. *California's Pictorial Letter Sheets, 1849–1869*. San Francisco: David Magee, 1967.

Baker, Hugh Sanford. "The Book Trade in California, 1849–59." *California Historical Society Quarterly* 30 (1951): 365.

Barth, Gunther. *Instant Cities: Urbanization and the Rise of San Francisco and Denver*. New York: Oxford University Press, 1975.

Belt, Betty C. "The Leland Stanford Home in San Francisco." Master's thesis, Stanford University, 1947.

Berglund, Barbara. *Making San Francisco American: Cultural Frontiers in the Urban West, 1846–1906*. Lawrence: University Press of Kansas, 2007.

Birt, Rodger C. "Envisioning the City: Photographs in the History of San Francisco." PhD diss., Yale University, 1985.

Boddam-Whetham, J. W. *Across Central America*. London: Hurst and Blackett, 1877.

Boruck, Marcus. "The Residence of Leland Stanford." *Centennial Spirit of the Times*, July 4, 1876, 31.

Braun, Marta. *Picturing Time: The Work of Étienne-Jules Marey (1830–1904)*. Chicago: University of Chicago Press, 1992.

Bright, Deborah. "The Machine in the Garden Revisited." *Art Journal* 51 (Summer 1992): 60–72.

Brookman, Philip. "To Pavements and Homesteads Here: Landscape, Photography, and the Transcendence of Space, 1850–1890." In *Expanding Horizons: Painting and Photography of American and Canadian Landscape, 1860–1918*, edited by Hilliard T. Goldfarb, 237–43. Montreal: Montreal Museum of Fine Arts; Barrington, IL: Art Stock Books, 2009.

Brown, Elspeth H. *The Corporate Eye: Photography and the Rationalization of American Commercial Culture, 1884–1929*. Baltimore: Johns Hopkins University Press, 2005.

Brown, Joshua. *Beyond the Lines: Pictorial Reporting, Everyday Life, and the Crisis of Gilded Age America*. Berkeley: University of California Press, 2002.

Brown, Julie K. *Contesting Images: Photography and the World's Columbian Exposition*. Tucson: University of Arizona Press, 1994.

______. *Making Culture Visible: The Public Display of Photography at Fairs, Expositions, and Exhibitions in the United States, 1847–1900*. New York: Routledge, 2001.

Buel, C. C. "Columbian Exposition Preliminary Glimpses of the Fair." *Century Illustrated Magazine* 45, no. 4 (February 1893): 615–25.

Cady, Theron G. "Peninsula—Birthplace of the Movie." *Peninsula Life* (August 1946).

Campbell, Robert. *In Darkest Alaska: Travel and Empire Along the Inside Passage*. Philadelphia: University of Pennsylvania Press, 2007.

Carden, Robert A. "Photography in California." *Photographic and Fine Art Journal*, no. 10 (April 1857): 112–13.

Clark, George T. *Leland Stanford: War Governor of California, Railroad Builder, and Founder of Stanford University*. Stanford, CA: Stanford University Press, 1931.

Conkling, Roscoe P., and Margaret B. Conkling. *The Butterfield Overland Mail, 1857–1869: Its Organization and Operation over the Southern Route to 1861; Subsequently over the Central Route to 1866; and under Wells, Fargo and Company in 1869*. Glendale, CA: Arthur H. Clark, 1947.

Conningham, Frederick A. *Currier & Ives Prints: An Illustrated Check List*. New York: Crown Publishing, 1949.

Coolidge, Susan. "A Few Hints on the California Journey." *Scribner's Monthly* 6, no. 1 (May 1873): 25–31.

Crary, Jonathan. *Suspensions of Perception: Attention, Spectacle, and Modern Culture*. Cambridge, MA: MIT Press, 1999.

______. "Unbinding Vision: Manet and the Attentitive Observer in the Late Nineteenth Century." In Leo Charney and Vanessa R. Schwartz, *Cinema and the Invention of Modern Life*, 46–71. Berkeley: University of California Press, 1995.

Cresswell, Tim. *On the Move: Mobility in the Modern Western World*. New York: Routledge, 2006.

Current, Karen, William R. Current, Margaret L. Kaplan, and Ellen Shultz. *Photography and the Old West*. New York: Harry N. Abrams, 1978.

Dagognet, François. *Étienne-Jules Marey: A Passion for the Trace*. Translated by Galeta Robert and Jeanine Herman. Cambridge, MA: MIT Press, 1992.

Danly, Susan, and Cheryl Leibold, eds. *Eakins and the Photograph: Works by Thomas Eakins and His Circle in the Collection of the Pennsylvania Academy of the Fine Arts*. Washington, DC: Published for the Pennsylvania Academy of the Fine Arts by the Smithsonian Institution Press, 1994.

Davis, Keith F. "The High-Speed Photographs of Francis Blake." *Massachusetts Historical Review* 2 (2000): 1–26.

Davis, Margo, and Roxanne Nilan. *The Stanford Album: A Photographic History, 1885–1945*. Stanford, CA: Stanford University Press, 1989.

De Duve, Thierry. "Time Exposure and Snapshot: The Photograph as Paradox." *October* 5 (Summer 1978): 113–25.

Doane, Mary Ann, *The Emergence of Cinematic Time: Modernity, Contingency, the Archive*. Cambridge, MA: Harvard University Press, 2002.

Duhousset, Émile. *Le cheval dans la nature et dans l'art*. Paris: Henri Laurens, 1901.

______. "Variétés: Reproduction instantanée des allures du cheval..." *L'Illustration*, January 25, 1879, 59.

Dupeux, Cécile, et al. *La passion du mouvement au XIXe siècle: Hommage à E. J. Marey*. Beaune: Musée Marey, 1991.

Edwards, Elizabeth, ed. *Anthropology and Photography 1860–1920*. New Haven, CT: Yale University Press, 1992.

Fardon, George Robinson, *San Francisco Album: Photographs of the Most Beautiful Views and Public Buildings*. San Francisco: Chronicle Books and Fraenkel Gallery, 1999.

______. *San Francisco in the 1850s: 33 Photographic Views by G. R. Fardon*. Introduction by Robert A. Sobieszek. Rochester: NY: International Museum of Photography at George Eastman House; New York: Dover Publications, 1977.

Fraser, W. Lewis. "At the Fair." *Century Magazine* 46, no. 1 (May 1893): 3–21.

Frizot, Michel, ed. *E. J. Marey, 1830/1904: La photographie du mouvement*. Paris: Centre Georges Pompidou, Musée National d'Art Moderne, 1977.

Fry, Walter, and John R. White. *Big Trees*. Stanford, CA: Stanford University Press, 1938.

Gmelch, Sharon. *The Tlingit Encounter with Photography*. Philadelphia: University of Pennsylvania Museum of Archaeology and Anthropology, 2008.

Gombrich, E. H. "Moment and Movement in Art." *Journal of the Warburg and Courtauld Institutes* 27 (1964): 293–306.

Gréard, Vallery K. O. *Meissonier: His Life and His Work*. New York: A. C. Armstrong and Son, 1897.

Greeley, Horace. *An Overland Journey: From New York to San Francisco in the Summer of 1859*. New York: Knopf, 1963.

Gregory, Thomas Jefferson. *History of Solano and Napa Counties, California*. Los Angeles: Historic Record Company, 1912.

Haas, Robert Bartlett. "William Herman Rulofson, Pioneer Daguerreotypist and Photographic Educator." *California Historical Society Quarterly* 34, no. 4 (1955): 289–301.

Haes, Frank. "The Early Days of Animal Photography." *Photographic News* 36 (January 15, 1892): 38–39.

Hales, Peter B. *Silver Cities: The Photography of American Urbanization, 1839–1915*. Philadelphia: Temple University Press, 1984.

Harmon, Albert A. "Gold Rush Panoramas: The View from Nob Hill." *Sea Letter* 2, no. 1 (September 1964).

Harte, Bret. *San Francisco in 1866; Being Letters to the "Springfield Republican"*. Edited by George R. Stewart and Edwin S. Fussell. San Francisco: Book Club of California, 1951.

Hendricks, Gordon. *Albert Bierstadt*. New York: Harrison House/Harry N. Abrams, 1988.

______. *The Life and Work of Thomas Eakins*. New York: Grossman Publishers, 1974.

______. "A May Morning in the Park." *Philadelphia Museum of Art Bulletin* 60 (Spring 1965): 56.

______. *The Photographs of Thomas Eakins*. New York: Grossman Publishers, 1972.

Hittell, John S. *Commerce and Industries of the Pacific Coast of North America*. San Francisco: A. L. Bancroft, 1882.

______. *The Resources of California: Comprising Agriculture, Mining, Geography, Climate, Commerce, etc., and the Past and Future Development of the State*. San Francisco: A. Roman, 1866.

______. *Yosemite: Its Wonders and Its Beauties*. San Francisco: H. H. Bancroft, 1868.

Hively, William, Lawrence Dinnean, and Lawton Kennedy. *Nine Classic California Photographers*. Berkeley: Friends of the Bancroft Library, University of California, 1980.

Hood, Mary V. "Charles L. Weed, Yosemite's First Photographer." *Yosemite Nature Notes* 38, no. 6 (June 1959): 76–87.

Hoyt, Edwin P. *Leland Stanford: A Biography of the Governor of California Who Built a Controversial Railroad and Founded a Great University*. New York: Abelard-Schuman, 1967.

Hungerford, Constance Cain. *Ernest Meissonier: Master in His Genre*. Cambridge: Cambridge University Press, 1999.

Hutchings, James M. *Scenes of Wonder and Curiosity in California. Illustrated with Over One Hundred Engravings. A Tourist's Guide to the Yo-Semite Valley*. New York: A. Roman, 1871.

Hutchinson, Elizabeth. "They Might Be Giants: Galen Clark, Carleton Watkins, and the Big Tree." In *A Keener Perception: Ecocritical Studies in American Art History*, edited by Alan C. Braddock and Christopher Irmscher, 110–128. Tuscaloosa: University of Alabama Press, 2009.

Hyde, Ralph. *Panoramania! The Art and Entertainment of the "All-Embracing" View*. London: Trefoil and Barbican Art Gallery, 1988.

Issel, William, and Robert W. Cherny. *San Francisco, 1865–1932: Politics, Power, and Urban Development*. Berkeley: University of California Press, 1986.

Jackson, Helen Hunt [H. H.]. *Ah-wah-ne days; a visit to the Yosemite Valley in 1872.* San Francisco: Book Club of California, 1971.

______. *Bits of Travel at Home.* Boston: Roberts Brothers, 1878.

Josephson, Matthew. *The Robber Barons.* New York: Harcourt, Brace, 1934.

Jussim, Estelle. *Frederic Remington: The Camera and the Old West.* Fort Worth, TX: Amon Carter Museum, 1983.

Kammen, Michael G. "How the West Was Seen: The Conquest of Space, Time, and Vision." *Reviews in American History* 31, no. 3 (September 2003): 406–13.

Kelsey, Robin. *Archive Style, Photographs, and Illustrations for U.S. Surveys, 1850–1890.* Berkeley: University of California Press, 2007.

Kibbey, Mead B. *The Railroad Photographs of Alfred A. Hart, Artist.* Sacramento: California State Library Foundation, 1996.

Klein, Maury. *The Life and Legend of Jay Gould.* Baltimore: Johns Hopkins University Press, 1997.

Kosinski, Dorothy. *The Artist and the Camera.* New Haven, CT: Yale University Press, 1999.

Krauss, Rosalind. "Photography's Discursive Spaces: Landscape/View." *Art Journal* 42, no. 4 (Winter 1982): 311–19.

Kurutz, Gary F. "California Books Illustrated with Original Photographs, 1856–1890." *Biblio-Cal. Notes* 7, no. 2 (Summer/Fall 1974): 3–16.

______. *The Triumph of Helios: Photographic Treasures of the California State Library.* Sacramento: California State Library Foundation, 2006.

Lang, Walter Barnes. *The First Overland Mail, Butterfield Trail, St. Louis to San Francisco, 1858–1861.* East Aurora, NY: W. B. Lang, 1940.

Mannoni, Laurent. *Étienne-Jules Marey: La mémoire de l'oeil.* Paris: Cinémathèque française, 1999.

Manthorne, Katherine. "Plantation Pictures in the Americas, circa 1880: Land, Power, and Resistance." *Nepantla: Views from the South* 2, no. 2, (2001): 317–53.

______. *Tropical Renaissance: North American Artists Exploring Latin America, 1839–1879.* Washington, DC: Smithsonian Institution Press, 1989.

Marey, Étienne-Jules. *Animal Mechanism: A Treatise on Terrestrial and Aerial Locomotion.* New York: D. Appleton, 1874.

______. "La chronophotographie: Nouvelle méthode pour analyser le mouvement dans les sciences physiques et naturelles." In *Revue Générale des Sciences* 2 (1891): 689–719. Reprinted as *La photographie du mouvement* (Paris: Levé, 1892).

______. *La machine animale: Locomotion terrestre et aérienne.* Paris: Baillière, 1873.

______. *Movement.* Translated by Eric Pritchard. New York: D. Appleton, 1895.

______. *Physiologie du mouvement: Le vol des oiseaux.* Paris: Masson 1890.

Margolis, David. *To Delight the Eye: Original Photographic Book Illustrations of the American West.* Dallas: DeGolyer Library, Southern Methodist University, 1994.

Martini, John Arturo. *Fort Point: Sentry at the Golden Gate.* San Francisco: Golden Gate National Park Association, 1991.

______. *Fortress Alcatraz: Guardian of the Golden Gate.* Berkeley: Ten Speed Press, 2004.

Mathews, Nancy Mowll, with Charles Musser. *Moving Pictures: American Art and Early Film, 1880–1910.* Manchester, VT: Hudson Hills Press in association with Williams College Museum of Art, 2005.

Mattison, David. "G. R. Fardon's Portraits on Patent Leather at the London International Exhibition, 1862: Newly Discovered at the Victoria and Albert Museum." *Photographic Canadiana* 25, no. 3 (November/December 1999): 47.

McCosker, M. J. "Philadelphia and the Genesis of the Motion Picture." *Pennsylvania Magazine of History and Biography* 65, no. 4 (October 1941): 401–19.

McDonald, Terrence J. *The Parameters of Urban Fiscal Policy: Socioeconomic Change and Political Culture in San Francisco, 1860–1905.* Berkeley: University of California Press, 1986.

Murray, Keith A. *The Modocs and Their War.* Norman: University of Oklahoma Press, 1959.

"Muybridge v. Stanford. Suit of the Artist for Damages for the Governor's Horse Book." *Salt Lake Daily Tribune,* January 30, 1883.

Naef, Weston. "Artists and Photographers of the Western Expeditions." *Art News* (December 1975): 62–64.

______. *Carleton Watkins in Yosemite.* Los Angeles: J. Paul Getty Museum, 2008.

Naef, Weston, ed. *In Focus: Carleton Watkins—Photographs from the J. Paul Getty Museum.* Los Angeles: Getty Publications, 1997.

Naef, Weston J., James N. Wood, and Therese Thau Heyman. *Era of Exploration: The Rise of Landscape Photography in the American West, 1860–1885.* Buffalo, NY: Albright-Knox Art Gallery; Boston: New York Graphic Society, 1975.

Nagel, Gunther W. "Monument to Elegance: Leland Stanford's Nob Hill Mansion." *American West* 7, no. 5 (September 1975): 18–25.

Newhall, Beaumont. "Photography and the Development of Kinetic Visualization." *Journal of the Warburg and Courtauld Institutes* 7 (1944): 40–45.

Nickel, Douglas R. *Carleton Watkins: The Art of Perception.* San Francisco: San Francisco Museum of Modern Art; New York: Harry N. Abrams, 1999.

Ogden, Kate Nearpass. "Sublime Vistas and Scenic Backdrops: Nineteenth-Century Painters and Photographers at Yosemite." *California History* 59 (Summer 1990): 134–53.

Olsen, Victoria. "Indelible Images: Leland Stanford, Eadweard Muybridge's Heart- (and Hoof-) Stopping Image of a Trotting Horse Not Only Answered a Hotly Debated Question but Revolutionized Photography." *Smithsonian* 35, no. 6 (2004): 20.

Orvell, Miles. *The Real Thing: Imitation and Authenticity in American Culture, 1880–1940.* Chapel Hill: University of North Carolina Press, 1989.

Osborne, Carol M. *Museum Builders in the West: The Stanfords as Collectors and Patrons of Art, 1870–1906.* Stanford, CA: Stanford University Museum of Art, 1980.

Palmquist, Peter E. *Carleton Watkins: Photographer of the American West.* Albuquerque: University of New Mexico Press, 1983.

______. *Carleton Watkins: Photographs, 1861–1874.* San Francisco: Bedford Arts Publishers, 1989.

______. "Charles Leander Weed, California's Peripatetic Photographer." *California History* 58, no. 3 (Fall 1979): 194–219.

______. "The Daguerreotype in San Francisco." *History of Photography* 4, no. 3 (July 1980): 207–38.

______. *Lawrence and Houseworth/Thomas Houseworth and Co.: A Unique View of the West, 1860–1886.* Columbus, OH: National Stereoscopic Association, 1980.

______. *Points of Interest: California Views, 1860–1870—The Lawrence and Houseworth Albums.* San Francisco: Berkeley Hills Books, 2002.

______. "Robert H. Vance: The Maine and Boston Years (1825–c. 1850)." *Daguerreian Annual 1991* (1991): 199–216.

______. "Robert Vance, Pioneer in Western Landscape Photography." *American West* 18, no. 5 (September–October 1981): 22–27.

______. "Watkins at Work." *History of Photography* 6, no. 4 (October 1982): 291–325.

______. "Yosemite's First Stereo Photographer, Charles Leander Weed (1824–1903)." *Stereo World* 6, no. 4 (September–October 1979): 4–11.

Palmquist, Peter E., and Thomas R. Kailbourn. *Pioneer Photographers of the Far West: A Biographical Dictionary, 1840–1865.* Stanford, CA: Stanford University Press, 2000.

Parry, Ellwood C., III. "Thomas Eakins's 'Naked Series' Reconsidered: Another Look at the Standing Nude Photographs Made for the Use of Eakins's Students." *American Art Journal* 20, no. 2 (1988): 53–77.

Perry, Claire. *Pacific Arcadia: Images of California, 1600–1915.* New York: Oxford University Press, 1999.

Pettigrew, J. Bell. *Animal Locomotion or Walking, Swimming, and Flying, with a Dissertation on Aeronautics (1874).* London: Kegan Paul, 1874.

Phillips, Sandra S., et al. *Taking Place: Photographs from the Prentice and Paul Sack Collection.* San Francisco: San Francisco Museum of Modern Art, 2005.

"The Photographic Gun." *Anthony's Photographic Bulletin* 13, no. 11 (June 1882): 170–71.

Rabinbach, Anson. *The Human Motor: Energy, Fatigue, and the Origins of Modernity.* Berkeley: University of California Press, 1992.

Rejlander, Oscar Gustave. "On Photographing Horses." *British Journal Photographic Almanac* (1872–1873): 115.

Riddle, Jefferson C. Davis. *The Indian History of the Modoc War.* With new introduction by Peter Cozzens. Mechanicsburg, PA: Stackpole Books, 2004. First published 1914 by Marnell.

Robertson, David. *West of Eden: A History of the Art and Literature of Yosemite.* Berkeley: Wilderness Press, 1984.

Rossell, Deac. *Ottomar Anschütz and His Electrical Wonder.* Hastings, UK: Projection Box, 1997.

Ruskin, John. *The Storm-Cloud of the Nineteenth Century: Two Lectures Delivered at the London Institution, February 4 and 11, 1884.* New York: John Wiley and Sons, 1884.

Sanborn, Helen J. *A Winter in Central America and Mexico.* Boston: Lee and Shepard; New York: C. T. Dillingham, 1886.

Scharf, Aaron. *Art and Photography.* New York: Penguin Books, 1968.

Schinto, Jeanne. "Deviled Ham Untouched by Hands: Food-Related Vintage Stereoviews." *Gastronomica: The Journal of Food and Culture* 2, no.4 (Fall 2002): 53–59.

Schwartz, Joan M. "G. R. Fardon, Photographer of Early Vancouver." *Afterimage* 6, no. 5 (December 1978): 5, 21.

Schwartz, Vanessa R. *Spectacular Realities: Early Mass Culture in Fin-de-Siècle Paris.* Berkeley: University of California Press, 1998.

______. "Cinematic Spectatorship Before the Apparatus: The Public Taste for Reality in Fin-de-Siècle Paris." In Leo Charney and Vanessa R. Schwartz, *Cinema and the Invention of Modern Life,* 297–319. Berkeley: University of California Press, 1995.

Scott, Amy, ed. *Yosemite: Art of an American Icon.* Berkeley: University of California Press, 2006.

Shaw, Jonathan. *Time/Motion: Photographs by Eadweard Muybridge, Harold Edgerton, and Jonathan Shaw.* Stockport, UK: Dewi Lewis Publishing, 2004.

Souriau, Paul. *The Aesthetics of Movement.* Amherst: University of Massachusetts Press, 1983.

Spiley, Louis Walton, *Photography's Great Inventors.* Philadelphia: American Museum of Photography, 1965.

Street, Richard Steven. *Photographing Farmworkers in California.* Stanford, CA: Stanford University Press, 2004.

Tissandier, Gaston. "Les allures du cheval." *La Nature,* December 14, 1878.

Tutorow, Norman E. *Leland Stanford: Man of Many Careers.* Menlo Park, CA: Pacific Coast Publishers, 1971.

Vance, Robert H. *Catalogue of Daguerreotype Panoramic Views in California.* New York: Baker, Godwin, 1851.

Ventura, Anita. "Gold Rush Panoramas, San Francisco—1850 to 1853." *Sea Letter* 2, nos. 2–3 (October 1964).

Weber, Eva. *Great Photographers of the American West.* Greenwich, CT: Brompton Books, 1993.

Westervelt, Ellsworth. "A Trip to Central America." *Scribner's Monthly* 15, no. 5 (March 1878): 609–23.

Williams, J. W. "The Butterfield Overland Mail Road Across Texas." *Southwestern Historical Society Quarterly* 61, no. 1 (July 1957): 14–18.

Williams, Linda. *Hard Core: Power, Pleasure, and the "Frenzy of the Visible."* Berkeley: University of California Press, 1999.

Williams, Robert Gregory. *States and Social Evolution: Coffee and the Rise of National Governments in Central America.* Chapel Hill: University of North Carolina Press, 1994.

"Wing Feathers. How They Move While the Wing Is Stationary." *New York Times,* August 20, 1885, 2.

Wolf, Daniel. *The American Space: Meaning in Nineteenth-Century Landscape Photography.* Middletown, CT: Wesleyan University Press, 1983.

Selection of Related Books and Articles about the History of Photography

Abney, Sir William de W. *Instantaneous Photography.* New York: Scoville and Adams, 1895.

Aubenas, Sylvie. *Gustave Le Gray, 1820–1884.* Los Angeles: Getty Trust Publications, 2002.

Aubenas, Sylvie, and André Gunthert. *La révolution de la photographie instantanée, 1880–1900.* Paris: Bibliothèque nationale de France and Société française de photographie, 1996.

Baldwin, Gordon, Malcolm Daniel, and Sarah Greenough. *All the Mighty World: The Photographs of Roger Fenton, 1852–1860.* New Haven, CT: Yale University Press, 2004.

Barthes, Roland. *Camera Lucida: Reflections on Photography.* Translated by Richard Howard. New York: Hill and Wang, 1981.

Bayer, Jonathan. "The Panoramic Image in Photography." In *The Panoramic Image,* 29–32. Southampton, UK: University of Southampton, John Hansard Gallery, 1981.

Blanchard, Valentine. "Instantaneous Photography Thirty Years Ago." *Photographic Quarterly* 2, no. 6 (1891): 147–57.

______. "On Instantaneous Photography." *British Journal of Photography* 10 (January 1, 1863): 5–6.

______. "Stereographs: Instantaneous Views of London Photographed by Valentine Blanchard." *British Journal of Photography* 9 (October 15, 1862): 381–82.

Blanchere, Henri de la. "Instantaneous Photography." *British Journal of Photography* 12 (May 26, 1865): 275–76.

Claudet, Antoine. "Instantaneous Photography a Quarter of a Century Ago." *Photographic News* 12 (May 22, 1868): 249.

Coe, Brian. *George Eastman and the Early Photographers.* London: Priory Press, 1973.

Crompton, Dennis, Richard Franklin, and Stephen Herbert, eds. *Servants of Light: The Book of the Lantern.* Leicester, UK: Magic Lantern Society, 1997.

Dallmeyer, J. H., Ltd. *The Early History of the House of Dallmeyer: Giving an Outline of the Progress Made in Photographic Optics by J. H. Dallmeyer and His Successors.* London: J. H. Dallmeyer, 1926.

Daniel, Malcolm R. *Edgar Degas, Photographer.* New York: Harry N. Abrams, 1998.

Darrah, William C. *The World of Stereographs.* 1977. Reprint, Chambersburg, PA: Alan C. Hood, 1998.

Dawson, George. "Notes on Instantaneous Shutters." *British Journal of Photography* 9 (December 1, 1862): 445–46.

______. "On Instantaneous Photography." *British Journal of Photography* 10 (April 1, 1863): 138–39.

Despratz, l'Abbé. "On Instantaneous Photography." *British Journal of Photography* 9 (November 15, 1861): 404.

Edkins, Diana. *Panoramic Photography.* New York: Grey Art Gallery and Study Center, 1977.

Edwards, Steve. *The Making of English Photography: Allegories.* University Park: Pennsylvania State University Press, 2006.

Ewing, James. "On the Manipulation, Developing, and Intensifying of Negatives." *British Journal of Photography* 10 (February 16, 1863): 71–73.

Fels, Thomas Weston. *Watkins to Weston: 101 Years of California Photography, 1849–1950.* Santa Barbara: Santa Barbara Museum of Art, 1992.

Fry, Samuel. "Instantaneous Photography." *Photographic News* 5 (October 25, 1861): 496–97.

Galassi, Peter. *Before Photography: Painting and the Invention of Photography.* Boston: New York Graphic Society, 1981.

Gernsheim, Helmut. *The History of Photography, from the Earliest Use of the Camera Obscura in the Eleventh Century up to 1914.* London: Oxford University Press, 1955.

______. *The Rise of Photography, 1850–1880: The Age of Collodion.* London: Thames and Hudson, 1988.

______. *Roger Fenton: Photographer of the Crimean War.* London: Secker and Warburg, 1954.

Gernsheim, Helmut and Alison. *Masterpieces of Victorian Photography, 1840–1900, from the Gernsheim Collection.* London: Phaidon Press, 1951

Green, David, ed. *Stillness and Time: Photography and the Moving Image.* Brighton, UK: Photoworks/Photoforum Press, 2006.

Green-Lewis, Jennifer. *Framing the Victorians: Photography and the Culture of Realism.* Ithaca, NY: Cornell University Press, 1996.

Grubb, Howard. "Instantaneous Pocket Cameras." *British Journal of Photography* 18 (March 3, 1871): 99.

Gunthert André. "La conquête de l'instantané: Archéologie de l'imaginaire photographique en France (1841–1895)." PhD diss., École des Hautes Études en Sciences Sociales, 1999.

Hannavy, John, ed. *Encyclopedia of Nineteenth-Century Photography.* New York: Taylor & Francis Group, 2008.

Herschel, Sir John F. W. "Instantaneous Photography." *Photographic News* 4 (May 1860): 1.

Highley, Samuel. "Instantaneous Shutters." *British Journal of Photography* 8 (March 15, 1861): 104–5, 128–30.

Hirsch, Robert. *Seizing the Light: A Social History of Photography.* New York: McGraw-Hill Higher Education, 2008.

"Instantaneous Photography v. Eyesight." *British Journal of Photography* 21 (July 1882).

Jenkins, Reese V. *Images and Enterprise: Technology and the American Photographic Industry, 1839 to 1925.* Baltimore: Johns Hopkins University Press, 1975.

Johnson, Drew Heath, ed. *Capturing Light: Masterpieces of California Photography, 1850 to the Present.* Oakland: Oakland Museum of California; New York: W. W. Norton, 2001.

Keene, Alfred. "A New Rapid Dry Process." *British Journal of Photography* 9 (December 1, 1862): 446–47.

Keller, Corey. *Brought to Light: Photography and the Invisible, 1840–1900.* San Francisco: San Francisco Museum of Modern Art, in association with Yale University Press, 2008.

Kelsey, Robin, and Blake Stimson, eds. *The Meaning of Photography.* New Haven, CT: Yale University Press, 2008.

Kibble, John. "On Instantaneous Photography." *British Journal of Photography* 9 (October 15, 1862): 446–47.

Lloyd, Valerie. *Roger Fenton: Photographer of the 1850s.* London: South Bank Board, 1988.

Londe, Albert. *La photographie instantanée: Théorie et pratique.* Paris: Gauthier-Villars, 1886.

Lothrop, Eaton S. *A Century of Cameras from the Collection of the International Museum of Photography at George Eastman House.* Dobbs Ferry, NY: Morgan and Morgan, 1973.

Mann, A. "A Description of an Instantaneous Shutter." *British Journal of Photography* 9 (July 1, 1862): 243–45.

Marder, William, Estelle Marder, and Robert G. Duncan, ed. *Anthony: The Man, the Company, the Cameras—An American*

Photographic Pioneer. Plantation, FL: Pine Ridge Publishing, 1982.

Marien, Mary Warner. *Photography: A Cultural History*. Upper Saddle River, NJ: Prentice Hall, 2006.

"Monsieur Tissot's Instantaneous Shutter." *British Journal of Photography* 10 (January 1, 1863): 449.

Morris, Richard. *John Dillwyn Llewelyn, 1810–1882: The First Photographer in Wales*. Cardiff: Welsh Arts Council, 1980.

Munson, Doug. "The Practice of Wet-Plate Photography." In *The Documentary Photograph as a Work of Art: American Photographs, 1860–1876*, 33–38. Chicago: David and Alfred Smart Gallery, University of Chicago, 1976.

Naef, Weston J. *Photographers of Genius at the Getty*. Los Angeles: J. Paul Getty Museum, 2004.

Newhall, Beaumont. *The History of Photography*. New York: Museum of Modern Art, 1949.

______. *Photography: Essays & Images—Illustrated Readings in the History of Photography*. New York: Museum of Modern Art, 1980.

Olsen, Victoria. *From Life: Julia Margaret Cameron & Victorian Photography*. New York: Palgrave Macmillan, 2003.

"On Instantaneous Exposures, and Improved Means for Effecting Them." *British Journal of Photography* 13 (January 12, 1866): 13–14.

Osterman, Mark. *The Wet-Plate Process: A Working Guide*. Rochester, NY: Scully & Osterman, 1996.

Ostroff, Eugene. *Western Views and Eastern Visions*. Washington, DC: Smithsonian Institution Traveling Exhibition Service with the cooperation of the United States Geological Survey, 1981.

Petschler, H. "Instantaneous Shutter." *British Journal of Photography* 9 (September 1, 1861): 330.

"Photographs by Moonlight, and Sunlight Effects." *Photographic News* 14 (August 19, 1870): 385–86.

"Photography and American Art." *The Century* 34, no. 3 (July 1887): 476–77.

"Rapid Sea Views and Moonlight Pictures." *Photographic News* 14 (September 2, 1870): 418.

Reilly, James M. *The Albumen & Salted Paper Book: The History and Practice of Photographic Printing, 1840–1895*. Rochester, NY: Light Impressions, 1980.

"The Relation of Photography to Art." *Scientific American* 68, no. 25 (June 24, 1893): 386.

Respini, Eva. *Into the Sunset: Photography's Image of the American West*. New York: Museum of Modern Art, 2009.

Robinson, David. *The Lantern Image—Iconography of the Magic Lantern, 1420–1880*. Nutley, UK: Magic Lantern Society, 1993.

Robinson, David, Stephen Herbert, and Richard Crangle, eds. *Encyclopedia of the Magic Lantern*. Kirkby Malzeard, UK: Magic Lantern Society, 2001.

Robinson, Henry Peach. "The Changes in Photography during the Last Few Years." *British Journal of Photography* 11 (February 1, 1864): 39.

Robinson, Henry Peach, and Cherrill Nelson King. "Photographs by Moonlight." *Photographic News* 14 (August 26, 1870): 407.

Ryan, James R. *Picturing Empire: Photography and the Visualization of the British Empire*. Chicago: University of Chicago Press, 1997.

Sandweiss, Martha A. *Photography in Nineteenth-Century America*. New York: Harry N. Abrams, 1991.

______. *Print the Legend: Photography and the American West*. New Haven, CT: Yale University Press, 2002.

Schaaf, Larry J. *Out of the Shadows: Herschel, Talbot, and the Invention of Photography*. New Haven, CT: Yale University Press, 1992.

______. "Sir John Herschel's 1839 Royal Society Paper on Photography." *History of Photography* 3, no. 1 (January 1979): 47–60.

Scharf, Aaron. *Pioneers of Photography: An Album of Pictures and Words*. New York: Harry N. Abrams, 1976.

Schimmelman, Janice G. *American Photographic Patents: The Daguerreotype and Wet Plate Era, 1840–1880*. Nevada City, NV: Carl Mautz Publishing, 2002.

Sobieszek, Robert. *British Masters of the Albumen Print: A Selection of Mid-Nineteenth Century Victorian Photography*. Chicago: University of Chicago Press, 1976.

Sutton, Thomas. "A New Wet Collodion Process." *British Journal of Photography* 18 (March 10, 1871): 109.

Taft, Robert. *Photography and the American Scene: A Social History, 1839–1889*. New York: MacMillan Company, 1938.

"Talk in the Studio: Photography and Moonlight." *Photographic News* 14 (February 25, 1870): 95.

Taylor, Roger. *George Washington Wilson: Artist and Photographer, 1823–93*. Aberdeen, UK: University of Aberdeen Press, 1981.

______. *Impressed by Light: British Photographs from Paper Negatives, 1840–1860*. New Haven, CT: Yale University Press, 2007.

Thomas, Ann, ed. *Beauty of Another Order: Photography in Science*. New Haven, CT: Yale University Press, 1997.

Thompson, Stephen. "Instantaneity." *Photographic News* 15 (March 3, 1871): 102–3.

Trachtenberg, Alan. *Classic Essays on Photography*. New Haven, CT: Leete's Island Books, 1980.

______. *The Incorporation of America: Culture and Society in the Gilded Age*. 1982. Reprint, New York: Hill and Wang, 2007.

______. *Reading American Photographs: Images as History, Mathew Brady to Walker Evans*. New York: Hill and Wang, 1990.

Tucker, Jennifer. *Nature Exposed: Photography as Eyewitness in Victorian Science*. Baltimore: Johns Hopkins University Press, 2005.

Wagstaff, Samuel J. *American Stereographs: A Selection from Private Collections*. New York: Grey Art Gallery, 1980.

Weaver, Mike. *British Photography in the Nineteenth Century: The Fine Art Tradition*. Cambridge: Cambridge University Press, 1989.

Welling, William. *Photography in America: The Formative Years, 1839–1900*. Albuquerque: University of New Mexico Press, 1987. First published 1978 by Crowell.

Wing, Paul. *Stereoscopes: The First One Hundred Years*. Nashua, NH: Transition Publishing, 1996.

Selection of Related Books and Articles about the History of Cinema

Auerback, Jonathan. *Body Shots: Early Cinema's Incar-nations*. Berkeley: University of California Press, 2007.

Barnes, John. *The Beginnings of the Cinema in England, 1894–1901*. Exeter: University of Exeter Press, 1998.

Barnouw, Erik. *The Magician and the Cinema*. Oxford: Oxford University Press, 1981.

Burch, Noël. *Life to Those Shadows*. Berkeley: University of California Press, 1990.

Campany, David. *Photography and Cinema*. London: Reaktion Books, 2008.

Card, James. "Problems of Film History." *Hollywood Quarterly* 4, no. 3 (April 1950): 279–88.

Ceram, C. W. *Archaeology of the Cinema*. New York: Harcourt, Brace and World, 1965.

Chanan, Michael. *The Dream That Kicks: The Prehistory and Early Years of the Cinema in Britain*. London: Routledge, 1995.

Charney, Leo. "In a Moment: Film and the Philosophy of Modernity." In Leo Charney and Vanessa R. Schwartz, *Cinema and the Invention of Modern Life*, 279–94. Berkeley: University of California Press, 1995.

Coe, Brian. *The History of Movie Photography*. London: Ash & Grant, 1981.

Coissac, G.-Michel. *Histoire du Cinématographe de ses origines à nos jours*. Paris: Cinéopse/Gauthier-Villars, 1925

Demenÿ, Georges. *Les origines du cinématographe*. Paris: Henry Pailin, 1909.

Fell, John L., ed. *Film before Griffith*. Berkeley: University of California Press, 1983.

Fescourt, Henri. *Le cinéma, dès origines à nos jours*. Paris: Éditions du Cygne, 1932.

Fielding, Raymond, ed. *A Technological History of Motion Pictures and Television*. Berkeley: University of California Press, 1967.

Gosser, H. Mark. *Selected Attempts at Stereoscopic Moving Pictures and Their Relationship to the Development of Motion Picture Technology, 1852–1903*. New York: Arno, 1977.

Gunning, Tom. "An Aesthetic of Astonishment: Early Film and the [In]Credulous Spectator." In *Viewing Positions*, edited by Linda Williams, 114–33. New Brunswick, NJ: Rutgers University Press, 1995.

______. "Doing for the Eye What the Phonograph Does for the Ear." In *The Sounds of Early Cinema*, edited by Richard Abel and Rick Altman, 13–31. Bloomington: Indiana University Press, 2001.

Harding, Colin, and Simon Popple, eds. *In the Kingdom of Shadows: A Companion to Early Cinema*. London: Cygnus Arts, 1996.

Hecht, Ann and Hermann. *Pre-cinema History: An Encyclopaedia and Annotated Bibliography of the Moving Image before 1896. Volume I*. London: Bowker-Saur/BFI, 1993.

Hendricks, Gordon. *The Edison Motion Picture Myth*. Berkeley: University of California Press, 1961.

______. *The Kinetoscope: America's First Commercially Successful Motion Picture Exhibitor*. New York: Beginnings of the American Film, 1966.

______. *Origins of the American Film.* Reprint, New York: Arno, 1972. Containing: *The Edison Motion Picture Myth* (University of California Press, 1961); *Beginnings of the Biograph* (Hendricks, 1964); and *The Kinetoscope* (Hendricks, 1966).

Herbert, Stephen, ed. *A History of Early Film.* 3 vols. London: Routledge, 2000.

______. *A History of Pre-cinema.* 3 vols. London: Routledge, 2000.

______. *When the Movies Began....* London: Projection Box, 1994.

Herbert, Stephen, and Luke McKernan, eds. *Who's Who of Victorian Cinema: A Worldwide Survey.* London: British Film Institute, 1996.

Jenkins, Charles Francis. *Animated Pictures: An Exposition of the Historical Development of Chronophotography.* Washington, DC: C. Francis Jenkins, 1898.

Leffman, Henry. "The Invention of the Motion Picture." *Journal of the Franklin Institute* 207, no. 6 (June 1929): 825–28.

Liesegang, Franz Paul. *Dates and Sources: A Contribution to the History of the Art of Projection and to Cinematography.* Translated and edited by Hermann Hecht. London: Magic Lantern Society of Great Britain, 1986.

Macgowan, Kenneth. *Behind the Screen: The History and Techniques of the Motion Picture.* New York: Delacorte Press, 1965.

Mannoni, Laurent. *The Great Art of Light and Shadow: Archaeology of the Cinema.* Translated and edited by Richard Crangle. Exeter: University of Exeter Press, 2000.

Mannoni, Laurent, Donata Pesenti Campagnoni, and David Robinson, eds. *Light and Movement: Incunabula of the Motion Picture, 1420–1896.* Pordenone: Le Giornate del Cinema Muto, 1995.

Musser, Charles. *Before the Nickelodeon: Edwin S. Porter and the Edison Manufacturing Company.* Berkeley: University of California Press, 1991.

______. *The Emergence of Cinema: The American Screen to 1907.* Berkeley: University of California Press, 1994.

Penley, Constance. "The Imaginary of the Photograph in Film Theory." *Photographies,* no. 4 (April 1984): 122–23.

Pfragner, Julius. *The Motion Picture from Magic Lantern to Sound Film.* Translated and edited by Theodore McClintock. Folkestone, UK: Bailey Bros. & Swinfen, 1974.

Quigley, Martin, Jr. *Magic Shadows.* Washington, DC: Georgetown University Press, 1948.

Ramsaye, Terry. *A Million and One Nights: A History of the Motion Picture.* New York: Simon and Schuster, 1926.

Robinson, David. *From Peep Show to Palace: The Birth of American Film.* New York: Columbia University Press, 1996.

Rosen, Philip. *Narrative, Apparatus, Ideology: A Film Theory Reader.* New York: Columbia University Press, 1986.

Rossell, Deac. *Living Pictures: The Origins of the Movies.* Albany: State University of New York Press, 1998.

______. "The New Thing with the Long Name and the Old Thing with the Name That Isn't Much Shorter: A Cinema Chronology, 1889–1896." *Film History* 7, no. 2 (Summer 1995).

Sadoul, Georges. *Histoire générale du cinéma: Part 1—L'invention du cinéma: 1832–1897.* Paris: Denoël, 1946.

Stewart, Garrett. *Between Film and Screen: Modernism's Photo Synthesis.* Chicago: University of Chicago Press, 2000.

Tosi, Virgilio. *Cinema before Cinema: The Origins of Scientific Cinematography.* London: British Universities Film & Video Council, 2005.

Von Zglinicki, Friedrich. *Der Weg des Films. V. 1: Textband; V. 2: Bildband.* Berlin: Rembrandt Verlag, 1956; reprint, Oms Press, 1979.

Waldekranz, Rune. *Sa foddes filmen: Ett massmediums uppkomst och genombrott.* Stockholm: Bokforlaget Pan/Norstedts, 1976.

Zone, Ray. *Stereoscopic Cinema and the Origins of 3-D Film, 1838–1952.* Lexington: University Press of Kentucky, 2007.

Selected Books and Articles Related to Theories of Time

Bergson, Henri. *Matter and Memory.* New York: Cosimo Classics, 2007.

______. *Time and Free Will: An Essay on the Immediate Data of Consciousness.* Mineola, NY: Dover Publications, 2001.

Borges, Jorge Luis. "A New Refutation of Time." In *Borges: Selected Non-Fictions.* London: Penguin, 2000.

Frampton, Hollis. "Incisions in History/Segments of Eternity." *Art Forum* (October 1974): 38–50.

Kern, Stephen. *The Culture of Time and Space, 1880–1918.* Cambridge, MA: Harvard University Press, 1983.

Newton-Smith, W. H. *Structure of Time.* London: Routledge and Kegan, 1980.

Selected Films about Eadweard Muybridge

Andersen, Thom, dir. *Eadweard Muybridge, Zoopraxographer.* 60 minutes. 1975.

Hanson, David, dir. *Homage to Eadweard Muybridge.* 2 minutes. Kino International, 1994.

Sheldon, James, dir. *Eadweard Muybridge.* 20 minutes. Voyager Company and Addison Gallery of American Art, 1991.

Selected Web Sites Devoted to Eadweard Muybridge

The Bancroft Library, University of California, Berkeley. *Lone Mountain College Collection of Stereographs by Eadweard Muybridge, 1867–1880.* http://www.oac.cdlib.org/findaid/ark:/13030/tf6t1nb6w7/

Herbert, Stephen. *The Compleat Eadweard Muybridge.* http://www.stephenherbert.co.uk/muybCOMPLEAT.htm

Kingston Museum, The Royal Borough of Kingston upon Thames. *Muybridge Collection.* http://www.kingston.gov.uk/museum/muybridge

Kingston Museum and Heritage Service, and Kingston University. *Eadweard Muybridge: Defining Modernities.* http://www.eadweardmuybridge.co.uk/

Museum of American History, Smithsonian Institution, Washington, DC. *Freeze Frame: Eadweard Muybridge's Photography of Motion.* http://americanhistory.si.edu/muybridge/index.htm

University Archives and Records Center, University of Pennsylvania, Philadelphia. *Guide to the Eadweard Muybridge, 1830–1904, Collection, 1870–1981.* http://www.archives.upenn.edu/faids/upt/upt50/muybridgee.html

List of Illustrations

All photographs are by Eadweard Muybridge unless otherwise stated. Most titles have been taken directly from the mounts of the prints, Muybridge's *Catalogue of Photographic Views* (published by Bradley and Rulofson in 1873), and from *Animal Locomotion, Prospectus and Catalogue of Plates* (1887).

Works exhibited at Tate Britain are marked with an asterisk (*). See p. 353 for a list of all works exhibited.

Andrew J. Russell
East and West shaking hands at laying last rail, 1869, from Andrew J. Russell, *Photographs taken during construction of the Union Pacific Railroad*, 1864–1869
Yale Collection of Western Americana, Beinecke Rare Book & Manuscript Library
[Page 49, fig. 35]

Shoshone Indians at Corinne (754), 1869 (*)
Two albumen silver prints on studio card
Collection of Leonard A. Walle
[Page 51, fig. 36]

Great American Overland Route—Yachting on San Francisco Bay, near San Rafael (380), 1873
Albumen silver print on printed card
Collection of Leonard A. Walle
[Page 51, fig. 37]

Chang Woo Gow. Celebrated Chinese giant. Virginia, Nevada, August 31, 1870
Two albumen silver prints on card
Special Collections, University of Nevada-Reno Library
[Page 52, fig. 38]

Art Gallery at Woodward's Gardens, 1870
Two albumen silver prints on studio card
Courtesy of The Bancroft Library, University of California, Berkeley, 1905.01213
[Page 53, fig. 39]

First-Order Light-house at Punta de los Reyes, Seacoast of California, 296 Feet Above Sea (4136), 1871 (*)
Albumen silver print
U.S. Coast Guard Historian's Office
[Page 54, fig. 40]

Grand Masonic Ceremony Laying the Cornerstone of the City Hall and Law Courts, 1872
Albumen silver print on cabinet card
Collection of California Historical Society, Virginia M. Storti Collection
[Page 57, fig. 41]

Base of Yowiye Fall (1318), 1872
Two albumen silver prints on studio card
Corcoran Gallery of Art, Washington, D.C., Anonymous Gift, 2009.025.001
[Page 57, fig. 42]

Timothy O'Sullivan
Ancient Ruins in the Cañon De Chelle, N.M., In a niche 50 feet above present Cañon bed, 1873, from *Geographical and Geological Explorations and Surveys West of the 100th Meridian, Seasons of 1871, 1872, and 1873*
Albumen silver print bound in album
Corcoran Gallery of Art, Washington, D.C., Gift of William Wilson Corcoran, 86.11a-xx
[Page 58, fig. 43]

Ancient Glacier Boulders at Lake Tenaya (1545), 1872
Two albumen silver prints on studio card
Robert N. Dennis Collection of Stereoscopic Views, Miriam and Ira D. Wallach Division of Art, Prints and Photographs, The New York Public Library, Astor, Lenox and Tilden Foundations
[Page 59, fig. 44]

Albert Bierstadt's Studio (1586), 1872 (*)
Two albumen silver prints on studio card
Collection of Leonard A. Walle
[Page 60, fig. 45]

Thomas Houseworth & Co.
Thomas Houseworth's store at No. 9 Montgomery Street, San Francisco, ca. 1873, from *California Scenery. A Series of Stereoscopic Views of most of the points of interest on the Pacific Coast*
Albumen silver print
Gift of Florence V. Flinn, The Society of California Pioneers
[Page 62, fig. 46]

The Domes, 1872, plate 52 from *Sun Pictures of the Yo-Semite Valley, Cal*, Coyne and Relyae, publishers, 1874
Huntington Library, Art Collections
[Page 63, fig. 47]

Looking Down Yo-Semite Valley, 1872, plate 59 from *Sun Pictures of the Yo-Semite Valley, Cal*, Coyne and Relaye, publishers, 1874
Huntington Library, Art Collections
[Page 63, fig. 48]

Carleton Watkins (attributed)
The Domes from the Merced River, Yosemite, ca. 1872
Albumen silver print
Courtesy of Fraenkel Gallery, San Francisco
[Page 64, fig. 49]

Yowiye, Nevada Fall, 600 feet fall (4201), 1872 (Verso)
Albumen silver print on cabinet card
Corcoran Gallery of Art, Washington, D.C., Museum Purchase, 2010.002.002
[Page 64, fig. 50; For Recto, see p.180, fig. 95]

Captain Jack's Cave in the Lava Beds (1602), 1873 (*)
Two albumen silver prints on studio card
Collection of Leonard A. Walle
[Page 65, fig. 51]

Panorama of Lava Beds from Signal Station at Tule Lake Camp, South (1604), 1873
Two albumen silver prints on studio card
Collection of Leonard A. Walle
[Page 65, fig. 52]

One-Eyed Dixie and Modoc Squaws (1625), 1873
Two albumen silver prints on studio card
Robert N. Dennis Collection of Stereoscopic Views, Miriam and Ira D. Wallach Division of Art, Prints and Photographs, The New York Public Library, Astor, Lenox and Tilden Foundations
[Page 66, fig. 53]

Pictures from the lava beds, from "The Last of the Modocs," *Harper's Weekly*, June 21, 1873 (*)
Engraving on newsprint
Corcoran Gallery of Art, Washington, D.C.
Image courtesy Picture Collection, The New York Public Library, Astor, Lenox, and Tilden Foundations
[Page 66, fig. 54]

Flora Muybridge, 1874 (*)
Two albumen silver prints on studio card
Collection of Leonard A. Walle
[Page 68, fig. 55]

The Brandenburg Album of Bradley & Rulofson "Celebrities" and Muybridge Photographs, page 98, 1874 (*)
Albumen silver prints mounted on paper
Iris & B. Gerald Cantor Center for Visual Arts at Stanford University, Museum Purchase Fund, 1972.9.98
[Page 69, fig. 56]

Timothy O'Sullivan
Historic Spanish Record of the Conquest, South Side of Inscription Rock, N.M. No. 3, 1873, from *Geographical and Geological Explorations and Surveys West of the 100th Meridian, Seasons of 1871, 1872, and 1873*
Albumen silver print bound in album
Corcoran Gallery of Art, Washington, D.C.
Gift of William Wilson Corcoran, 86.11a-xx
[Page 70, fig. 57]

Coffee Harvesting at San Isidro, elevation 1000 feet, Guatemala, 1875 (*)
Albumen silver print
Smithsonian American Art Museum, Gift of Jeffrey A. Lichtenberg in honor of his parents Laura and Harold Lichtenberg
[Page 73, fig. 58]

Title page, from *The Pacific Coast of Central America and Mexico; the Isthmus of Panama; Guatemala; and the cultivation and shipment of coffee, illustrated by Muybridge*, 1876
Albumen silver print
Courtesy Special Collections, Stanford University Libraries
[Page 74, fig. 59]

Panorama of San Francisco from California Street Hill, 1877 (*)
Albumen silver print mounted on boudoir card
Wilson Centre for Photography
[Page 75, fig. 60]

View Looking South and West from the Windows of Leland Stanford's House, 1877
Albumen silver print mounted on India-tinted board
The J. Paul Getty Museum, Los Angeles
[Page 75, fig. 61]

Kate and Robert Johnson Photograph Album, 1880 (*)
84 albumen silver prints in album
William L. Clements Library, University of Michigan
[Page 76, fig. 62; see also pages 245–250, plates 119–124]

Detail from Plate 7 of *Panorama of San Francisco from California Street Hill, San Francisco, California*, 1878
For full details, see pages 240–244, plate 118 below
[Page 77, fig. 63]

Unknown photographer (possibly Eadweard Muybridge)
Harness Racing, ca. 1876
Courtesy of The Bancroft Library, University of California, Berkeley
[Page 79, fig. 64]

Currier & Ives
The California Wonder Occident, owned by Gov. L. Stanford, 1873
Lithograph
Prints & Photographs Division, Library of Congress, Washington, D.C.
[Page 79, fig 65]

John Koch
Occident Trotting, 1877 (*)
Watercolor with gouache
Iris & B. Gerald Cantor Center for Visual Arts at Stanford University, Stanford Family Collections, 1983.48
[Page 80, fig. 66]

"Occident," Owned by Leland Stanford; trotting at a 2:30 gait over the Sacramento track, in July 1877, 1877
Albumen silver print
Iris & B. Gerald Cantor Center for Visual Arts at Stanford University, Stanford Family Collections, 19737
[Page 80, fig. 67]

Detail of the Track at Leland Stanford's Horse Farm in Palo Alto, 1878–79
Albumen silver print
Corcoran Gallery of Art, Washington, D.C.
Gift of Charles Isaacs
[Page 82, fig. 68]

Unknown photographer
Interior of the Camera Shed at Leland Stanford's Horse Farm in Palo Alto, no date
Albumen silver print
Corcoran Gallery of Art, Washington, D.C.
Gift of Charles Isaacs
[Page 83, fig. 69]

Patent Model for Method and Apparatus for Photographing Objects in Motion, 1879
Wood, metal
Smithsonian Institution, National Museum of American History, Kenneth E. Behring Center
[Page 84, fig. 70]

The Horse in Motion, illus. by Muybridge. "Sallie Gardner," owned by Leland Stanford, running at a 1:40 gait over the Palo Alto track, 19th June 1878: 2 frames showing diagram of foot movements, 1878 (*)
Albumen silver print on cabinet card
Prints & Photographs Division, Library of Congress, Washington, D.C.
[Page 85, fig. 71]

The Stride of the Trotting Horse, "Abe Edgington," owned by Leland Stanford; being driven by Marvin, trotting at a 2:24 gait over Mr. Stanford's Race Track at Menlo Park, on 11th June, 1878, 1878 (*)
12 albumen silver prints on cabinet card
Wilson Centre for Photography
[Page 85, fig. 72]

Detail of the Track at Leland Stanford's Horse Farm in Palo Alto, 1879
Albumen silver print
Corcoran Gallery of Art, Washington, D.C.
[Page 86, fig. 73]

Horses. Leaping a 3 Ft. 6 Hurdle. Phryne L. Plate 49, 1879, from *The Attitudes of Animals in Motion*, 1881 (*)
Albumen silver print
Courtesy Special Collections, Stanford University Libraries
[Page 87, fig. 74]

Studies of Foreshortenings. Horses. Running. Mahomet. Plates 143–144, 1879, from *The Attitudes of Animals in Motion*, 1881 (*)
Albumen silver print
Chrysler Museum of Art, Norfolk, Virginia, Museum Purchase
[Page 87, fig. 75]

Thomas Eakins
A May Morning in the Park (The Fairman Rogers Four-in-Hand), 1879–80
Oil on canvas
Philadelphia Museum of Art
Gift of William Alexander Dick, 1930
[Page 88, fig. 76]

Zoopraxiscope, 1879–85 (*)
Mixed media projector
By permission of Kingston Museum and Heritage Service
[Page 89, fig. 77]

Mule Bucking and Kicking, 12 phases, (verso), 1893
Glass zoopraxiscope disc, silver gelatin emulsion, paint, metal
Courtesy of George Eastman House, International Museum of Photography and Film
[Page 90, fig. 78]

Mule Bucking and Kicking, 12 phases, No. 29, 1893
Glass zoopraxiscope disc negative, silver gelatin emulsion
Smithsonian Institution, National Museum of American History, Kenneth E. Behring Center
[Page 91, fig. 79]

The Attitudes of Animals in Motion: A Series of Photographs Illustrating the Consecutive Positions Assumed by Animals in Performing Various Movements, 1881 (*)
174 albumen silver prints mounted on paper in bound volume
Partial Gift of The Beinecke Foundation, Inc. Addison Gallery of American Art, Phillips Academy, Andover, Massachusetts
[Page 92, fig. 80]

Horses. Running. Phryne L. Plate 40, 1879, from *The Attitudes of Animals in Motion*, 1881 (*)
Albumen silver print
National Gallery of Art, Washington, Gift of Mary and Dan Solomon 2006.131.7
Image Courtesy of The Board of Trustees
[Page 92, fig. 81]

Phryne L—Running. Stride. 19 ft. 9 in. Plate XVI, from J. D. B. Stillman, *The Horse in Motion as Shown by Instantaneous Photography, with a Study on Animal Mechanics Founded on Anatomy and the Revelations of the Camera, in Which Is Demonstrated the Theory of Quadrupedal Locomotion*, 1881–82
Lithograph in book
National Gallery of Art, Gift of Mary and Dan Solomon
Image courtesy of the Board of Trustees, National Gallery of Art, Washington, D.C.
[Page 93, fig. 82]

Thomas Eakins
Marey Wheel Photographs of Unidentified Model, with Eadweard Muybridge Notations, 1884
Albumen silver print on paper
Hirshhorn Museum and Sculpture Garden, Smithsonian Institution, Gift of Joseph H. Hirshhorn, 1966
Photography by Lee Stalsworth
[Page 94, fig. 83]

A, striking a blow; B, throwing disk; C, heaving a 75-lb. stone; D, throwing a ball; E, throwing disk; F, heaving 75-lb. stone. Plate 523, 1887
Collotype on paper
Corcoran Gallery of Art, Washington, D.C., Museum Purchase, 87.7.479
[Page 94, fig. 84]

Electronic Timing Device Used by Eadweard Muybridge at University of Pennsylvania for Animal Locomotion, 1885
Metal, wire
Smithsonian Institution, National Museum of American History, Kenneth E. Behring Center
[Page 95, fig. 85]

Eadweard Muybridge's outdoor camera house, 36th and Pine Streets, Philadelphia, ca. 1885
Gelatin silver print
Collections of the University of Pennsylvania Archives
[Page 96, fig. 86]

Infantile paralysis; child, walking on hands and feet. Plate 539, 1887 (*)
Collotype on paper
Corcoran Gallery of Art, Washington, D.C.
Museum Purchase, 87.7.491
[Page 96, fig. 87]

Boxing; 53 knocking 52 down (shoes). Plate 333, 1887
Collotype on paper
Corcoran Gallery of Art, Washington, D.C.
Museum Purchase,
87.7.318
[Page 97, fig. 88]

Jean-Louis-Ernest Meissonier
Campagne de France, 1814, 1864
Oil on wood
Musée d'Orsay, Paris, Courtesy Réunion des Musées Nationaux / Art Resource, NY
[Page 98, fig. 89]

Frederic Remington
Off the Range (Coming Through the Rye), 1902
Bronze with green patina
Corcoran Gallery of Art, Washington, D.C.
Museum Purchase, 05.7
[Page 98, fig. 90]

Descriptive Zoopraxography, or the Science of Animal Locomotion Made Popular (frontispiece), 1893 (*)
Book
Museum Purchase. Addison Gallery of American Art, Phillips Academy, Andover, Massachusetts
[Page 100, fig. 91]

Father of the Forest (Est). 450 feet high, 112 feet circumference (130), 1867
Two albumen silver prints on studio card
Collection of California Historical Society
[Page 110, plate 6]

Yosemite Tourists, 1867
Two albumen silver prints on studio card
Collection of California Historical Society
[Page 112, plate 7]

Yosemite Hotel (59), 1867
Two albumen silver prints on studio card
Corcoran Gallery of Art, Washington, D.C., Anonymous Gift, 2009.025.003
[Page 113, plate 8]

Pioneers' Cabin, 318 ft. high, 73 ft. circumference (125), 1867
Two albumen silver prints on studio card
Collection of Leonard A. Walle
[Page 113, plate 9]

Fissure in Eagle Rock. 1100 ft. deep, 1867 (*)
Two albumen silver prints on studio card
Collection of Leonard A. Walle
[Page 114, plate 10]

California—Yosemite Valley. Iris from the Summit of Pi-wy-ack (297), 1867 (*)
Two albumen silver prints on studio card
Collection of Leonard A. Walle
[Page 115, plate 11]

Summit of Third Fall of the Yosemite (4035), 1867
Albumen silver print
Iris & B. Gerald Cantor Center for Visual Arts at Stanford University, Elizabeth K. Raymond Fund, 19747
[Page 116, plate 12]

Kahchoomah, Wild Cat Fall, 30 feet high, on the main fork of the Merced (4061), 1867
Albumen silver print
Iris & B. Gerald Cantor Center for Visual Arts at Stanford University, Elizabeth K. Raymond Fund, 19742
[Page 117, plate 13]

An Hour Before Sunset, Bay of San Francisco, 1868
Two albumen silver prints on studio card
Museum Purchase. Addison Gallery of American Art, Phillips Academy, Andover, Massachusetts
[Page 118, plate 14]

First ME Church, Powell Street, San Francisco, 1871
Albumen silver print
Fine Arts Museums of San Francisco, Gift of William A. Trubody
[Page 119, plate 15]

Panorama, from Rincon Hill (428), 1868
Two albumen silver prints on studio card
Collection of Leonard A. Walle
[Page 120, plate 16]

Panorama, from Rincon Hill (434), 1868
Two albumen silver prints on studio card
Courtesy of The Bancroft Library, University of California, Berkeley, 1971.055, Series 2
[Page 121, plate 17]

Military Post at Goat Island (4108), 1870
Albumen silver print
Collection of California Historical Society, Virginia M. Storth Collection
[Page 122, plate 18]

The Steamer "Golden City" on the California Dry Dock (4123), 1868
Albumen silver print
Courtesy of The Bancroft Library
University of California, Berkeley
[Page 123, plate 19]

Fort Wrangle, 1868
Albumen silver print
Courtesy of The Bancroft Library, University of California, Berkeley, 1905.17137:146–PIC
[Page 124, plate 20]

Fort Tongass, 1868
Albumen silver print
Courtesy of The Bancroft Library, University of California, Berkeley, 1905.17137:143–PIC
[Page 125, plate 21]

Group of Indians, Alaska (479), 1868 (*)
Two albumen silver prints on studio card
Collection of Leonard A. Walle
[Page 126, plate 22]

Alaska Ter.—Sitka. Russo Greek Priests, Alaska (472), 1868 (*)
Two albumen silver prints on studio card
Collection of Leonard A. Walle
[Page 127, plate 23]

Cut at Alta, looking East (787), 1869 (*)
Two albumen silver prints on studio card
California History Section, California State Library
[Page 128, plate 24]

Long Ravine Trestle and Bridge—113 feet high, 878 feet long, looking East (766), 1869 (*)
Two albumen silver prints on studio card
Collection of Leonard A. Walle
[Page 128, plate 25]

View from Cape Horn, looking South, Mineral Bar Bridge 2,500 feet below (779), 1869
Two albumen silver prints on studio card
Robert N. Dennis Collection of Stereoscopic Views, Miriam and Ira D. Wallach Division of Art, Prints and Photographs, The New York Public Library, Astor, Lenox and Tilden Foundations
[Page 129, plate 26]

Coming storm—Wasatch Mountains, from the Weber River, near Ogden (756), 1869
Two albumen silver prints on studio card
Collection of Leonard A. Walle
[Page 129, plate 27]

William Herman Rulofson and Family Group at His San Francisco Home, 1871
Two albumen silver prints on studio card
Collection of California Historical Society, Gift of Alfred Currie Rulofson, Jr.
[Page 130, plate 28]

Progress of Construction, U.S. Branch Mint, 1870
Albumen silver print
Collection of California Historical Society
[Page 131, plate 29]

Progress of Construction, U.S. Branch Mint, 1871
Albumen silver print
Collection of California Historical Society
[Page 132, plate 30]

Progress of Construction, U.S. Branch Mint, 1871
Albumen silver print
Collection of California Historical Society
[Page 133, plate 31]

Geysers of Cal., 1870, from *Photo Album No. 9*, page 27, ca. 1873
72 albumen silver prints in album
Collection of California Historical Society
[Page 134, plate 32]

The "Heathen Chinee" Prospecting (848), 1868 (*)
Two albumen silver prints on studio card
Collection of Leonard A. Walle
[Page 135, plate 33]

Photo Album No. 9, page 36, ca. 1873
72 albumen silver prints in album
Collection of California Historical Society
[Page 136, plate 34]

Photo Album No. 9, page 43, ca. 1873
72 albumen silver prints in album
Collection of California Historical Society
[Page 137, plate 35]

Animals at Woodward's Gardens (589), 1870 (*)
Two albumen silver prints on studio card
Collection of Leonard A. Walle
[Page 138, plate 36]

Views in Woodward's Gardens, 1870 (*)
Two albumen prints mounted on studio card
Museum Purchase. Addison Gallery of American Art, Phillips Academy, Andover, Massachusetts
[Page 139, plate 37]

Point Shubrick, Tower Head, Lighthouse, Parrot Rock, and Gull Peak from the Hole in the Wall, South Farallon Island, 1871 (*)
Albumen silver print
U.S. Coast Guard Historian's Office
[Page 140, plate 38]

The Ramparts, Funnel Rock, Hole in the Wall, Pyramid, Sugar Loaf, Oil House, and Landing Cove on Fisherman's Bay, South Farallon Island, 1871 (*)
Albumen silver print
U.S. Coast Guard Historian's Office
[Page 141, plate 39]

Cape Disappointment Lighthouse, Mouth of Columbia River, First Order Fixed Light, 232 feet above Sea Level (4164), 1871
Albumen silver print
U.S. Coast Guard Historian's Office
[Page 142, plate 40]

Cape Blanco, Lighthouse and Dwelling (4161), 1871
Albumen silver print
U.S. Coast Guard Historian's Office
[Page 143, plate 41]

Point Loma Lighthouse, 1871
Albumen silver print
National Archives, Washington, D.C.
[Page 144, plate 42]

Light House at Punta de los Reyes, Coast of California (8074), 1871
Albumen silver print
National Archives, Washington, D.C.
[Page 145, plate 43]

Pigeon Point Lighthouse, 1873 (*)
Albumen silver print
California History Section, California State Library
[Page 146, plate 44]

Pigeon Point Lighthouse, 1873 (*)
Albumen silver print
California History Section, California State Library
[Page 147, plate 45]

City Hall Under Construction, 1873
Albumen silver print
California History Section, California State Library
[Page 148, plate 46]

Mills Seminary, Seminary Park, Alameda Co., Cal., 1873
Albumen silver print
Collection Centre Canadien d'Architecture
Canadian Centre for Architecture, Montreal
[Page 149, plate 47]

View of South Park, San Francisco, ca. 1873
Half-stereo albumen silver print on original mount
San Francisco Museum of Modern Art, Gift of Mr. and Mrs. John B. Ritchie
[Page 150, plate 48]

Saucilito from the N.P.C.R.R. Wharf, looking South (1664), 1873
Albumen silver print (half stereo)
Lent by The Metropolitan Museum of Art, Purchase, The Horace W. Goldsmith Foundation Gift, through Joyce and Robert Menschel, 1991 (1991.1203)
Image copyright © The Metropolitan Museum of Art, Art Resource, NY
[Page 151, plate 49]

Buena Vista Vineyard, Sonoma. Fertilizing the Soil (1091), 1871–72
Two albumen silver prints on studio card
Courtesy of The Bancroft Library, University of California, Berkeley, 1988.103.1
[Page 152, plate 50]

Buena Vista Vineyard, Sonoma. Plowing (1092), 1871–72
Two albumen silver prints on studio card
Courtesy of The Bancroft Library, University of California, Berkeley, 1988.103.2
[Page 152, plate 51]

Buena Vista Vineyard, Sonoma (4167), 1871–72, from Lone Mountain College Collection, Album 3, 1873
Albumen silver prints mounted in album
Courtesy of The Bancroft Library, University of California, Berkeley, 1971.055, Series 1, Volume 3
[Page 153, plate 52]

An Octogenarian (1580), 1872
Two albumen silver prints on studio card
Robert N. Dennis Collection of Stereoscopic Views, Miriam and Ira D. Wallach Division of Art, Prints and Photographs, The New York Public Library, Astor, Lenox and Tilden Foundations
[Page 154, plate 53]

Valley of the Yosemite, from Mariposa Trail. No. 3, 1872 (*)
Albumen silver print
From the American Geographical Society Library
University of Wisconsin-Milwaukee Libraries, 79:446
[Page 155, plate 54]

Tutokanula, "The Great Chief El Capitan" from Rocky Ford (4178), 1872
Albumen silver print on cabinet card
Corcoran Gallery of Art, Washington, D.C.
Museum Purchase, 2010.002.001
[Page 156, plate 55]

Valley of the Yosemite, from Rocky Ford. No. 4, 1872
Albumen silver print
From the American Geographical Society Library
University of Wisconsin-Milwaukee Libraries, 79:447
[Page 157, plate 56]

Tutokanula. Valley of the Yosemite. The Great Chief "El Capitan." 3,300 feet high. No. 9, 1872 (*)
Albumen silver print
Collection of George Eastman House (Gift of Virginia Adams) 1981:2375:0005
Courtesy of George Eastman House, International Museum of Photography and Film
[Page 158, plate 57]

Tutokanula. Valley of the Yosemite. No. 11, from *Yosemite Views*, 1872 (*)
40 albumen silver prints mounted in album
California History Section, California State Library
[Page 159, plate 58; see also page 172, plate 71]

Valley of the Yosemite. Confluence of the Merced, and Yosemite Creek. No. 21, 1872
Albumen silver print
The Society of California Pioneers
[Page 160, plate 59]

Cottonwood Bend. Valley of the Yo-semite. No. 15, 1872
Albumen silver print

Collection of California Historical Society
[Page 161, plate 60]

Mirror Lake, 1872, from *Photo Album No. 9*, page 10, ca. 1873
72 albumen silver prints in album
Collection of California Historical Society
[Page 162, plate 61]

Mirror Lake, (Reflection of Mount Waiya). "Pine Mountain" No. 25, 1872
Albumen silver print
Lent by the Metropolitan Museum of Art, David Hunter McAlpin Fund, 1966 (66.724.2)
Image copyright © The Metropolitan Museum of Art / Art Resource, NY
[Page 163, plate 62]

Falls of the Yosemite. From Glacier Rock. (Great Grizzly Bear) 2600 feet fall. No. 36, 1872
Albumen silver print
Courtesy of The Bancroft Library, University of California, Berkeley, 1962.019:36—ffALB
[Page 164, plate 63]

Valley of the Yosemite. From Mosquito Camp. No. 22, 1872 (*)
Albumen silver print
Corcoran Gallery of Art, Washington, D.C.
Museum Purchase, 2007.003
[Page 165, plate 64]

Vernal Fall, 1872, from *Photo Album No. 9*, page 11, ca. 1873
72 albumen silver prints in album
Collection of California Historical Society
[Page 166, plate 65]

Pi-Wi-Ack. Valley of the Yosemite. (Shower of Stars) "Vernal Fall." 400 Feet Fall. No. 29, 1872 (*)
Albumen silver print
San Francisco Museum of Modern Art. Accessions Committee Fund and gift of Jeffrey Fraenkel and Frish Brandt
[Page 167, plate 66]

Piwayack (Shower of Stars) Early Morning (1272), 1872 (*)
Two albumen silver prints on studio card
Collection of California Historical Society
[Page 168, plate 67]

Wild Cat Fall. Valley of the Yosemite. No. 30, 1872
Albumen silver print
The Society of California Pioneers
[Page 169, plate 68]

Valley of the Yosemite from Union Point. No. 33, 1872
Albumen silver print
Museum of Fine Arts, Boston. Gift of Charles T. and Alma A. Isaacs 1983.462
Photograph © March 2010 Museum of Fine Arts, Boston
[Page 170, plate 69]

Tenaya Canyon. Valley of the Yosemite. From Union Point. No. 35, 1872
Albumen silver print
National Gallery of Art, Washington, Horace W. Goldsmith Foundation through Robert and Joyce Menschel 1997.31.1
Image Courtesy of The Board of Trustees
[Page 171, plate 70]

Ancient Glacier Channel. Lake Tenaya, No. 47, 1872, from *Yosemite Views*, page 38, 1873
40 albumen silver prints mounted in album
California History Section, California State Library
[Page 172, plate 71]

Glacier Channels. Valley of the Yosemite. From Panorama Rock. No. 41, 1872 (*)
Albumen silver print
The Society of California Pioneers
[Page 173, plate 72]

Wm. H. Seward. 85 feet in Circumference, 268 feet high. Mariposa Grove of Mammoth Trees. No. 51, 1872
Albumen silver print on board
Fine Arts Museums of San Francisco, Gift of Mrs. Raymond Perkins
[Page 174, plate 73]

Charles Leander Weed (attributed)
U.S. Grant, Mariposa Grove, portrait of Eadweard Muybridge seated at base of tree, 1872 (*)
Albumen silver print
The J. Paul Getty Museum, Los Angeles
[Page 175, plate 74]

Emptying bucket of water. Plate 401, 1887
Collotype on paper
Corcoran Gallery of Art, Washington, D.C.
Museum Purchase, 87.7.385
[Page 176, fig. 92]

Emptying bucket of water. Plate 399, 1887 (*)
Collotype on paper
Corcoran Gallery of Art, Washington, D.C.
Museum Purchase, 87.7.383
[Page 178, fig. 93]

Helios Flying Studio, 1870 (*)
Albumen silver print on carte-de-visite
Courtesy of the California History Room, California State Library, Sacramento, California
[Page 179, fig. 94]

Yowiye, Nevada Fall, 600 feet fall (4201), 1872
Albumen silver print on cabinet card
Corcoran Gallery of Art, Washington, D.C., Museum Purchase, 2010.002.002
[Page 180, figs. 95; for verso, see page 64, fig. 50]

San Rafael, Ross' Landing (1723), 1873
Two albumen silver prints on studio card
Courtesy of The Bancroft Library, University of California, Berkeley, 1971.055, Series 2
[Page 181, fig. 96]

Caspar David Friedrich
Moonrise over the Sea, 1822
Oil on canvas
Nationalgalerie, Staatliche Museen zu Berlin, Courtesy Bildarchiv Preussischer Kulturbesitz / Art Resource, NY
[Page 182, fig. 97]

Lake Tenaya. Sierra Nevada Mountains. No. 46, 1872
Albumen silver print
Courtesy of The Bancroft Library, University of California, Berkeley, BANC PIC 1962.019:46—ffALB
[Page 186, fig. 98]

United States Appraisers' Building, 1874 (*)
Albumen silver print
Iris & B. Gerald Cantor Center for Visual Arts at Stanford University, Museum Purchase Fund, 1974.121.1
[Page 188, fig. 99]

United States Appraisers' Building, 1875
Albumen silver print
San Francisco History Center, San Francisco Public Library
[Page 188, fig. 100]

United States Appraisers' Building, 1876 (*)
Albumen silver print
Iris & B. Gerald Cantor Center for Visual Arts at Stanford University, Museum Purchase Fund, 1974.121.2
[Page 188, fig. 101]

United States Appraisers' Building, 1876 (*)
Albumen silver print
Iris & B. Gerald Cantor Center for Visual Arts at Stanford University, Museum Purchase Fund, 1974.121.3
[Page 188, fig. 102]

United States Appraisers' Building, 1876 (*)
Albumen silver print
San Francisco History Center, San Francisco Public Library
[Page 188, fig. 103]

United States Appraisers' Building, 1876
Albumen silver print
San Francisco History Center, San Francisco Public Library
[Page 188, fig. 104]

United States Appraisers' Building, 1877
Gelatin silver print
San Francisco History Center, San Francisco Public Library
[Page 188, fig. 105]

United States Appraisers' Building, 1877
Gelatin silver print
San Francisco History Center, San Francisco Public Library
[Page 188, fig. 106]

The Brandenburg Album of Bradley & Rulofson "Celebrities" and Muybridge Photographs, page 45, 1874 (*)
Albumen silver prints mounted on paper
Iris & B. Gerald Cantor Center for Visual Arts at Stanford University, Museum Purchase Fund, 1972.9.45
[Page 190, plate 75]

The Brandenburg Album of Bradley & Rulofson "Celebrities" and Muybridge Photographs, page 104, 1874 (*)
Albumen silver prints mounted to paper
Iris & B. Gerald Cantor Center for Visual Arts at Stanford University, Museum Purchase Fund, 1972.9.104
[Page 191, plate 76]

The Lava Beds, No. 8 (1619), (verso), 1873
One of 12 studio cards
National Archives, Washington, D.C.
[Page 192, plate 77; see pages 194-196, plates 79-89]

A Modoc Brave on the War Path (1627), 1873 (*)
Two albumen silver prints on studio card
Wilson Centre for Photography
[Page 193, plate 78]

The Lava Beds, Nos. 1–12 (1619), 1873
12 studio cards, two albumen silver prints each
National Archives, Washington, D.C.
[Pages 194–196, plates 79–90]

Donald Mc Ky, the celebrated Warm Springs Indian Scout and his chief men (1623), 1873 (*)
Two albumen silver prints on studio card
Collection of Leonard A. Walle
[Page 197, plate 91]

Generals Jeff C. Davis, Hardie, Gillem, and Officers of the Modoc Campaign (1621), 1873 (*)
Two albumen silver prints on studio card
Collection of Leonard A. Walle
[Page 198, plate 92]

The Peace Commission Tent, and Stone on which Gen. Canby was sitting when Shot by Capt. Jack (1601), 1873 (*)
Two albumen silver prints on studio card
Collection of Leonard A. Walle
[Page 199, plate 93]

Panorama of Lava Beds from Signal Station at Tule Lake Camp, South (Nos. 1605-1607), 1873
3 studio cards, two albumen silver prints on each
Collection of Leonard A. Walle
[Pages 200-201, plates 94-96]

Warm Spring Indian Scouts on Picket Duty (1630), 1873 (*)
Two albumen silver prints on studio card
Collection of Leonard A. Walle
[Page 202, plate 97]

Bringing in the Wounded after an Engagement (1617), 1873 (*)
Two albumen silver prints on studio card
Collection of Leonard A. Walle
[Page 203, plate 98]

Trade card of Muybridge for Bradley & Rulofson, 1874 (*)
Albumen silver print on carte-de-visite
Collection of Leonard A. Walle
[Page 204, plate 99]

Reception of the Artist at Panama (4262), 1875, from *The Pacific Coast of Central America and Mexico, The Isthmus of Panama, Guatemala, and the Cultivation and Shipment of Coffee*, 1876 (*)
144 albumen silver prints mounted in album
California History Section, California State Library
[Page 205, plate 100]

Church of San Filipe (4243), 1875
Albumen silver print
Courtesy of The Bancroft Library, University of California, Berkeley, 1971.055, Series 3
[Page 206, plate 101]

Burial of a Native—Panama, 1875
Albumen silver print
Smithsonian American Art Museum,
Gift of Mitchell and Nancy Steir
[Page 207, plate 102]

Ruins of a Church, Antigua, Guatemala, 1875 (*)
Albumen silver print
Collection Centre Canadien d'Architecture
Canadian Centre for Architecture, Montreal
[Page 208, plate 103]

Ruins of the Church of San Domingo, Panama, 1875 (*)
Albumen silver print
National Gallery of Art, Washington
Gift of Mitchell and Nancy Steir 2004.163.10
Image Courtesy of The Board of Trustees, National Gallery of Art, Washington
[Page 209, plate 104]

Laundry of Calvario, Guatemala (4347), 1875
Albumen silver print
Courtesy of The Bancroft Library, University of California, Berkeley, 1971.055, Series 3
[Page 210, plate 105]

Bridge on the Puerto Bello, Panama (4267), 1875 (*)
Albumen silver print
Department of Special Collections, Charles E. Young Research Library, UCLA
[Page 211, plate 106]

Solola, Guatemala, 1875 (*)
Albumen silver print
Smithsonian American Art Museum
Gift of Mitchell and Nancy Steir
[Page 212, plate 107]

Quetzaltenango, from the Crater of a Volcano, 1875 (*)
Albumen silver print
Wilson Centre for Photography
[Page 213, plate 108]

Machinery for Pulping Coffee, Las Nubes, 1875 (*)
Albumen silver print
Wilson Centre for Photography
[Page 214, plate 109]

Reception of the Artist, Quetzaltenango, Guatemala, 1875 (*)
Albumen silver print
Smithsonian American Art Museum, Gift of Jeffrey A. Lichtenberg in honor of his parents Laura and Harold Lichtenberg
[Page 215, plate 110]

J. Walter Wilson
Mr. Muybridge Shows His Instantaneous Photographs of Animal Motion at the Royal Society, Illustrated London News,
May 25, 1889 (*)
Engraving on newsprint
Corcoran Gallery of Art, Washington, D.C.
[Page 216, fig. 107]

Savings and Loan Society, Clay street (340), 1869
Two albumen silver prints on studio card
Collection of Leonard A. Walle
[Page 218, fig. 108]

Chinese Women Outside, Woodward's Gardens, 1870
Albumen silver print on carte-de-visite
Yale Collection of Western Americana, Beinecke Rare Book & Manuscript Library
[Page 219, fig. 109]

A, walking; B, ascending step; C, throwing disk; D, using shovel; E, F, using pick. Plate 521, 1887 (*)
Collotype on paper
Corcoran Gallery of Art, Washington, D.C.
Museum Purchase, 87.7.477
[Page 220, fig. 110]

Carleton Watkins
Primitive Mining; The Rocker, Calaveras Co., Cal. (self portrait), ca. 1876
Albumen silver print
The Society of California Pioneers
[Page 221, fig. 111]

Carleton Watkins
Cathedral Rock—Yosemite. No. 20, 1861
Albumen silver print
The J. Paul Getty Museum, Los Angeles
[Page 222, fig. 112]

Cathedral Rocks, Valley of the Yosemite, 2,600 Feet High. No. 20, 1872
Albumen silver print
The Nelson-Atkins Museum of Art, Kansas City, Missouri. Gift of the Hall Family Foundation, 2005.37.43.
Photograph by Thomas Palmer
[Page 223, fig. 113]

Edgington—Fast (Phase 1), 1878 (*)
Lantern slide, collodion positive on glass
Wilson Centre for Photography
[Page 230, plate 111]

Thomas Day, Importer of Gas Fixtures, French Clocks, Fine Bronzes, Etc. 122 and 124 Sutter Street, San Francisco, 1880
Two albumen silver prints on single card mount
California History Section, California State Library
[Page 231, plate 112]

Leland Stanford's Residence Photograph Album, 1878 (*)
Album of 62 albumen silver prints mounted on 62 leaves of white woven paper
San Francisco History Center, San Francisco Public Library
[Pages 232-233, plates 113-114]

View from Windows Looking South, 1877
Albumen silver print mounted on India-tinted board
The J. Paul Getty Museum, Los Angeles
[Page 234, plate 115]

View from Windows Looking South-west, 1877
Albumen silver print mounted on India-tinted board
The J. Paul Getty Museum, Los Angeles
[Page 235, plate 116]

Panorama of San Francisco from California Street Hill, 1877 (*)
11 albumen silver prints mounted on linen in quarto album
Wilson Centre for Photography
[Pages 236–239, plate 117]

Panorama of San Francisco from California Street Hill, San Francisco, California, 1878
13 albumen silver prints mounted on linen in elephant album
Reproduction taken from the copy of the panorama in the Robert N. Dennis Collection of Stereoscopic Views, Miriam and Ira D. Wallach Division of Art, Prints and Photographs, The New York Public Library, Astor, Lenox and Tilden Foundations
[Pages 240–244, plate 118; see also detail on page 77, fig. 63]

Kate and Robert Johnson Photograph Album, 1880 (*)
84 albumen silver prints in album
William L. Clements Library, University of Michigan
[Pages 245–250, plates 119–124]

Untitled (landscape near Watsonville, California), ca. 1879
Albumen silver print
The Cleveland Museum of Art
John L. Severance Fund, 1994.187
[Page 251, plate 125]

Leland Stanford, Jr. on his Pony "Gypsy"—Phases of a Stride by a Pony While Cantering, 1879 (*)
Collodion positive on glass
Wilson Centre for Photography
[Page 252, plate 126, and Cover]

General view of experiment track, background and cameras. Plate F, 1879, from *The Attitudes of Animals in Motion*, 1881 (*)
Albumen silver print
Courtesy Special Collections, Stanford University Libraries
[Page 253, plate 127]

Camera and shutter (Side View). Plate B, 1879–80, from *The Attitudes of Animals in Motion*, 1881 (*)
Albumen silver print
Courtesy Special Collections, Stanford University Libraries
[Page 254, plate 128]

The Horse in Motion. "Abe Edgington," owned by Leland Stanford; driven by C. Marvin, trotting at a 2:24 gait over the Palo Alto track, 15th June 1878, 1878 (*)
Albumen silver print on card
Prints & Photographs Division, Library of Congress, Washington, D.C.
[Page 255, plate 129]

Horses. Trotting. Occident. Plate 32, 1879, from *The Attitudes of Animals in Motion*, 1881 (*)
Albumen silver print
Courtesy Special Collections
Stanford University Libraries
[Page 256, plate 130]

Horses. Cantering. Mahomet. Plate 18, 1879, from *The Attitudes of Animals in Motion*, 1881 (*)
Albumen silver print
Courtesy Special Collections
Stanford University Libraries
[Page 257, plate 131]

Various Animals, Boar, Trotting. Plate 91, 1879, from *The Attitudes of Animals in Motion*, 1881 (*)
Albumen silver print
Courtesy Special Collections
Stanford University Libraries
[Page 258, plate 132]

Various Animals, Greyhound, Running. Plate 76, 1879, from *The Attitudes of Animals in Motion*, 1881 (*)
Albumen silver print
Courtesy Special Collections
Stanford University Libraries
[Page 259, plate 133]

Athletes. Flip Flap. Plate 109, 1879, from *The Attitudes of Animals in Motion*, 1881 (*)
Albumen silver print

Courtesy Special Collections
Stanford University Libraries
[Page 260, plate 134]

Athletes. Posturing. Plate 115, 1879, from *The Attitudes of Animals in Motion*, 1881 (*)
Albumen silver print
Courtesy Special Collections
Stanford University Libraries
[Page 261, plate 135]

Athletes. Swinging Pick. Plate 110, 1879, from *The Attitudes of Animals in Motion*, 1881 (*)
Albumen silver print mounted on heavy weight paper
Partial Gift of The Beinecke Foundation, Inc.
Addison Gallery of American Art, Phillips Academy, Andover, Massachusetts
[Page 262, plate 136]

Studies of Foreshortenings. Horses. Running. Mahomet. Plates 131–132, 1879, from *The Attitudes of Animals in Motion*, 1881 (*)
Albumen silver print
The Museum of Modern Art, New York. The Family of Man Fund. Accession Number: 223.1986
Digital image © The Museum of Modern Art / Licensed by SCALA / Art Resource, NY
[Page 263, plate 137]

Nimrod Pacing, maquette for a zoopraxiscope disc, ca. 1879 (*)
Wet plate collodion on glass, positive lantern slides
Iris & B. Gerald Cantor Center for Visual Arts at Stanford University, Stanford Family Collections,1994.46.1-21
[Page 264, plate 138]

Mule Bucking and Kicking, 12 phases, (recto), 1893
Glass zoopraxiscope disc, silver gelatin emulsion, paint, metal
Courtesy of George Eastman House
International Museum of Photography and Film
[Page 265, plate 139]

Athlete, Running Long Jump, 14 phases, 1893 (*)
Glass zoopraxiscope disc, paint, metal
By permission of Kingston Museum and Heritage Service
[Page 266, plate 140]

Four Horses with Riders, Dogs, and Spectators, 20/21 phases, No. 35, ca. 1879–80
Glass zoopraxiscope disc, paint
By permission of Kingston Museum and Heritage Service
[Page 267, plate 141]

Woman Dancing, 1893
Glass zoopraxiscope disc negative
Smithsonian Institution, National Museum of American History, Kenneth E. Behring Center
[Page 268, plate 142]

Woman Dancing, 1893 (*)
Glass zoopraxiscope disc, paint, metal
By permission of Kingston Museum and Heritage Service
[Page 269, plate 143]

Eadweard Muybridge self-portrait, 1883
Three albumen silver prints mounted on board
Courtesy of the Thomas Anshutz papers, ca. 1870–1942, Archives of American Art
Smithsonian Institution
[Page 270, fig. 114]

A, sitting down; B, sitting down; C, sprinkling water; D, stooping for cup and drinking. Plate 490, 1887 (*)
Collotype on paper
Corcoran Gallery of Art, Washington, D.C.
Museum Purchase, 87.7.450
[Page 274, fig. 115]

Base-ball; throwing. Plate 286, 1887
Collotype on paper
Corcoran Gallery of Art, Washington, D.C.
Museum Purchase, 87.7.272
[Page 276, fig. 116]

Cyanotype proof for Animal Locomotion. Ascending and Descending Stairs (Blanche Epler). Plate 504, 1885 (*)
Cyanotype prints on Masonite, paper and ink
Smithsonian Institution, National Museum of American History, Kenneth E. Behring Center
[Page 277, fig. 117]

Glass positive for Animal Locomotion. Ascending and Descending Stairs. Plate 504, 1887
Glass dry plate positives, paper
Smithsonian Institution, National Museum of American History, Kenneth E. Behring Center
[Page 278, fig. 118]

Ascending and descending stairs. (Movements. Female). Plate 504, 1887 (*)
Collotype on paper
Corcoran Gallery of Art, Washington, D.C.
Museum Purchase, 87.7.460
[Page 279, fig. 119]

Toilet; two models, 1 disrobing 8. Plate 428, 1887
Collotype on paper
Corcoran Gallery of Art, Washington, D.C.
Museum Purchase, 87.7.401
[Page 281, fig. 120]

Frances Benjamin Johnston
Portrait of Eadweard Muybridge, 1890 (*)
Gelatin silver print
Smithsonian Institution, National Museum of American History, Kenneth E. Behring Center
[Page 283, fig. 121]

Sequenced image of a rotating sulky wheel with Muybridge climbing steps, 1887
Glass dry plate positives, tape
Courtesy of George Eastman House
International Museum of Photography and Film
[Page 284, plate 144]

A, throwing water from bucket; B, descending step; C, ascending step; D, lawn tennis. Plate 524, 1887
Collotype on paper
Corcoran Gallery of Art, Washington, D.C.
Museum Purchase, 87.7.480
[Page 286, plate 145]

Heaving a 20-lb. rock. Plate 315, 1887
Collotype on paper
Corcoran Gallery of Art, Washington, D.C.
Museum Purchase
[Page 287, plate 146]

A, 98 and 100, wrestling; B, 98 and 100, wrestling; C, 99 and 101, wrestling; D, 99 and 101, sparring without gloves. Plate 520, 1887
Collotype on paper
Corcoran Gallery of Art, Washington, D.C.
Museum Purchase, 87.7.476
[Page 288, plate 147]

A, 97, jumping; B, 98, hand-spring; C, 98, somersault; D, 99, somersault; E, 99, spring over man's back. Plate 522, 1887 (*)
Collotype on paper
Corcoran Gallery of Art, Washington, D.C.
Museum Purchase, 87.7.478
[Page 289, plate 148]

Cyanotype proof for Animal Locomotion. Blacksmiths hammering an anvil, males, pelvis cloth (Messrs. Redinger and Breen). Plate 374, 1885 (*)
Cyanotype prints on Masonite, paper and ink
Smithsonian Institution, National Museum of American History, Kenneth E. Behring Center
[Page 290, plate 149]

Blacksmiths, two models, hammering on anvil. (Movements. Male). Plate 374, 1887 (*)
Collotype on paper
Corcoran Gallery of Art, Washington, D.C.
Museum Purchase, 87.7.358
[Page 291, plate 150]

Cyanotype proof for Animal Locomotion. Jumping, running broad jump. Plate 160, 1885 (*)
Cyanotype prints on Masonite, paper and ink
Smithsonian Institution, National Museum of American History, Kenneth E. Behring Center
[Page 292, plate 151]

Jumping; running broad jump. (Movements. Male). Plate 160, 1887 (*)
Collotype on paper
Corcoran Gallery of Art, Washington, D.C.
Museum Purchase, 87.7.160
[Page 293, plate 152]

Walking. (Movements. Male). Plate 7, 1887 (*)
Collotype on paper
Corcoran Gallery of Art, Washington, D.C.
Museum Purchase, 1887, 87.7.7
[Page 294, plate 153]

Running full speed. (Movements. Male). Plate 63, 1887 (*)
Collotype on paper
Corcoran Gallery of Art, Washington, D.C.
Museum Purchase, 87.7.63
[Page 295, plate 154]

Turning around in surprise and running away. (Movements. Female). Plate 73, 1887 (*)
Collotype on paper
Corcoran Gallery of Art, Washington, D.C.
Museum Purchase, 87.7.73
[Page 296, plate 155]

Ascending stairs. (Movements. Male). Plate 91, 1887 (*)
Collotype on paper
Corcoran Gallery of Art, Washington, D.C.
Museum Purchase, 87.7.91
[Page 297, plate 156]

Glass Positive for Animal Locomotion. Plate 143. Descending Stairs, a cup and saucer in right hand, 1887
Glass dry plate positives, paper
Courtesy of George Eastman House
International Museum of Photography and Film
[Page 298, plate 157]

Descending stairs, turning, cup and saucer in right hand. Plate 143, 1887
Collotype on paper
Corcoran Gallery of Art, Washington, D.C.
Museum Purchase, 87.7.143
[Page 299, plate 158]

Descending stairs with hands clasped. Plate 128, 1887 (*)
Collotype on paper
Corcoran Gallery of Art, Washington, D.C.
Museum Purchase, 87.7.128
[Page 300, plate 159]

Jumping; running straight high jump. Plate 155, 1887
Collotype on paper
Corcoran Gallery of Art, Washington, D.C.
Museum Purchase, 87.7.155
[Page 301, plate 160]

Emptying bucket of water. Plate 404, 1887
Collotype on paper
Corcoran Gallery of Art, Washington, D.C.
Museum Purchase, 87.7.387
[Page 302, plate 161]

Emptying bucket of water. Plate 403, 1887
Collotype on paper
Corcoran Gallery of Art, Washington, D.C.

Museum Purchase, 87.7.386
[Page 303, plate 162]

Ironing clothes. Plate 435, 1887
Collotype on paper
Corcoran Gallery of Art, Washington, D.C.
Museum Purchase, 87.7.405
[Page 304, plate 163]

Wringing clothes. Plate 433, 1887
Collotype on paper
Corcoran Gallery of Art, Washington, D.C.
Museum Purchase, 87.7.403
[Page 305, plate 164]

Getting into bed. Plate 263, 1887 (*)
Collotype on paper
Corcoran Gallery of Art, Washington, D.C.
Museum Purchase, 87.7.250
[Page 306, plate 165]

Emptying bucket of water. Plate 400, 1887
Collotype on paper
Corcoran Gallery of Art, Washington, D.C.
Museum Purchase, 87.7.384
[Page 307, plate 166]

Contortions on the ground. Plate 510, 1887 (*)
Collotype on paper
Corcoran Gallery of Art, Washington, D.C.
Museum Purchase, 87.7.466
[Page 308, plate 167]

Fencing. (Movements. Male). Plate 349, 1887 (*)
Collotype on paper
Corcoran Gallery of Art, Washington, D.C.
Museum Purchase, 87.7.334
[Page 309, plate 168]

Wrestling; Graeco-Roman. Plate 347, 1887 (*)
Collotype on paper
Corcoran Gallery of Art, Washington, D.C.
Museum Purchase, 87.7.332
[Page 310, plate 169]

Boxing; open hand. Plate 340, 1887
Collotype on paper
Corcoran Gallery of Art, Washington, D.C.
Museum Purchase, 87.7.325
[Page 311, plate 170]

Two models; one standing, the other sitting, crossing legs. Plate 239, 1887
Collotype on paper
Corcoran Gallery of Art, Washington, D.C.
Museum Purchase, 87.7.231
[Page 312, plate 171]

Crossing brook on step-stones with fishing-pole and can. Plate 175, 1887 (*)
Collotype on paper
Corcoran Gallery of Art, Washington, D.C.
Museum Purchase, 87.7.170
[Page 313, plate 172]

Jumping; running straight high jump. (Movements. Female). Plate 156, 1887 (*)
Collotype on paper
Corcoran Gallery of Art, Washington, D.C.
Museum Purchase, 87.7.156
[Page 314, plate 173]

Walking, two models meeting, and partly turning. Plate 45, 1887 (*)
Collotype on paper
Corcoran Gallery of Art, Washington, D.C.
Museum Purchase, 87.7.45
[Page 315, plate 174]

Child, running. Plate 469, 1887 (*)
Collotype on paper
Corcoran Gallery of Art, Washington, D.C.
Museum Purchase, 87.7.430
[Page 316, plate 175]

Two models, child, 70 N, bringing bouquet to 12. Plate 465, 1887 (*)
Collotype on paper
Corcoran Gallery of Art, Washington, D.C.
Museum Purchase, 87.7.426
[Page 317, plate 176]

Cyanotype proof for Animal Locomotion. Dancing (Fancy), Woman Draped (Miss Larrigan). Plate 188, 1885 (*)
Cyanotype prints on Masonite, paper and ink
Smithsonian Institution, National Museum of American History, Kenneth E. Behring Center
[Page 318, plate 177]

Dancing (fancy.) (Movements. Female). Plate 188, 1887 (*)
Collotype on paper
Corcoran Gallery of Art, Washington, D.C.
Museum Purchase, 87.7.188
[Page 319, plate 178]

Cyanotype proof for Animal Locomotion. Movement of the Hand, Lifting a Ball (Mr. Tadd). Plate 534, 1885 (*)
Cyanotype prints on Masonite, paper and ink
Smithsonian Institution, National Museum of American History, Kenneth E. Behring Center
[Page 320, plate 179]

Movement of the hand; lifting a ball. Plate 534, 1887 (*)
Collotype on paper
Corcoran Gallery of Art, Washington, D.C.
Museum Purchase, 87.7.487
[Page 321, plate 180]

Glass Positive for Animal Locomotion. Plate 169. Jumping over boy's back (leapfrog), 1887
Glass dry plate positives, paper
Courtesy of George Eastman House, International Museum of Photography and Film
[Page 322, plate 181]

Jumping over boy's back (leap-frog). Plate 169, 1887 (*)
Collotype on paper
Corcoran Gallery of Art, Washington, D.C.
Museum Purchase, 87.7.169
[Page 323, plate 182]

Movement of the hand; beating time. Plate 535, 1887
Collotype on paper
Corcoran Gallery of Art, Washington, D.C.
Museum Purchase, 87.7.488
[Page 324, plate 183]

Head-spring, a flying pigeon interfering. Plate 365, 1887 (*)
Collotype on paper
Corcoran Gallery of Art, Washington, D.C.
Museum Purchase, 87.7.350
[Page 325, plate 184]

Trotting; bareback; dark-gray Belgian horse. Dusel. Plate 599, 1887
Collotype on paper
Corcoran Gallery of Art, Washington, D.C.
Museum Purchase, 87.7.545
[Page 326, plate 185]

Trotting; sulky; bay mare Lizzie M. Plate 609, 1887
Collotype on paper
Corcoran Gallery of Art, Washington, D.C.
Museum Purchase, 87.7.555
[Page 327, plate 186]

Gallop; saddle; thoroughbred bay horse Bouquet. Plate 632, 1887
Collotype on paper
Corcoran Gallery of Art, Washington, D.C.
Museum Purchase, 87.7.578
[Page 328, plate 187]

Jumping a hurdle; saddle; bay horse Daisy. Plate 640, 1887 (*)
Collotype on paper
Corcoran Gallery of Art, Washington, D.C.
Museum Purchase, 87.7.586
[Page 329, plate 188]

Fallow deer; doe; galloping; and kid; jumping. Plate 691, 1887
Collotype on paper
Corcoran Gallery of Art, Washington, D.C.
Museum Purchase, 87.7.615
[Page 330, plate 189]

Canter; bareback; rider, 106, nude; white horse Clinton. Plate 617, 1887 (*)
Collotype on paper
Corcoran Gallery of Art, Washington, D.C.
Museum Purchase, 87.7.563
[Page 331, plate 190]

Baboon; walking on all-fours. Plate 748, 1887 (*)
Collotype on paper
Corcoran Gallery of Art, Washington, D.C.
Museum Purchase, 87.7.656
[Page 332, plate 191]

Elephant; walking. Plate 733, 1887 (*)
Collotype on paper
Corcoran Gallery of Art, Washington, D.C.
Museum Purchase, 87.7.641
[Page 333, plate 192]

Dog; Walking; Interrupted; Mastiff-Dread. Plate 703, 1887 (*)
Collotype on paper
Gift of the Edwin J. Beinecke Trust. Addison Gallery of American Art, Phillips Academy, Andover, Massachusetts
[Page 334, plate 193]

Cockatoo; flying. Plate 759, 1887 (*)
Collotype on paper
Corcoran Gallery of Art, Washington, D.C.
Museum Purchase, 87.7.667
[Page 335, plate 194]

Exhibited Works

Works are listed in chronological order, and then in plate order within specific projects.

Base of First Fall of the Yosemite, from cave of the Spirit's voice (40), 1867
Two albumen silver prints on studio card
Collection of Leonard A. Walle
[Page 20, plate 4]

The Flying Studio, Photographer's Equipment in the Field (114), 1867
Two albumen silver prints on studio card
Collection of Leonard A. Walle
[Page 18, plate 2]

California—Yosemite Valley. Iris from the Summit of Pi-wy-ack (297), 1867
Two albumen silver prints on studio card
Collection of Leonard A. Walle
[Page 115, plate 11]

Fissure in Eagle Rock. 1100 ft. deep, 1867
Two albumen silver prints on studio card
Collection of Leonard A. Walle
[Page 114, plate 10]

Kekootoyem (Water Asleep) Mirror Lake from the Western Bank, 1867
Albumen silver print
Iris & B. Gerald Cantor Center for Visual Arts at Stanford University, Elizabeth K. Raymond Fund, 19735

Tolookweack Fall in South Canon, Yosemite (19741), 1867
Albumen silver print
Iris & B. Gerald Cantor Center for Visual Arts at Stanford University, Elizabeth K. Raymond Fund, 19741

Alaska Ter.—Sitka. Russo Greek Priests, Alaska (472), 1868
Two albumen silver prints on studio card
Collection of Leonard A. Walle
[Page 127, plate 23]

Group of Indians, Alaska (479), 1868
Two albumen silver prints on studio card
Collection of Leonard A. Walle
[Page 126, plate 22]

Group of Indians (489), 1868
Two albumen silver prints on studio card
Collection of Leonard A. Walle
[Page 46, fig. 30]

Steamers Zealous and Scout in Esquimault Harbor (511), 1868
Two albumen silver prints on studio card
Collection of Leonard A. Walle
[Page 46, fig. 29]

The "Heathen Chinee" Prospecting (848), 1868
Two albumen silver prints on studio card
Collection of Leonard A. Walle
[Page 135, plate 33]

Effect of Earthquake in San Francisco, California St. East of Montgomery St., 1868
Albumen silver print on carte-de-visite
Collection of California Historical Society, Gift of Mrs. Augustus M. Lord
[Page 47, fig. 31]

Snow Sheds at Emigrant Gap, looking East (721), 1869
Two albumen silver prints on studio card
Collection of Leonard A.Walle

Shoshone Indians at Corinne (754), 1869
Two albumen silver prints on studio card
Collection of Leonard A. Walle
[Page 51, fig. 36]

Long Ravine Trestle and Bridge—113 feet high, 878 feet long, looking East (766), 1869
Two albumen silver prints on studio card
Collection of Leonard A. Walle
[Page 128, plate 25]

Cut at Alta, looking East (787), 1869
Two albumen silver prints on studio card
California History Section, California State Library
[Page 128, plate 24]

Helios Flying Studio, 1870
Albumen silver print on carte-de-visite
Courtesy of the California History Room, California State Library, Sacramento, California
[Page 179, fig. 94]

Views in Woodward's Gardens, 1870
Two albumen prints mounted on studio card
Museum Purchase. Addison Gallery of American Art, Phillips Academy, Andover, Massachusetts
[Page 139, plate 37]

Bust of California (Powers) at Woodward's Gardens, 1870
Two albumen silver prints on studio card
California History Sections, California State Library

Animals at Woodward's Gardens (589), 1870
Two albumen silver prints on studio card
Collection of Leonard A. Walle
[Page 138, plate 36]

Chinee Bottle Washer, Buena Vista Vineyard (1115), 1871
Two albumen silver prints on studio card
Collection of Leonard A.Walle

First-Order Light-house at Punta de los Reyes, Seacoast of California, 296 Feet Above Sea (4136), 1871
Albumen silver print
U.S. Coast Guard Historian's Office
[Page 54, fig. 40]

Lighthouse at Punta de los Reyes, Coast of California (8075), 1871
Albumen silver print
U.S. Coast Guard Historian's Office

Point Shubrick, Tower Head, Lighthouse, Parrot Rock, and Gull Peak from the Hole in the Wall, South Farallon Island, 1871
Albumen silver print
U.S. Coast Guard Historian's Office
[Page 140, plate 38]

The Ramparts, Funnel Rock, Hole in the Wall, Pyramid, Sugar Loaf, Oil House, and Landing Cove on Fisherman's Bay, South Farallon Island, 1871
Albumen silver print
U.S. Coast Guard Historian's Office
[Page 141, plate 39]

Sacramento–Residence of Leland Stanford, Mr. Stanford's Horses (17166), 1872
Two albumen silver prints on studio card
Wilson Centre for Photography

Piwayack (Shower of Stars) Early Morning (1272), 1872
Two albumen silver prints on studio card
Collection of California Historical Society
[Page 168, plate 67]

Contemplation Rock, Glacier Point (1385), 1872
Two albumen silver prints on studio card
Collection of California Historical Society
[Page 22, fig. 6]

A Log of (Piute) Bucks (1577), 1872
Two albumen silver prints on studio card
Robert N.Dennis Collection of Stereoscopic Views, Miriam and Ira D.Wallach Division of Art, Prints and Photographs, The New York Public Library, Astor, Lenox and Tilden Foundations

Albert Bierstadt's Studio (1586), 1872
Two albumen silver prints on studio card
Collection of Leonard A. Walle
[Page 60, fig. 45]

Yosemite Views, 1872
40 albumen silver prints mounted in album
California History Section, California State Library
[Page 159, plate 58; see also page 172, plate 71]

Valley of the Yosemite, from Mariposa Trail. No. 3, 1872
Albumen silver print
From the American Geographical Society Library
University of Wisconsin–Milwaukee Libraries, 79:446
[Page 155, plate 54]

Tutokanula. Valley of the Yosemite. The Great Chief "El Capitan." 3,300 feet high. No. 9, 1872
Albumen silver print
Collection of George Eastman House (Gift of Virginia Adams) 1981:2375:0005.
Courtesy of George Eastman House,
International Museum of Photography and Film
[Page 158, plate 57]

Valley of the Yosemite. From Mosquito Camp. No. 22, 1872
Albumen silver print
Corcoran Gallery of Art, Washington, D.C.
Museum Purchase, 2007.003
[Page 165, plate 64]

Tokoya and Hunto. Valley of the Yosemite. (The Basket,) (The Watching Eye,) 3,568 feet high. No. 24, 1872
Albumen silver print
The Society of California Pioneers
[Page 42, fig. 25]

Pi-Wi-Ack. Valley of the Yosemite. (Shower of Stars) "Vernal Fall." 400 Feet Fall. No. 29, 1872
Albumen silver print
San Francisco Museum of Modern Art. Accessions Committee Fund and gift of Jeffrey Fraenkel and Frish Brandt
[Page 167, plate 66]

The High Sierra. Valley of the Yosemite. From Glacier Point, No. 38, 1872
Albumen silver print
From the American Geographical Society Library, University of Wisconsin-Milwaukee Libraries

Glacier Channels. Valley of the Yosemite. From Panorama Rock. No. 41, 1872
Albumen silver print
The Society of California Pioneers
[Page 173, plate 72]

Mount Hoffman, Sierra Nevada Mountains. From Lake Tenaya. No. 48, 1872
Albumen silver print
Collection of George Eastman House
(Gift of Harvard University) 1981:2377:0011

Charles Leander Weed (attributed)
U.S. Grant, Mariposa Grove, portrait of Eadweard Muybridge seated at base of tree, 1872
Albumen silver print
The J. Paul Getty Museum, Los Angeles
[Page 175, plate 74]

Pigeon Point Lighthouse, 1873
Albumen silver print
California History Section, California State Library
[Page 146, plate 44]

Pigeon Point Lighthouse, 1873
Albumen silver print
California History Section, California State Library
[Page 147, plate 45]

The Peace Commission Tent, and Stone on which Gen. Canby was sitting when Shot by Capt. Jack (1601), 1873
Two albumen silver prints on studio card
Collection of Leonard A. Walle
[Page 199, plate 93]

Captain Jack's Cave in the Lava Beds (1602), 1873
Two albumen silver prints on studio card
Collection of Leonard A. Walle
[Page 65, fig. 51]

Bringing in the Wounded after an Engagement (1617), 1873
Two albumen silver prints on studio card
Collection of Leonard A. Walle
[Page 203, plate 98]

Generals Jeff C. Davis, Hardie, Gillem, and Officers of the Modoc Campaign (1621), 1873
Two albumen silver prints on studio card
Collection of Leonard A. Walle
[Page 198, plate 92]

Donald Mc Ky, the celebrated Warm Springs Indian Scout and his chief men (1623), 1873
Two albumen silver prints on studio card
Collection of Leonard A. Walle
[Page 197, plate 91]

Toby, (the squaw who warned General Canby of his impending fate), and four old Modoc squaws (1624), 1873
Two albumen silver prints on studio card
Collection of Leonard A. Walle

A Modoc Brave on the War Path (1627), 1873
Two albumen silver prints on studio card
Wilson Centre for Photography
[Page 193, plate 78]

Warm Spring Indian Scouts on Picket Duty (1630), 1873
Two albumen silver prints on studio card
Collection of Leonard A. Walle
[Page 202, plate 97]

Pictures from the lava beds, from "The Last of the Modocs," Harper's Weekly, June 21, 1873
Engraving on newsprint
Corcoran Gallery of Art, Washington, D.C.
[Page 66, fig. 54]

Flora Muybridge, 1874
Two albumen silver prints on studio card
Collection of Leonard A. Walle
[Page 68, fig. 55]

The Brandenburg Album of Bradley & Rulofson "Celebrities" and Muybridge Photographs, page 45, 1874
Albumen silver prints mounted on paper
Iris & B. Gerald Cantor Center for Visual Arts at Stanford University, Museum Purchase Fund, 1972.9.45
[Page 190, plate 75]

The Brandenburg Album of Bradley & Rulofson "Celebrities" and Muybridge Photographs, page 98, 1874
Albumen silver prints mounted on paper
Iris & B. Gerald Cantor Center for Visual Arts at Stanford University, Museum Purchase Fund, 1972.9.98
[Page 69, fig. 56]

The Brandenburg Album of Bradley & Rulofson "Celebrities" and Muybridge Photographs, page 104, 1874
Albumen silver prints mounted to paper
Iris & B. Gerald Cantor Center for Visual Arts at Stanford University, Museum Purchase Fund, 1972.9.104
[Page 191, plate 76]

United States Appraisers' Building, 1874
Albumen silver print
Iris & B. Gerald Cantor Center for Visual Arts at Stanford University, Museum Purchase Fund, 1974.121.1
[Page 188, fig. 99]

Trade card of Muybridge for Bradley & Rulofson, 1874
Albumen silver print on carte-de-visite
Collection of Leonard A. Walle
[Page 204, plate 99]

The Pacific Coast of Central America and Mexico, The Isthmus of Panama, Guatemala, and the Cultivation and Shipment of Coffee, 1876
144 albumen silver prints mounted in album
California History Section, California State Library
[See page 54, fig.59 for Title Page; and page 205, plate 100, *Reception of the Artist at Panama (4262)*, 1875]

Bridge on the Puerto Bello, Panama (4267), 1875
Albumen silver print
Department of Special Collections, Charles E. Young Research Library, UCLA
[Page 211, plate 106]

Ruins of the Church of San Domingo, Panama, 1875
Albumen silver print
National Gallery of Art, Washington
Gift of Mitchell and Nancy Steir 2004.163.10
[Page 209, plate 104]

Ancient Sacrificial Stone, Naranjo, Guatemala, 1875
Albumen silver print
Wilson Centre for Photography

Ruins of a Church, Antigua, Guatemala, 1875
Albumen silver print
Collection Centre Canadien d'Architecture
Canadian Centre for Architecture, Montreal
[Page 208, plate 103]

Solola, Guatemala, 1875
Albumen silver print
Smithsonian American Art Museum
Gift of Mitchell and Nancy Steir
[Page 212, plate 107]

Quetzaltenango, from the Crater of a Volcano, 1875
Albumen silver print
Wilson Centre for Photography
[Page 213, plate 108]

Reception of the Artist, Quetzaltenango, Guatemala, 1875
Albumen silver print
Smithsonian American Art Museum, Gift of Jeffrey A. Lichtenberg in honor of his parents Laura and Harold Lichtenberg
[Page 215, plate 110]

Setting Out a Coffee Plantation at Antigua de Guatemala, 1875
Albumen silver print
National Gallery of Art, Washington, Gift of Mary and Dan Solomon and Patrons' Permanent Fund 2006.133.109

Coffee Harvesting at San Isidro, elevation 1000 feet, Guatemala, 1875
Albumen silver print
Smithsonian American Art Museum, Gift of Jeffrey A. Lichtenberg in honor of his parents Laura and Harold Lichtenberg
[Page 73, fig. 58]

Machinery for Pulping Coffee, Las Nubes, 1875
Albumen silver print
Wilson Centre for Photography
[Page 214, plate 109]

Loading Coffee for the Port, Las Nubes, 1875
Albumen silver print
Wilson Centre for Photography
G. Andrews, after Muybridge

Coffee Cultivation in Guatemala, Central America, from *The Graphic, October 20*, 1877
Wood Engravings
Wellcome Library, London (25225i)

United States Appraisers' Building, 1876
Albumen silver print
Iris & B. Gerald Cantor Center for Visual Arts at Stanford University, Museum Purchase Fund, 1974.121.2
[Page 188, fig. 101]

United States Appraisers' Building, 1876
Albumen silver print
Iris & B. Gerald Cantor Center for Visual Arts at Stanford University, Museum Purchase Fund, 1974.121.3
[Page 188, fig. 102]

United States Appraisers' Building, 1876
Albumen silver print
San Francisco History Center, San Francisco Public Library
[Page 188, fig. 103]

Occident. Owned by Leland Stanford. Driven by Jas. Tennant, 1877, 1877
Albumen silver print on cabinet card
Prints & Photographs Division
Library of Congress, Washington, D.C.
[See page 80, fig. 67]

John Koch
Occident Trotting, 1877
Watercolor with gouache
Iris & B. Gerald Cantor Center for Visual Arts at Stanford University, Stanford Family Collections, 1983.48
[Page 80, fig. 66]

Occident Trotting, 1877
Albumen silver print
Iris & B. Gerald Cantor Center for Visual Arts at Stanford University, Stanford Family Collections, 13924

Panorama of San Francisco from California Street Hill, 1877
Albumen silver print mounted on boudoir card
Wilson Centre for Photography
[Page 75, fig. 60]

Panorama of San Francisco from California Street Hill, 1877
11 albumen silver prints mounted on linen in quarto album
Wilson Centre for Photography
[Pages 236–239, plate 117]

Panorama of San Francisco from California Street Hill (Key). Morse's Gallery, San Francisco, 1877
Albumen silver print
Prints & Photographs Division, Library of Congress, Washington, D.C.

Panorama of San Francisco from California Street Hill, San Francisco, California, 1878
13 albumen silver prints mounted on linen in elephant album
By permission of Kingston Museum and Heritage Service
[For reproductions of the version in the New York Public Library, see pages 240–244, plate 118; see also detail on page 77, fig. 63]

Leland Stanford's Residence Photograph Album, 1878
Album of 62 albumen silver prints mounted on 62 leaves of white woven paper
San Francisco History Center, San Francisco Public Library
[Pages 232–233, plates 113–114]

California Street Railroad, 1878
Albumen silver print mounted on board
California History Section, California State Library

A very rare set of Muybridge proof prints for The Horse in Motion, 1878
Four albumen silver prints on cards
Wilson Centre for Photography

The Stride of the Trotting Horse, "Abe Edgington," owned by Leland Stanford; being driven by Marvin, trotting at a 2:24 gait over Mr. Stanford's Race Track at Menlo Park, on 11th June, 1878, 1878
12 albumen silver prints on cabinet card
Wilson Centre for Photography
[Page 85, fig. 72]

The Horse in Motion, Illustrated by Muybridge. "Mahomet," owned by Leland Stanford; ridden by G.Domm, cantering at an 8 minute gait over the Palo Alto track, 18th June 1878, 1878
Albumen silver print on cabinet card
San Francisco Museum of Modern Art, Purchase through a gift of Elizabeth Gates and Accessions Committee Fund

The Horse in Motion, illus. by Muybridge. "Sallie Gardner," owned by Leland Stanford, running at a 1:40 gait over the Palo Alto track, 19th June 1878, 1878
Albumen silver print on cabinet card
Prints & Photographs Division, Library of Congress, Washington, D.C.

The Horse in Motion, illus. by Muybridge. "Sallie Gardner," owned by Leland Stanford, running at a 1:40 gait over the Palo Alto track, 19th June 1878: 2 frames showing diagram of foot movements, 1878
Albumen silver print on cabinet card
Prints & Photographs Division, Library of Congress, Washington, D.C.
[Page 85, fig. 71]

The Horse in Motion, Illustrated by Muybridge, "Occident" owned by Leland Stanford; driven by C.Marvin, trotting at a 2.20 gait over the Palo Alto track, 20th June 1878, 1878
Albumen silver print on cabinet card
Iris & B. Gerald Cantor Center for Visual Arts at Stanford University, Stanford Family Collections, 13928

Kate and Robert Johnson Photograph Album, 1880
84 albumen silver prints in album
William L. Clements Library, University of Michigan
[Page 76, fig. 62; see also pages 245–250, plates 119–124]

The Attitudes of Animals in Motion: A Series of Photographs Illustrating the Consecutive Positions Assumed by Animals in Performing Various Movements, 1881
174 albumen silver prints mounted on paper in bound volume
Department of Special Collections
Charles E. Young Research Library, UCLA
[Page 24, fig. 7: *Athletes. Twisting Summersault. Plate 106*, 1879]

The Attitudes of Animals in Motion: A Series of Photographs Illustrating the Consecutive Positions Assumed by Animals in Performing Various Movements, 1881
174 albumen silver prints mounted on paper in bound volume
Partial Gift of The Beinecke Foundation, Inc. Addison Gallery of American Art, Phillips Academy, Andover, Massachusetts
[Page 92, fig. 80]

Camera and shutter (Side View). Plate B, 1879–80, from *The Attitudes of Animals in Motion*, 1881
Albumen silver print
Courtesy Special Collections, Stanford University Libraries
[Page 254, plate 128]

Camera and shutter (Front view). Plate C, 1879–80, from *The Attitudes of Animals in Motion*, 1881
Albumen silver print
Courtesy Special Collections, Stanford University Libraries

General view of experiment track, background and cameras. Plate F, 1879, from *The Attitudes of Animals in Motion*, 1881
Albumen silver print
Courtesy Special Collections, Stanford University Libraries
[Page 253, plate 127]

Horses. Cantering. Mahomet. Plate 18, 1879, from *The Attitudes of Animals in Motion*, 1881
Albumen silver print
Courtesy Special Collections, Stanford University Libraries
[Page 257, plate 131]

Horses. Trotting. Occident. Plate 32, 1879, from *The Attitudes of Animals in Motion*, 1881
Albumen silver print
Courtesy Special Collections, Stanford University Libraries
[Page 256, plate 130]

Horses. Running. Phryne L. Plate 40, 1879, from *The Attitudes of Animals in Motion*, 1881
Albumen silver print
National Gallery of Art, Washington, Gift of Mary and Dan Solomon 2006.131.7
Image Courtesy of The Board of Trustees
[Page 92, fig. 81]

Horses. Leaping a 3 Ft. 6 Hurdle. Phryne L. Plate 49, 1879, from *The Attitudes of Animals in Motion*, 1881
Albumen silver print
Courtesy Special Collections, Stanford University Libraries
[Page 87, fig. 74]

Various Animals, Greyhound, Running. Plate 76, 1879, from *The Attitudes of Animals in Motion*, 1881
Albumen silver print
Courtesy Special Collections, Stanford University Libraries
[Page 259, plate 133]

Various Animals, Ox, Trotting. Plate 79, 1879, from *The Attitudes of Animals in Motion*, 1881
Albumen silver print
Courtesy Special Collections, Stanford University Libraries

Various Animals, Boar, Trotting. Plate 91, 1879, from *The Attitudes of Animals in Motion*, 1881
Albumen silver print
Courtesy Special Collections, Stanford University Libraries
[Page 258, plate 132]

Athletes. Running. Plate 99, 1879, from *The Attitudes of Animals in Motion*, 1881
Albumen silver print
Courtesy Special Collections, Stanford University Libraries

Athletes. Flip Flap. Plate 109, 1879, from *The Attitudes of Animals in Motion*, 1881
Albumen silver print
Courtesy Special Collections, Stanford University Libraries
[Page 260, plate 134]

Athletes. Swinging Pick. Plate 110, 1879, from *The Attitudes of Animals in Motion*, 1881
Albumen silver print mounted on heavy weight paper
Partial Gift of The Beinecke Foundation, Inc.
Addison Gallery of American Art, Phillips Academy, Andover, Massachusetts
[Page 262, plate 136]

Athletes. Boxing. Plate 111, 1879, from *The Attitudes of Animals in Motion*, 1881
Albumen silver print
Courtesy Special Collections, Stanford University Libraries

Athletes. Posturing. Plate 115, 1879, from *The Attitudes of Animals in Motion*, 1881
Albumen silver print
Courtesy Special Collections, Stanford University Libraries
[Page 261, plate 135]

Studies of Foreshortenings. Horses. Walking. Mahomet. Plates 119–120, 1879, from *The Attitudes of Animals in Motion*, 1881
Albumen silver print
National Gallery of Art, Washington, Gift of Mary and Dan Solomon and Patrons' Permanent Fund 2006.133.106

Studies of Foreshortenings. Horses. Running. Mahomet. Plates 131–132, 1879, from *The Attitudes of Animals in Motion*, 1881
Albumen silver print
The Museum of Modern Art, New York. The Family of Man Fund. Accession Number: 223.1986
[Page 263, plate 137]

Studies of Foreshortenings. Horses. Running. Mahomet. Plates 143–144, 1879, from *The Attitudes of Animals in Motion*, 1881
Albumen silver print
Chrysler Museum of Art, Norfolk, Virginia, Museum Purchase
[Page 87, fig. 75]

Skeleton of Horse.Leaping. Leaving the Ground. Plate 201, 1879, from *The Attitudes of Animals in Motion*, 1881
Albumen silver print
Courtesy Special Collections, Stanford University Libraries

Edgington—Fast (Phase 1), 1878
Lantern slide, collodion positive on glass
Wilson Centre for Photography
[Page 230, plate 111]

"Sallie Gardner" Galloping (Phases 1, 2, 5, 6, 7, 8, 10, 11), 1878
Eight lantern slides, collodion positives on glass
Iris & B. Gerald Cantor Center for Visual Arts at Stanford University, Stanford Family Collections, 13867-74

Horses. Cantering. Gypsy. Plate 15 (Phase 12), 1879
Lantern slide, collodion positive on glass
By permission of Kingston Museum and Heritage Service

Leland Stanford, Jr. on his Pony "Gypsy"—Phases of a Stride by a Pony While Cantering, 1879
Collodion positive on glass
Wilson Centre for Photography
[Page 252, plate 126, and Cover]

Nimrod Pacing, maquette for a zoopraxiscope disc, ca. 1879
Wet plate collodion on glass, positive lantern slides
Iris & B. Gerald Cantor Center for Visual Arts at Stanford University, Stanford Family Collections, 1994.46.1–21
[Page 264, plate 138]

Horse trotting. 21 phases. Disc '22', ca. 1879
Glass zoopraxiscope disc, paint, metal
By permission of Kingston Museum and Heritage Service

Zoopraxiscope, 1879–85
Mixed media projector
By permission of Kingston Museum and Heritage Service
[Page 89, fig. 77]

J. D. B. Stillman, *The Horse in Motion as Shown by Instantaneous Photography, with a Study on Animal Mechanics Founded on Anatomy and the Revelations of the Camera, in Which Is Demonstrated the Theory of Quadrupedal Locomotion*, 1881–82
Lithograph in book
Wilson Centre for Photography
[Page 93, fig. 82: *Phryne L—Running. Stride. 19 ft. 9 in. Plate XVI*]

Muybridge Laboratory Notebook #3, 1885
Ink on notebook paper
Collection of Menschel Library at George Eastman House (Gift of Eastman Kodak Company; ex collection of Dr. George E. Nitzsche), 1982:2603:0001

Cyanotype proof for Animal Locomotion. *Jumping, running broad jump. Plate 160*, 1885
Cyanotype prints on Masonite, paper and ink
Smithsonian Institution, National Museum of American History, Kenneth E. Behring Center
[Page 292, plate 151]

Cyanotype proof for Animal Locomotion. *Dancing (Fancy), Woman Draped (Miss Larrigan). Plate 188*, 1885
Cyanotype prints on Masonite, paper and ink
Smithsonian Institution, National Museum of American History, Kenneth E. Behring Center
[Page 318, plate 177]

Cyanotype proof for Animal Locomotion. *Blacksmiths hammering an anvil, males, pelvis cloth (Messrs. Redinger and Breen). Plate 374*, 1885
Cyanotype prints on Masonite, paper and ink
Smithsonian Institution, National Museum of American History, Kenneth E. Behring Center
[Page 290, plate 149]

Cyanotype proof for Animal Locomotion. *Ascending and Descending Stairs (Blanche Epler). Plate 504*, 1885
Cyanotype prints on Masonite, paper and ink
Smithsonian Institution, National Museum of American History, Kenneth E. Behring Center
[Page 277, fig. 117]

Cyanotype proof for Animal Locomotion. *Movement of the Hand, Lifting a Ball (Mr. Tadd). Plate 534*, 1885
Cyanotype prints on Masonite, paper and ink
Smithsonian Institution, National Museum of American History, Kenneth E. Behring Center
[Page 320, plate 179]

Animal Locomotion. An Electro-Photographic Investigation of Consecutive Phases of Animal Movements. Published under the Auspices of the University of Pennsylvania. Prospectus and Catalogue of Plates, Philadelphia 1887
Book
By permission of Kingston Museum and Heritage Service

Walking. (Movements. Male). Plate 7, 1887
Collotype on paper
Corcoran Gallery of Art, Washington, D.C.
Museum Purchase, 1887, 87.7.7
[Page 294, plate 153]

Walking, two models meeting, and partly turning. Plate 45, 1887
Collotype on paper
Corcoran Gallery of Art, Washington, D.C.
Museum Purchase, 87.7.45
[Page 315, plate 174]

Running full speed. (Movements. Male). Plate 63, 1887
Collotype on paper
Corcoran Gallery of Art, Washington, D.C.
Museum Purchase, 87.7.63
[Page 295, plate 154]

Turning around in surprise and running away. (Movements. Female). Plate 73, 1887
Collotype on paper
Corcoran Gallery of Art, Washington, D.C.
Museum Purchase, 87.7.73
[Page 296, plate 155]

Ascending stairs. (Movements. Male). Plate 91, 1887
Collotype on paper
Corcoran Gallery of Art, Washington, D.C.
Museum Purchase, 87.7.91
[Page 297, plate 156]

Descending incline, right hand on breast (shoes). Plate 119, 1887
Collotype on paper
Corcoran Gallery of Art, Washington, D.C.
Museum Purchase, 87.7.119

Descending stairs with hands clasped. Plate 128, 1887
Collotype on paper
Corcoran Gallery of Art, Washington, D.C.
Museum Purchase, 87.7.128
[Page 300, plate 159]

Jumping; running straight high jump. (Movements. Female). Plate 156, 1887
Collotype on paper
Corcoran Gallery of Art, Washington, D.C.
Museum Purchase, 87.7.156
[Page 314, plate 173]

Jumping; running broad jump. (Movements. Male). Plate 160, 1887
Collotype on paper
Corcoran Gallery of Art, Washington, D.C.
Museum Purchase, 87.7.160
[Page 293, plate 152]

Jumping over boy's back (leap-frog). Plate 169, 1887
Collotype on paper
Corcoran Gallery of Art, Washington, D.C.
Museum Purchase, 87.7.169
[Page 323, plate 182]

Crossing brook on step-stones with fishing-pole and can. Plate 175, 1887
Collotype on paper
Corcoran Gallery of Art, Washington, D.C.
Museum Purchase, 87.7.170
[Page 313, plate 172]

Hopping on right foot. Plate 186, 1887
Collotype on paper
Corcoran Gallery of Art, Washington, D.C.
Museum Purchase, 87.7.186

Dancing (fancy.) (Movements. Female). Plate 188, 1887
Collotype on paper
Corcoran Gallery of Art, Washington, D.C.
Museum Purchase, 87.7.188
[Page 319, plate 178]

Sitting down, placing feet on chair. Plate 247, 1887
Collotype on paper
Wellcome Library, London (27760i)

Getting into bed. Plate 263, 1887
Collotype on paper
Corcoran Gallery of Art, Washington, D.C.
Museum Purchase, 87.7.250
[Page 306, plate 165]

Wrestling; Graeco-Roman. Plate 347, 1887
Collotype on paper
Corcoran Gallery of Art, Washington, D.C.
Museum Purchase, 87.7.332
[Page 310, plate 169]

Fencing. (Movements. Male). Plate 349, 1887
Collotype on paper
Corcoran Gallery of Art, Washington, D.C.
Museum Purchase, 87.7.334
[Page 309, plate 168]

Back somersault. Plate 362, 1887
Collotype on paper
Victoria and Albert Museum, London: 876-'89

Head-spring, a flying pigeon interfering. Plate 365, 1887
Collotype on paper
Corcoran Gallery of Art, Washington, D.C.
Museum Purchase, 87.7.350
[Page 325, plate 184]

Blacksmiths, two models, hammering on anvil. (Movements. Male). Plate 374, 1887
Collotype on paper
Corcoran Gallery of Art, Washington, D.C.
Museum Purchase, 87.7.358
[Page 291, plate 150]

Emptying bucket of water. Plate 399, 1887
Collotype on paper
Corcoran Gallery of Art, Washington, D.C.
Museum Purchase, 87.7.383
[Page 178, fig. 93]

Two models, 8 pouring bucket of water over 1. Plate 406, 1887
Collotype on paper
Corcoran Gallery of Art, Washington, D.C.
Museum Purchase, 87.7.389

Two models, child, 70 N, bringing bouquet to 12. Plate 465, 1887
Collotype on paper
Corcoran Gallery of Art, Washington, D.C.
Museum Purchase, 87.7.426
[Page 317, plate 176]

Child running. Plate 469, 1887
Collotype on paper
Corcoran Gallery of Art, Washington, D.C.
Museum Purchase, 87.7.430
[Page 316, plate 175]

A, sitting down; B, sitting down; C, sprinkling water; D, stooping for cup and drinking. Plate 490, 1887
Collotype on paper
Corcoran Gallery of Art, Washington, D.C.
Museum Purchase, 87.7.450
[Page 274, fig. 115]

Ascending and descending stairs. (Movements. Female). Plate 504, 1887
Collotype on paper
Corcoran Gallery of Art, Washington, D.C.
Museum Purchase, 87.7.460
[Page 279, fig. 119]

Contortions on the ground. Plate 510, 1887
Collotype on paper
Corcoran Gallery of Art, Washington, D.C.
Museum Purchase, 87.7.466
[Page 308, plate 167]

A, walking; B, ascending step; C, throwing disk; D, using shovel; E, F, using pick. Plate 521, 1887
Collotype on paper
Corcoran Gallery of Art, Washington, D.C.
Museum Purchase, 87.7.477
[Page 220, fig. 110]

A, 97, jumping; B, 98, hand-spring; C, 98, somersault; D, 99, somersault; E, 99, spring over man's back. Plate 522, 1887
Collotype on paper
Corcoran Gallery of Art, Washington, D.C.
Museum Purchase, 87.7.478
[Page 289, plate 148]

Movement of the hand; lifting a ball. Plate 534, 1887
Collotype on paper
Corcoran Gallery of Art, Washington, D.C.
Museum Purchase, 87.7.487
[Page 321, plate 180]

Infantile paralysis; child, walking on hands and feet. Plate 539, 1887
Collotype on paper
Corcoran Gallery of Art, Washington, D.C.
Museum Purchase, 87.7.491
[Page 96, fig. 87]

Hemiplegia; walking with a cane. Plate 552, 1887
Collotype on paper
Wellcome Library, London (572329i)

Trotting; free; light-grey horse. Eagle. Plate 596, 1887
Collotype on paper
Corcoran Gallery of Art, Washington, D.C.
Museum Purchase, 87.7.542

Canter; bareback; rider, 106, nude; white horse Clinton. Plate 617, 1887
Collotype on paper
Corcoran Gallery of Art, Washington, D.C.
Museum Purchase, 87.7.563
[Page 331, plate 190]

Gallop; bay horse. Daisy. Plate 628, 1887
Collotype on paper
Victoria and Albert Museum, London: 1142-'89

Jumping a hurdle; saddle; bay horse Daisy. Plate 640, 1887
Collotype on paper
Corcoran Gallery of Art, Washington, D.C.
Museum Purchase, 87.7.586
[Page 329, plate 188]

Buffalo; galloping. Plate 700, 1887
Collotype on paper
Corcoran Gallery of Art, Washington, D.C.
Museum Purchase, 87.7.624

Dog; Walking; Interrupted; Mastiff-Dread. Plate 703, 1887
Collotype on paper
Gift of the Edwin J. Beinecke Trust. Addison Gallery of American Art, Phillips Academy, Andover, Massachusetts
[Page 334, plate 193]

Elephant; walking. Plate 733, 1887
Collotype on paper
Corcoran Gallery of Art, Washington, D.C.
Museum Purchase, 87.7.641
[Page 333, plate 192]

Elephants; two walking. Plate 735, 1887
Collotype on paper
Corcoran Gallery of Art, Washington, D.C.
Museum Purchase, 87.7.643

Baboon; walking on all-fours. Plate 748, 1887
Collotype on paper
Corcoran Gallery of Art, Washington, D.C.
Museum Purchase, 87.7.656
[Page 332, plate 191]

Cockatoo; flying. Plate 759, 1887
Collotype on paper
Corcoran Gallery of Art, Washington, D.C.
Museum Purchase, 87.7.667
[Page 335, plate 194]

American eagle; flying. Plate 769, 1887
Collotype on paper
Corcoran Gallery of Art, Washington, D.C.
Museum Purchase, 87.7.677

Chickens; scared by torpedo. Plate 781, 1887
Collotype on paper
Corcoran Gallery of Art, Washington, D.C.
Museum Purchase, 87.7.689

Walking, scattering flowers and turning around. Phase 8, Lateral, from *Animal Locomotion, Plate 53*, 1887
Lantern slide, collodion positive on glass
By permission of Kingston Museum and Heritage Service (6273)

Descending incline, throwing handkerchief over shoulders. Phase 1, Lateral, from *Animal Locomotion, Plate 133*, 1887
Lantern slide, collodion positive on glass
By permission of Kingston Museum and Heritage Service (8161)

Cricket; batting; drive. Phase 9, Front foreshortening, from *Animal Locomotion, Plate 291*, 1887
Lantern slide, collodion positive on glass
By permission of Kingston Museum and Heritage Service (6101)

Cricket; batting; drive. Phase 6, Rear foreshortening, from *Animal Locomotion, Plate 291*, 1887
Lantern slide, collodion positive on glass
By permission of Kingston Museum and Heritage Service (6099)

Cricket; batting; drive. Phase 11, Rear foreshortening, from *Animal Locomotion, Plate 291*, 1887
Lantern slide, collodion positive on glass
By permission of Kingston Museum and Heritage Service (6110 b)

Back Somersault. Phase 3, from *Animal Locomotion, Plate 362*, 1887
Lantern slide, collodion positive on glass
By permission of Kingston Museum and Heritage Service (6127)

Canter. Phase 10, from *Animal Locomotion, Plate 617*, 1887
Lantern slide, collodion positive on glass
By permission of Kingston Museum and Heritage Service

Cockatoo; flying. Phase 12 (uncropped), from *Animal Locomotion, Plate 760*, 1887
Lantern slide, collodion positive on glass
By permission of Kingston Museum and Heritage Service (8185)

J. Walter Wilson
Mr. Muybridge Shows His Instantaneous Photographs of Animal Motion at the Royal Society, Illustrated London News, May 25, 1889
Engraving on newsprint
Corcoran Gallery of Art, Washington, D.C.
[Page 216, fig. 107]

Athlete, Running Long Jump, 14 phases, ca. 1893
Glass zoopraxiscope disc, paint, metal
By permission of Kingston Museum and Heritage Service
[Page 266, plate 140]

Woman Dancing, 1893
Glass zoopraxiscope disc, paint, metal
By permission of Kingston Museum and Heritage Service
[Page 269, plate 143]

Descriptive Zoopraxography, or the Science of Animal Locomotion Made Popular (frontispiece), 1893
Book
Museum Purchase. Addison Gallery of American Art, Phillips Academy, Andover, Massachusetts
[Page 100, fig. 91]

Unknown photographer
Zoopraxographical Hall, World's Columbian Exposition, Chicago (1009), 1893
Two albumen silver prints on studio card
California History Section, California State Library
[Page 28, fig. 10]

Frances Benjamin Johnston
Portrait of Eadweard Muybridge, 1890
Gelatin silver print
Smithsonian Institution, National Museum of American History, Kenneth E. Behring Center
[Page 283, fig. 121]

Eadweard Muybridge
The Human Figure in Motion. An Electro-Photographic Investigation of Consecutive Phases of Muscular Actions,
London 1901
Book
Tate Library

Acknowledgments

Philip Brookman, Carolyn Kerr, and Ian Warrell

Eadweard Muybridge is in the broadest sense a collaborative effort between many colleagues, a research project intended to bring together for the first time the full array of Eadweard Muybridge's art and explore its tantalizing relationships between art and science, photography, and cinema. A project of this sort could not be accomplished without significant assistance from many people. From the start, numerous institutions and individuals have supported the Corcoran Gallery of Art's efforts, and more recently those of Tate Britain, in pursuing a better understanding of Muybridge's world, and we are indebted to everyone involved in this process.

The Corcoran Gallery of Art has supported the research for this exhibition over a number of years, permitting it to draw on a wide range of collections, and to include much unfamiliar material. Many scholars, curators, artists, and others have also lent their expertise. Most immediately, Marta Braun, Andy Grundberg, Corey Keller, and Rebecca Solnit deserve special recognition for their insightful contributions to this catalogue. In addition to these new perspectives, the following individuals have all helped to build our current understanding of Muybridge: Robert Bartlett Haas, Gordon Hendricks, Stephen Herbert, Anita Ventura Mozley, and Phillip Prodger, among others too numerous to list.

As well as a number of private collectors, colleagues at museums, libraries, and galleries have generously agreed to lend fragile objects for presentation in the exhibition. Others have given their time and experience to assist the project. For this we wish to thank: Iglika Avramova; Dick Balzer; Martin Barnes; Bernard Barryte; Christopher Baruth; Toni Booth; Marisa Bourgoin; Jeffrey Bowdoin; Frish Brandt; Susan Cary; Marina Chao; Katie Clifford; John Cohen; Peta Cook; Verna Curtis; Malcolm Daniel; Emily Darragh; Michelle Anne Delaney; Rowan De Saulles; Mary Anne Doane; Heidi Donner; James Eason; David E. Falkner; Nancy Ferguson; Polly Fleury; Jeffrey Fraenkel; Robert Frank; Betsy Fryberger; Leanda Gahegan; Peter Galassi; Jim Ganz; Jim Goldberg; Susan Goldstein; Frank Goodyear; Sarah Gordon; Maria Grandinette; Sarah Greenough; Genie Guerard; Todd Gustavson; Violet Hamilton; Jeff Harrison; Michael Harvey; Christopher Havern; Ann Havinga; George Hemphill; Ruth Hibbard; Tom Hinson; Hogan and Hartson, LLP; Charles Isaacs; Carol Johnson; Drew Johnson; David Julyan; Toby Jurovics; Alan Jutzi; Betsy and Frank Karel; Patricia Keats; Corey Keller; Robin Kelsey; Allison Kemmerer; Sarah Kennel; Margaret Kimball; Hope Kingsley; Gary Kurutz; Jill Lamb; June Leaf; Clayton Lewis; Connie McCabe; Peter MacGill; Paul Martineau; Cheryl Maslin; Brian May; George Miles; Aimee Morgan; Christina Moretta; Mary Morganti; Charles Musser; Weston Naef; Alison Nordström; Mark Osterman; Shannon Perich; Susan Peschel; Sandra Phillips; Stephen Pinson; Jock Reynolds; Yancey Richardson; Jovanka Ristic; Jeff Rosenheim; Emma Rummins; Paul Sack; William Schupbach; James Sheldon; Rebecca Short; Andrew Smith; Stephanie Smith; Aimee Soubier; James Sousa; Susan Roberts-Manganelli; Joe Strubble; Emma Stuart; Vicente Todolí; Anne Tucker; Elena Vidal; Jack Von Euw; Diane Waggoner; Rachel Waldron; Leonard Walle; Jenny Watts; Wendy Welker; Glenn Willumson; Michael Wilson; James Zeender.

At the Corcoran, where this exhibition began its tour, the project was made possible through the hard work of many colleagues: Paul Greenhalgh, Fred Bollerer, Paul Roth, Amanda Maddox, Elizabeth Parr, Dana Gildenhorn, Nancy Swallow, Dare Hartwell, Ken Ashton, Andrea Jain, Ila Furman, Cory Hixson, Michael Baltzer; Charles Sthreshley, Sarah Durkee, Linda Powell, Michelle Clair, Kristin Guiter, Mariana Nork, Janice Marks, Kate Earnest, David Julyan, Darci Spasojevich, Steve Brown, David Dorsey, Kirk Pillow, Bernard Welt, Mario Ascencio, and Pat Reid, as well as research interns Kate Phillips, Josh Brilliant, Kelly Montana, and Katie Clausen. Kaitlin Booher and Larissa Leclair served at different times as curatorial associates, assisting greatly with logistics, research, and editing. Sarah Wagner and Martin Salazar provided conservation support, and Linda McNamara designed the exhibition. The development and organisation of the exhibition at the Corcoran Gallery of Art was funded from its inception by American Express, the Trellis Fund, the National Endowment for the Arts, and Deane and Paul Shatz, all of whom have been extremely generous.

At Tate Britain, the idea for a Muybridge exhibition was embraced with enthusiasm by former Director, Stephen Deuchar. Since his departure to the Art Fund at the end of 2009, the practicalities of seeing the realization of the project have been led by Judith Nesbitt, Chief Curator, working with Carolyn Kerr, Ian Warrell, Sofia Karamani, and Emma Baker. Additional support has been provided by Juleigh Gordon-Orr, Hanne Dahl, Alice Teng, Amy Thompson, Jessie Turner, and Shoair Mavlian. The invaluable role of registrar for this project has been performed by Gillian Buttimer, and the installation of the exhibition has been overseen by Andy Shiel and Mikei Hall.

It has been a pleasure to collaborate once again with Oliver Salway, of SOFT-ROOM, who also designed *How We Are: Photographing Britain* for Tate Britain in 2007. The graphics for the exhibition have been created by Micha Weidmann and Will Martin. The crucial job of lighting everything has been undertaken by John Johnson and Lightwaves.

The task of bringing Muybridge's art to Tate Britain's various audiences has been performed across the gallery's many departments by the following members of staff: Piers Townshend; Vanessa Griffiths; Gina Koutsika; Jennifer Batchelor; Madeleine Keep; Louise Butler; Duncan Holden; Marilena Reina; Celeste Menich; Caroline Arno; Andrew Tullis; Selina Jones; Alex O'Neill; Claire Eva; Simon Grant; Mariko Finch; Will Gompertz; Kirstie Beaven; Kate Vogel; Dan Crompton; Duncan Mackie; Shuja Rahman; Silaja Suntharalingam. Roger Thorp, Anna Ridley, and Beth Thomas have collaborated with Steidl on the production of this book for Tate. We are also grateful to Simon Baker, Chris Stephens, Melanie Grant, and Marek Pytel.

After its showing in London, the exhibition travels back across the Atlantic to the San Francisco Museum of Modern Art. We are grateful to Neal Benezra, Sandra Phillips, and Corey Keller, who have been enormously helpful in coordinating our collaborations.

This book could not have been produced without the constant and steadfast assistance of Gerhard Steidl and his terrific staff at Steidl in Göttingen and London. For this we are indebted to Michael Mack, Bernard Fischer, Monte Packham, Julia Braun, Jonas Wettre, and Duncan Whyte. Michelle Piranio edited the extensive text with great skill and good humor throughout, selflessly lending her time when it was most needed. Mark Gulezian and Ben Blackwell assisted with photography.

In addition, this publication was made possible, in part, by the Paul Mellon Centre for Studies in British Art and Furthermore, a program of the J. M. Kaplan Fund.

First published in the United Kingdom and Ireland 2010 by order of the Tate Trustees
by Tate Publishing, a division of Tate Enterprises Ltd,
Millbank, London SW1P 4RG
www.tate.org.uk/publishing

on the occasion of *Eadweard Muybridge*

Corcoran Gallery of Art, Washington
10 April–18 July 2010

Tate Britain, London
8 September 2010–16 January 2011

San Francisco Museum of Modern Art
26 February–7 June 2011

The exhibition at Tate Britain is supported by the Terra Foundation for American Art

The accompanying catalogue was made possible, in part, by the Paul Melon Centre for Studies
in British Art and Furthermore, a program of the J. M. Kaplan Fund

A catalogue record for this book is available from the British Library

ISBN 978 1 85437 837 8

Edited by Michelle Piranio with Philip Brookman
Editorial assistance by Kaitlin Booher
Designed by Bernard Fischer and Gerhard Steidl
Steidl's digital darkroom: Julia Braun, Jonas Wettre, Duncan Whyte
Printed in Germany by Steidl

Steidl Publishers
Düstere Str. 4
37073 Göttingen, Germany
Phone +49 551-49 60 60, Fax +49 551-49 60 649
E-mail mail@steidl.de
www.steidlville.com, www.steidl.de

Front cover: *Leland Stanford, Jr. on his Pony "Gypsy"—Phases of a Stride by a Pony While Cantering* (detail),
1879, Wilson Centre for Photography
Back cover: *Athletes. Posturing. Plate 115, (detail)*, 1879, from *The Attitudes of Animals in Motion*, 1881,
Courtesy Special Collections, Stanford University Libraries

Lenders to the Exhibition

Addison Gallery of American Art, Phillips Academy, Andover, Massachusetts
American Geographical Society Library, University of Wisconsin-Milwaukee Libraries
California Historical Society, San Francisco
California State Library, Sacramento
Iris & B. Gerald Cantor Center for Visual Arts at Stanford University
Centre Canadien d'Architecture/Canadian Centre for Architecture, Montreal
Chrysler Museum of Art, Norfolk, Virginia
William L. Clements Library, University of Michigan, Ann Arbor
Corcoran Gallery of Art, Washington, D.C.
George Eastman House, Rochester, New York
The J. Paul Getty Museum, Los Angeles
Kingston Museum and Heritage Service, Kingston, United Kingdom
Library of Congress, Washington, D.C.
The Museum of Modern Art, New York
National Gallery of Art, Washington, D.C.
The New York Public Library
San Francisco History Center, San Francisco Public Library
San Francisco Museum of Modern Art
Smithsonian American Art Museum, Washington, D.C.
Smithsonian Institution, National Museum of American History, Behring Center, Washington, D.C.
The Society of California Pioneers, San Francisco
Special Collections, Stanford University Libraries
Tate Library and Archive
U.S. Coast Guard Historian's Office, Washington, D.C.
Victoria and Albert Museum, London
Leonard A. Walle
Wellcome Library, London
Wilson Centre for Photography, London
Charles E. Young Research Library, University of California, Los Angeles, Department of Special Collections